D0867367

Methods of
Experimental Physics

VOLUME 27

SCANNING TUNNELING MICROSCOPY

METHODS OF EXPERIMENTAL PHYSICS

Robert Celotta and Thomas Lucatorto, *Editors-in-Chief*

Founding Editors

L. MARTON
C. MARTON

Volume 27

Scanning Tunneling Microscopy

Edited by

Joseph A. Stroscio
National Institute of Standards and Technology
Gaithersburg, Maryland

and

William J. Kaiser
Jet Propulsion Laboratory
Pasadena, California

ACADEMIC PRESS, INC.
Harcourt Brace Jovanovich, Publishers

Boston San Diego New York London Sydney Tokyo Toronto

This book is printed on acid-free paper. ♾

ACADEMIC PRESS, INC.
1250 Sixth Avenue, San Diego, CA 92101-4311

United Kingdom Edition published by
ACADEMIC PRESS LIMITED
24–28 Oval Road, London NW1 7DX

ISBN 0-12-475972-6

Printed in the United States of America

93 94 95 96 BC 9 8 7 6 5 4 3 2 1

CONTENTS

CONTRIBUTORS ix

PREFACE xi

LIST OF VOLUMES IN TREATISE xv

1. Theory of Scanning Tunneling Microscopy
by J. TERSOFF and N. D. LANG

1.1. Basic Principles 1

1.2. Theory of STM Imaging 5

1.3. Metal Surfaces: STM as Surface Topography 8

1.4. Semiconducting Surfaces: Role of Surface Electronic Structure 11

1.5. Adsorbates on Metal Surfaces 14

1.6. Close Approach of the Tip: The Strong-Coupling Regime. . 16

1.7. Tunneling Spectroscopy 22

1.8. Mechanical Tip–Sample Interactions 24

References 27

2. Design Considerations for an STM System
by SANG-IL PARK and ROBERT C. BARRETT

2.1. Introduction 31

2.2. Theoretical Considerations 33

2.3. Mechanical Structure and Components 51

2.4. Control Electronics 60

2.5. Common Problems and Further Improvements 66

Acknowledgments 75
References 75

3. Extensions of STM
by H. Kumar Wickramasinghe

3.1. Introduction 77
3.2. Historical 77
3.3. STM and Some Extensions 78
3.4. Near-Field Thermal Microscopy and Extensions 83
3.5. Scanning Force Microscopy and Applications 86
3.6. Conclusion 92
References 93

4. Methods of Tunneling Spectroscopy
by Joseph A. Stroscio and R. M. Feenstra

4.1. Instrumentation 96
4.2. General Current versus Voltage Characteristics 100
4.3. Voltage-Dependent Imaging Measurements 104
4.4. Fixed Separation I–V Measurements 112
4.5. Variable Separation Measurements 134
References 145

5. Semiconductor Surfaces
5.1. Silicon 149
by Russell Becker and Robert Wolkow
References 220

5.2. Germanium 225
by Russell Becker
References 248

5.3. Gallium Arsenide. 251
by R. M. FEENSTRA and JOSEPH A. STROSCIO
References 275

6. Metal Surfaces
by YOUNG KUK

6.1. Introduction 277

6.2. Corrugation Amplitudes and the Tunneling Tip 278

6.3. Tunneling Spectroscopy of Metal Surfaces 280

6.4. Clean Metal Surfaces 286

6.5. Adsorbate on Metal Surfaces 293

6.6. Conclusion 303

Acknowledgment 303

References 303

7. Ballistic Electron Emission Microscopy
by L. D. BELL, W. J. KAISER, M. H. HECHT, and L. C. DAVIS

7.1. Introduction 307

7.2. Theory 311

7.3. Experimental Details 326

7.4. Results 330

7.5. Conclusions 346

Acknowledgment 347

References 348

8. Charge-Density Waves
by R. V. COLEMAN, ZHENXI DAI, W. W. MCNAIRY, C. G. SLOUGH, and CHEN WANG

8.1. Transition Metal Chalcogenides 349

8.2. Charge-Density Wave Formation 350
8.3. Charge-Density Waves in Transition Metal Chalcogenides 352
8.4. Experimental STM and AFM Response to CDW Structures 352
8.5. Experimental Techniques. 355
8.6. 1T Phase Transition Metal Dichalcogenides 358
8.7. 2H Phase Transition Metal Dichalcogenides 385
8.8. 4Hb Phase Transition Metal Dichalcogenides. 393
8.9. Linear Chain Transition Metal Trichalcogenides 402
8.10. Conclusions 421
Acknowledgments 423
References 423

9. Superconductors
by Harald F. Hess
9.1. Introduction 427
9.2. The Superconducting State 427
9.3. Experimental Techniques for STM on Superconductors. . . 430
9.4. Spectrum in Zero Field 432
9.5. Vortex Data 434
9.6. Interpretation 441
9.7. Conclusion 448
References. 448

Index 451

CONTRIBUTORS

Numbers in parentheses indicate the pages on which the authors' contributions begin.

Robert C. Barrett (31), *Department of Applied Physics, Stanford University, Stanford, CA 94305*

Russell Becker (149), *AT&T Bell Laboratories, 600 Mountain Avenue, Murray Hill, NJ 07974*

L. D. Bell (307), *Center for Space Microelectronics Technology, Jet Propulsion Laboratory, California Institute of Technology, Pasedena, CA 91109*

R. V. Coleman (349), *Department of Physics, University of Virginia, Charlottesville, Virginia 22901*

Zhenxi Dai (349), *Department of Physics, University of Virginia, Charlottesville, Virginia 22901*

L. C. Davis (307), *Scientific Research Laboratory, Ford Motor Company, Dearborn, MI 48121*

R. M. Feenstra (95), *IBM Research Division, T. J. Watson Research Center, Yorktown Heights, NY 10598*

M. H. Hecht (307), *Center for Space Microelectronics Technology, Jet Propulsion Laboratory, California Institute of Technology, Pasadena, CA 91109*

Harald F. Hess (427), *AT&T Bell Laboratories, Murray Hill, NJ 07974*

W. J. Kaiser (307), *Center for Space Microelectronics Technology, Jet Propulsion Laboratory, California Institute of Technology, Pasadena, CA 91109*

Young Kuk (277), *Seoul National University, Seoul, Korea*

N. D. Lang (1), *IBM Research Division, T. J. Watson Research Center, Yorktown Heights, NY 10598*

W. W. McNairy (349), *Department of Physics, University of Virginia, Charlottesville, Virginia 22901*

Sang-il Park (31), *Park Scientific Instruments, Sunnyvale, CA 94040*

C. G. Slough (349), *Department of Physics, University of Virginia, Charlottesville, Virginia 22901*

Joseph A. Stroscio (95), *Electron and Optical Physics Division, National Institute of Standards and Technology, Gaithersburg, MD 20899*

J. Tersoff (1), *IBM Research Division, T. J. Watson Research Center, Yorktown Heights, NY 10598*

CHEN WANG (349), *Department of Physics, University of Virginia, Charlottesville, Virginia 22901*

H. KUMAR WICKRAMASINGHE (77), *IBM Research Division, T. J. Watson Research Center, P.O. Box 218, Yorktown Heights, NY 10598*

ROBERT WOLKOW (149), *AT&T Bell Laboratories, 600 Mountain Avenue, Murray Hill, NJ 07974*

PREFACE

Understanding the atomic structure of materials, the behavior and reactions of atoms at surfaces, and the nature of electronic properties at the atomic scale have been the goals of fundamental and applied research for many decades. Until recently, our imagination and intuition for physics on the atomic scale was verified by only a few often unwieldy experimental probes that provided only an indirect or incomplete glimpse of atomic structure at surfaces. The development of scanning tunneling microscopy (STM) has revolutionized our approach to the investigation of many aspects of material properties at the atomic scale. The STM, and the techniques it has inspired, have provided methods to measure accurately the structure and electronic properties of single surface atoms. Extensions of the STM technique, such as atomic force microscopy, have allowed various other properties to be measured on a nanometer scale. As STM methods have been refined, several experiments have demonstrated the ability to manipulate individual atoms accurately. These recent STM demonstrations are not only accomplishments of modern science, but also fulfill the aspirations of those who first imagined the existence of atoms.

Electron tunneling was first demonstrated by the elegant experiments of Giaver during the early 1960s. The following decade showed dynamic growth in electron tunneling, resulting in a powerful method for characterizing materials and structures. This work led to the achievement of vacuum tunneling by R. Young and co-workers in the early 1970s. Young additionally combined three-dimensional scanning and electron tunneling in the field emission regime to produce the "topographiner." A decade passed until the development of the scanning tunneling microscope by Binnig and Rohrer in the early 1980s. Their STM produced beautiful, detailed images of surface atoms, which led to today's renaissance in electron tunneling, as seen in this present book.

This volume of *Experimental Methods in Physics* presents many landmark studies and techniques of STM by leading researchers in the field. The aim of the book is to bring together in one volume those "historical" pieces of work that have defined the STM field over roughly the past 10 years. STM is introduced, in this volume, with a chapter on the theoretical understanding of STM by Tersoff and Lang. These authors show exactly what the STM measures, with the concept of measuring an energy-dependent local state density at the surface. An experimental methods volume would not be complete without a detailed dicussion of design criteria and instrumentation.

Chapter 2 of this volume, by Park and Barett, discusses this topic in detail with attention to the important topics of servo loops, vibration isolation, and scanner designs. STM instrumentation has inspired an entire spectrum of nanometer-scale instrumentation that probe physical properties ranging from thermal coefficients to forces. While a complete description of these STM extensions would require an entire volume (or volumes), Wichramasinghe reviews the more widely used STM extensions in Chapter 3.

The ability to vary the tunneling bias over several volts was exploited early in STM development and led to the important application of scanning tunneling spectroscopy. Because there are a variety of variables under control in the STM, various spectroscopic methods have been developed, including voltage dependent imaging, fixed and variable tip-sample distance current-voltage spectroscopy, and current imaging spectroscopy. Chapter 4, by Stroscio and Feenstra, describes tunneling spectroscopy with the STM, with extensive examples taken from measurements on Si and GaAs surfaces.

The remainder of this volume deals with applications of STM of specific materials or areas. The largest group of activity in the STM field has been the investigation of semiconductor surfaces. Chapter 5 describes those studies pertaining to silicon by Becker and Wolkow, followed by studies on germanium surfaces by Becker, and finally gallium arsenide studies are described by Feenstra, and Stroscio. The application of STM to metal surfaces are reviewed by Kuk in Chapter 6. Historically, metal surfaces have been less studied due to the small signals and the resulting increased demand on instrumentation, but Chapter 6 shows that there has been an increase in applications to metal surfaces as instrumentation has matured. Chapter 7 by Bell, Kaiser, Hecht, and Davis, describes the new method of ballistic electron emission microscopy to the study of buried interfaces. The application of STM to charge density wave phenomena is a very rich scientific area, and shows one of the best examples of the STM sensitivity to local state density and not merely atomic positions. Coleman, Dai, McNairy, and Slough describe in detail their pioneering work in this field in Chapter 8. The related low temperature application of the STM in the field of superconductivity is described by Hess in Chapter 9. As STM low temperature technology spreads, we expect the very rich combination of low temperature measurement and an atomic scale probe to lead to exciting future scientific studies.

The restriction of the present volume to the aforementioned areas is simply imposed by space and time requirements, and we regret not being able to include all of the wonderful developments in other areas, such as biological, electrochemical, and various other extensions of the STM technique. I (JAS) would like to thank Bob Celotta for inviting me to edit this volume and Randall Feenstra for introducing me to the STM field. By wife, Frances, deserves the greatest acknowledgement for her unending support in my work.

I (WJK) would also like to thank Bob Celotta for his valuable support of our efforts. Also, I would like to thank R. C. Jaklevic, L. Douglas Bell, and Michael Hecht, for many years of collaboration on exciting STM research. In particular, I owe a large debt to my wife, Catherine, for her constant encouragement and enthusiasm.

Joseph A. Stroscio
Gaithersburg, Maryland

W. J. Kaiser
Pasadena, California

METHODS OF EXPERIMENTAL PHYSICS

Editors-in-Chief
Robert Celotta and Thomas Lucatorto

Volume 1. Classical Methods
Edited by Immanuel Estermann

Volume 2. Electronic Methods, Second Edition (in two parts)
Edited by E. Bleuler and R. O. Haxby

Volume 3. Molecular Physics, Second Edition (in two parts)
Edited by Dudley Williams

Volume 4. Atomic and Electron Physics—Part A: Atomic Sources and Detectors; Part B: Free Atoms
Edited by Vernon W. Hughes and Howard L. Schultz

Volume 5. Nuclear Physics (in two parts)
Edited by Luke C. L. Yuan and Chien-Shiung Wu

Volume 6. Solid State Physics—Part A: Preparation, Structure, Mechanical and Thermal Properties; Part B: Electrical, Magnetic and Optical Properties
Edited by K. Lark-Horovitz and Vivian A. Johnson

Volume 7. Atomic and Electron Physics—Atomic Interactions (in two parts)
Edited by Benjamin Bederson and Wade L. Fite

Volume 8. Problems and Solutions for Students
Edited by L. Marton and W. F. Hornyak

Volume 9. Plasma Physics (in two parts)
Edited by Hans R. Griem and Ralph H. Lovberg

Volume 10. Physical Principles of Far-Infrared Radiation
By L. C. Robinson

Volume 11. Solid State Physics
Edited by R. V. Coleman

Volume 12. Astrophysics—Part A: Optical and Infrared Astronomy
Edited by N. Carleton
Part B: Radio Telescopes; Part C: Radio Observations
Edited by M. L. Meeks

Volume 13. Spectroscopy (in two parts)
Edited by Dudley Williams

Volume 14. Vacuum Physics and Technology
Edited by G. L. Weissler and R. W. Carlson

Volume 15. Quantum Electronics (in two parts)
Edited by C. L. Tang

Volume 16. Polymers—Part A: Molecular Structure and Dynamics; Part B: Crystal Structure and Morphology; Part C: Physical Properties
Edited by R. A. Fava

Volume 17. Accelerators in Atomic Physics
Edited by P. Richard

Volume 18. Fluid Dynamics (in two parts)
Edited by R. J. Emrich

Volume 19. Ultrasonics
Edited by Peter D. Edmonds

Volume 20. Biophysics
Edited by Gerald Ehrenstein and Harold Lecar

Volume 21. Solid State: Nuclear Methods
Edited by J. N. Mundy, S. J. Rothman, M. J. Fluss, and L. C. Smedskjaer

Volume 22. Solid State Physics: Surfaces
Edited by Robert L. Park and Max G. Lagally

Volume 23. Neutron Scattering (in three parts)
Edited by K. Sköld and D. L. Price

Volume 24. Geophysics—Part A: Laboratory Measurements; Part B: Field Measurements
Edited by C. G. Sammis and T. L. Henyey

Volume 25. Geometrical and Instrumental Optics
Edited by Daniel Malacara

Volume 26. Physical Optics and Light Measurements
Edited by Daniel Malacara

Volume 27. Scanning Tunneling Microscopy
Edited by Joseph A. Stroscio and William Kaiser

1. THEORY OF SCANNING TUNNELING MICROSCOPY

J. Tersoff and N. D. Lang

IBM Research Division, T. J. Watson Research Center,
Yorktown Heights, New York

Scanning tunneling microscopy[1] (STM) has proven to be a powerful and unique tool for the determination of the structural and electronic properties of surfaces. While STM was originally introduced as a method for topographic imaging of surfaces, many related techniques have since grown out of it. These include local spectroscopy,[2,3] scanning potentiometry,[4] and a host of other local probes.

This chapter reviews the current understanding of STM and tunneling spectroscopy.

1.1. Basic Principles

1.1.1 Vacuum Tunneling

In vacuum tunneling, the potential in the vacuum region acts as a barrier to electrons between the two metal electrodes, in this case the surface and the tip. This barrier is shown schematically in Fig. 1. The transmission probability for a wave incident on a barrier in one dimension is easily calculated. For STM we typically need consider only the limit of weak transmission, corresponding to the most common range of barrier heights and widths. This limit gives a very simple behavior.

The solutions of Schrödinger's equation inside a rectangular barrier in one dimension have the form

$$\psi = e^{\pm \kappa z}. \tag{1}$$

Thus the crucial parameter is κ, where

$$\kappa^2 = 2m(V_B - E)/\hbar^2. \tag{2}$$

Here E is the energy of the state, and V_B is the potential in the barrier. In general, as shown in Fig. 1, V_B may not be constant across the gap; but for the moment in our discussion, it will be adequate to replace the potential in the barrier with its average value, so that we need only consider a rectangular barrier. In the simplest case V_B is simply the vacuum level; so for states at the Fermi level, $V_B - E$ is just the work function.

METHODS OF EXPERIMENTAL PHYSICS
Vol. 27

ISBN 0-12-475972-6

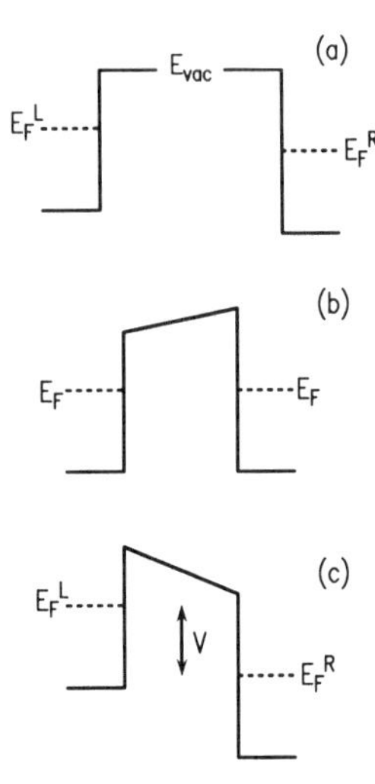

FIG. 1. Schematic of potential barrier between electrodes for vacuum tunneling. (a) Two non-interacting metal electrodes, separated by vacuum. The Fermi levels E_F of the two materials differ by an amount equal to the work function difference. (E_F^L and E_F^R denote the Fermi levels of the left and right electrode respectively, in the cases (a) and (c) where the two are not in equilibrium.) (b) The two electrodes are allowed to come into electrical equilibrium, so that there is a unique common Fermi level. The difference in work functions is now manifested as an electric field in the vacuum region. (c) A voltage is applied. There is a voltage drop V across the gap, i.e., the Fermi levels differ by eV. The field in the barrier includes contributions from both the applied voltage and the work function difference. The arrows indicate the range of energy over which tunneling can occur. At higher energies, there are no electrons to tunnel, while at lower energies, there are no empty states to tunnel into.

The transmission probability, or the tunneling current, thus decays exponentially with barrier width d as

$$I \propto e^{-2\kappa d}. \tag{3}$$

The generalization to a real three-dimensional surface is given below.

For tunneling between two metals with a voltage difference V across the gap, only the states within V above or below the Fermi level can contribute to tunneling, with electrons in states within V below the Fermi level on the negative side tunneling into empty states within V above the Fermi level on the positive side. As shown in Fig. 1, other states cannot contribute either because there are no electrons to tunnel at higher energy, or because of the exclusion principle at lower energy. (Of course we should say "within energy eV above or below the Fermi level," where e is the electron charge. But throughout the article, we will write V in such instances, to avoid confusion with the abbreviation for "electron volts.")

Since most work functions are around 4–5 eV, from Eq. (2) we find that

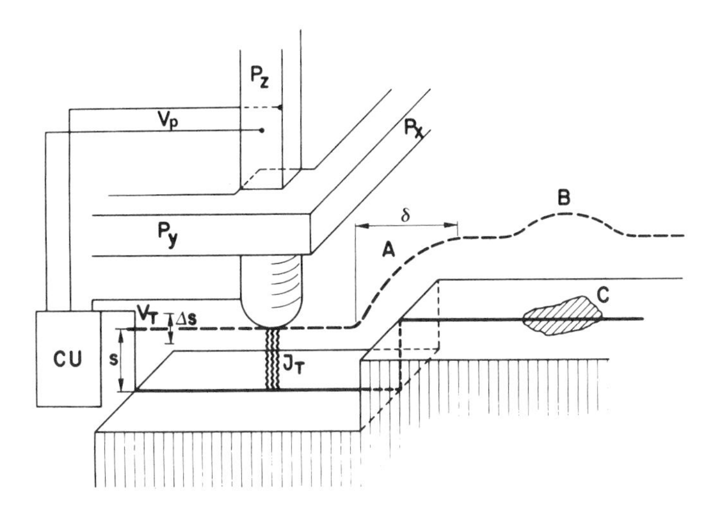

FIG. 2. Principle of operation of the scanning tunneling microscope (schematic, not to scale). The piezodrives P_x and P_y scan the metal tip over the surface. The control unit (CU) applies the necessary voltage V_P to the piezodrive P_z to maintain constant tunnel current J_T at bias voltage V_T. The broken line indicates the z displacement in a y scan at (A) a surface step, and (B) a spot C with lower work function. (From Ref. 1.)

typically $2\kappa \sim 2\,\text{Å}^{-1}$. Thus the tunneling current drops by nearly an order of magnitude for every 1 Å of vacuum between the electrodes. Such tunneling can therefore only be observed in practice for very small separations. Achieving such small separations, and keeping the current even moderately stable, requires very precise control of the positions of the electrodes, limiting vibrations to much less than an angstrom.

While such vacuum tunneling has long been understood in principle, the first reports[5] of the direction observation of vacuum tunneling did not come until the 1970s, and in 1982 Binnig *et al.*[5] demonstrated well-controlled vacuum tunneling in their first step towards STM. They used a piezoelectric driver to accurately control the height of a metal tip above a surface, and confirmed the expected exponential behavior. Tunneling experiments before these were largely restricted to tunneling through a static barrier, consisting of a layer of oxide sandwiched between metal electrodes.[6]

1.1.2 Scanning Tunneling Microscopy

The basic idea behind STM is quite simple,[1] as illustrated in Fig. 2. A sharp metal tip is brought close enough to the sample surface that electrons can tunnel quantum mechanically through the vacuum barrier separating tip and sample. As discussed earlier, this tunneling current is extremely sensitive to the gap, i.e., to the height of the tip above the surface.

The position of the tip in three dimensions is accurately controlled by piezoelectric drivers. The tip is scanned in the two lateral dimensions, while a feedback circuit constantly adjusts the tip height, to keep the current constant. A constant current yields roughly a constant tip height, so the shape of the surface is reproduced by the path of the tip, which can be inferred directly from the voltage supplied to the piezoelectric drivers.

It is also possible to use a slower feedback for the tip height, so that the height remains constant above the average surface, and small features are reflected in fluctuations of the current, rather than in the tip height.[7] However, this "constant height" mode of imaging is only practical in special cases where the surface is extremely flat, and is not fundamentally different from the usual "constant current" mode, so it is not discussed separately here.

Yet another mode of imaging is occasionally used with interesting results.[8] By modulating the tip height slightly (at a frequency above the response of the feedback), and measuring the resulting current modulation, one can obtain the local value of $d \ln I/dz$, and hence κ (or equivalently what is sometimes called the apparent barrier height or "effective work function," $\hbar^2\kappa^2/2m$) across the whole surface. This yields a picture in which the local variations of κ presumably reflect differences in chemical composition or such, rather than topography.

Of course, since the tip has a finite radius, the surface topography is determined only with a finite resolution. One can make a simple estimate of the resolution as follows.[1] At a given lateral position x, relative to the center of the tip, the height of the corresponding point on the tip is $d + x^2/2R$, assuming a parabolic tip with radius of curvature R, and distance of closest approach d. The corresponding current in a one-dimensional model is $I(x) \propto \exp(-\kappa x^2/R)$. Thus the current has a Gaussian profile, with root-mean-square width $\sim 0.7(R/\kappa)^{1/2}$. Since κ is typically ~ 1 Å^{-1}, even a large tip radius such as 1000 Å gives a rather sharp (though not atomic) resolution of 50 Å.

It is possible to make metal tips with a radius of curvature of a few hundred Å, but not much less. However, because the tunneling current is so sensitive to distance, if the tip is a bit rough, most of the current will go to whatever atomic-scale asperity approaches closest to the surface. It is generally believed that the best STM images result from tunneling to a single atom, or at most a few atoms, on the tip.

1.2. Theory of STM Imaging

1.2.1 Beyond Topography

So long as the features resolved are on the nanometer scale or larger, interpretation of the STM image as a surface topograph (complicated by local variations in barrier height) is generally adequate. But soon after the invention of STM,[1] Binnig *et al.* reported the first atomic-resolution images.[9] On the atomic scale, it is not even clear what one would mean by a topograph.

The most reasonable definition would be that a topograph is a contour of constant surface charge density. However, there is no reason why STM should yield precisely a contour of constant charge density, since only the electrons near the Fermi level contribute to tunneling, whereas all electrons below the Fermi level contribute to the charge density. Thus on some level, the interpretation of STM images as surface topographs must be inadequate. The following sections describe a more precise interpretation of STM images, applicable even in the case of atomic resolution.

One can calculate directly the transmission coefficient for an electron incident on the vacuum barrier between a surface and tip.[10] (We discuss such calculations later.) However, such a calculation is fairly complex for a realistic model of the surface. Fortunately, for typical tip–sample separations (of order 9 Å nucleus-to-nucleus[11]) the coupling between tip and sample is weak, and the tunneling can be treated with first-order perturbation theory. Since the problem is otherwise rather intractable except for simple models of the surface,[10–12] we restrict ourselves to this weak-coupling limit in the first part of our discussion.

1.2.2 Tunneling Hamiltonian Approach

In first-order perturbation theory, the current is

$$I = \frac{2\pi e}{\hbar} \sum_{\mu,\nu} \{ f(E_\mu)[1 - f(E_\nu)] - f(E_\nu)[1 - f(E_\mu)] \} |M_{\mu\nu}|^2 \delta(E_\nu + V - E_\mu), \tag{4}$$

where $f(E)$ is the Fermi function, V is again the applied voltage, $M_{\mu\nu}$ is the tunneling matrix element between states ψ_μ and ψ_ν of the respective electrodes, and E_μ is the energy of ψ_μ. For most purposes, the Fermi functions can be replaced by their zero-temperature values, i.e. unit step functions, in which case one of the two terms in braces becomes zero. In the limit of small voltage, this expression further simplifies to

$$I = \frac{2\pi}{\hbar} e^2 V \sum_{\mu,\nu} |M_{\mu\nu}|^2 \delta(E_\mu - E_F)\delta(E_\nu - E_F). \tag{5}$$

These equations are quite simple. The only real difficulty is in evaluating the tunneling matrix elements. Bardeen[13] showed that, under certain assumptions, these can be expressed as

$$M_{\mu\nu} = \frac{\hbar^2}{2m} \int d\mathbf{S} \cdot (\psi_\mu^* \nabla \psi_\nu - \psi_\nu \nabla \psi_\mu^*), \tag{6}$$

where the integral is over any surface lying entirely within the barrier region and separating the two half-spaces. If we choose a plane for the surface of integration, and neglect the variation of the potential in the region of integration, then the surface wave function at this plane can be conveniently expanded in the generalized plane-wave form

$$\psi = \int d\mathbf{q} a_\mathbf{q} e^{-\kappa_q z} e^{i\mathbf{q}\cdot\mathbf{x}}, \tag{7}$$

where z is height measured from a convenient origin at the surface, and

$$\kappa_q^2 = \kappa^2 + |\mathbf{q}|^2. \tag{8}$$

A similar expansion applies for the other electrode, replacing $a_\mathbf{q}$ with $b_\mathbf{q}$, z with $z_t - z$, and $\mathbf{x}$ with $\mathbf{x} - \mathbf{x}_t$. Here $\mathbf{x}_t$ and z_t are the lateral and vertical components of the position of the tip. Then, substituting these wave functions into Eq. (6), one obtains

$$M_{\mu\nu} = -\frac{4\pi^2\hbar^2}{m} \int d\mathbf{q} a_\mathbf{q} b_\mathbf{q}^* \kappa_q e^{-\kappa_q z_t} e^{i\mathbf{q}\cdot\mathbf{x}_t}. \tag{9}$$

Thus, given the wave functions of the surface and tip separately, i.e., $a_\mathbf{q}$ and $b_\mathbf{q}$, one has a reasonably simple expression for the matrix element and tunneling current.

1.2.3 Modeling the Tip

In order to calculate the tunneling current, and hence the STM image or spectrum, it is first necessary to have explicitly the wave functions of the surface and tip, for example in the form of Eq. (7) for use in Eq. (9). Unfortunately, the actual atomic structure of the tip is generally not known.[14] Even if it were known, the very low symmetry makes accurate calculation of the tip wave functions difficult.

One can therefore adopt a reasonable but somewhat arbitrary model for the tip. To motivate the simplest possible model for the tip, consider what would be the ideal STM.[15] First, one wants the maximum possible resolution, and therefore the smallest possible tip. Second, one wants to measure the properties of the bare surface, not of the more complex interacting system of surface and tip. Therefore, the ideal STM tip would consist of a mathematical point source of current, whose position we denote $\mathbf{r}_t$. In that case, Eq. (5) for the current at small voltage reduces to[15]

$$I \propto \sum_{\nu} |\psi_\nu(\mathbf{r}_t)|^2 \delta(E_\nu - E_F) \equiv \rho(\mathbf{r}_t, E_F). \tag{10}$$

Thus the ideal STM would simply measure $\rho(\mathbf{r}_t, E_F)$. This is a familiar quantity, the local density of states at E_F (LDOS), i.e., the charge density from states at the Fermi level. Note that the LDOS is evaluated for the bare surface, i.e., in the absence of the tip, but at the position which the tip will occupy. Thus within this model, STM has quite a simple interpretation as measuring a property of the bare surface, without reference to the complex tip–sample system.

It is important to see how far this interpretation can be applied for more realistic models of the tip. Reference 15 showed that Eq. (10) remains valid, regardless of tip size, so long as the tunneling matrix elements can be adequately approximated by those for an *s*-wave tip wave function. The tip position r_t must then be interpreted as the effective center of curvature of the tip, i.e., the origin of the *s*-wave which best approximates the tip wave functions.

One can also to some extent go beyond the *s*-wave tip approximation, while still getting useful analytical results. A discussion of the contribution of wave function components of higher angular momentum was given by Tersoff and Hamann,[15] who showed that these made little difference for the observable Fourier components of typical STM images. This issue was raised again by Chung *et al.*,[16] and by Chen,[17] who extended the analysis of Ref. 15.

Chen noted that in more recent images of close-packed metals,[18,19] the relevant Fourier components are high enough that higher angular momentum components of the tip wave function could indeed affect the image substantially. However, it is important to recognize that such deviations from the behavior expected for an *s*-wave tip would be large only on this special class of surfaces. If a tip were to have a purely d_z wave function, then the corrugation of Al(111), for example, would be drastically increased; but the effect on surfaces observable with typical STM resolution would be relatively modest. Whether such tips exist is in any case an open question.

Sacks *et al.*[20] have also introduced a model that includes all components of the wave functions of both tip and surface. The essential approximation there was to treat both electrodes as rather flat, so that any deviations from planarity could be treated in perturbation theory.

Most recently, it was shown[21] that for a free-electron-like tip, the *s*-wave tip model should accurately describe STM images except in the case of tunneling to band-edge states, e.g., to semiconducting surfaces at low voltage; and that even then, none but the lowest Fourier component of the image should differ much from the *s*-wave result. This of course neglects the obvious effects of tip geometry, e.g., double tips.[22]

To model the tip more realistically, one must turn to numerical calculations of wave functions for a specific tip. Several studies in this vein, which support the simple LDOS interpretation of Eq. (10), are described in a later section.[23-26]

1.3. Metal Surfaces: STM as Surface Topography

1.3.1 Calculation of the LDOS

In order to interpret STM images quantitatively, just as with other experimental techniques, it is often necessary to calculate the image for a proposed structure or set of structures, and compare it with the actual image. Such calculations for STM are still rather rare. We will discuss some examples and general features of such calculations, as well as the qualitative interpretation of STM images without detailed calculations.

For simple metals, there is typically no strong variation of the local density of states or wave functions with energy near the Fermi level. For purposes of STM, the same is presumably true for noble and even transition metals, since the *d* shell apparently does not contribute significantly to the tunneling current.[25] It is therefore convenient in the case of metals to ignore the voltage dependence, and consider the limit of small voltage of Eq. (10). (Effects of finite voltage are discussed in Sections 1.4 and 1.7.)

This is particularly convenient, since we then require only the calculation of the LDOS $\rho(\mathbf{r}, E_F)$, a property of the bare surface. Nevertheless, even this calculation is quite demanding numerically. One case however of the STM image being calculated for a real metal surface, and compared with experiment, is that of Au(110) 2×1 and 3×1, treated in Ref. 15. The LDOS calculated for these Au surfaces is shown in Fig. 3.

1.3.2 Atom-Superposition Modeling

The real strength of STM is that, unlike diffraction, it is a local probe, and so can be readily applied to large, complex unit cells, or even to disordered surfaces or to isolated features such as defects. However, while the accurate calculation of $\rho(\mathbf{r}, E_F)$ is difficult even for Au(110) 3×1, it is out of the question for surfaces with very large unit cells, and *a fortiori* for disordered surfaces or defects. It is therefore highly desirable to have a method, however approximate, for calculating STM images in these important but intractable cases.

Such a method has been suggested and tested in Ref. 15. It consists of

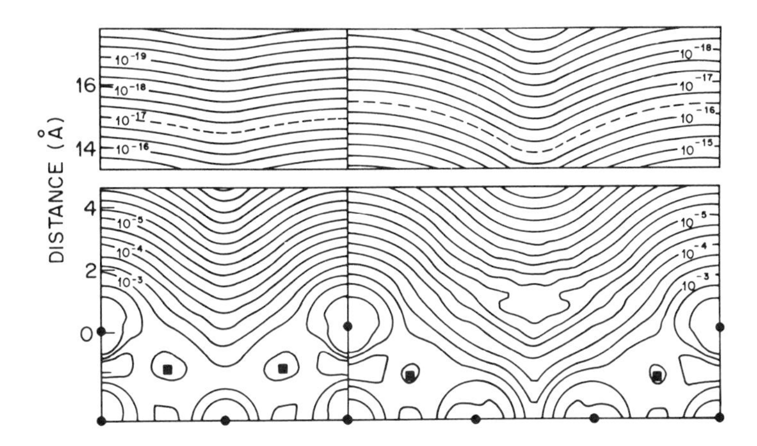

FIG. 3. Calculated $\rho(\mathbf{r}, E_F)$ for Au(110) 2 × 1 (left) and 3 × 1 (right) surfaces. This figure shows (1$\bar{1}$0) plane through outermost atoms. Positions of nuclei are indicated by circles (in plane) and squares (out of plane). Contours of constant ρ are labeled in units of a.u.$^{-3}$eV^{-1}. Note break in vertical distance scale. Assuming a 9 Å tip radius in the s-wave tip model, the center of curvature of the tip is calculated to follow the dashed line. (From Ref. 15.)

approximating the LDOS (10) by a superposition of spherical atomic-like densities,

$$\rho(\mathbf{r}, E_F) \propto \sum_{\mathbf{R}} e^{-2\kappa|\mathbf{r}-\mathbf{R}|}/(E_0\kappa|\mathbf{r}-\mathbf{R}|). \tag{11}$$

Here each term is an atomic-like density centered on the atom site. The choice of an s-wave Hankel function allows convenient analytical manipulations, and provides an accurate description even at large distances.[15] E_0 is an energy which relates the charge to the density of states, and is typically of order 0.5–1.0 eV.[15]

This approach is expected to work very well for simple and noble metals, and was tested in detail for Au(110).[15] The success of the method relies on the fact that the model density, by construction, has the same analytical properties as the true density for a constant potential. Thus if the model is accurate near the surface, it will automatically describe accurately the decay with distance.

As one example of where this approach can be useful, the image expected for Au(110) 3 × 1 was calculated for two plausible models of the structure, differing only in the presence or absence of a missing row in the second layer.[15] The similarity of the model images at distances of interest suggested that the structure in the second layer could not be reliably inferred from experimental images. Quantifying the limits of valid interpretation in this way is an essential part of the analysis of STM data.

While this method is intended primarily for metals, Tromp *et al.*[27] applied

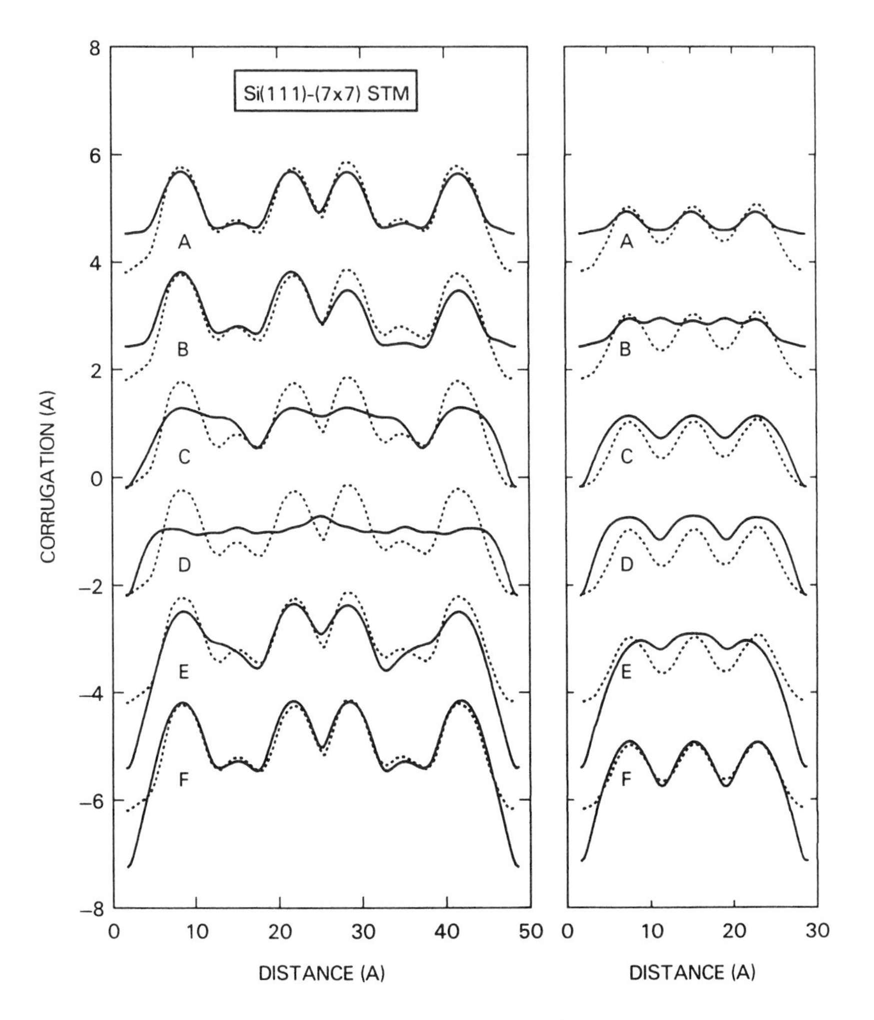

FIG. 4. Comparison between experimental STM images of Si(111) 7 × 7 (dotted line), and the images calculated with the atom-superposition model (Eq. 11), for six different proposed structures of this surface. Left panel shows line scan along long diagonal of 7 × 7 cell, right panel along short diagonal. Curve A is for the adatom model proposed by Binnig *et al.*,[9] Curve F is for the model of Takayanagi *et al.*,[28] other curves are discussed in Ref. 27. (From Ref. 27.)

it to Si(111) 7 × 7 with remarkable success. They simulated the images expected for a number of different proposed models of this surface, and compared them with an experimental image. This comparison is shown in Fig. 4. The so-called "Dimer-Adatom-Stacking Fault" model of Takayanagi[28] gives an image which agrees almost perfectly with experiment, and a simple adatom model[9] is also rather close. Other models, although intended to be consistent with the STM measurements, lead to images with little similarity to experiment. Thus the usefulness of such image simulations must not be underestimated, although few such applications have been made to date.

1.4. Semiconducting Surfaces: Role of Surface Electronic Structure

1.4.1 Voltage Dependence of Images

At very small voltages, the *s*-wave approximation for the tip led to the very simple result in Eq. (10). At larger voltages, one might hope that this could be easily generalized to give a simple expression such as

$$I \sim \int_{E_F}^{E_F+V} \rho(\mathbf{r}_t, E)\, dE. \tag{12}$$

This is not strictly correct for two reasons. First, the matrix elements and the tip density of states are somewhat energy dependent, and any such dependence is neglected in Eq. (12). Second, the finite voltage changes the potential, and hence the wave functions, outside the surface. A more careful discussion is given in Section 1.7. Nevertheless, Eq. (12) is a reasonable approximation for many purposes,[26] so long as the voltage is much smaller than the work function. We shall therefore use Eq. (12) in discussing STM images of semiconductors at modest voltages.

Unlike metals, semiconductors show a very strong variation of $\rho(\mathbf{r}, E_F + V)$ with voltage. In particular, this quantity changes discontinuously at the band edges. With negative sample voltage, current tunnels out of the valence band, while for positive voltage, current tunnels into the conduction band. The corresponding images, reflecting the spatial distribution of valence and conduction-band wave functions respectively, may be qualitatively different.

A particularly simple and illustrative example, which has been studied in great detail, is GaAs (110). There, it was proposed[15] that since the valence states are preferentially localized on the As atoms, and the conduction states on the Ga atoms, STM images of GaAs (110) at negative and positive sample bias should reveal the As and Ga atoms respectively. Such atom-selective imaging was confirmed[15] by direct calculation of Eq. (12), and was subsequently observed experimentally.[29] (See Section 5.3 for further discussion.)

In a single image of GaAs (110), whether at positive or negative voltage, one simply sees a single "bump" per unit cell, as shown in Fig. 5. In fact, the images at opposite voltage look quite similar. It is therefore crucial to obtain both images *simultaneously*, so that the dependence of the absolute position of the "bump" on voltage can be determined. While neither image alone is very informative, by overlaying the two images the zig-zag rows of the (110)

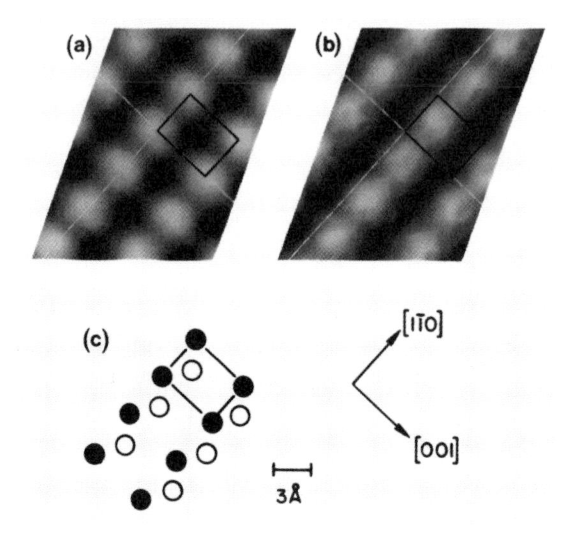

FIG. 5. Grey-scale STM images of GaAs(110) acquired at sample voltages of (a) + 1.9 and (b) − 1.9 V. (c) Top view of GaAs surface structure. The As and Ga atoms are shown as open and filled circles respectively. The rectangle indicates a unit cell, whose position is the same in all three figures. (From Ref. 29.)

surface can be clearly seen.[29] Thus in this case, as with many non-metallic surfaces, voltage-dependent imaging is essential for the meaningful interpretation of STM images on an atomic scale.

Even in this simple case, however, the interpretation of the voltage-dependent images as revealing As or Ga atoms directly is a bit simplistic. Figure 6 shows a line-scan from a measured GaAs image, as well as theoretical "images" (i.e., contours of constant LDOS) for two cases: the ideal surface formed by rigid truncation of the bulk; and the real surface, where the As atom buckles upward, and the Ga downward. In each case, two curves are shown, corresponding to positive and negative bias.

For the ideal surface, at both biases the maxima in Fig. 6a are almost directly over the respective atoms, supporting the simple interpretation of the image. For the buckled surface, though, the apparent positions of the atoms in the images deviate significantly from the actual positions. The separation between the Ga and As atoms in the (001) direction after buckling is less than 1.3 Å, while the separation in this direction of the maxima in the image (Fig. 6b) is 2.0 Å. Thus, the distance between maxima in the image differs from that between atoms by over 50%. Qualitatively, one might say that the maxima correspond to the positions of the respective dangling bonds; but such an interpretation is difficult to quantify.

This deviation of the bumps from the atom positions could be viewed as

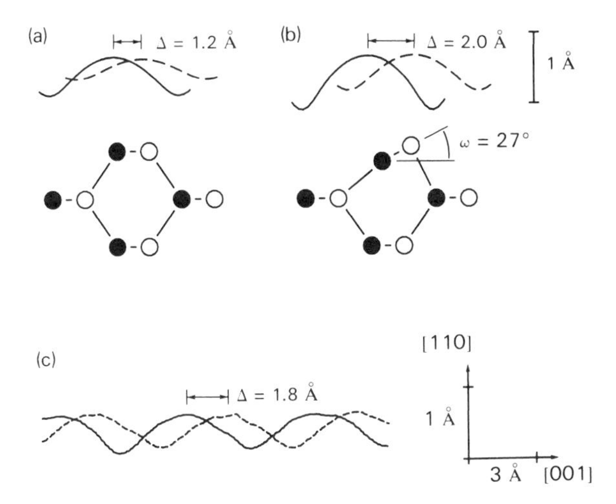

FIG. 6. Contour of constant LDOS in $(1\bar{1}0)$ cross-section, for occupied (dashed line) and unoccupied (solid line) states. Absolute vertical positions are arbitrary. Δ is the lateral distance between peak positions for occupied and unoccupied states. (a) Theoretical results for ideal (unbuckled) surface. (b) Theoretical results for surface with 27° buckling. Side view of atomic structure is also shown, with open and filled circles indicating As and Ga. (c) Experimental results. (From Ref. 29.)

an undesirable complication, since it makes the image even less like a topograph. Alternatively, it is possible to take advantage of this deviation. The apparent positions of the atoms turn out to be rather sensitive to the degree of buckling associated with the (110) surface reconstruction. As a result, it is possible to infer the surface buckling quantitatively from the apparent atom positions.[29] Thus the images are actually quite rich in information; but the quantitative interpretation requires a more detailed analysis than is often feasible. Even for a qualitative analysis, we cannot sufficiently emphasize the importance of voltage-dependent imaging, to help separate electronic from topographic features.

In tunneling to semiconductors, there are additional effects not present in metals. There may be a large voltage drop associated with band-bending in the semiconductor, in addition to the voltage drop across the gap.[30] This means that the tunneling voltage may be substantially less than the applied voltage, complicating the interpretation. More interestingly, local band-bending associated with defects or adsorbates on the surface can lead to striking non-topographic effects in the image,[30] and localized states in the band gap lead to fascinating voltage-dependent images.[31] While beyond the scope of this review, these effects are of great interest for anyone involved in STM of semiconductor surfaces. Chapter 5 gives a more detailed account of the rich field of STM on semiconductor surfaces.

1.4.2 Imaging Band-Edge States

A particularly interesting situation can arise in tunneling to semiconducting surfaces at low voltages, and also to some semi-metals such as graphite.[32,33] At the lowest possible voltages, only states at the band edge participate in tunneling. These band-edge states typically (though not necessarily) fall at a symmetry point at the edge of the surface Brillouin zone. In this case, the states which are imaged have the character of a standing wave on the surface.

This standing-wave character leads to an image with striking and peculiar properties.[33,34] The corrugation is anomalously large; and unlike the normal case, it does not decrease rapidly with distance from the surface. This gives the effect of unusually sharp resolution.[33,34] For example, in the case of graphite the unit cell is easily resolved despite the fact that it is only 2 Å across. This effect was also seen on Si(111) 2 × 1.[32]

In most such cases the image can be adequately described by a universal form,[33] consisting of an array of dips with the periodicity of the lattice. (The dips however are weakened by a variety of effects, and in any case may not be well resolved because of the limits of instrumental response.[33])

Such an image can be extremely misleading. When there is one "bump" per unit cell, it is tempting to infer that there is one topographic feature per unit cell. However, in the case of Si(111) 2 × 1, like GaAs(110), it is necessary to combine images from both positive and negative voltage to get a reasonable picture of the surface. In extreme cases, such as certain images of charge–density–wave materials,[35] the image may in fact carry no information whatever regarding the distribution of atoms within the unit cell.[33,34]

1.5. Adsorbates on Metal Surfaces

1.5.1 Topography

There have been several studies of adsorbates on metals[23-26] modeling the tip (and the sample) as an atom adsorbed on a model free-electron metal substrate (i.e., the so-called "jellium" model).[36] Results of one such calculation for the STM images of a sodium and a sulfur atom on jellium (with a sodium atom tip) are shown in Figs. 7 and 8.

In the case of Na, where everything is highly metallic, the theoretical image in Fig. 7 is almost indistinguishable from the contour of constant local density of states, in striking confirmation of the *s*-wave tip model. In addition, both of these look like the contour of total charge density, which we used above as the definition of a surface topograph. For sulfur, Fig. 8, the image is again well reproduced by the LDOS, but these are distinctly different from the total charge density. So while the simple LDOS interpretation of Eq.

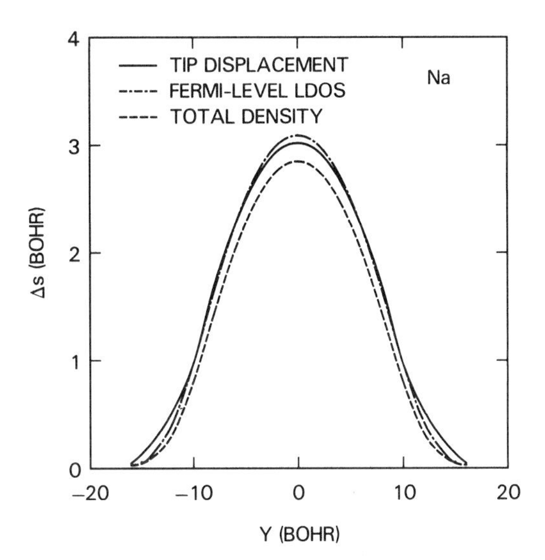

FIG. 7. Comparison of theoretical STM image for Na adatom sample and Na adatom tip, with contours of constant Fermi-level local density of states and constant total charge density. (From Ref. 23.)

(10) remains valid, the image does not correspond as closely to a topograph in this case.

It was found also in these studies that the local density of states at the Fermi level for certain adsorbed atoms like helium and oxygen adsorbed on

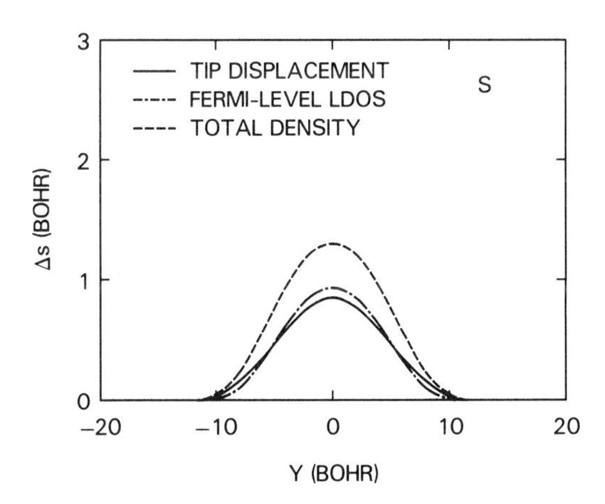

FIG. 8. Comparison of theoretical STM image for S adatom sample and Na adatom tip, with contours of constant Fermi-level local density of states and constant total charge density. (From Ref. 23.)

free-electron metals was lower than that for the bare metal; that is, the presence of the atom makes a negative contribution to the LDOS at this energy. This of course would cause these atoms to look like holes or depressions in the surface in a topographic image; there is evidence of this in the studies of Wintterlin *et al.*[37] for oxygen on Al(111).

1.5.2 Voltage Dependence of Images: Apparent Size of an Adatom

The tip displacement in constant-current mode is studied in Ref. 25 for an isolated adatom for different bias voltages. We call the maximum vertical tip displacement due to an adatom the apparent vertical size of the adatom, Δs; thus this reference studies $\Delta s(V)$.

Curves of $\Delta s(V)$ for Na and Mo sample adatoms atoms are shown in Figs. 9 and 10 for V in the range -2 to $+2$ eV. Positive V again corresponds to the polarity in which electrons tunnel into *empty* states of the sample. At the top of each figure, the corresponding sample state-density curve for energies in the range -2 to $+2$ eV is given for comparison.

Figure 9 shows the total additional state density (i.e., the total state density minus that of the bare metal) and the $\Delta s(V)$ curve for the case of the Na sample atom. The 3*s* resonance peak in the density of states is mostly above the Fermi level; the $\Delta s(V)$ curve clearly reflects this resonance. The curvature of $\Delta s(V)$ up toward the left in the negative bias region is simply an effect of the exponential barrier penetration factor. It would be present even if the state density of the sample, and the tip as well, had no structure whatsoever.

In the case of Mo shown in Fig. 10, the large state-density peak just below the Fermi level corresponds to the 4*d* state of the free atom, and the smaller peak about 1 eV above the Fermi level corresponds to the 5*s* state. The graph of $\Delta s(V)$ is very similar to that of Na, with the large peak at positive bias associated with the 5*s* states of the adsorbed atom. It is striking how little evidence there is of the 4*d* state in this graph.[25] The reason for this is that the valence *d* orbitals in the transition elements are in general quite localized relative to the valence *s* orbitals, and will thus have a much smaller amplitude at the tip. The relative unimportance of the contribution from *d* states in tunneling spectroscopy has been confirmed experimentally for Au and Fe by Kuk and Silverman.[38]

1.6. Close Approach of the Tip: The Strong-Coupling Regime

1.6.1 From Tunneling to Point Contact

Let us now consider what happens when we start to reduce the tip–sample distance, and pass into the region of transition from tunneling to point-

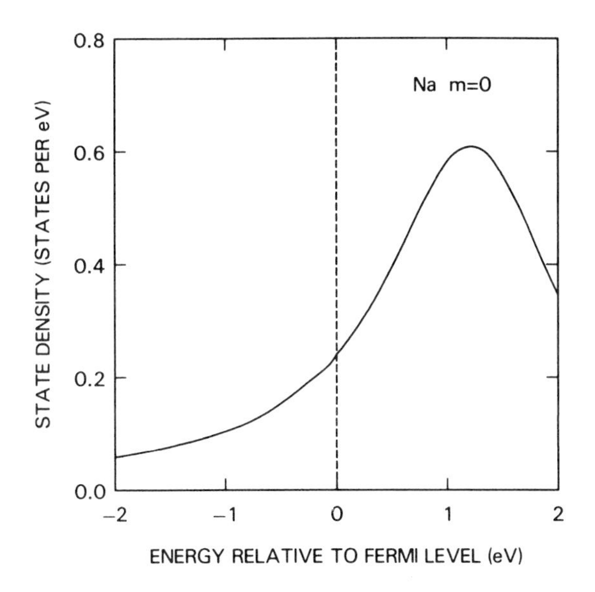

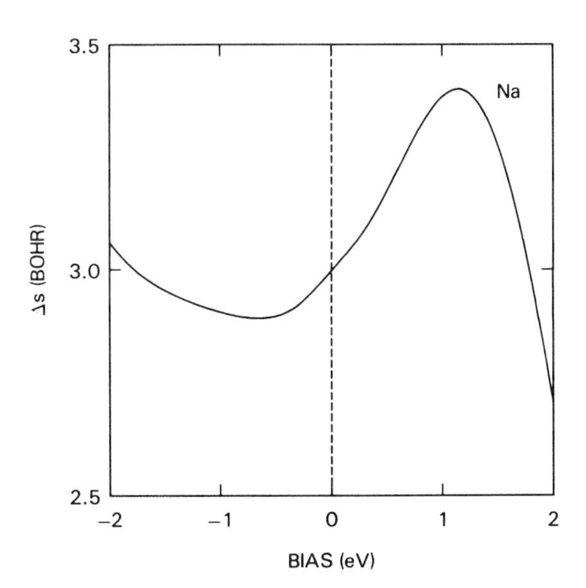

FIG. 9. Top: Difference in eigenstate density between the metal-adatom system and the bare metal for adsorbed Na. Only the $m = 0$ component is shown. The $3s$ resonance is clearly evident. Bottom: $\Delta s(V)$ for Na sample adatom. (From Ref. 25.)

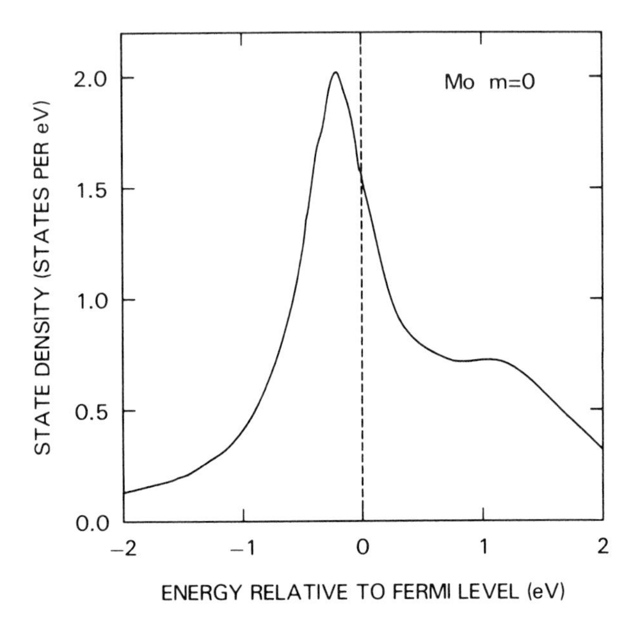

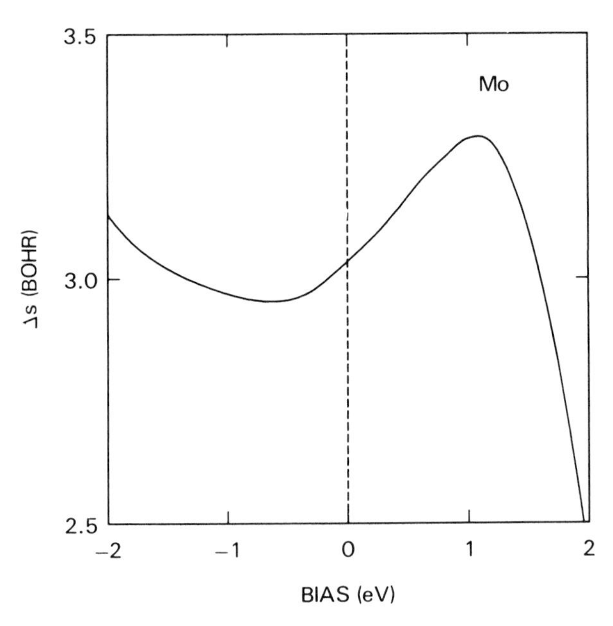

FIG. 10. Top: Mo eigenstate density difference. The lower-energy Mo peak correspo
the upper smaller peak to 5s. Bottom: $\Delta s(V)$ for Mo sample adatom. (From Ref. 25

contact. The initial contact takes place, in the experiment we discuss, between a single tip atom and the sample surface.[11,38,39] We study the current that flows between the two flat metallic electrodes in the jellium model, one of which here has an adsorbed atom (the tip electrode), as a function of distance between them. For simplicity we consider the case of a small applied bias voltage.

Now in contrast to the studies described earlier, it is not possible here to use a tunneling-Hamiltonian formalism, since the overlap of the wave functions of the two electrodes is no longer smaller. It is necessary therefore to treat the tip and sample together as a single system in computing the wave functions.

We are interested in the additional current density due to the presence of the atom, $\delta\mathbf{j}(\mathbf{r}) = \mathbf{j}(\mathbf{r}) - \mathbf{j}_0$, where $\mathbf{j}(\mathbf{r})$ is the current density in the presence of the atom and $\mathbf{j}_0$ that in its absence, and in the total additional current δI which is obtained by integration of $\delta\mathbf{j}$ over an appropriate surface. We can define the additional conductance due to the presence of the atom as $\delta G = \delta I/V$, and it is convenient to define an associated resistance $R \equiv 1/\delta G$.

Consider as before the simple case of a Na tip atom. The tip–sample separation s is measured from the nucleus of the tip atom to the positive-background edge of the sample; the distance d between the center of the Na atom and the positive-background edge of the tip electrode is held at 3 bohrs, the equilibrium value for $s \rightarrow \infty$.

In Fig. 11, the resistance R defined above is shown as a function of separation s. At large separations, R changes exponentially with s. As s is decreased toward d, the resistance levels out at a value of 32,000 Ω. (For $s = d$, the atom is midway between the two metal surfaces, so this can in some sense be taken to define contact between the tip atom and the sample surface.) We can understand this leveling out, including the order of magnitude of the resistance, from discussions of Imry and Landauer.[40] These authors point out that there will be a "constriction" resistance $\pi\hbar/e^2 = 12{,}900\,\Omega$ associated with an ideal conduction channel, sufficiently narrow to be regarded as 1-dimensional, which connects two large reservoirs. Our atom, in the instance in which it is midway between the two electrodes, contacting both, forms a rough approximation to this.

Just such a plateau in the resistance was found in the experiments of Gimzewski and Möller[39] using a Ag sample surface and an Ir tip (though the identity of the tip atom itself was not determined). These authors fix the voltage on the tip at some small value, and measure the current as a function of distance as the tip is moved toward the sample. They find a plateau in the current at about the same distance from the surface that it is found here, with the resistance at the plateau $\sim 35\,\mathrm{k\Omega}$. Analogous results were found by Kuk and Silverman[38] with a minimum resistance of $\sim 24\,\mathrm{k\Omega}$.

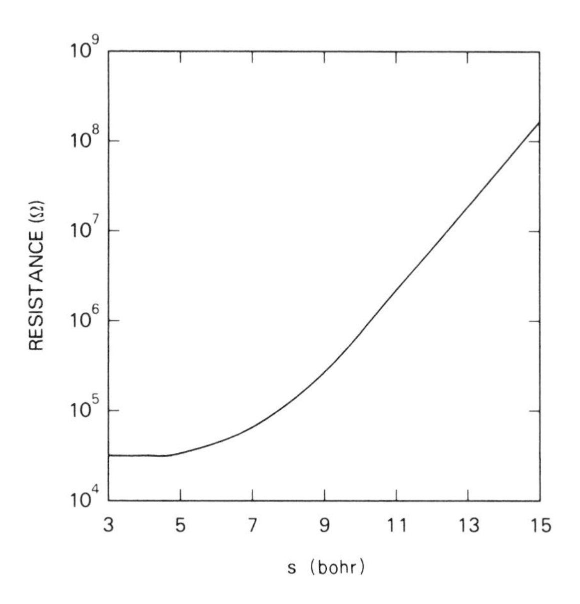

FIG. 11. Calculated resistance $R \equiv 1/\delta G$ as a function of tip–sample separation s for a Na tip atom. (From Ref. 11.)

1.6.2 Measuring the Tunneling Barrier

Now consider the height of the tunneling barrier as it is commonly measured in the STM.[41] For the simple case of a one-dimensional square barrier of height φ above the Fermi level, the tunneling current I at a small bias V is proportional to $V\exp(-2s\sqrt{2m\varphi}/\hbar)$, where s is the barrier thickness; thus for constant bias, $\varphi = (\hbar^2/8m)(d\ln I/ds)^2$. This relationship is often used to *define* an *apparent* barrier height (or local "effective work function")

$$\varphi_A \equiv \frac{\hbar^2}{8m}\left(\frac{d\ln I}{ds}\right)^2. \tag{13}$$

A number of authors, as noted previously, present "local barrier-height" images of surfaces (that is, images of φ_A) as an alternative to the more commonly shown "topographic" images.[8]

Reference 41 studies apparent barrier height φ_A as a function of separation for a single-atom tip and a flat sample surface, using exactly the same model employed previously to discuss the resistance. An experimental study of the dependence of apparent barrier height on separation has been reported by Kuk and Silverman.[38] Now it is clear that for tip–sample separation $s \to \infty$, $\varphi_A(s) \to \Phi$, the sample surface work function. The question of interest is

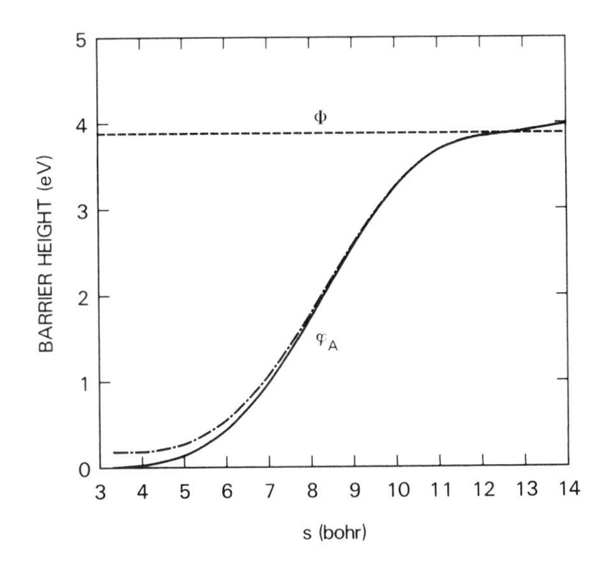

FIG. 12. Apparent barrier height φ_A as a function of tip–sample separation s for two electrodes, one representing the sample, the other, with an adsorbed Na atom, representing the tip. See text for distinction between solid and dot-dash curves. The work function Φ for the sample electrode by itself is shown for comparison. (From Ref. 41.)

whether or not $\varphi_A(s)$ will be close to Φ for the range of tip–sample separations commonly used in scanning tunneling microscopy.

The apparent barrier height $\varphi_A(s)$ obtained in Ref. 41 from $\delta I(s)$ via Eq. (13) is shown in Fig. 12. Note that φ_A is well below the sample work function Φ except at the largest distances shown in the graph, where in fact it goes slightly above (it reaches a maximum of 4.3 eV at $s \sim 19$ bohrs). That it is below Φ for intermediate distances results from the long range of both the exchange-correlation part of the surface potential (see the discussion in Ref. 41), and the electrostatic potential due to the charge-transfer dipole that forms at the Na tip atom. That it is even further below at the shortest distances results from the complete absence at these separations of a potential barrier to electron tunneling, as we will see. Near the point of nominal contact ($s = 3$ bohrs), the tunneling current reaches a plateau as we have seen and thus φ_A is zero. (It is strictly zero only when all tunneling to the substrate upon which the atom is adsorbed is neglected; including such tunneling in a physically reasonable manner gives the dashed curve in Fig. 12.)

Figure 13 gives contour maps of the total potential seen by an electron, v_{eff}, for several distances s. It shows the contour $v_{\text{eff}}(\mathbf{r}) = E_F$ (the one closest to the atom), with E_F the Fermi energy, as well as a number of other contours for values above ε_F; thus the contour-filled areas represent regions where a

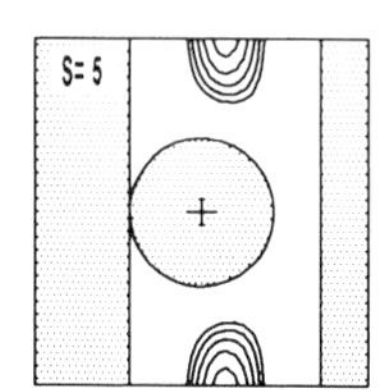

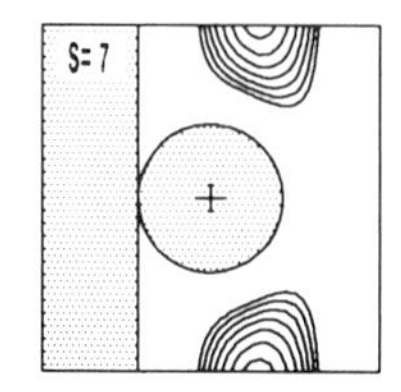

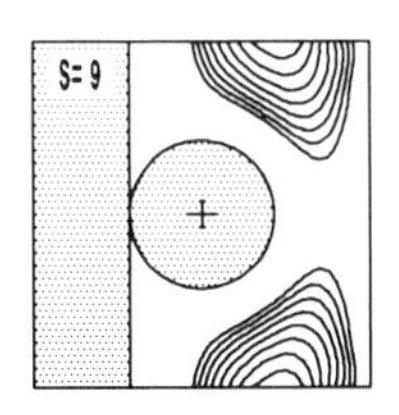

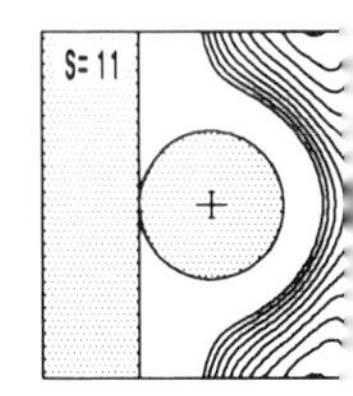

FIG. 13. Contour maps of the potential v_{eff} for two-electrode case with an adsorbed Na atom (tip). The presence of the atom is represented by a shaded circle with a cross at the position of the nucleus; the positive background regions of both tip and sample electrodes are shaded also. Maps are shown for four values of s (given in bohrs; 1 bohr = 0.529 Å), which is the distance between the nucleus of the tip atom and the positive background edge of the sample electrode. The nucleus is at the center of each box, so that the sample electrode in fact lies entirely outside of the box for all but the smallest of the s values shown ($s = 5$ bohrs). The contour closest to the atom in each case is that for $v_{\text{eff}} = E_F$; the other contours correspond to higher energy values. (From Ref. 41.)

Fermi-level electron encounters a potential barrier. At $s = 5$ bohrs, for example, the electrons moving from the single-atom tip to the sample electrode encounter essentially no barrier whatsoever, while at $s = 11$ bohrs, there is a barrier for electrons tunneling along all directions.

For $s = 9$ bohrs, there is a barrier for tunneling in most directions, but not for tunneling nearly normal to the surface. However, even for tunneling in the normal direction there will be an *effective* barrier. The reason for this is that although there is a small opening in the barrier, whose transverse size we denote by a, electrons moving through this opening will, by the uncertainty principle, have a minimum transverse momentum of $O(\hbar/a)$. This in turn decreases the energy available for motion along the direction of the surface normal, and thus a Fermi-level electron will in fact have to tunnel through a barrier even in the region of this "opening."

1.7. Tunneling Spectroscopy

1.7.1 Qualitative Theory

Tunneling spectroscopy in planar junctions was studied long before STM.[6] However, the advent with STM of spatially-resolved spectroscopy has led to a resurgence of interest in this area. Because of the difficulty of calculating $I(\mathbf{r}_t, V)$ in general, most detailed analyses have instead focused simply on $I(V)$, without regard to its detailed tip-position dependence.

Selloni *et al.*[42] suggested that the results of Tersoff and Hamann[15] could be *qualitatively* generalized for modest voltages as

$$I(V) \propto \int_{E_F}^{E_F+V} \rho(E)T(E, V)\,dE, \tag{14}$$

where $\rho(E)$ is the local density of states given by Eq. (10) at or very near the surface, and *assuming* a constant density of states for the tip. This is similar to Eq. (12), except that the effect of the voltage on the surface wave functions is included through a barrier transmission coefficient $T(E, V)$.

Unfortunately, this simple model still does not lend itself to a straightforward interpretation of the tunneling spectrum.[26] In particular, the derivative dI/dV has no simple relationship to the density of states $\rho(E_F + V)$, as might have been hoped. At best, one can say that a sharp feature in the density of states of the sample (or tip), at an energy $E_F + V$, will lead to a feature in $I(V)$ or its derivatives at voltage V.

Even this rather weak statement may prove unreliable in practice, where spectral features have considerable widths. The reason for this problem is that $T(E, V)$ is very strongly V-dependent when the voltage becomes an appreciable fraction of the work function. Thus the V-dependence of $T(E, V)$ may distort features in the spectrum.[26,32]

Stroscio, Feenstra, and coworkers[32,43] proposed a simple but effective solution to this problem. They normalize[32] dI/dV by dividing it by I/V. This yields $d\ln I/d\ln V$, and so effectively cancels out the exponential dependence of $T(E, V)$ on V. At semiconductor band edges, where the current goes to zero, a slight smoothing of I/V eliminates the singular behavior.[43]

This normalization is, however, both unnecessary and undesirable at small voltages; in that case, I/V is well behaved, whereas $(dI/dV)/(I/V)$ is identically equal to unity for ohmic systems, and so carries no information. Thus the appropriate way of displaying spectroscopy results depends on the problem at hand, and a variety of approaches for collecting and displaying data have been considered.[30,32,43] Chapter 4 gives a variety of examples of tunneling spectroscopy.

1.7.2 Quantitative Theory

A rigorous treatment of the tunneling spectrum in STM requires calculation of the wave functions of surface and tip at finite voltage. This is a difficult problem, which is not yet tractable except in simplified models. A natural approximation[26] is therefore to use the zero-voltage wave functions, but to shift all the surface wave functions in energy relative to the tip, by an amount corresponding to the applied voltage V. Unfortunately, the result then depends on the position of the surface of integration for Eq. (6).

However, by positioning the surface of integration half-way between the two planar electrodes, the resulting error is second order in the voltage, and rather small as long as the voltage is much less than the work function.[26] This result assumes that surface and tip have equal work functions, and is derived for a one-dimensional model only. Nevertheless, it seems safe to assume that the conclusion is more generally valid. With this approximation, calculation of the tunneling spectrum becomes relatively straightforward.

Such calculations[26] confirm the qualitative applicability of simple models such as Eq. (14). As an example, the solid curve in Fig. 14 (bottom) gives $(dI/dV)/(I/V)$ for the case of a Ca sample adatom and a Na tip atom. The positions of the two Ca peaks and one Na peak in the spectrum correspond reasonably well to the features of the state-density curve shown at the top. These calculations thus confirm that the spectrum $(dI/dV)/(I/V)$ mimics the density of states reasonably well, as proposed by Stroscio *et al.*[32] Note also the negative values of $(dI/dV)/(I/V)$ for biases near $+2\,\text{eV}$. Such "negative resistance" effects have been discussed by Esaki and Stiles,[44] and have now been observed in the STM context by Bedrossian *et al.*[45] and Lyo and Avouris.[46]

In an experiment, it would be most convenient if the tip state density could be taken as relatively featureless, and thus be omitted from consideration. In some cases, tips are used which are purposely blunt (and probably disordered), and it is found that the tip state density appears to play no significant role in the results.[32] This seems quite reasonable, in view of the expectation that most of the sharp features in the density of states of such a tip would be washed out. Even if the tip were very sharp (a single atom), its state-density structure should be similar to the often broad resonances in the cases studied here; it would certainly not exhibit the complex surface-state structure that may be characteristic of an extended ordered surface made of these same atoms.

Calculations of STM spectra including true voltage-dependent wave functions have not yet been reported. But methods do exist now for performing such calculations "exactly" for a simple model of surface and tip,[12] or more approximately for a real surface.[47]

There is also great interest in inelastic tunneling spectroscopy. This typically involves energy exchange with phonon modes, so the interesting energy scale (and hence voltage) is quite small. Thus $T(E, V)$ in Eq. (14) can in this case be taken as constant. The primary theoretical problem is then to understand the electron–phonon coupling, which has been discussed in several papers.[48] However, in practice the primary issue in inelastic tunneling is simply the experimental problem of achieving very low noise, necessary for observing this weak effect.

1.8. Mechanical Tip–Sample Interactions

Ideally, in STM the tip and surface are separated by a vacuum gap, and are mechanically non-interacting. However, sometimes anomalies are

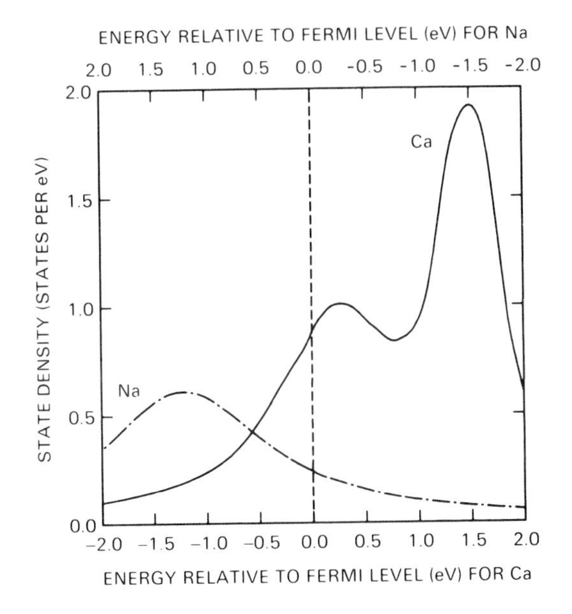

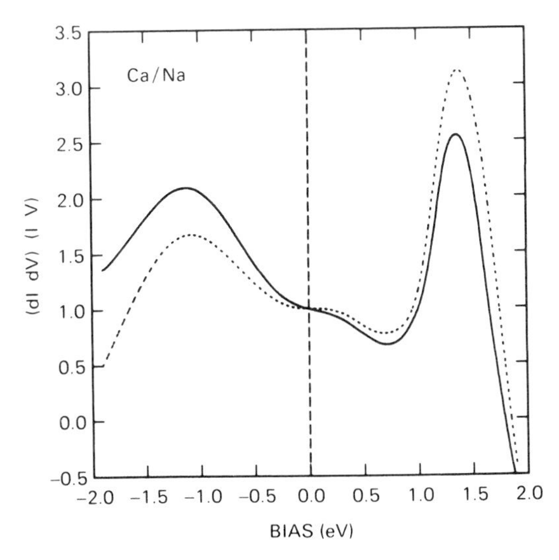

FIG. 14. Top: Curves of the difference in eigenstate density between the metal-adatom system and the bare metal for adsorbed Ca and Na. Note that the energy scale for Na (top) is reversed. The $3s$ resonance for Na is clearly evident. The lower-energy Ca peak corresponds to $4s$, the upper to $3d$ (and some $4p$). (Only $m = 0$ is shown.) Bottom: Solid line is calculated curve of $(dI/dV)/(I/V)$ versus V for Ca/Na; dotted line is same quantity evaluated using a simple analytical model discussed in Ref. 26. Center-to-center distance between the atoms is held fixed at 18 bohrs (1 bohr = 0.529 Å). (From Ref. 26.)

observed, which are most easily explained by assuming a mechanical interaction between the tip and surface. In particular, since the earliest vacuum tunneling experiments of Binnig *et al.*,[5] it has been observed that for dirty surfaces, the current varies less rapidly than expected with vertical displacement of the tip. Coombs and Pethica[49] pointed out that this behavior can be explained by assuming that the dirt mediates a mechanical interaction between surface and tip.

As discussed in Section 1.6.2, the apparent barrier height is defined in terms of $d \ln I/ds$. For dirty surfaces, $d \ln I/ds$ is smaller than expected, leading to an inferred barrier height which is unphysically small. For graphite in air, this inferred barrier height is often less than 0.1 eV.

The suggestion of Coombs and Pethica is that some insulating dirt (e.g., oxide) is squeezed between the surface and tip, acting in effect as a spring. The actual point of tunneling could be a nearby part of the tip which is free of oxide, or an asperity poking through the dirt. As the tip is lowered, the dirt becomes compressed, pushing down the surface or compressing the tip if it is sufficiently soft. As a result, the surface–tip separation does not really decrease as much as expected from the nominal lowering of the tip, and so the current variation is correspondingly less.

This issue gained renewed importance with the observation of giant corrugations in STM of graphite.[50,51] Ridiculously large corrugations were sometimes observed, up to 10 Å or more vertically within the 2 Å wide unit cell of graphite. Soler *et al.*[50] at first attributed these corrugations to direct interaction of the tip and surface, but a detailed study by Mamin *et al.*[51] suggests that in fact the interaction is mediated by dirt, consistent with the earlier proposal.[49] Very clean UHV measurements[52] apparently give corrugations of at most a small fraction of an Å, consistent with recent theoretical calculations for realistic single-atom tips.[53]

For the model of direct interaction,[50] there is a complex nonlinear behavior. But for the dirt-mediated interaction model, a linear treatment is appropriate.[49] In this case, assuming the mechanical interaction has negligible corrugation (e.g., because of the large interaction area), the image seen is simply the ideal image, with the vertical axis distorted by a constant scale factor.

It seems surprising that it is possible to obtain any image at all, when there is dirt between surface and tip, since some scraping might be expected as the tip is scanned. For graphite, one could imagine that the dirt (e.g., metal oxide) moves with the tip, sliding nicely along the inert graphite surface. Or the "dirt" might be a liquid such as water. However, any such model is based on indirect inference, and this phenomenon must be considered as not really understood at present.

This lack of a complete understanding is pointed up by the possible role of forces even in certain clean situations. In the recent observation of close-packed metal surfaces, in which individual atoms were resolved,[18,19] it was suggested[19] that these cases like graphite may represent an enhancement of the corrugation by mechanical interactions between tip and surface. Apparent barrier heights as small as 1 eV are not unusual even in ultra-clean UHV experiments,[54] where atomic-resolution images would seem to be incompatible with direct mechanical contact. Such effects may be associated with a very deformable tip (i.e., a "whisker");[19,54] or they might be unrelated, reflecting e.g., the small apparent barrier heights discussed above for small tip–surface separation. However, the question of what forces exist in clean vacuum experiments, and their effect on STM images, remains largely unexplored at this time.

References

1. G. Binnig, H. Rohrer, Ch. Gerber, and E. Weibel, *Phys. Rev. Lett.* **49**, 57 (1982). Some general reviews of STM include G. Binnig and H. Rohrer, *Rev. Mod. Phys.* **59**, 615 (1987); P. K. Hansma and J. Tersoff, *J. Appl. Phys.* **61**, R1 (1987). For a previous review specifically of the theory of STM which touched on some more technical issues, see J. Tersoff, p. 77 in *Scanning Tunneling Microscopy and Related Techniques*, (R. J. Behm, N. Garcia, and H. Rohrer, eds.) NATO ASI Series Vol. 184, Kluwer Academic Publishers, 1990.
2. R. M. Feenstra, W. A. Thompson, and A. P. Fein, *Phys. Rev. Lett.* **56**, 608 (1986); J. A. Stroscio, R. M. Feenstra, D. M. Newns, and A. P. Fein, *J. Vac. Sci. Technol. A* **6**, 499 (1988).
3. R. J. Hamers, R. M. Tromp, and J. E. Demuth, *Phys. Rev. Lett.* **56**, 1972 (1986).
4. P. Muralt and D. W. Pohl, *Appl. Phys. Lett.* **48**, 514 (1986).
5. G. Binnig, H. Rohrer, Ch. Gerber, and E. Weibel, *Appl. Phys. Lett.* **40**, 178 (1982). For earlier reports of vacuum tunneling see E. C. Teague, Ph.D. Thesis, North Texas State University, 1978, reprinted in *J. Res. National Bureau of Standards* **91**, 171 (1986); R. Young, J. Ward, and F. Scire, *Phys. Rev. Lett.* **27**, 922 (1971); and W. A. Thompson, *Rev. Sci. Instr.* **47**, 1303 (1976).
6. See for example C. B. Duke, p. 1 in *Tunneling in Solids*, Suppl. 10 of *Solid State Physics*, (F. Seitz and D. Turnbull, eds.) Academic Press, New York, 1969, and references therein.
7. A. Bryant, D. P. E. Smith, and C. F. Quate, *Appl. Phys. Lett.* **48**, 832 (1986).
8. G. Binnig and H. Rohrer, *Surf. Sci.* **126**, 236 (1983); R. Wiesendanger, L. Eng, H. R. Hidber, P. Oelhafen, L. Rosenthaler, U. Staufer, and H.-J. Güntherodt, *Surf. Sci.* **189/190**, 24 (1987); B. Marchon, P. Bernhardt, M. E. Bussell, G. A. Somorjai, M. Salmeron, and W. Siekhaus, *Phys. Rev. Lett.* **60**, 1166 (1988).
9. G. Binnig, H. Rohrer, Ch. Gerber, and E. Weibel, *Phys. Rev. Lett.* **50**, 120 (1983).
10. N. Garcia, C. Ocal, and F. Flores, *Phys. Rev. Lett.* **50**, 2002 (1983); E. Stoll, A. Baratoff, A. Selloni, and P. Carnevali, *J. Phys. C* **17**, 3073 (1984).
11. N. D. Lang, *Phys. Rev. B* **36**, 8173 (1987).
12. N. D. Lang, A. Yacoby, and Y. Imry, *Phys. Rev. Lett.* **63**, 1499 (1989).
13. J. Bardeen, *Phys. Rev. Lett.* **6**, 57 (1961).
14. For a unique exception, see Y. Kuk, P. J. Silverman, and H. Q. Nguyen, *J. Vac. Sci. Technol. A* **6**, 524 (1988).

15. J. Tersoff and D. R. Hamann, *Phys. Rev. B* **31**, 805 (1985); and *Phys. Rev. Lett.* **50**, 1998 (1983).
16. M. S. Chung, T. E. Feuchtwang, and P. H. Cutler, *Surf. Sci.* **187**, 559 (1987).
17. C. J. Chen, *J. Vac. Sci. Technol. A* **6**, 319 (1988), and *Phys. Rev. Lett.* **65**, 448 (1990).
18. V. M. Hallmark, S. Chiang, J. F. Rabolt, J. D. Swalen, and R. J. Wilson, *Phys. Rev. Lett.* **59**, 2879 (1987).
19. J. Wintterlin, J. Wiechers, H. Brune, T. Gritsch, H. Höfer, and R. J. Behm, *Phys. Rev. Lett.* **62**, 59 (1989).
20. W. Sacks, S. Gauthier, S. Rousset, and J. Klein, *Phys. Rev. B* **37**, 4489 (1988).
21. J. Tersoff, *Phys. Rev. B* **41**, 1235 (1990).
22. S. Park, J. Nogami, and C. F. Quate, *Phys. Rev. B* **36**, 2863 (1987); H. A. Mizes, S. Park, and W. A. Harrison, *Phys. Rev. B* **36**, 4491 (1987).
23. N. D. Lang, *Phys. Rev. Lett.* **56**, 1164 (1986).
24. N. D. Lang, *Phys. Rev. Lett.* **55**, 230 and 2925 (E) (1985).
25. N. D. Lang, *Phys. Rev. Lett.* **58**, 45 (1987).
26. N. D. Lang, *Phys. Rev. B* **34**, 5947 (1986).
27. R. M. Tromp, R. J. Hamers, and J. E. Demuth, *Phys. Rev. B* **34**, 1388 (1986).
28. K. Takayanagi, Y. Tanishiro, M. Takahashi, and S. Takahashi, *J. Vac. Sci. Technol. A* **3**, 1502 (1985), and references therein.
29. R. M. Feenstra, J. A. Stroscio, J. Tersoff, and A. P. Fein, *Phys. Rev. Lett.* **58**, 1192 (1987).
30. R. M. Feenstra and J. A. Stroscio, *J. Vac. Sci. Technol. B* **5**, 923 (1987); J. A. Stroscio and R. M. Feenstra, *J. Vac. Sci. Technol. B* **6**, 1472 (1988); P. Mårtensson and R. M. Feenstra, *Phys. Rev. B* **39**, 7744 (1989).
31. R. M. Feenstra, *Phys. Rev. Lett.* **63**, 1412 (1989); P. N. First, J. A. Stroscio, R. A. Dragoset, D. T. Pierce, and R. J. Celotta, *Phys. Rev. Lett.* **63**, 1416 (1989).
32. J. A. Stroscio, R. M. Feenstra, and A. P. Fein, *Phys. Rev. Lett.* **57**, 2579 (1986); but regarding the role of tip electronic structure in spectroscopy, see by contrast R. M. Tromp, E. J. Van Loenen, J. E. Demuth, and N. D. Lang, *Phys. Rev. B* **37**, 9042 (1988), where an important tip contribution was observed.
33. J. Tersoff, *Phys. Rev. Lett.* **57**, 440 (1986).
34. J. Tersoff, *Phys. Rev. B* **39**, 1052 (1989).
35. R. V. Coleman, B. Giambattista, P. K. Hansma, A. Johnson, W. W. McNairy, and C. G. Slough, *Adv. Phys.* **37**, 559 (1988).
36. For a discussion of the jellium model, see e.g., N. D. Lang, p. 309, in *Theory of the Inhomogeneous Electron Gas*, (S. Lundqvist and N. H. March, eds.) Plenum Press, New York, 1983.
37. J. Wintterlin, H. Brune, H. Höfer, and R. J. Behm, *Appl. Phys. A* **47**, 99 (1988).
38. Y. Kuk and P. J. Silverman, *J. Vac. Sci. Technol. A* **8**, 289 (1990). Regarding the role of *d* states, see also discussion in J. E. Demuth, U. Koehler, and R. J. Hamers, *J. Microscopy* **152**, 299 (1988).
39. J. K. Gimzewski and R. Möller, *Phys. Rev. B* **36**, 1284 (1987); and data of Gimzewski and Möller reproduced in Ref. 11.
40. Y. Imry, p. 101 in *Directions in Condensed Matter Physics: Memorial Volume in Honor of Shang-keng Ma*, (G. Grinstein and G. Mazenko, eds.) World Scientific, Singapore, 1986; R. Landauer, *Z. Phys. B* **68**, 217 (1987).
41. N. D. Lang, *Phys. Rev. B* **37**, 10395 (1988).
42. A. Selloni, P. Carnevali, E. Tosatti, and C. D. Chen, *Phys. Rev. B* **31**, 2602 (1985).
43. R. M. Feenstra and P. Mårtensson, *Phys. Rev. Lett.* **61**, 447 (1988).
44. L. Esaki and P. J. Stiles, *Phys. Rev. Lett.* **16**, 1108 (1966).
45. P. Bedrossian, D. M. Chen, K. Mortensen, and J. A. Golovchenko, *Nature* **342**, 258 (1989).

46. I.-W. Lyo and Ph. Avouris, *Science* **245**, 1369 (1989).
47. J. Tersoff, *Phys. Rev. B* **40**, 11990 (1989).
48. B. N. J. Persson and A. Baratoff, *Phys. Rev. Lett.* **59**, 339 (1987); B. N. J. Persson and J. E. Demuth, *Solid State Comm.* **57**, 769 (1986); G. Binnig, N. Garcia, and H. Rohrer, *Phys. Rev. B* **32**, 1336 (1985).
49. J. H. Coombs and J. B. Pethica, *IBM J. Res. Develop.* **30**, 455 (1986).
50. J. M. Soler, A. M. Baro, N. Garcia, and H. Rohrer, *Phys. Rev. Lett.* **57**, 444 (1986).
51. H. J. Mamin, E. Ganz, D. W. Abraham, R. E. Thomson, and J. Clarke, *Phys. Rev. B* **34**, 9015 (1986).
52. R. J. Hamers, in *Physics of Solid Surfaces*, (G. Chiarotti, ed.) Landolt-Börnstein, (in press).
53. J. Tersoff and N. D. Lang, *Phys. Rev. Lett.* **65**, 1132 (1990).
54. R. M. Feenstra, private communication.

2. DESIGN CONSIDERATIONS FOR AN STM SYSTEM

Sang-il Park

Park Scientific Instruments, Sunnyvale, California

Robert C. Barrett

Department of Applied Physics, Stanford University, Stanford, California

2.1. Introduction

In 1982 Binnig *et al.*[1] published the first STM image of the Si(111) surface reconstructed in the 7 × 7 pattern. It was amazing to learn that they could image individual atoms of silicon. At the time, and for the next few years, it seemed almost unbelievable. The STM image of the 7 × 7 pattern, with its display of individual atoms, answered questions that had baffled investigators for years. That result was the stimulus that initiated many STM projects. In the early years the rate of success in the groups that pioneered the new era with the STM was limited by the technical difficulties associated with the new instrument. The validity of the initial results was in question and skepticism was widespread until 1985 when selected groups[2,3,4] were able to reproduce the beautiful images on the silicon surface. They were able to confirm that the STM is a viable instrument for studying conducting surfaces with atomic resolution. Since that time the field has experienced a tremendous increase in popularity.

The number of groups building STMs continued to grow in size. They tried a diversity of designs but images with respectable resolution came from only a very few. At first glance the design and structure of the STM seems simple, but there are subtle aspects that are quite difficult to control. The exacting tolerances for instruments with atomic resolution place a number of constraints on the design. Low frequency vibrations, small thermal drifts, and electrical noise are major obstacles to the accurate placement of the tip. The spacing between tip and sample must be controlled with an accuracy that is better than ±0.05 Å. Bringing the tip into tunneling range (about 6 Å separation), and scanning the tip in a raster pattern without crashing it into the sample, requires a design that is both innovative and precise. Electrical and mechanical feedback loops must be capable of maintaining a constant tip–sample separation to better than the width of one atom. Stable tips with a

METHODS OF EXPERIMENTAL PHYSICS
Vol. 27

ISBN 0-12-475972-6

single atom at the end must be reproduced easily and economically. Different samples have different surface characteristics, and the STM must be flexible enough to adjust to the differences in topography, conductivity, and volatility. In addition the sample preparation technique is critical if one is to achieve atomic resolution. STM experiments are challenging; if the microscope is untested, it is difficult to determine from the images alone which of the above problems may exist.

This chapter is devoted to STM instrumentation. Related review articles on applications and instrumentation appear in references.[5–8] Here we will restrict our attention to the fundamental application of the STM: imaging and spectroscopy in both ultra-high vacuum (UHV) and ambient environments. Instrumentation for other applications and other environments are found in corresponding chapters of this book.

The most convenient STM mode is operation in air—the Air-STM is often used for inert samples. The most stringent STM mode is operation in ultra-high vacuum—the UHV-STM is required for accurate and reliable studies of clean surfaces. The design concepts of the two instruments are quite similar, but the construction and operation of the Air-STM is simpler by far. Acoustic coupling through the air is always present in the Air-STM; however the Air-STM, with its compact and rigid form, is less sensitive to mechanical vibrations. In contrast, the UHV-STM contains rather large devices for mounting and changing the tip and sample. These reduce the mechanical rigidity of the microscope and lower the mechanical resonant frequencies; therefore the systems used for isolating the vibrations are more complex in the UHV-STM than in the Air-STM. The design of a UHV-STM is further complicated by the bakeout that is required for outgassing the entire vacuum system. Finally, since the tunneling current is easily affected by contaminants from the ambient atmosphere, the sample must be cleaned with a rigorous procedure that usually involves heating it to a very high temperature in UHV.

In the next section, we present our theoretical analysis of the systems for feedback control and vibration isolation as applied to STM. It is possible to design the STM without this rigorous calculation of the behavior of the feedback and isolation systems, but the analysis alllows us to improve the performance of the STM when it is constrained to operate under specific conditions. The control electronics and the data acquisition system will also be discussed in order to show how all of the components fit together. In the last section of this chapter, we will discuss typical problems and difficulties that are encountered in operating an STM together with possible remedies and improvements.

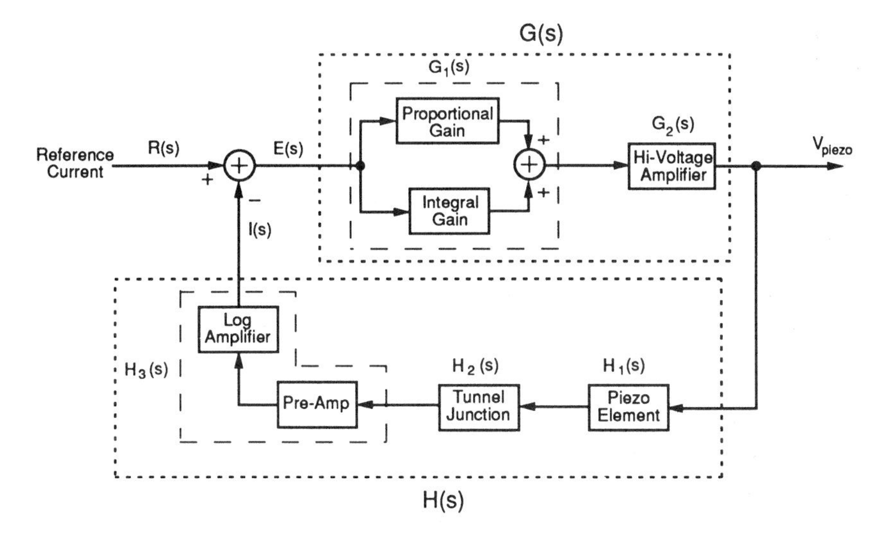

FIG. 1. Block diagram for the feedback control system of an STM.

2.2 Theoretical Considerations

2.2.1 Feedback Theory

In scanning tunneling microscopy, a feedback system is used to control the tip–sample spacing in order to maintain a stable tunneling junction. A fixed tip–sample bias voltage is applied and the desired tunneling current is selected by the operator. The control system is used to adjust the gap between the tip and sample until that tunneling current is achieved. As the tip is scanned across the surface, variations in the sample topography and electronic structure affect the tunneling current. The control system must react to bring the current back to the desired value. Ideally, the adjustments should be made instantaneously and exactly. However, such demands are unrealistic. This section will describe some of the theory behind the design of an adequate STM feedback control system. More general discussions of control theory can be found in appropriate textbooks.[9,10]

Figure 1 illustrates a typical block diagram for the *z*-motion of an STM (i.e., the motion of the tip perpendicular to the plane of the sample). A straightforward analysis of the system may be performed if the system is assumed to be linear. Most of the components behave in a fairly linear fashion except for the tunneling gap. The tunneling gap output is an exponentially varying current for linear variations in the gap distance. This current can be linearized with a log-amplifier, or it can be treated as linear if variations in the gap distance are small. In either case, we will limit our analysis to the case of a linear system. Linear systems such as this one are best analyzed in the Laplace transform domain. The Laplace transform, $F(s)$, of a function, $f(t)$, is defined to be:

$$F(s) \equiv \int_0^\infty f(t)\, e^{-st}\, dt. \tag{1}$$

This transformation is a generalized Fourier transform. The time variable t is replaced with a complex frequency s. In the Fourier transform the basis functions are simple sinusoids, while in the Laplace transform the basis functions are sinusoids with amplitudes that vary exponentially in time. The imaginary part of s is the frequency of the sinusoid. The real part of s is the rate of growth (for a positive real part s) or decay (for a negative real part of s) of the amplitude of the oscillation. The Laplace transform is good for analyzing linear systems because the response of real systems is very often well modeled by exponentially damped sinusoids.

In order to begin the analysis of the STM system, each block of the system must be replaced by its transfer function. The transfer function of a component is the Laplace transform of its impulse response (the output of the component when excited by a unit impulse at time $t = 0$). For example, the high-voltage amplifier will typically have a transfer function given by:

$$G_2(s) = \frac{k_{\text{hv}}\, \omega_{\text{hv}}}{s + \omega_{\text{hv}}}. \tag{2}$$

This transfer function describes an amplifier with a DC gain of k_{hv} and a high frequency gain rolloff with a cutoff frequency of $f_c = \omega_{\text{hv}}/2\pi$. In cases where the cutoff frequency is much higher than other frequencies of interest, this transfer function can be simplified to be that of an ideal amplifier:

$$G_2(s) = k_{\text{hv}} \qquad (\omega_{\text{hv}} \to \infty). \tag{3}$$

The piezoelectric element is more complicated and will typically have a second-order response. This means that its transfer function will have two dominant poles which result in an oscillatory impulse response. For example, a piezoelectric actuator might have a transfer function given by:

$$H_1(s) = k_p \omega_p^2 \frac{\alpha s + 1}{s^2 + \omega_p s/Q + \omega_p^2}, \tag{4}$$

where k_p is the DC response of the piezoelectric (Å/V), ω_p is the dominant resonant frequency, Q is the quality factor, and α determines the phase of the output.

The goal of the feedback system design is to minimize the error signal E. The designer must develop a control block with transfer function $G_1(s)$ such that the system maintains the desired tunneling current with the necessary accuracy. The constraints put on the system response are:

1. The resulting system must be stable, i.e., the error signal must not increase in an unbounded fashion as a function of time. This is equivalent to saying that the expression for $E(s)$ must not have any poles that are in the right half of the s-plane.
2. The steady-state error of the system must be small.
3. The transient response of the system must be fast and any oscillations must decay quickly.

Examining the block diagram, we see that the error signal may be written as

$$E(s) = \frac{R(s)}{1 + G_1(s)G_2(s)H_1(s)H_2(s)H_3(s)}, \tag{5}$$

where $R(s)$ is the reference signal which defines the desired tunneling current. The response of the control system to changes in tunneling current, for example because of a scan over changing topography, can be studied by examining this error signal. To simplify the analysis, such changes can be modeled as changes in the reference current, R.

Consider an STM control system which has a simple proportional control function, $G_1(s) = k_1\omega_c/(s + \omega_c)$, with a high frequency cutoff at $f_c = \omega_c/2\pi$. We see that the resulting error signal is

$$E(s) = \frac{R(s)}{1 + \gamma k_1}, \tag{6}$$

where $\gamma = [\omega_c/(s + \omega_c)]G_2(s)H_1(s)H_2(s)H_3(s)$. That is, the error signal is the reference signal divided by $(1 + \gamma k_1)$. If the system is stable, the steady-state error is given by $sE(s)$ in the limit when s goes to zero. In this limit, γ becomes a constant equal to the product of the DC gains of the other components in the loop. If the proportional gain k_1 is large, the steady-state error signal is small, but non-zero. A more complete analysis reveals that for sufficiently high gain the system becomes unstable in the manner described later in this study. Thus, there is a trade-off between reducing the error signal and maintaining system stability. The system stability is improved by including a low-pass filter in the proportional gain stage, such that the cutoff frequency (ω_c) is much lower than the piezoelectric element's resonant frequency. This filter prevents the proportional gain stage from amplifying the resonance.

The finite steady-state error can be eliminated by including an integrator in the control block. The proportional and integral signals are then added together. The resulting transfer function is

$$G_1(s) = \frac{k_1\omega_1}{s + \omega_1} + \frac{k_2}{s}. \tag{7}$$

The first term is the proportional gain with an upper cutoff frequency ω_1, while the second term is an integrator with gain k_2. The integrator causes a small error signal to be integrated in time, producing more and more control action until the error is eliminated. Thus, the steady-state error of this system is zero for a reference that changes in a step-like manner.

If a fast transient response is necessary, as with fast imaging speeds, then a derivative control signal can be included. This type of signal "anticipates" errors by correcting for the direction in which errors are moving before they grow large. A much faster transient response can be achieved in this way, but at the cost of decreased system stability. Since differentiators amplify signals proportionately to their frequency, they must be followed by a low-pass filter to prevent the amplification of high-frequency noise. The transfer function of a differentiator is

$$G_{1,\text{diff}}(s) = k_3 \omega_3 \frac{s}{s + \omega_3}, \tag{8}$$

where k_3 is the gain and ω_3 is the upper cutoff frequency.

In order to see how these different control methods work, we will analyze a simplified form of a real STM system. Analytical or computer modeling of an STM control system can produce accurate predictions for the response of the system to disturbances, such as reference current changes or changes in topography. All of the standard control system procedures (e.g., root-locus analysis, frequency domain analysis, and optimal control system design) can be applied to this system. To simplify the calculations, the system is reduced to a minimal form which only contains the dynamic response of the mechanical components of the microscope $H(s)$ and the control electronics $G(s)$.

To evaluate the response of the system, we can apply a unit step input to the reference current $R(s)$ and evaluate the resulting tunneling current $I(s)$. Ideally, the response $I(s)$ will be identical to the reference $R(s)$, but in reality this is not the case. This step response provides the steady-state error of the system ($R(s) - I(s)$ for large times) as well as the transient response of the system. The response of this simplified system is given by:

$$I(s) = R(s) - E(s) = \frac{G(s)H(s)}{1 + G(s)H(s)} R(s). \tag{9}$$

Even a simplified system is difficult to analyze analytically because of the complexity of finding the roots of the polynomials of s as a function of different control gains. So, in order to evaluate the system, we calculate its response numerically. This calculation is performed by having a computer algebraically evaluate the rational polynomial expression for $I(s)$ from Eq. (9). Then it does a partial fraction decomposition of the expression and evaluates the inverse Laplace transform of each term of the decomposition. This calculation yields the time-domain step response of the system (Figs.

3–6). Commercial software packages such as MATLAB may be used for this analysis.

As a model for the mechanical response of the microscope, we assume a general second-order impulse response:

$$H(s) = \frac{\alpha s + 1}{s^2 + \omega_0 s/Q + \omega_0^2}, \tag{10}$$

where ω_0 is the undamped resonant frequency of the system, Q is the quality factor, and α determines the phase of the response. These parameters can be measured on a real STM by applying a step input to the piezoelectric element and measuring the response of the tunneling current with a digital storage oscilloscope. The Fourier transform of the measured step response will consist of several peaks corresponding to resonances of the mechanical system. Each peak can be fit to a resonant term, which, in the time-domain, is written

$$f(t) = Ae^{-Bt} \sin(Ct + D), \tag{11}$$

where A, B, C, and D are fitting parameters. These parameters can be determined from the peak amplitude, center frequency, phase at the center frequency, and full width at half maximum. The impulse response is then found by differentiating the step response with respect to time. By comparing the Laplace transform of the resulting expression with Eq. (10), ω_0, Q, α, and the overall amplitude can be determined. Although most systems will have multiple peaks, one is often dominant and can be used to adequately model the system response. Figure 2 illustrates the fit obtained on one of our air microscopes for matching one peak. The fit is seen to be remarkably good for the dominant peak. A better fit can be obtained by including more terms, one for each peak. However, for the purposes of modeling our system, the one-resonance approximation is acceptable. The parameters are found to be approximately:

$$\omega_0 = 7 \times 10^3 \text{ rad/sec}, \tag{12}$$

$$Q = 20, \tag{13}$$

$$\alpha = 2.6 \times 10^{-5} \text{ sec}. \tag{14}$$

These values result in an impulse response given by:

$$H(s) = \frac{2.6 \times 10^{-5} s + 1}{s^2 + 350s + 4.9 \times 10^7}. \tag{15}$$

Note that we are ignoring the absolute value of the DC response of the piezoelectric element in this example in order to simplify the calculation. That value is absorbed into the gain of the control function.

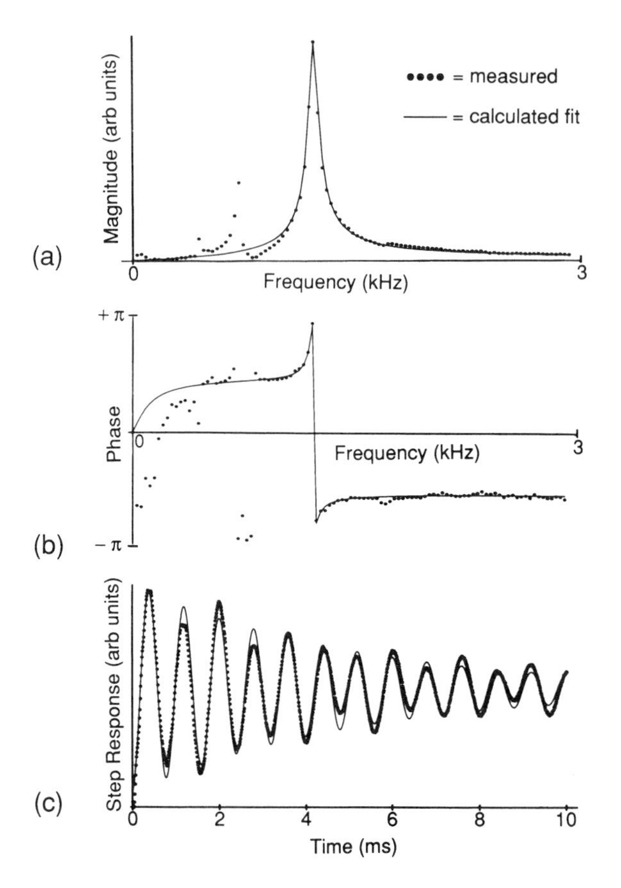

FIG. 2. Step response of the mechanical system as measured and with a fit to the dominant peak. (a) Magnitude of the frequency response. (b) Phase of the frequency response. (c) Time response.

Now, we can examine various control functions for producing a system that responds quickly and accurately to variations in the reference signal. The first and simplest to try is proportional control. For this case, we let

$$G(s) = A\,\frac{\omega_c}{s + \omega_c}, \tag{16}$$

where A is the proportional gain and ω_c is the high frequency cutoff of the amplifier. If a high frequency cutoff were not included in our model, the calculations would predict a stable system for arbitrarily high gain. This is not realistic and shows that our model is incomplete. When modeling systems, it is important to recognize unrealistic behavior in order to avoid producing meaningless results that come from oversimplification.

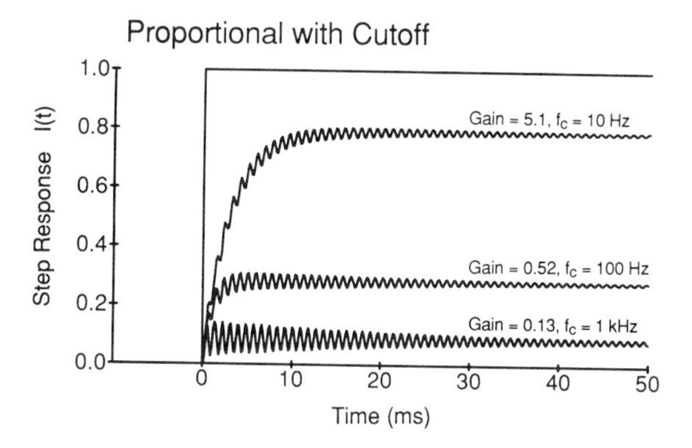

FIG. 3. Response of the tunneling current to a unit change in the reference current. Each curve corresponds to the response using a proportional control function followed by a low-pass filter with the indicated cutoff frequency. The gain was set to 75% of the maximum stable gain with the given cutoff frequency.

In order to evaluate the proportional gain system, the system response was examined with cutoff frequencies of 10, 100, and 1000 Hz. First, the maximum allowable gain for a stable system with each of these cutoff frequencies was determined. This calculation is easily done by solving for the poles of the response function $I(s)$ as a function of the gain. Whenever all of the poles are in the left half of the s-plane, the system is stable. When one of the poles moves to the right half of the s-plane, then the response will have a term which grows exponentially with time. This implies an unstable system. After the maximum allowable gain was determined, the step response of the system was evaluated with the gain set to 75% of the maximum value for the given cutoff frequency. Figure 3 shows these three different step responses. At these gain settings, the step input excites the resonance of the microscope, which then slowly decays. The steady-state response indicates a finite error, as expected for a proportional gain system. Since a lower cutoff frequency allows a higher gain before the instability sets in, the steady-state error is less with the lower cutoff. The transient response of the system, however, is slowed because of that low cutoff frequency. As we go to lower cutoff frequencies and higher gains (maintaining the product $A\omega_c$ constant), the control begins to look similar to an integrator:

$$G_{\text{prop}}(s) = A\,\frac{\omega_c}{s+\omega_c} \rightarrow \frac{A\omega_c}{s} = G_{\text{int}}(s), \tag{17}$$

for small ω_c. In essence the control function is an integrator with finite DC gain. In general, it is advantageous to have infinite DC gain in the loop; therefore it makes sense to use an integral control function.

With an integral gain component the error is integrated in time, thus the

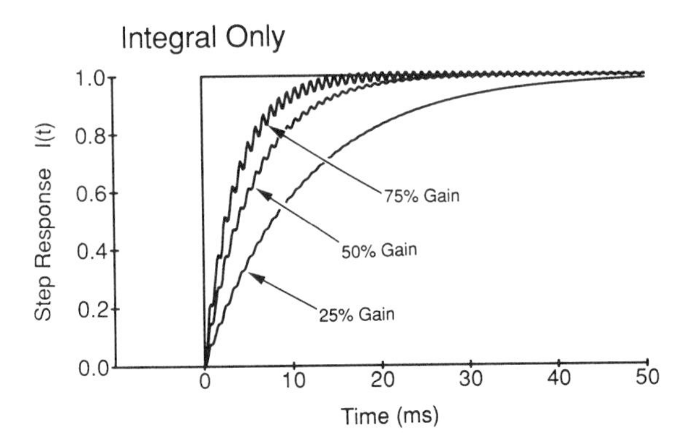

FIG. 4. Response of the tunneling current to a unit change in the reference current using an integral control function. The indicated gains are a fraction of the maximum stable gain.

steady-state error is reduced to zero. We will first consider a control system which is only an integrator:

$$G(s) = \frac{A}{s}, \tag{18}$$

where A is the integrator gain. The step response of the system is again computed and we find that the system remains stable up to $A_{max} = 1.65 \times 10^{10}\,\text{sec}^{-1}$. In Fig. 4 we show three plots for the step response of the system for A set to 25%, 50%, and 75% of A_{max}. Several observations can be made. First, for any stable, non-zero gain the steady-state error is zero. Second, increasing the gain increases the system's speed of response as well as exciting the mechanical resonance of the instrument. With a gain equal to 75% of A_{max}, the integrator gives a very similar response to the best proportional-only system, with a response time of about 8 msec, but with the complete elimination of the steady-state error. Therefore, the addition of proportional gain with the integral gain will not significantly improve the transient response of the system.

What can be done if an even faster transient response is required? For example, for fast constant-current imaging with an STM, the 8 msec response time may be inadequate. The best method for improving the response would be to build a mechanical system with a higher resonant frequency, thus removing the limitations of the electromechanical components. However, if that is impractical, a derivative control signal can be included to improve the response time. The derivative signal responds to rates of change in the error signal, instead of the error value itself. Therefore, it can compensate for disturbances before the errors grow very large. Since the derivative cannot measure the absolute value of the error, another control signal must be added to compensate for that. A good solution is to combine the derivative and integral control systems. We model this system with a control function of the form:

$$G(s) = \frac{A_1}{s} + A_2\omega_c \frac{s}{s + \omega_c}. \tag{19}$$

Here the first term is the integrator with a gain of A_1 and the second term is the differentiator with a gain of A_2 and an upper cutoff frequency of ω_c. This low-pass filter must be included to prevent the differentiator from giving high gains to high frequency noise in the system. This linear model is not completely realistic because differentiators tend to drive the system with very large signals, producing nonlinearities. When the tunneling current changes rapidly, or when noise is present, a differentiator can push the amplifiers into a nonlinear response regime. Thus, the predictions from this model can only be used qualitatively. These simulations do give a good intuitive feel, however, for how the derivative signal improves the system response.

Essentially, the derivative signal improves the transient response of the system by allowing larger control gains while still maintaining system stability. The other gains of the system can be set much higher without making the system unstable. To illustrate this, we choose a moderate derivative signal, $A_2 = 1 \times 10^8$ sec, $\omega_c = 2\pi \times 100$ Hz, and increase the integral gain. (These values are moderate in the sense that the closed-loop poles have moved in between the open-loop poles and the open-loop zeros.) We then simulate the step response of the system for three different values of integral gain, $5\times$, $10\times$, and $20 \times A_{max}$, the value that previously drove the system unstable without the derivative gain. These plots are shown in Fig. 5. The derivative signal now increases the ringing of the response and significant overshoot occurs. As before, the system has a zero steady-state error because of the integrator signal. Finally, we note that the transient response has been improved significantly. The system reaches the reference value for the first time after only 0.15 msec and, with the gain set to $10\times$, comes permanently to within 10% of the final value after 2 msec.

As a final comparison, we show in Fig. 6 the step response for the three different control systems: proportional, integral, and integral plus derivative. The proportional has an 8 msec response time to within 10% of the final value and a steady-state error of about 20%. The integral performs quite well, with an 8 msec response time and no steady-state error. If we include a derivative signal with the integrator, there is a much faster transient response with about 20% overshoot and zero steady-state error. For most applications, the integral-only control system is adequate and simple. For highly demanding applications, the derivative–plus–integral system provides a faster but less stable response.

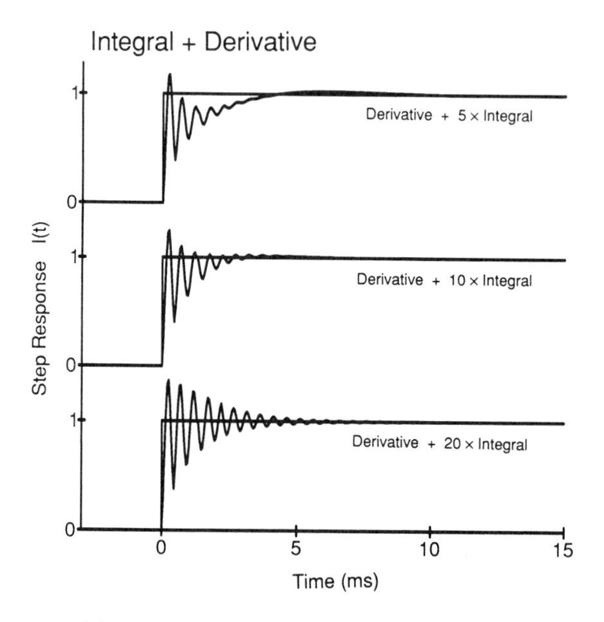

FIG. 5. Response of the tunneling current to a unit change in the reference current using an integral plus derivative control signal. The gain of the derivative signal is fixed and the integral gain is to set to various multiples of the maximum allowable gain using an integral-only control function.

In conclusion, theoretical models for the control system of an STM can be most helpful in determining the type of system that will be needed for adequate response in the instrument. The models provide reliable results which compare well with actual measurements when the system is operated in the linear regime. However, when the control gains are increased to

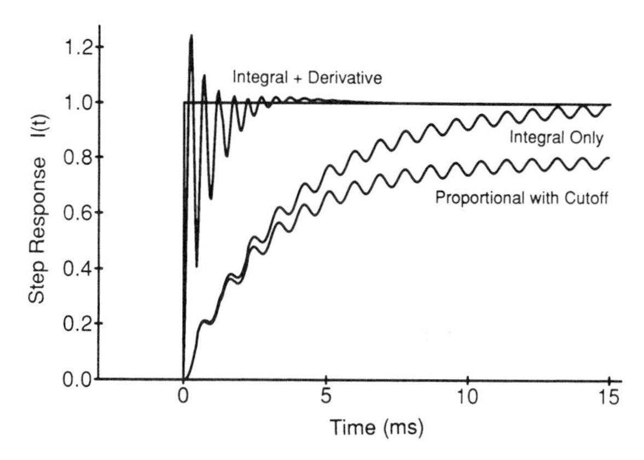

FIG. 6. A comparison of the step responses for each of the three different control functions.

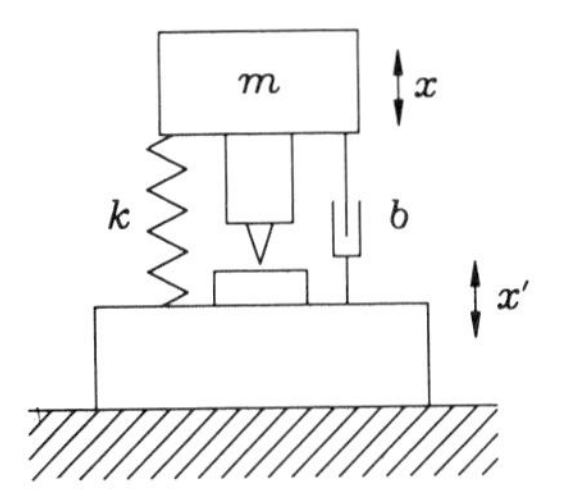

FIG. 7. Model of the tunneling unit. The system has been reduced to a simple damped harmonic oscillator.

optimize the transient response of the system, the simplest models usually fail. Generally, the simple models predict a much better system response than can be achieved in practice. Complications in the system that have been neglected in the model (such as amplifier and tunneling junction nonlinearities, neglected resonances in other mechanical components, and signal noise) will limit the final system performance.

2.2.2 Vibration Isolation Theory

Vibration isolation is another important area of STM where design parameters should be carefully considered. It is obvious that environmental vibrations are much greater than the scale that we want to measure; they are large enough to ruin tunneling conditions. Since the STM is measuring and controlling the tip position to less than an atomic diameter, the vibration isolation has to damp out noise to the same level.[11] While it is possible to design a dynamic control system,[12] we will restrict the discussion here to simple, passive damping systems. Furthermore, each component has six degrees of freedom, and we should include six coordinates in order to describe its complete motion. However, it is difficult to trace all of these motions and, very often, analyzing only one or two motions of the components will describe the whole system effectively. We will, therefore, reduce the system to a few rigid objects with several characteristic parameters, and consider only their vertical motion. This is appropriate as floor vibration is mostly vertical. The following discussion can be applied to both spring suspension stages and stacked metal plate isolators, even though the physical description is closer to spring suspension stages. The analysis of multiple stacked plates was discussed by Okano *et al.*[13] Other points to be considered in designing vibration isolation systems for an STM will be discussed in section 2.3.4. More general vibration theory can be found in Ref. 14.

In order to understand the tolerable level of vibration and how much isolation is required, we first examine the mechanical nature of the tunneling device. This is illustrated in Fig. 7 and includes the sample, sample holder, sample positioner, scanner, and tip. In order to simplify the analysis, we define the overall resonant frequency (ω_0) and quality factor (Q) of this unit as we did in the previous section. We can write the equation of motion in the form:

$$m\ddot{x} + b(\dot{x} - \dot{x}') + k(x - x') = 0, \tag{20}$$

where m is the effective mass, x is its vertical position, b is the damping factor, k is the spring constant of this unit, and x' is the vertical displacement of the STM frame. Using the relation of

$$\omega_0 = \sqrt{\frac{k}{m}}, \tag{21}$$

$$\gamma = \frac{b}{2m}, \tag{22}$$

we can rewrite Eq. (20) as

$$\ddot{x} + 2\gamma(\dot{x} - \dot{x}') + \omega_0^2(x - x') = 0. \tag{23}$$

When the STM frame is driven sinusoidally, i.e., $x'(t) = x_0' e^{i\omega t}$, we can use the steady state harmonic solution $x(t) = x_0 e^{i\omega t}$ to obtain the relationship between the variation in tunneling gap distance ($x_0 - x_0'$) and external vibration amplitude (x_0'). Or, equivalently, we can take the Laplace transform of Eq. (23) and replace $s = -i\omega$ to see the frequency dependence. The transfer function then becomes

$$|T_1(\omega)| = \left|\frac{x_0 - x_0'}{x_0'}\right| = \left|\frac{\omega^2}{\omega_0^2 - \omega^2 + 2i\gamma\omega}\right|. \tag{24}$$

$T_1(\omega)$ is the response of the tunneling gap spacing to external vibration just as $H(s)$ in Eq. (10) is the response of the gap spacing to the electrical driving signal. The quality factor Q is equal to $\omega_0/2\gamma$. $|T_1(\omega)|$ is plotted in Fig. 8 for $\gamma = 0.025\omega_0$. In the lower frequency range, T_1 becomes ω^2/ω_0^2. Therefore, as the driving frequency ω decreases by a factor of 10, the system response becomes smaller by a factor of 100. If the resonant frequency ω_0 is 2 kHz, the system response is less than 10^{-6} for driving frequencies below 2 Hz. This is sufficiently small for an STM considering that the typical floor vibration amplitude is a few thousand angstroms. For vibration frequencies above 10 Hz, we obviously need an isolation system.

The vibration isolation stage, as shown in Fig. 9, can be described in a similar manner. In this model, the mass m' represents the tunneling device of Fig. 7, which is now assumed to be rigid. The equation of motion is equivalent to eq. (20).

$$m'\ddot{x}' + b'(\dot{x}' - \dot{x}'') + k'(x' - x'') = 0. \tag{25}$$

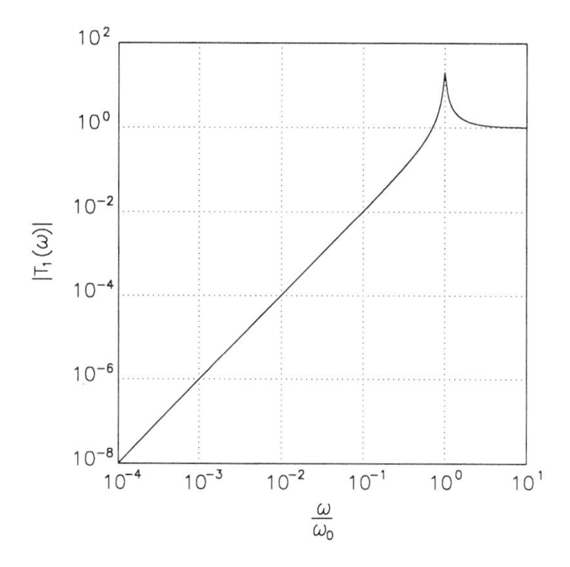

FIG. 8. Frequency response of the tunneling unit to external mechanical vibrations. Calculated from Eq. (24) for the case of $\gamma = 0.025\omega_0$.

We are interested in the amplitude ratio of x'_0 and x''_0.

$$|T_2(\omega)| = \left|\frac{x'_0}{x''_0}\right| = \left|\frac{\omega_1^2 + 2i\gamma'\omega}{\omega_1^2 - \omega^2 + 2i\gamma'\omega}\right|, \tag{26}$$

where $\omega_1 = \sqrt{k'/m'}$, $\gamma' = b'/2m'$. This function is plotted in Fig. 10 for several values of γ'. This curve is almost mirror symmetric to the one in Fig. 8. If there is little damping ($\gamma' \ll \omega_1$), the vibration isolation is more efficient at high frequencies but the oscillation at the resonant frequency is large. Heavy damping (near critical damping $\gamma' = \omega_1$) will reduce the amplitude of the resonance oscillation, but it does not attenuate the high frequency vibrations as effectively.

To evaluate the response of the tunneling gap to external vibration, including the details of both the tunneling unit and the isolation stage, we examine

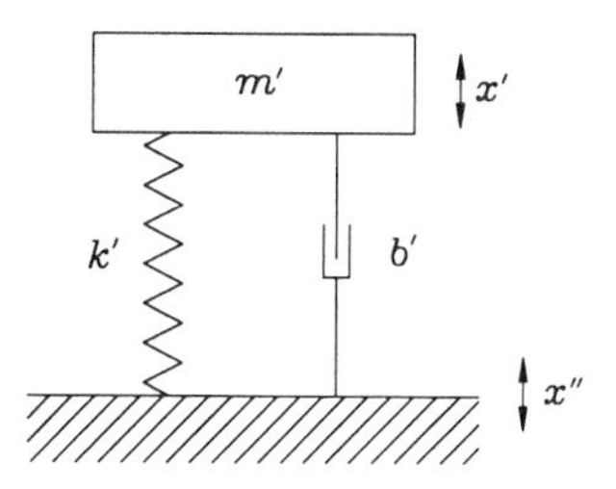

FIG. 9. Model of a single stage vibration isolation system.

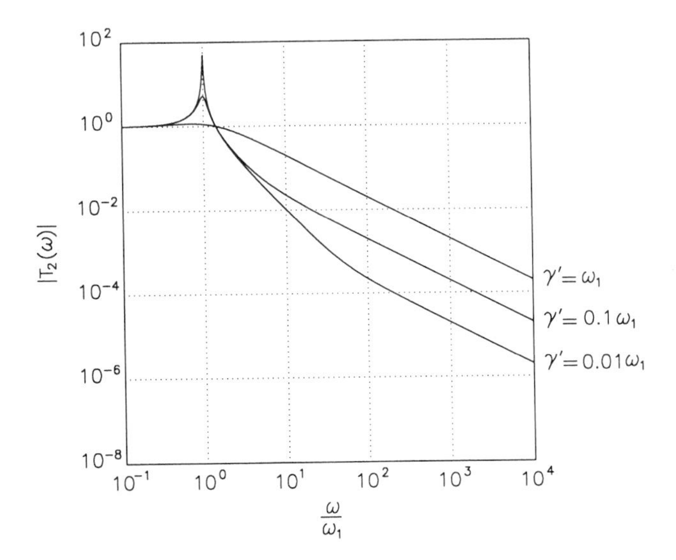

FIG. 10. Vibration attenuation of a single stage vibration isolation system for three different damping factors. Calculated from Eq. (26).

the magnitude of the product of T_1 and T_2. This is plotted in Fig. 11 for the case when $\omega_0/\omega_1 = 1000$, that is, when the resonant frequency of the tunneling unit is 1000 times greater than that of the isolation stage. In the intermediate frequency range ($\omega_1 \ll \omega \ll \omega_0$), Eq. (24) and Eq. (26) can be ap-

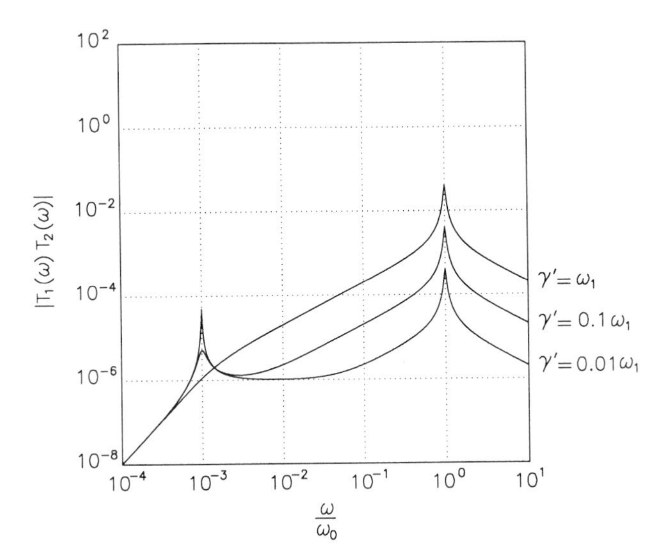

FIG. 11. Vibration attenuation for the tunneling unit combined with a single isolation stage.

proximated and the noise spectrum becomes

$$|T_1(\omega)T_2(\omega)| = \frac{\omega_1^2}{\omega_0^2} \tag{27}$$

for small damping, and

$$|T_1(\omega)T_2(\omega)| = \frac{2\gamma'\omega}{\omega_0^2} \tag{28}$$

for large damping. The damping factor should be adjusted such that the overall vibrational noise is minimized. If the tunneling unit has a sufficiently high resonant frequency, lower damping should be selected. If the low frequency vibration is more problematic, heavier damping will be helpful. However, increased damping will degrade the vibration isolation at high frequencies and additional isolation stages will be required.

It is interesting to note that the best vibration isolation is at the intermediate frequency, and its efficiency depends solely on the ratio of ω_1 and ω_0. Therefore, it is desirable to have a low resonant frequency (ω_1) for the vibration isolation stage, and a high resonant frequency (ω_0) for the tunneling unit. However, there is a limit to how well we can satisfy these requirements. The resonant frequency is determined by the physical dimensions of the microscope. This is a serious restriction. The spring constant k is related to the stretch length of the spring Δl by

$$m'g = k'\Delta l, \tag{29}$$

and we can write the resonant frequency ω as

$$\omega_1 = \sqrt{\frac{g}{\Delta l}}. \tag{30}$$

Thus, the resonant frequency is determined by the stretched length of spring only. Even if we enlarge the system by four times, the resonant frequency will be reduced only by half. Also, increasing the resonant frequency of the tunneling unit (ω_0) is not easy. The size of tunneling unit cannot be reduced arbitrarily since we need a reasonable space to accommodate the samples.

We can extend the analysis to the case of the double spring stage (Fig. 12). Here x'' denotes the vertical displacement of the second stage and x''' denotes that of the outer STM frame. We will use Eq. (25) for the first stage. The additional equation of motion is

$$m''\ddot{x}'' + b''(\dot{x}'' - \dot{x}''') + k''(x'' - x''') + k'(x'' - x') = 0. \tag{31}$$

Here m'' is the mass, b'' is the damping factor, and k'' is the spring constant of the second stage.

For simplicity, we neglect the damping ($b' = b'' = 0$) and consider the

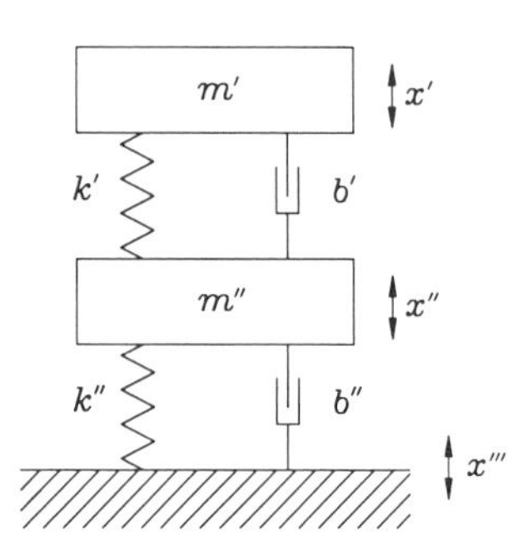

FIG. 12. Model of a double stage vibration isolation system.

situation where the outer frame is driven by $x'''(t) = x_0'''e^{i\omega t}$. Using the same method as before, we obtain

$$-\omega_2^2 x_0''' + (\omega_2^2 + \omega_3^2 - \omega^2)x_0'' - \omega_3^2 x_0' = 0. \tag{32}$$

We have defined $\omega_2 = \sqrt{k''/m''}$ and $\omega_3 = \sqrt{k'/m''}$. We use $x_0'' = x_0'(\omega_1^2 - \omega^2)/\omega_1^2$ from Eq. (25) to rewrite Eq. (32).

$$\frac{x_0'}{x_0'''} = \frac{\omega_1^2\omega_2^2}{\omega^4 - (\omega_1^2 + \omega_2^2 + \omega_3^2)\omega^2 + \omega_1^2\omega_2^2}. \tag{33}$$

This formula gives the ratio between the vibration amplitudes of the first stage and outer frame.

There is a constraint that specifies the relation between ω_1, ω_2, and ω_3. Since the physical size of the microscope is finite, the total stretched length of the springs is fixed. That is,

$$\Delta l + \Delta l' = \text{constant}, \tag{34}$$

where Δl and $\Delta l'$ denote respectively the stretch length of the first and second stage springs.

$$\Delta l = \frac{m'g}{k'} = \frac{1}{\omega_1^2}g$$

$$\Delta l' = \frac{(m' + m'')g}{k''} = \left(\frac{1}{\omega_2^2} + \frac{\omega_3^2}{\omega_1^2\omega_2^2}\right)g$$

Equation (34) can be rewritten as

$$\frac{\omega_1^2 + \omega_2^2 + \omega_3^2}{\omega_1^2\omega_2^2} = C, \tag{35}$$

where C is a constant. Cg denotes the total stretched length of springs.

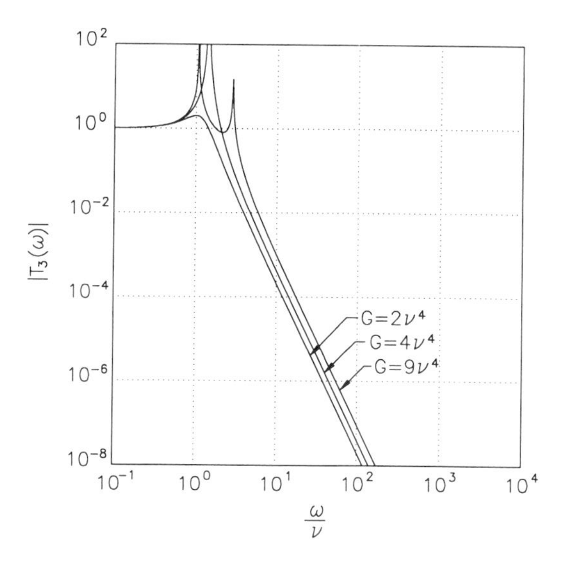

FIG. 13. Vibration attenuation of a double stage vibration isolation system for three different spring configurations. Calculated from Eq. (36). The frequency unit is defined as $\nu^2 = 1/C$. See text.

Equation (33) becomes

$$|T_3(\omega)| = \left|\frac{x_0'}{x_0'''}\right| = \frac{G}{|\omega^4 - CG\omega^2 + G|}, \tag{36}$$

where

$$G = \omega_1^2\omega_2^2. \tag{37}$$

The values of $|T_3(\omega)|$ are plotted in Fig. 13 for several values of G. In order to compare the efficiency of the double stage vibration isolation system to that of the single stage system, we have defined a new reference frequency $\nu = \sqrt{1/C}$. This new reference frequency is the same as ω_1 used for the single spring stage calculation (Fig. 10) when the total stretched length of springs is the same. It is obvious that the double stage system is much more efficient at high frequencies than a single stage system with the same physical dimension. We also notice that a smaller value of G gives better vibration isolation.

From Eq. (35), we find that G has a positive minimum value when

$$\omega_1^2 = \omega_2^2 = \frac{1 + \sqrt{1 + C\omega_3^2}}{C}. \tag{38}$$

This value becomes smaller as ω_3 decreases and C increases. This result has several implications for an optimal design.

1. k'/m' should be equal to k''/m''.

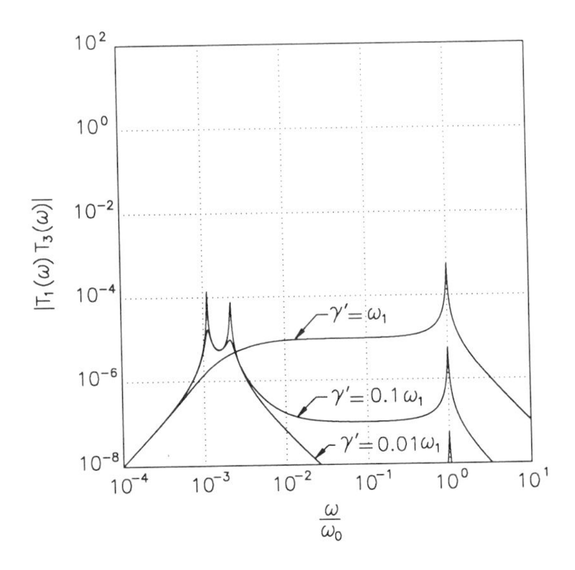

FIG. 14. Vibration attenuation for the tunneling unit combined with a double isolation stage.

2. k'/m'' $(=\omega_3^2)$ should be reduced as much as possible.
3. The total stretched length of springs Cg should be allowed as large as possible.
4. For a given k'/m'' and Cg, k'/m' and k''/m'' should be determined by Eq. (38).

In practice, the third condition is determined by the physical size of the STM. The fourth condition may be too strict. Fortunately the value of G does not increase rapidly as ω_1 and ω_2 deviate from the value of Eq. (38), so this condition is not as critical. We should keep in mind the first and second conditions. A heavier mass m'' of the second stage is preferable.

It is interesting to examine the efficiency of the double stage vibration isolation system at its optimal condition. At the best, the minimum value of ω_1^2 is $2v^2$ when $\omega_3 = 0$. This gives $G = 4v^4$. We can expect the maximum value of ω_3 to be ω_2, since the outer spring should be stiffer than the inner one. By setting $\omega_3 = \omega_2$, we get $\omega_1^2 = \omega_2^2 = \omega_3^2 = 3v^2$ and $G = 9v^4$ from Eq. (38) as the worst case. The vibrational noise spectrum with the double stage, $|T_1(\omega)T_3(\omega)|$, is plotted for the case of $G = 6v^4$ and $\gamma' = \gamma'' = \omega_1$, $0.1\omega_1$, $0.01\omega_1$ in Fig. 14. We notice that they exhibit much better vibration response than with a single stage system of the same total spring length. If the springs of the two stages are designed to overlap, the unit v in Fig. 13 will be smaller than the unit in Fig. 10, and the efficiency of double stage system will be further improved.

In a real STM, the double stage system has additional advantages. The suspension spring is not ideal and the spring wire itself can transmit high frequency vibrations. The electrical wires that connect the tunneling unit on the first stage to the outside world can transmit low frequency vibrations. In the double stage vibration isolation system, such coupling can be efficiently eliminated by clamping the wires on the intermediate stage.

2.3 Mechanical Structure and Components

2.3.1 Tip

The tunneling tip is the most crucial component of the STM.[15] The geometry and chemical identity of the tip influences both topographic and spectroscopic measurements. The best images are obtained when tunneling is limited to a single metallic atom at the end of the tip. Anomalous imaging artifacts will appear when simultaneous tunneling occurs through multiple atoms on the tip. This is commonly referred to as *double-tip* imaging. When tunneling involves a group of atoms, the STM image appears as the summation of multiple images. If the primary tunneling current from the closest atom is similar in magnitude to that of the background tunneling current from neighboring atoms, atomic scale features of the sample will not be resolved. If the atom at the end of the tip is nonmetallic, the STM tunneling spectrum will not represent the true electronic structure of the sample surface. Nonmetallic tip atoms may also make the tunneling current unstable; the feedback will move the tip forward to compensate for the lack of tunneling current, and the tip may crash into the sample. Therefore, a good tip is required not only for high resolution imaging but also for elimination of tip artifacts due to variations in conductance. We will return to a discussion of tip artifacts in Section 2.5.2.

It is difficult to control the shape of a tip down to the final atom, although its shape can be monitored by field ion microscopy.[16–20] Fortunately, nature is kind to the STM researcher. It turns out that if certain metal wires are fractured, or cut, the result is a rugged surface with a single atom as the endpoint. There is a strong dependence of the tunneling current on the tip–sample separation, and the STM will "find" the atom that is protruding toward the sample. As will be developed in the theory chapter, the tunneling current varies exponentially with the gap distance. With a typical work function of 4 eV, the tunneling current increases by a factor of 10 when the gap distance decreases by 1 Å. That is, if one atom at the apex of the tip is 1 Å closer to the sample than all the other atoms of tip, most of the tunneling current will flow through the apex atom and we can expect atomic resolution. In this way, even with a blunt tip one can obtain images of atoms (see Fig.

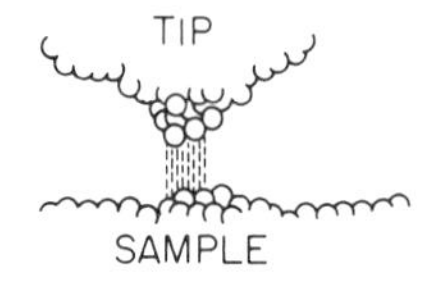

FIG. 15. Naturally occurring protuberances reduce the area of the tunneling current, giving the STM its atomic resolution.

15). In fact, the first STM image of silicon 7×7 by Binnig *et al.*[1] was achieved with a mechanically ground tungsten tip. The macroscopic radius of mechanically prepared tips is large and rarely produces atomic resolution images. Naturally the probability of having a single atomic tip with small background tunneling increases as the radius at the end of the tip becomes smaller. The relationship between the size of the cluster on the end of the tip and the measured corrugation height has been studied quantitatively by Kuk *et al.*[21]

The most common types of STM tips are electrochemically etched tungsten wires and cut platinum-iridium wires. Use of materials that do not oxidize in air such as Pt-Ir, Pt, or Au might be advantageous as far as the stability of tunneling in air. However, tungsten is most commonly used in UHV, where oxidation is not a factor. Tungsten has the advantages of being relatively inexpensive and easily etched to produce the desired macroscopic tip shapes. Furthermore, tungsten wires have been used in the past for field ion microscopy, where it has been possible to observe and to control the tip geometry on a microscopic scale.

Several methods have been developed to prepare sharp and "clean" tungsten STM tips. Electrochemical etching, similar to the sample preparation method of field-ion microscopy, has been widely used. This is illustrated in Fig. 16. In this method, we use 10% potassium hydroxide (KOH) solution with 10–20 V AC bias to etch 0.2–0.5 mm diameter tungsten wire (either polycrystalline or single crystal). The typical current is 2–3 A. The side of the wire will etch more quickly than the bottom, causing the wire to "neck" into a sharp tip shape. The aspect ratio can be controlled by the exposed length of wire to the solution and the time that the applied voltage is shut off.

Tungsten tips made with this procedure have a "wet" oxide layer covering

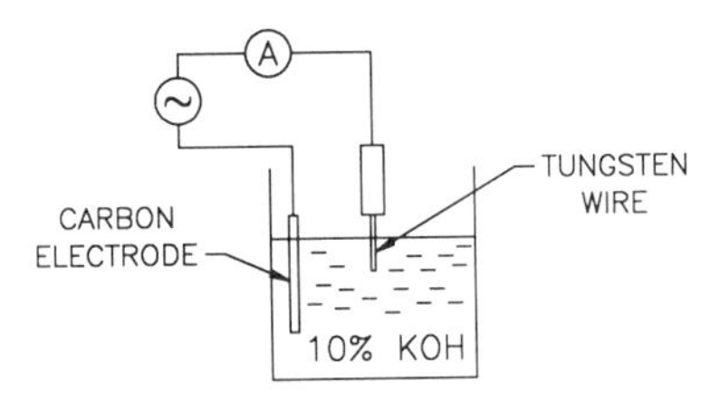

FIG. 16. Schematic drawing of an electrochemical tip-etching apparatus.

the surface. If the oxide layer is not removed, the tunneling current is unstable and the image quality is poor. Tunneling under these conditions can cause the tip to crash into the sample. This might seem to be disastrous, but the crash occasionally improves the tip quality, possibly by breaking through the oxide layer or by depositing a conducting atom on the end of the tip. It is preferable to avoid these lucky circumstances, by removing the oxide with a more controlled method such as ion milling, annealing, or field evaporation from the tip.

In ion milling, an ion beam of inert gas (e.g., Ne or Ar) is aimed at the tip while the tip is rotated.[22] If single crystal (111) tungsten wire is used, it has been reported that a steady-state faceted tip, which is sharp on a near atomic scale, can be produced by extensive ion milling.[23] The tip can be annealed by electron beam heating[24] or by attaching the tip to a thick filament wire in the presence of a high electric field.[17,19] The tips treated by ion milling or annealing can be exposed to air for a short time while being transferred into the STM chamber without degrading the tip quality. Field evaporation may be performed with the tip in the STM by applying a few hundred volts of DC or AC bias between the tip and sample (tip negative for DC bias) with a tip–sample spacing of about 10 μm. A few micro-amperes of field emission current can be achieved by adjusting the gap spacing and applied voltage. This method can be repeatedly performed during STM experiments. One has to be careful to limit the high voltage current to protect the preamplifier circuits of the STM. The sample can also be damaged during field emission and it is advisable to use this procedure with the tip far from the area of the sample that is to be imaged.

Tunneling tips prepared by each of the described methods will produce high resolution STM images that are reproducible. It is hard to judge which of these is superior as there is no direct way to test the condition of tip other than by analyzing the images. The preferred method depends on the nature of the sample and the instrumentation available. For example, the annealing method tends to make the tip rounded at the end. Such tips perform well on flat samples, but the method is less useful on samples with many steps and protrusions.

During STM operation it is inevitable that the tip will crash into the sample, even with well-designed control electronics and with extreme care by the operator. Sometimes crashing can improve the tip condition, but in general the tip becomes less sharp. Therefore it is desirable to have a facility for exchanging the tips in vacuum.

2.3.2 Scanner

The scanner is the device that moves the tunneling tip across the sample

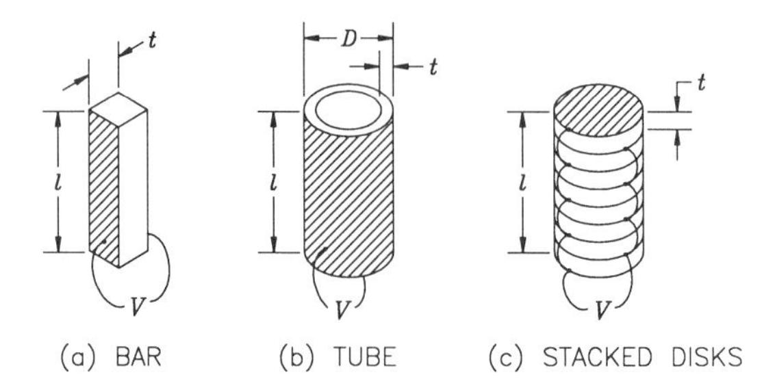

FIG. 17. Geometries of piezoelectric actuator elements.

surface and controls the tip–sample separation. The requirements of a good scanner are: 1) high resolution—the necessary resolution is less than 1 Å in the lateral direction (x, y) and 0.05 Å in the vertical direction (z); 2) orthogonality—movement of each of the three axes should be independent; 3) linearity—the amount of movement should be proportional to the applied voltage; 4) mechanical rigidity—a rigid scanner will have a high resonant frequency, which is desirable for both vibration isolation and feedback performance; and 5) large range—it is desirable to cover as large a sample area as possible.

Scanners made from piezoelectric actuators meet these design specifications. The commercially available piezoelectric ceramic PZT [$Pb(Zr,Ti)O_3$] is suitable and is used in most STM scanners. Common problems with piezoelectric actuators include non-linearity, hysteresis, and creep. Piezoelectric ceramics are ferroelectric materials and for this reason their response to applied electric fields is nonlinear. This leads to hysteresis and creep. Such effects become increasingly noticeable at higher electric field strengths and with higher piezoelectric sensitivity materials. The electric fields applied for atomic-scale scanning are small, and the nonlinear effects are minimal. For larger scan areas (10–100 μm), the nonlinearity becomes more noticeable due to the high fields required to drive the scanner. Piezoelectric creep is the slow motion of the actuator after a large change in the DC level of the applied field. These nonlinearity and creep effects can be minimized by controlling the total charge applied to the piezoelectric rather than the voltage.[25] Inserting a capacitor in series with the piezo accomplishes a similar effect.[26] However, even with these methods, the actual three-dimensional scanner has non-ideal motion due to stresses created during scanning. More sophisticated methods are required to remove these nonlinear effects. This problem will be discussed in more detail in Section 2.5.5.

In Fig. 17 three types of piezoelectric actuators are shown. The bar and tube geometries are used in the lateral mode (the displacement is perpendicular to the applied electric field), while stacked disks are used in parallel mode (the displacement is in the same direction as the applied electric field). The stacked disks are wired in parallel while mechanically connected in series. Their displacement Δl can be written as

$$\Delta l_l = d_{31} \frac{V}{t} l, \tag{39}$$

for lateral mode (Fig. 17a, b), and

$$\Delta l_p = d_{33} \frac{V}{t} l = d_{33} Vn, \tag{40}$$

for parallel mode (Fig. 17c). For PZT the typical value of d_{31} is $(-1$ to $-2) \times 10^{-10}$ m/V and the typical value of d_{33} is $(+2$ to $+5) \times 10^{-10}$ m/V. As we increase the voltage in the poling direction, the length becomes shorter in lateral mode and longer in parallel mode. The actual displacement will be less than the free displacement since stress is developed. In an STM scanner the stalling force (defined as the force required to keep the piezoelectric from contracting or expanding with a given applied electric field) of a typical piezoelectric is much greater than the stress load, so we can use the free displacement formulas as a good approximation. The polarity of the applied voltage V should be in the same direction as the original poling field. A small negative field may be applied, but without careful control depoling or partial depoling can result. A negative voltage that is too high may force the poling in the opposite direction. The negative field must not exceed the point where the piezo response $\Delta l/\Delta V$ changes significantly with the applied voltage.

For comparable mechanical strength, the thickness of the tube wall can be thinner than the thickness of a bar. Therefore, a smaller voltage is necessary for the same piezo displacement with a tube than with a bar. Similarly, the stacked disk geometry needs less voltage than either the tube or bar for a given displacement. However, the maximum field that can be applied to an element is constant because of electrostatic breakdown. Therefore, the maximum displacement with a bar is similar to that with a tube. Similarly, the maximum displacement of a stacked disk actuator cannot be increased by increasing the number of disks in the stacks n. For a given total length, reducing the thickness of each disk and increasing the number of disks merely decreases the voltage required for a given displacement (i.e., increases the sensitivity). The maximum displacement of an actuator is ultimately limited by the piezoelectric material and the size of the actuator. The desired sensitivity for the piezoelectric elements should be determined by considering the desired scan range and resolution, together with the output range and noise level of driving circuits. Higher sensitivity is not always desirable. The driving signal has a certain amount of noise, and higher sensitivity means more uncertainty in the tip position. Of course the length of the piezoelectric

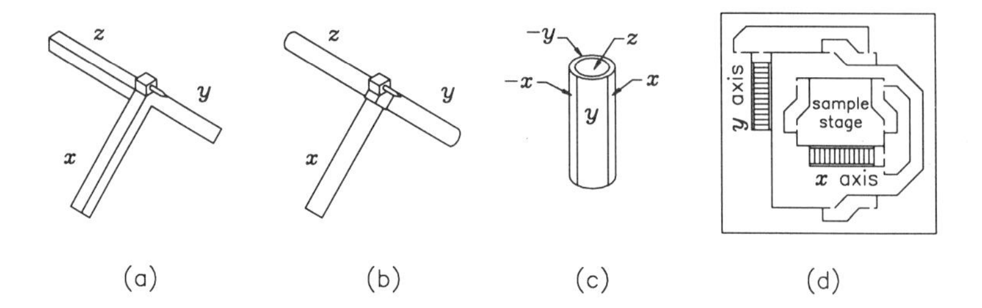

FIG. 18. Several geometries for STM scanners. (a) bar tripod, (b) tube tripod, (c) single tube, (d) stacked disks with mechanical amplifier.

element can be reduced if it is more sensitive, but orthogonality and linearity may suffer.

In Fig. 18, four typical STM scanners are shown: (a) bar tripod, (b) tube tripod, (c) single tube, and (d) stacked disks. A tripod scanner with three piezoelectric ceramic bars was the type first used in STM. It has a reasonably high resonant frequency, relatively low sensitivity (Å/V) and is reasonably orthogonal. The tripod scanner with three piezoelectric tubes has a higher sensitivity and can be driven with low-voltage, integrated operational amplifiers for scans of a few thousand angstroms. It is also appropriate for large area scans using a high-voltage driver.

The single piezoelectric tube scanner[27] has become popular due to its compact and simple structure, high sensitivity, and high resonant frequency. This scanner has the outside electrode split into four sections of equal area. By applying mirror symmetric driving signals across two diagonally opposite electrodes, the tube is bent perpendicular to its axis, producing lateral motion. If the driving signal for the lateral motion is applied to only one section of the electrode, the orthogonality is degraded significantly. This is partially due to the asymmetric stress developed in the tube and partially due to the change of the charge on the inner electrode which is not cancelled by the electrode on the opposite section. The amount of lateral displacement is proportional to $(l^2/Dt)V$, where l is the length, D is the tube diameter, t is the wall thickness, and V is the applied voltage (the common-mode component of the voltage applied to the four outer electrodes). The vertical z motion is controlled by the voltage applied to the inner wall electrode and its displacement is represented as Eq. (39). A more detailed analysis of the single tube scanner has been done by Carr.[28]

Scanners have been made with stacked piezoelectric disks coupled with mechanical amplifiers which are suitable for scanning large objects over large areas (hundreds of microns),[29,30] yet most of these scanners have low mechanical resonant frequencies (< 100 Hz).

2.3.3 Sample Positioner

A variety of devices have been developed for the purpose of bringing the tip close to the sample in the STM. Such a device needs to be able to move the sample far enough away from the tip to allow sample transfers and then approach the sample to within the range of the *z*-axis of the scanner. The approach must be smooth and predictable with minimal backlash in order to avoid crashing the tip into the sample. To simplify the sample approach procedure, a computer can be used to step the sample a small distance toward the tip, with the tip fully retracted. After each step the tip is extended to see if any tunneling current can be detected. In this way, the sample is automatically brought to within tunneling range of the tip. There are several factors to be considered in selecting a sample approach mechanism. First of all, the reliability of the mechanism. Second, the geometry and rigidity of the device may be important for special sample shapes and sizes. For certain applications, two- or three-dimensional motion may be required to move the sample laterally in addition to approach in the *z*-direction. The speed of the approach mechanism may also be important if a sample approach over a long distance needs to be repeated frequently.

The devices that have been used can be divided into three categories: screw, clamp-step, and stick-slip. Screws can be used with lever, gear, or spring reduction mechanisms.[31–35] The screws can be turned by either a stepper motor mounted on the microscope, or with an external motor, or by hand using a retractable shaft. Using an externally controlled screw simplifies the part of the mechanism mounted on the tunneling stage and eliminates the power dissipation problem of the stepper motor, but at the cost of mechanical coupling and vibration. Mounting a stepper motor on the tunneling stage decouples mechanical vibrations. In this case, the stepper motor power dissipation should be minimized and the moving stage has to be smooth, otherwise the heat generated by the mechanism can cause severe thermal drift. Both types of screw devices are rigid and they provide reliable steps of a known size and, therefore, can relocate the sample precisely.

Clamp-step devices "walk" like an inchworm: clamping one foot to a base, expanding the body, clamping the other foot to the base while releasing the first foot, and then contracting the body. The "louse" walker used in the first STM was of this type,[36] using electrostatically clamped feet and a piezoelectric body. Using three feet this device is able to move in two dimensions. The clamp can cause problems, however, as the UHV environment may cause the feet to strongly adhere to the base and prevent the positioner from moving. Another sample positioner of this type is the piezoelectric inchworm.[37,38] This devices uses piezoelectric clamps as well as a piezoelectric body. The geometry is altered to be two concentric tubes with piezoelectric rings in between them which are used for clamping. The inner tube is fixed and the outer tube walks by a clamp-step action as before. This device is limited to one-dimensional motion but is rigid and reliable in UHV.

Various types of stick-slip inertial stages have been used successfully by several groups for sample positioning.[39–43] These devices are usually configured as a sample holder resting on a piezoelectric plate which is poled for shear motion. A sawtooth signal is applied to the piezoelectric plate so that it slowly moves the sample horizontally and then the piezoelectric quickly retracts so that it slips across the base. Thus the sample is moved one step across the base. Three-legged piezoelectric stick-slip devices can walk in two dimensions as well as rotate.[41,43] If the sample holder is just resting on the base instead of being actively clamped, the system is less rigid and it can only be used horizontally since the device depends on gravity for the friction needed for motion. In order to remove this limitation, the moving stage can be tightly clamped and use stronger piezoelectric force. By adjusting the clamping force appropriately, the stick-slip stage can be made more rigid and can also climb up and down vertically. Other variants of this design include the use of external magnetic solenoids for the locomotive power, [44,45] in which case heat dissipation problems must be considered.

Most of the sample positioning devices mentioned allow a wide range of step sizes, from less than 1000 Å to more than a micron. However, very small step sizes tend to be unreliable because of the surface roughness of the mechanical components. Performance can be improved with polished surfaces and/or using sapphire bearings. In general, care must be taken to prevent mechanical overshoot and backlash from causing tip crashes into the sample. All of these sample positioning devices have been proven in working microscopes and the selection of one should be based upon the unique requirements of the particular microscope design.

2.3.4 Vibration Isolation Stage

Considering the resolution and sensitivity of STM, the problem of isolating the microscope from vibrations in the external world should never be underestimated. This is especially true for UHV-STM, since atomic scale images at slow scan rates are the primary interest. There are two typical vibration isolation systems currently used in STM: the first is coiled spring suspension with magnetic damping, and the second is a stack of metal plates with viton dampers between each pair of steel plates.[46] Both systems have been used in STM instruments capable of recording images with atomic resolution. The first system is more delicate and complicated to build but in general provides more effective vibration isolation. The second system is easy to build and relatively stiff so that it is much easier to manipulate the internal components;

on the other hand, some additional pneumatic suspension systems may be required to improve the vibration isolation at low frequencies. A detailed comparison of the two systems is developed in Ref. 13.

As we discussed in Section 2.2.2, rigid and compact STMs are more immune to vibration. However in UHV-STMs, the sample and tip transfer mechanisms often require a large size for the microscope. Although the scanner itself has a high resonant frequency, what is important is the lowest resonant frequency of the mechanical structure which supports the sample and the tip. This includes the sample holder, the sample positioner, the base plate, and the scanner, as well as the sample and tip. The typical lowest resonant frequency of a UHV STM is between 1 and 2 kHz. If the floor vibration amplitude is a few thousand angstroms, we need to have a vibration isolation factor of at least 10^{-5} in order to attenuate the vibration level below 0.05 Å. Considering the relationship of vibration isolation efficiency to the resonant frequencies of the tunneling unit and the vibration isolation stage, Eq. (27), the vibration isolation stage must have a resonant frequency of less than a few Hertz.

A spring suspension stage can meet this requirement easily by allowing for enough spring extension. There are three parameters in designing an extension spring with a given material: the wire diameter, the winding diameter, and the free length. With a given mass load and extended length $(\Delta l + l)$, the parameters should be selected such that the stretch length (Δl) of the spring can be maximized while the spring is within its elastic limit. For a given spring constant per unit length, a larger winding diameter with thicker wire diameter allows more expansion $(\Delta l/l)$ than a smaller winding diameter with thinner wire diameter. Therefore the optimum spring can be determined as follows: 1) select the maximum winding diameter that can fit in the allowed space; 2) select the finest wire diameter that can stand the given mass load (retaining the spring within the elastic limit); and 3) adjust the length to fit the given span of suspension. There is an exception to the first condition: even when space permits, the winding diameter should not be too large, otherwise the wire diameter has to be too thick and it can transmit high frequency vibrations. The material for the spring should be selected so that it does not deform (sag) during bakeout (such as Inconel).

When two stages of spring suspension are used, it is desirable to make the springs of the two stages overlap each other rather than simply attaching the inner stage spring to the bottom of the outer stage. This overlap permits a longer possible span for the springs and therefore lowers the resonant frequencies. One requirement for this scheme is that the center of mass of each stage must be lower than the point from which the stage is suspended. An additional consideration for choosing the spring suspension points is that that lateral separation of the springs should be large compared to the spring length. Otherwise, there will be other eigenmodes which can be excited easily, such as swinging or torsional motions.

As a damping mechanism, permanent magnets and copper blocks are used. This eddy current damping provides a damping force which is proportional to the relative speed of the stages. This system is compatible with UHV and its damping factor can be adjusted easily by varying the strength of the magnets, the size of the copper blocks, and the spacing between them. The damping coefficient γ is given by

$$\gamma = B^2 S t C_0 / \rho, \tag{41}$$

where B is the magnetic field (weber/m^2), S is the cross sectional area of the magnet (m^2), t is the thickness of the copper block (m), and ρ is the resistivity of the copper block ($\Omega \cdot$ m). The dimensionless constant C_0 can be calculated by solving the Maxwell equations. The typical value of C_0 is 0.1–0.5, depending on the geometry of the copper block.[13,47]

In the stacked metal plate vibration isolation system, the stiffness of the rubber is higher than that of the coiled springs. Since the compressible rubber length is limited, the eigen-frequencies are higher (usually 10–100 Hz) and the performance is poorer than spring suspension stages. As the number of stacked plates increases, the eigen-frequency reduces and the vibration isolation improves. However it is impractical to achieve a resonant frequency below 10 Hz, and extra pneumatic support is usually necessary for atomic scale imaging. The pneumatic vibration isolation support has a surge tank and capillary flow resistance. It can provide a low resonant frequency (< 1 Hz) and the amplification near its resonant frequency is small. The pneumatic support is helpful in both spring suspension and stacked plate vibration isolation systems.

In order to prevent mechanical coupling between isolation stages, the electrical wires connecting the tunneling area to the outside electronics should be thin (36 gauge or higher) and clamped on the intermediate stage(s).

2.4 Control Electronics

In this section, we discuss the basic electronics that are required to control an STM. None of the components are particularly unusual, but the need for control of the instrument to sub-angstrom accuracy dictates that good design and construction practices must be followed.

2.4.1 Preamplifier Electronics

A current-to-voltage (transimpedance) preamplifier is used to detect the tunneling current and amplify it in order to limit the effects of interference

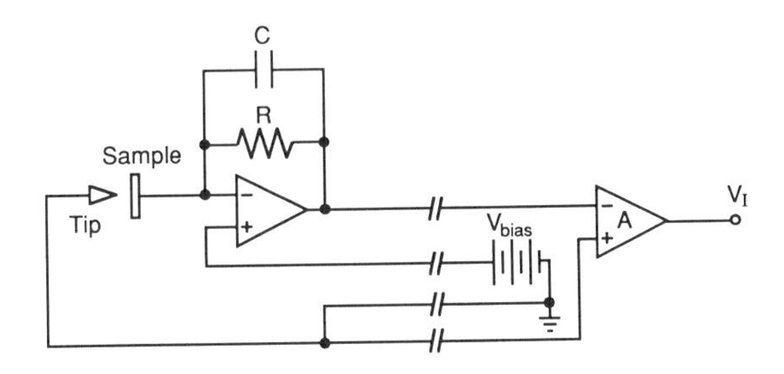

FIG. 19. Schematic for a current preamplifier.

and noise. In a typical STM application, the tunneling bias is of the order of 1 V and the tunneling current is of the order of 1 nA. This yields a junction resistance of 1 GΩ. Special care must be taken in the design of the preamplifier, since the high source resistance makes the circuit particularly susceptible to electrostatic coupling. To limit interference, the current wire should be kept short (ideally with the amplifier mounted close to the tunneling junction). The capacitance of the current wire also needs to be minimized to avoid a long RC time constant, with this large source resistance.

The amplifier itself can either be a commercial unit (such as from Keithley or Ithaco) or built around an operational amplifier. To limit the input current noise as well as the input offset error, an FET amplifier should be used with a low bias current and high input impedance. Commonly available amplifiers (such as the AD549[48] or OPA128[49]), have noise currents below 0.01 pA. The gain of the amplifier is typically set to 10^6 to 10^9 V/A. For low current applications (< 1 nA), proper input guarding of the amplifier is needed to realize the full input impedance of the amplifier. A suitable circuit is shown in Figure 19. Note that this amplifier can be used to apply the bias to the tunneling junction as well. Interference on the wires between the preamplifier and the rest of the STM electronics can be minimized by reading the output of the preamplifier with a differential amplifier in order to cancel common-mode noise.

2.4.2 Feedback Electronics

The feedback circuit is normally built using operational amplifiers. It starts with the logarithm of the output from the current preamplifier, and subtracts this value from the reference current setting to obtain the error signal. The error signal is then processed by the loop filter and amplified for adjusting the tunneling gap spacing. For spectroscopy applications, it is desirable to have a sample-and-hold circuit in the feedback that freezes the z-output at a certain voltage. This allows I–V measurements of the tunneling junction to be made without the feedback system responding. Care needs to be taken to

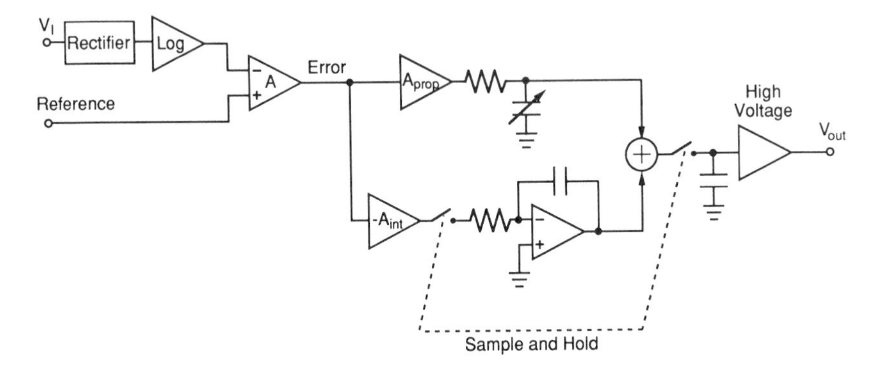

FIG. 20. Block diagram for a feedback control circuit with both proportional and integral control functions.

make sure that the system responds smoothly when the hold is turned off. This requires that the values of all integrating feedback elements be held as well as the whole system output. It is also convenient to include an approach circuit which can fully withdraw the tip and slowly allow it to approach until it reaches the desired tunneling conditions. A block diagram for a typical analog feedback circuit is shown in Fig. 20. A typical STM application uses proportional and integral control functions; these signals are summed to generate the feedback output. A wide range of gains is desirable for these control functions to enable the operator to tune the STM to a variety of operating conditions. In a typical STM, we use a current preamplifier with a gain of 10 MΩ. The output of the feedback circuit goes to a high-voltage amplifier and a piezoelectric tube, which combined give an effective sensitivity at the output of the feedback circuit of 1000 Å/V. For this system, the proportional signal has a gain adjustment that varies from 0.01 to 100 and a variable cutoff frequency that can be adjusted from 10 Hz to 10 kHz. The integral gain can be adjusted from 0.1 to 10^4 V $(\text{V} \cdot \text{sec})^{-1}$.

Some STMs use a digital feedback approach built around a digital signal processor (DSP) integrated circuit. With this method, the desired control function is converted into a finite impulse response (FIR) filter function and programmed into the DSP. The input to the control system (the tunneling current) is digitized by an analog-to-digital converter, processed by the DSP, and converted back to analog before going to the piezoelectric element. This approach has advantages and disadvantages. First, after the initial complication of building the DSP-based system, it is very easy to change the control system by reprogramming the DSP. This avoids having to build new hardware every time a new control idea needs to be implemented. Also, the reference signal as well as the whole control system is computer controlled to

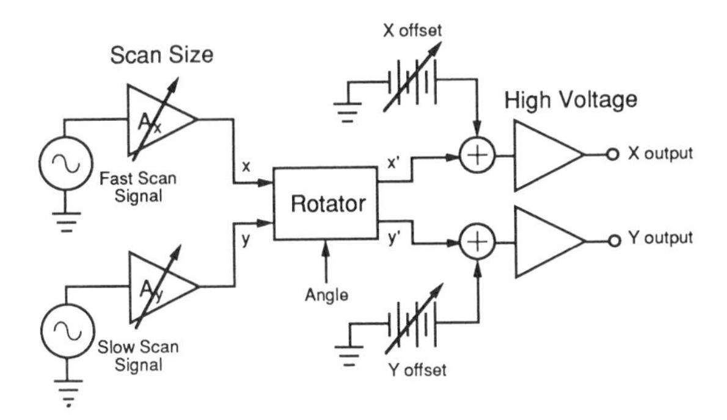

FIG. 21. Block diagram for a scan controller which allows the operator to adjust the scan size, rotation, and position.

simplify the user interface. Disadvantages include the added complexities of quantization error in the analog-to-digital converter and the finite sampling speed of the system, which may degrade the control system response. Finally, the DSP system removes some level of operator interaction with the feedback portion of the microscope. In an analog system, an experienced operator can easily optimize the control system by adjusting the various feedback gains while in the DSP system such adjustment is not as simple.

2.4.3 Scan Control Electronics

The circuitry for scan control (Fig. 21) allows the operator to adjust the region of the sample that is scanned when an image is recorded. The necessary adjustments are x and y scan size and x and y position offset, while other useful adjustments are x and y slope and scan rotation. Slope adjustments are used to compensate for any consistent tilt in the sample or scanner. A small fraction of the x and y signals are added to z to automatically follow such tilts. Generally a fundamental scan signal is generated, which is a raster scan with one fast-scan signal (the horizontal lines of the image) and one slow-scan signal (stepping vertically across the image). These fundamental scan signals can be either triangle waves or sine waves. For slow scans, triangle waves are simpler, but they contain many harmonics of their fundamental frequency and excite resonances of the microscope when used at high frequencies. Arithmetic and trigonometric functions are applied to these basic scan signals to generate outputs which are proportional to the desired x–y location of the scanner for each point of the image. Some thought should be given to the order in which these functions are applied in the circuit. In a typical system, the order of these functions is: size adjustment, rotation, offset, and slope. Following this order, for example, the rotation of the image is around the center of the image and not about the center of the scanner. Likewise, slope adjustments should be made after rotation since they are functions of the scanner/sample and not the image orientation.

Finally, these x and y signals, as well as the z signal from the feedback electronics, are amplified to the high-voltage levels required for the piezoelectric elements. High-voltage integrated operational amplifiers such as the PA83, PA85[50] work well for this application. In the tripod scanner the signals are simply scaled up to the required levels and applied to each of the three legs of the tripod. In the tube scanners, it is desirable to generate differential signals for x and y that are applied to opposite sides of the tube scanner to produce a symmetric drive arrangement. Likewise in the tube scanner, the z signal can be added to all of the outer electrodes, allowing the inner electrode to be held at one of the rail potentials. This arrangement ensures that the applied field is never in the depoling direction for the piezoelectric. If the z-voltage is applied to the inner electrode instead, the output range of the voltages should be limited so that they do not apply too high a reverse field.

2.4.4 Computer Interface and Data Acquisition

It is possible to operate an STM without any computer interface and control. The early STM experiments by Binnig *et al.*[1,36] were performed with only function generators for the x–y scan signals and a pen plotter to record data. However, as the applications of STM become more complex and digital image processing becomes more useful for both image distortion correction and data analysis, a computer interface becomes attractive for STM control. The simplest system can be built around a personal computer and commercial data acquisition boards. In this case, the main CPU bears the burden of controlling each step of the data collection process, and performance can be somewhat slow, especially for more complex functions of STM such as spectroscopy. Assembly level programming may be required for reasonable performance. A more versatile interface can be built using a digital signal process (DSP) board[51] or a general purpose interface such as the Hewlett-Packard multiprogrammer.[34] In this case each step of control and data acquisition is controlled by a dedicated microprocessor. The host CPU can then be free to control the data transfers and graphics.

The minimum computer interface should generate a raster scan via digital-to-analog convertors and read the z-data with an analog-to-digital converter. For advanced STM operation, the tunneling voltage, feedback on/off signal, and z output can also be generated by the computer. It is desirable to choose the number of data points in an image to be a power of two so that we can use the fast Fourier transform algorithm efficiently for image processing. Depending on the scan size, 128×128, 256×256, and 512×512 data points are often used.

While topographic image acquisition is straightforward, several innovative methods have been developed for spectroscopic data acquisition. The basic tunneling spectrum can be obained by the following sequence: move the tip to the desired location while the feedback is active, then freeze the feedback, and record the *I–V* curve by ramping the tunneling voltage. For a better signal to noise ratio, the bias voltage can be modulated and lock-in detection used to obtain the differential conductivity, dI/dV as a function of the bias voltage. During this measurement, the feedback can remain active so that the dynamic range of the signal can be increased. In this case, the modulation frequency should be much higher than the feedback response frequency so that the feedback maintains only the DC level of the tunneling current. The tunneling voltage must not be reduced to the point where the tip to sample distance becomes too small (contact). In order to overcome this limitation while achieving a wide dynamic range for the spectrum, Feenstra *et al.*[52] freeze the feedback and move the probe tip toward the sample during the voltage sweep, thereby amplifying the current and conductivity. A continuous ramp was added in the *z* scanner, moving the tip towards the surface as the magnitude of the voltage is reduced.

Spectroscopic measurements have some uncertainty in the lateral position of the tip due to thermal drifts. Hamers *et al.*[53] have devised a clever method to eliminate this uncertainty, named current image tunneling spectroscopy (CITS). This method generates a topographic image and an *I–V* spectrum at each point of the *x–y* image. While the size of the data is quite large, the data contains complete electronic and geometric information of the surface. In this method the feedback is activated only 30–50% of the time during the scan, and frozen for the rest of the time. When the feedback is active, a constant voltage is applied to the tunneling gap. When the feedback is frozen the bias voltage is ramped and the current is measured to produce an *I–V* curve. By repeating this sequence around 1 kHz, a constant sample to tip separation is maintained during *I–V* measurements and the *I–V* curve at each point along the raster scan is determined simultaneously with the topography.

Another method used to acquire spectroscopic information with exact lateral positioning is dual-bias voltage imaging, developed by Feenstra *et al.*[54] In this method, each line of the *x* scan is measured twice, once at each of two different bias voltages. Since the two bias voltage images are acquired in parallel, the registry of the two images is maintained even when there is considerable thermal drift. One example of this method is shown in Figure 23.

Other spectroscopic techniques include conductivity imaging and work function imaging. The conductivity image was first demonstrated by Becker *et al.*[55] A small AC voltage is added to the tunneling voltage for the conductivity image and the z-piezoelectric element for the work function image. The variation in the tunneling current at the modulating frequency is measured by a lock-in amplifier. Image data are taken from both the z feedback output (topographic image) and the lock-in amplifier output (conductivity or work function image). Further details of spectroscopic data acquisition are given in Chapter 4.

2.5 Common Problems and Further Improvements

2.5.1 Troubleshooting STM System

Operation of scanning tunneling microscopes in UHV is much more difficult than operation in air. Even with a well designed STM there are the added problems of tip and sample preparation and precise control of the vacuum. In the UHV environment, the instrument itself is quite delicate with many moving parts. Since access to the microscope in the vacuum chamber is limited, any problems that occur while under vacuum require that the system be opened to ambient for repair. Thus the pumpdown and bakeout procedure must be repeated. Moreover, when the experiment is unsuccessful, the symptoms are often vague and this makes the diagnosis of the problems very difficult. It is helpful to develop a test procedure for pinpointing the source of the problems.

It is good practice to test the STM system thoroughly in air before mounting it in the vacuum chamber. A simple and common test in air is to take images of highly oriented pyrolytic graphite (HOPG). This material is inert in air and provides a stable tunneling current. For this test, constant height mode imaging is adequate, where the image data are taken from variations in the tunneling current. In this mode, the scan speed should be high (> 20 Hz in x for 20–30 Å scan size) so that the data rate is in the higher frequency range where there is less mechanical noise. Feedback should be adjusted so that the tip does not follow atomic corrugations of the samples, but rather follows the low frequency acoustic noise and mechanical vibrations. Even with a blunt tip and a noisy environment, it is not difficult to get images showing atomic corrugations if the instrument is in good condition. If one can get clear images of the two-dimensional array of the graphite lattice, it can be assumed that the tunneling current detection system, x–y scanner and drive signals are in good condition. However, this test does not guarantee good tip condition, proper feedback operation, or the effectiveness of the vibration isolation.

In order to test the feedback performance, we have to use other samples that are inert and have some height variation (e.g., a gold film). In this test we should use the conventional constant current mode for imaging. By decreasing the x-scan rate and increasing the scan size, the feedback should be able to follow the large features of the sample. It is very useful to monitor the z motion of the scanner along the x-scan direction with the x–y mode of an oscilloscope. If the feedback parameters are correctly set, the forward and reverse trace should be similar except for some horizontal offset (due to the hysteresis of the piezoelectric). If there is a feedback oscillation, the gain of the feedback should be reduced or the cutoff frequency reduced. If there is a large hysteresis loop and the features do not appear sharp, the gain should be increased or the cutoff frequency increased. If the trace is not stable, it can be due to a bad tunneling junction, poor vibration isolation, or an inadequate feedback circuit. With a gold sample, the tunneling should be reasonably stable even with a poor tip. If the vibration isolation is a problem, the relative size of the noise from vibrations should decrease as the scan size is increased. If good images of gold are obtained at small scan sizes (around 100 × 100 Å), the system is in good condition overall. At this time, we should confirm that the noise level of the preamplifier and scan drive outputs are acceptable for atomic scale images.

During the test in air, a certain amount of noise is to be expected from acoustic coupling to the tunneling region. Therefore, it may be difficult to obtain images of individual atoms in the constant current mode in air. Without atomic images in constant current mode, it is difficult to judge the tip condition and the efficiency of the vibration isolation stages. In UHV, there is no noise from acoustic coupling through the air. So such acoustic pick-up can be ignored in these preliminary tests.

Once installed in vacuum, the STM can be tested on a real sample. If the sample is not well prepared or if it does not have a well ordered atomic lattice, the STM will still show some surface features. Poor images can appear even though the microscope is optimized and the sample properly prepared. If the images have fine features on the atomic scale without other noise, one can assume that the STM is in good condition. But the sample and tip preparation may still need improvement. If the image has low frequency noise, which appears as fluctuations in the vertical direction of the image, it is due to insufficient vibration isolation. If the image has high frequency noise which varies with the feedback gain and cutoff frequency, one should suspect the feedback circuit. When the STM system has power line noise (60 Hz or 50 Hz), it will appear as parallel lines in the image. The spacing of these lines does not depend on the scan size but depends on the scan frequency. We will discuss these problems in more detail in the following sections. Basically, the best diagnostic tool for locating the source of problems with an STM are the images themselves. An experienced user can see immediately from a few images whether problems are due to poor vibration isolation, circuitry, or tip or sample preparation. Alternatively, a spectral analysis of the tunneling current is often useful for discovering the sources of noise currents.

2.5.2 Sample and Tip Condition

The sample and the tip form the two electrodes of the tunneling junction. This tunneling junction is the source of the signal and, at the same time, the uncertainty in the operation of the STM. If our tests confirm the functions of other components of the STM, we can assume they are fixed in their functionality. On the other hand, the tunneling tip and the sample are frequently replaced and their condition may change even during the measurement. If the sample and tip condition is not good, the STM image will be filled with random noise. With such noisy data it is hard to identify the source of problems, and to distinguish between bad tip or bad sample preparation. Therefore, it is important to test the STM with a clean tip and sample, such as gold and platinum. Once the system demonstrates reasonable performance in the scanning mechanism, feedback, and vibration isolation, it is wise to invest more effort in improving the sample and tip preparation.

For semiconductors, sample preparation for STM imaging may be more stringent than it is for other surface analysis instruments. For example, a surface free of impurities is important for photoemission experiments, but for the STM it is necessary to have a well ordered and atomically flat surface. The measured region of the sample is also different. Typical surface science measurements such as LEED or Auger are averaged over a wide area and over several atomic layers below the surface. The STM measures the properties of only the top layer on an atom–by–atom basis. A good LEED pattern of a desired lattice does not guarantee that the surface is well prepared for STM. When the tip is blunt, the sample surface has to be flat over hundreds or thousands of angstroms to guarantee that the tip will land in a region showing the lattice. Detailed sample preparation techniques for specific experiments can be found in other chapters of this book.

The importance of good tip condition was discussed in an earlier section. STM images correctly reflect the sample geometry only when the current flows through a small portion of the STM tip where the density of states near the Fermi energy is constant. Transition metals are used as the tip material since they have this property and allow the energy dependence of only the local density of sample states to influence the STM image. However, the structure of the tip is not controlled at the atomic level and anomalous tips may occasionally appear. These non-ideal tips can modify the STM topographs in unusual ways, such as distorting or doubling the surface atoms. One must be careful to distinguish these anomalies from actual properties of the sample.

Non-ideal tips and their effects on STM image are most commonly observed with graphite.[56] The geometric and electronic structures of graphite are well known where one expects a centered hexagonal lattice. However, a variety of STM images can be obtained for the graphite surface as shown in Fig. 22. Many of these images do not reflect the three-fold symmetry of the graphite surface and we must assume that the image depends on the tip. A semiquantitative explanation for the variety of images can be obtained by assuming the tunneling takes place through two or more atoms at the end of the tip. The superposition of the tunneling current through each tip atom distorts the shape of the sample atoms. Multiple atomic tips separated by more than 100 Å have been observed from STM topographs of regions near a grain boundary of graphite.[57]

In the 7×7 reconstruction of the Si(111) surface, the symmetry is much more complex and the adatom spacing is larger than the graphite–atom spacing, so multiple atomic tips cause a doubling of atoms in the STM topographs. The pairing of defects that can appear in the images indicate that the tip atoms can be as close as 3 Å, while images of step boundaries indicate that they can be more than 20 Å apart.[58] The fact that the apparent spacing between the double tips can be as small as 3 Å indicates that a monatomic tip is required for sharp and definite imaging.

The relative contribution of each atom on the tip to the total current is proportional to the atoms' local density of states. When the two protruding tip atoms are chemically different, the contribution of each tip, and the doubling that appears in the image, can be voltage dependent. One example is shown in Fig. 23.[59] Figure 23(a) is an STM image of Si(111) 7×7 taken with a bias voltage of -2 V on the tip, while Fig. 23(b) is taken with a bias voltage of $+2$ V. These dual-polarity images are taken simultaneously by alternating the bias voltage for each scan line. On the bottom half of the image of Fig. 23(b) we see a doubling of Si adatoms; this pattern can be generated by superimposing two 7×7 images spaced by half the distance between two Si adatoms, which is 0.5×7.68 Å $= 3.8$ Å. The doubling of atoms disappears on the upper half of the image, indicating that one of the two tip atoms moved during the acquisition of the image. This effect does not appear in Fig. 23(a). The asymmetry of the lower halves of the images is impossible for identical tip atoms. It can be explained if we assume that one atom is metallic and the other atom has a higher density of states above than below the Fermi level. When an electron tunnels from the sample to the tip, both tip atoms contribute similar amounts to the tunneling current. However, when the electron tunnels from the tip to the sample, the nonmetallic atom does not have enough filled states and its contribution to tunneling is much smaller than the metallic atom.

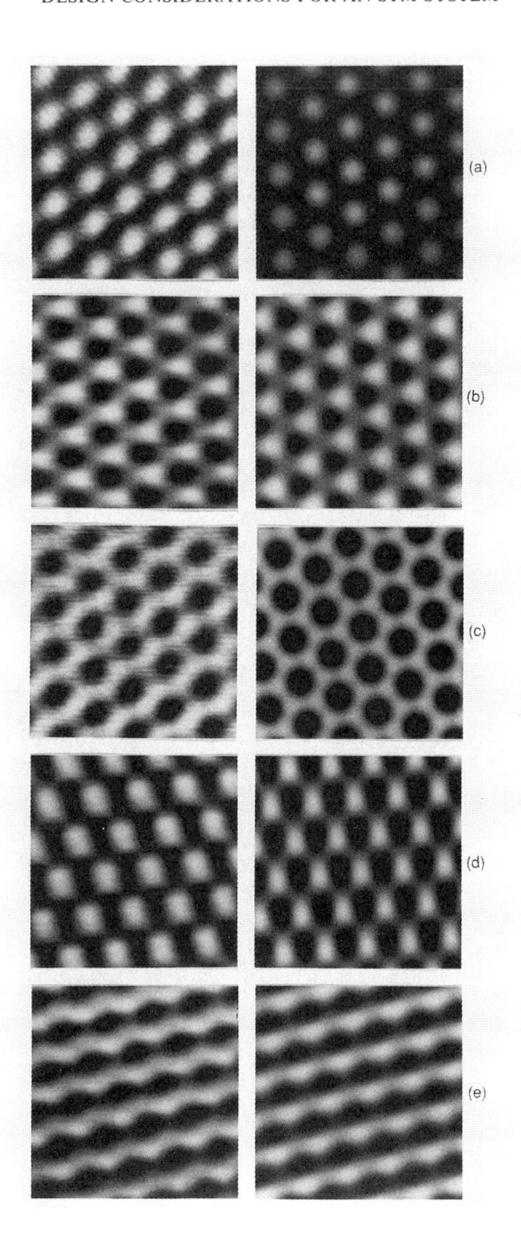

FIG. 22. Anomalous STM images of graphite. The experimental images are displayed in the left column. The computer-generated images corresponding to the experimental data are displayed in the right column. They are just a linear combination of three sine waves, whose amplitude and phases have been adjusted to match the experimental data. From Ref. 56.

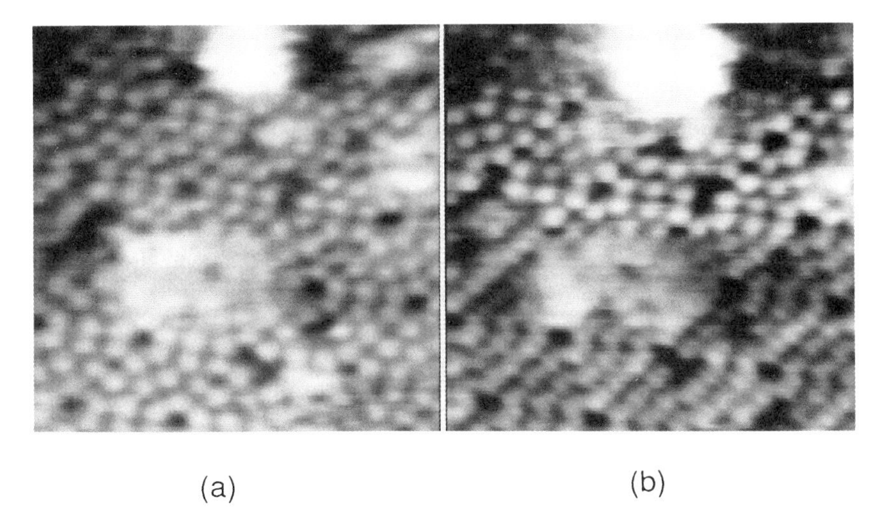

FIG. 23. A dual-polarity image of the Si(111) 7 × 7 surface obtained with (a) −2 V and (b) +2 V bias on the tip. The doubling of adatoms appears only in the lower half of the occupied states, (b). From Ref. 59.

2.5.3 Noise

In STM, we use a tunneling current in the range of 10 pA to 1 nA. This corresponds to a source impedance greater than 1 GΩ. When we amplify such a small tunneling signal, different sources of noise are amplified as well. Thermal noise (Johnson noise) and shot noise (due to the finite size of the charge quantum) are irreducible, but flicker noise ($1/f$ noise due to resistance fluctuations) and interference noise (cross-talk noise, microphonic pickup noise, 60 Hz line noise) can be reduced by careful design. Noise in vacuum tunneling was discussed by Möller *et al.*[60] Abraham *et al.*[61] showed how to reduce noise differentially by modulating the x position of the tip. This may be a good solution in certain applications where severe noise is inevitable, but the image obtained by this method is the derivative of the z height along the x scan direction rather than the z height itself. Here we will discuss instrumental noise problems, which may occur in common applications of STM.

Most of the noise originates from imperfections in the preamplifier and the tunneling junction. The tunneling region should be appropriately shielded from stray electromagnetic fields. The preamplifier should be a low noise type, as suggested in an earlier section, and its location should be close to the tunneling region. The wire connecting the tunneling signal to the preamplifier is sensitive to vibration; therefore it should be short and well supported mechanically. The wire may also induce low frequency noise in the tunneling signal as a result of changes in capacitance and magnetic flux. The feedback resistor of the preamplifier is responsible for much of the flicker noise. This noise largely depends on the material in the resistor. In general, metal film resistors exhibit better noise characteristics than carbon resistors. Special low noise metal film resistors designed for UHV are commercially available.[62]

Sixty Hz line noise is quite common, and it is difficult to remove when the signal is weak. This is not because it is hard to fix a component exhibiting line noise; rather it is hard to find the source. Frequently 60 Hz noise comes from improper grounding of the control electronics. We should keep in mind that the electromagnetic field of fluorescent lamps, stray magnetic field from power transformers, and mechanical vibrations from fans and other equipment can all cause 60 Hz noise. Once the source is identified, the solution is obvious. In order to find the source, other instruments, such as ion pumps, ion gauges, fluorescent lamps, motors, and other nearby equipment, should be turned off one by one. Another thing to check before tracing the electronic circuits is the power line itself. It is strongly advised to use a single power outlet for all instruments including the vacuum chamber controller, STM electronics, and computer. If more than one power supply has to be used, one should make sure that there is no potential difference between the different ground terminals. Sometimes a computer network or printers that share cables can cause line noise problems. In such cases, turning off each instrument may not eliminate the noise. One has to disconnect the power cord or network cable to verify that the ground loop is through the network.

If the 60 Hz noise is a problem internal to the STM system, we must trace the signal path in the electronics. In most cases, improper grounding causes 60 Hz noise. A general rule is that the signal ground should not carry current; it should be used just as a reference point to a signal, and meet with power ground or chassis ground only at one point (star ground). In order to narrow down the source of the 60 Hz noise, the actual signal path of the circuits should be traced. For this purpose, one might think it is reasonable to withdraw the tunneling tip from the sample and follow the signal path. However by doing this, it is hard to judge exactly where the 60 Hz originates. Even if the input signal contains a 60 Hz component, it may be too small to be detected. As the signal is amplified, the 60 Hz will emerge more clearly but we cannot tell whether the 60 Hz component came from the input signal or was added during the amplification. If the feedback circuit has a log-amp, its gain becomes very large when the tunneling current is near zero. Such an amplifier amplifies noise near zero input to such a point that this renders the test impractical. A good way to find the source of 60 Hz noise is to maintain the tunneling current at a finite value and examine the z output to the piezo while changing the feedback gain. If the 60 Hz component of the output increases as the feedback gain increases, then the source is before the gain setting point, most likely in the tunneling region and the preamplifier. If the

60 Hz component of the output decreases as the feedback gain is increased, then the 60 Hz noise source is after the gain setting point, most likely in the high voltage piezo driver amplifier.

2.5.4 Feedback Oscillation

Control system oscillation in the STM can be quite damaging as the oscillation normally grows exponentially until some component saturates. Unfortunately, many times the oscillation growth is only stopped by the tip being driven into the sample! Several things can cause this sort of oscillation. The first and most obvious is an improper feedback setting. In a simple control system, this problem can be alleviated by reducing the gains of the feedback signals, or reducing the cutoff frequency of any lowpass filters (such as on the proportional gain). However, in complex control systems things are not this simple. If derivative feedback is being used, the derivative signal is necessary to maintain system stability with the high gains in the other elements of the system (such as in the integral signal). Therefore, reducing the derivative gain can drive the system into oscillation. In such cases, it is necessary to turn off the control system (preferably by applying a voltage to withdraw the tip far away from the sample), turn down all the gains of the control system, and approach again by slowly turning up the appropriate gains.

Oscillations can build up for reasons other than high control gain. A tunneling junction with a contaminated tip or sample often exhibits an unstable tunneling current, and thus can result in oscillations. Such cases can be identified by observing how well the tunneling current is maintained at low control gains. If the tip-sample junction is unstable, the tunneling current will be unstable even when the control system is inactive. Finally, mechanical vibrations in the microscope can appear as feedback oscillations. These vibrations are normally identified by applying a light mechanical shock to the microscope and looking for vibrations in the tunneling current at the oscillation frequency.

2.5.5 Thermal Drift and Piezoelectric Hysteresis

STM images can be distorted because of non-ideal mechanical motions in the microscope. Varying temperatures across the microscope cause different components to move relative to one another. On an atomic scale these motions are significant. In order to reduce thermal drift effects, several versions of symmetrical design have been developed,[63–65] in which thermal drift is cancelled to a certain extent. Materials such as Invar, which have coefficients of low thermal expansion, can also be helpful. A more elaborate method for compensating the thermal drift vector is to add a slow ramp signal to the x and y axis control signals.[66] These ramps offset the drifts in the microscope. It is easy to set the ramps by taking two images over the same area of the sample and measuring the motion of a selected feature between the two images. The drift rate is given by the displacement vector divided by the time between the two images. These drifts can also be corrected by post-image processing software. In this method, the amount of drift in a particular image is determined (e.g., by measuring known crystal angles in the image). The thermal drift can then be removed by stretching and/or compressing the image in the appropriate directions to remove the distortion. This method is only possible for images of surfaces with a known structure unless the drift is measurable between successive images with a common identifying feature.

Other non-ideal motions in the microscope are due to the nonlinear characteristics of the piezoelectric elements. We can divide this problem into two distinct phenomena: 1) piezoelectric creep and 2) nonlinear displacement of the piezoelectric element with a linear applied electric field. When a control voltage is suddenly applied to a piezoelectric element, the element will immediately move approximately 95% of the full extent of the desired movement (limited by the mechanical resonant frequency of the element). The element will slowly move (creep) the remaining 5% over a period of several minutes. For atomic-scale imaging, this effect is not serious in the z direction. However, in the x–y direction the creep appears as an extra drift that remains for several minutes after any large motion across the sample has occurred. The most practical solution to the problem is to wait a suitable period of time before imaging after a large change in position across the sample.

The nonlinear displacement effect is more serious for large area scans, where the electric field is large. There are several methods to correct this nonlinearity. One is to take an image of a standard calibration grating to find the amount of distortion and re-map the image with software in a way that removes the distortion. This method is easy to implement but has some limitations. The amount and shape of the nonlinearity depend not only on the operating parameters such as scan size and scan speed, but also in the individual piezoelectric elements and time (aging). Therefore, to re-map an image of an arbitrary sample using parameters derived from a calibration grating, the image must be taken under the same conditions used for the calibration image. A more elaborate method is to use independent position sensors for the x and y axes, such as an optical sensor or a capacitance probe.[67,68] If we record the actual x–y position for each data point, the image can be processed and restored to its original shape. Alternatively, closed-loop feedback can be used to position the x–y scanner to an exact position at each data point. This method is independent of the operating conditions and the sensitivity of the piezoelectric. A difficulty arises from building an accurate position sensor. The complexity of the controlling hardware is a second problem. Nevertheless, this method serves as the ultimate solution for large area scanning with a piezoelectric actuator.

Acknowledgments

We are grateful to Prof. Calvin Quate and Dr. Jun Nogami at Stanford University, and Dr. Michael Kirk and Dr. Ian Smith at Park Scientific Instruments for valuable discussions and suggestions for this chapter.

References

1. G. Binnig, H. Rohrer, Ch. Gerber, and E. Weibel, *Phys. Rev. Lett.* **50**, 120 (1983).
2. R. S. Becker, J. A. Golovchenko, and B. S. Swartzentruber, *Phys. Rev. Lett.* **54**, 2678 (1985).
3. R. M. Tromp, R. J. Hamers, and J. E. Demuth, *Phys. Rev. Lett.* **55**, 1303 (1985).
4. R. M. Feenstra and A. P. Fein, *Phys. Rev. B* **32**, 1394 (1985).
5. J. Golovchenko, *Science* **232**, 48 (1986).
6. C. F. Quate, *Phys. Today* **39**(8), 26 (Aug. 1986).
7. P. K. Hansma and J. Tersoff, *J. Appl. Phys.* **61**, R1 (1987).
8. Y. Kuk and P. J. Silverman, *Rev. Sci. Instrum.* **60**, 165 (1989).
9. J. J. D'Azzo and C. H. Houpis, *Linear Control System Analysis and Design*, McGraw-Hill, New York, 1981.
10. J. R. Rowland, *Linear Control Systems: Modeling, Analysis and Design*, Wiley, New York, 1986.
11. D. W. Pohl, *IBM J. Res. Develop.* **30**, 417 (1986).
12. J. R. Sandercock, M. Tgetgel, and E. Meier, *RCA Review* **46**, 70 (1985).
13. M. Okano, K. Kajimura, S. Wakiyama, F. Sakai, W. Mizutani, and M. Ono, *J. Vac. Sci. Technol. A* **5**, 3313 (1987).
14. *Shock and Vibration Handbook*, 3rd ed. (C. M. Harris, ed.), McGraw-Hill, New York, 1988.
15. J. E. Demuth, U. Koehler, and R. J. Hamers, *J. Microscopy* **152**, 299 (1988).
16. H. W. Fink, *IBM J. Res. Develop.* **30**, 460 (1986).
17. Y. Kuk and P. J. Silverman, *Appl. Phys. Lett.* **48**, 1597 (1986).
18. Th. Michely, K. H. Besocke, and M. Teske, *J. Microscopy* **152**, 77 (1988).
19. H. Neddermeyer and M. Drechsler, *J. Microscopy* **152**, 459 (1988).
20. O. Nishikawa, M. Tomitori, and F. Katsuki, *J. Microscopy* **152**, 637 (1988).
21. Y. Kuk, P. J. Silverman, and H. Q. Nguyen, *J. Vac. Sci. Technol. A* **6**, 524 (1988).
22. D. K. Biegelsen, F. A. Ponce, J. C. Tramontana, and S. M. Koch, *Appl. Phys. Lett.* **50**, 696 (1987).
23. D. K. Biegelsen, F. A. Ponce, and J. C. Tramontana, *Appl. Phys. Lett.* **54**, 1223 (1989).
24. J. A. Stroscio, R. M. Feenstra, D. M. Newns, and A. P. Fein, *J. Vac. Sci. Technol. A* **6**, 499 (1988).
25. C. V. Newcomb and I. Flinn, *Electron. Lett.* **18**, 442 (1982).
26. H. Kaizuka, *Rev. Sci. Instrum.* **60**, 3119 (1989).
27. G. Binnig and D. P. E. Smith, *Rev. Sci. Instrum.* **57**, 1688 (1986).
28. R. G. Carr, *J. Microscopy* **152**, 379 (1988).

29. F. E. Scire and E. C. Teague, *Rev. Sci. Instrum.* **49**, 1735 (1978).
30. Commercially available from Wye Creek Instruments, Frederick, Maryland.
31. J. E. Demuth, R. J. Hamers, R. M. Tromp, and M. E. Welland, *J. Vac. Sci. Technol. A* **4**, 1320 (1986); *IBM J. Res. Dev.* **30**, 397 (1986).
32. D. P. E. Smith, A. Bryant, C. F. Quate, J. P. Rabe, Ch. Gerber, and J. D. Swalen, *Proc. Natl. Acad. Sci. USA* **84**, 969 (1987).
33. A. P. Fein, J. R. Kirtley, and R. M. Feenstra, *Rev. Sci. Instrum.* **58**, 1806 (1987).
34. Sang-il Park and C. F. Quate, *Rev. Sci. Instrum.* **58**, 2010 (1987).
35. B. Mitchel and G. Travaglini, *J. Microscopy* **152**, 681 (1988).
36. G. Binnig and H. Rohrer, *Helv. Phys. Acta* **55**, 726 (1982).
37. Commercially available from Burleigh Instruments, Inc., Fisher, New York.
38. A. Okumura, K. Miyamura, and Y. Gohshi, *J. Microscopy* **152**, 631 (1988).
39. D. W. Pohl, *Rev. Sci. Instrum.* **58**, 54 (1987).
40. Ph. Niedermann, R. Emch, and P. Descouts, *Rev. Sci. Instrum.* **59**, 368 (1988).
41. J. Frohn, J. F. Wolf, K. Besocke, and M. Teske, *Rev. Sci. Instrum.* **60**, 1200 (1989).
42. B. L. Blackford and M. H. Jericho, *Rev. Sci. Instrum.* **61**, 182 (1990).
43. K. Besocke, *Surf. Sci.* **181**, 145 (1987).
44. D. P. E. Smith and S. A. Elrod, *Rev. Sci. Instrum.* **56**, 1970 (1985).
45. B. W. Corb, M. Ringger, and H.-J. Güntherodt, *J. Appl. Phys.* **58**, 3947 (1985).
46. Ch. Gerber, G. Binnig, H. Fuchs, O. Marti, and H. Rohrer, *Rev. Sci. Instrum.* **57**, 221 (1986).
47. K. Nagaya and J. Dynam, *Sys. Meas. Control.* **106**, 52 (1984).
48. Analog Devices, Norwood, MA.
49. Burr Brown, Tucson, AZ.
50. Apex Microtechnology Corporation, Tucson, AZ.
51. R. D. Cutkosky, *Rev. Sci. Instrum.* **61**, 960 (1990).
52. R. M. Feenstra and P. Mårtensson, *Phys. Rev. Lett.* **61**, 447 (1988).
53. R. J. Hamers, R. M. Tromp, and J. E. Demuth, *Phys. Rev. Lett.* **56**, 1972 (1986).
54. R. M. Feenstra, W. A. Thompson, and A. P. Fein, *Phys. Rev. Lett.* **56**, 608 (1986); *J. Vac. Sci. Technol. A* **4**, 1315 (1986).
55. R. S. Becker, J. A. Golovchenko, D. R. Hamann, and B. S. Swartzentruber, *Phys. Rev. Lett.* **55**, 2032 (1985).
56. H. A. Mizes, Sang-il Park, and W. A. Harrison, *Phys. Rev. B* **36**, 4491 (1987).
57. T. R. Albrecht, H. A. Mizes, J. Nogami, Sang-il Park, and C. F. Quate, *Appl. Phys. Lett.* **52**, 362 (1988).
58. Sang-il Park, J. Nogami, and C. F. Quate, *Phys. Rev. B* **36**, 2863 (1987).
59. Sang-il Park, J. Nogami, H. A. Mizes, and C. F. Quate, *Phys. Rev. B* **38**, 4269 (1988).
60. R. Möller, A. Esslinger, and B. Koslowski, *Appl. Phys. Lett.* **55**, 2360 (1989).
61. D. W. Abraham, C. C. Williams, and H. K. Wickramasinghe, *Appl. Phys. Lett.* **53**, 1503 (1988); *J. Microscopy* **152**, 599 (1988).
62. PME and PVV series resistors from KDI Electronics, Whippany, New Jersey.
63. B. Drake, R. Sonnenfeld, J. Schneir, P. K. Hansma, G. Slough, and R. V. Coleman, *Rev. Sci. Instrum.* **57**, 441 (1986).
64. S. Gregory and C. T. Rogers, *J. Vac. Sci. Technol. A* **6**, 390 (1988).
65. S. Okayama, M. Komura, W. Mizuani, H. Tokumoto, M. Okano, K. Shimizu, Y. Kobayashi, F. Matsumoto, S. Wakiyama, M. Shigeno, F. Sakai, S. Fujiwara, O. Kitamura, M. Ono, and K. Kajimura, *J. Vac. Sci. Technol. A* **6**, 440 (1988).
66. R. M. Feenstra, *J. Vac. Sci. Technol. B* **7**, 925 (1989).
67. J. E. Griffith, G. L. Miller, C. A. Green, D. A. Grigg, and P. E. Russell, *J. Vac. Sci. Technol.* **B8**, 2023 (1990).
68. R. C. Barrett and C. F. Quate, *Rev. Sci. Instrum.* **62**, 1393 (1991).

3. EXTENSIONS OF STM

H. Kumar Wickramasinghe

IBM Research Division, T. J. Watson Research Center, Yorktown Heights, New York

3.1 Introduction

The STM has demonstrated that it is possible to stabilize and scan a fine probe tip over a sample surface to nm accuracies in (x,y,z) by using piezoelectric scanners coupled with electronic feedback techniques.[1] In order to achieve such precise control of the tip–sample spacing, it is necessary to derive an electronic feedback signal that varies rapidly as the tip–sample distance is varied. In the STM, we achieve this by monitoring the (almost exponential) decrease in tunnel current with increasing tip–sample spacing. We will see in this chapter, that the same scanning and feedback principles can also be applied to a range of other types of interactions than the tunnel current between tip and sample. All these new microscopes have the characteristic that their resolution is not determined by the wavelength that is used for the interaction as in conventional microscopy (the so called Abbe limit[2]) but rather by the size of the interacting probe that hovers over the sample surface to scan the image. As the resolution achieved is far superior to the wavelengths involved, these microscopies come under the general class of super-resolution microscopes.

So far, a number of interactions between probe tip and sample have been investigated. In this chapter, we review some of the work done in these related scanning probe techniques. Following a short historical survey of the development of super-resolution microscopy, we describe the basic principles of the scanning tunneling microscope and mention some of its applications to electrical and optical measurements. We then go on to discuss the work done in our group in the area of near-field thermal microscopy and its extensions and in the area of scanning force microscopy and its applications —in particular to electrostatic measurements such as charge, capacitance, and potentiometry. Finally, we offer some brief concluding remarks.

3.2 Historical

The first suggestion for a super-resolution microscope can be traced back

METHODS OF EXPERIMENTAL PHYSICS
Vol. 27

ISBN 0-12-475972-6

to the British scientist Synge[3,4] as far back as 1928. He suggested that one could build a tiny aperture at the end of a glass tip and raster scan this over an illuminated sample surface in order to detect sequentially, the light transmitted through sub-wavelength size regions. A picture would then be built up by using the detected signal to brightness-modulate a scan-synchronized CRT display. His later paper[4] suggested piezoelectric scanning, electronic magnification control and contrast enhancement. For apparently no good reason, this suggestion remained unnoticed until 1956 when O'Keefe[5] re-investigated the same ideas. He calculated the light transmitted through a small aperture 100 Å in size and predicted that it should be possible to achieve 100 Å resolution. He concluded, wrongly, that the technology for scanning and positioning was not available at that time and therefore no work was carried out. Following O'Keefe's paper, Baez[6] attempted to verify this basic concept by resorting to 2.4 KHz sound waves in air (14 cm wavelength) and a 1.5 cm diameter aperture.

The first demonstration of a near-field super-resolving scanning microscope was performed by Ash[7] in 1972 using microwave radiation at 3 cm wavelength; he achieved a resolution of 150 microns (that is $\lambda/200$). The STM is of course a supreme example of a super-resolution microscope; the wavelength of the electrons that scan the sample is on the order of 1 nm and atomic 0.2 nm resolution images are routinely obtained. Following the demonstration of the STM, several novel scanned probes have emerged. Table I lists some of the key ones.

3.3 STM and Some Extensions

3.3.1 Basic Principles of STM

As all the probe systems we shall talk about rely on the same piezoelectric scanning, electronic feedback, and display techniques, in this section we will explain the basic operating principle of the STM. The schematic of the STM is shown in Fig.1. In the STM, a fine tungsten (or other noble metal) tip is brought within a nanometer from a conducting surface while a voltage is applied between them. The gap separation between tip and sample is so small that electrons from the tip can tunnel from the atom at the very end of the tip to the nearest atom on the sample surface and generate a current. The tunnel current (around one nano-amp) decreases to 1/10 of its initial value for every 0.1 nm increase in gap separation. This current is compared with a reference current and the error signal so generated is applied to a gap control, z-piezo, which moves the tip up or down in order to maintain a constant tunnel current (equal to the reference value) as the tip is rastered across the sample to record an image. Typically, the signal that modulates the brightness of such an image is the variation in the voltage across the gap control piezo, which in turn is proportional to the variations in the up and down motion of the tip, as the tip is rastered across the sample. Although for a sample such as a gold surface, the STM image can be related to topography, in general such an image does not represent pure topography; also superimposed are variations from point to point of the value of the controlled parameter (in this case current) due to spatial variations in any property of the sample that could alter its value—topography being just one such property. Thus, for example, an image of a molybdenum disulphide or gallium arsenide crystal surface in general will not represent pure topography due to the variation in tunnel probability from atom to atom on its surface. With *a priori* knowledge, the different atoms can be distinguished in this way[8].

As we shall see in later sections, the basic piezoelectric scanning, feedback control, and display concepts described here can be directly transferred to other probe interactions. For example, when imaging a flat magnetic surface with a magnetic force microscope (MFM), the variations in the z-position signal can be related to variations in the magnetic interaction between tip and sample, in direct analogy to the variations in tunneling probability.

3.3.2 Electrical Extensions

Several extensions of the STM have been demonstrated, both in relation to electrical measurements and optical measurements. The scanning noise microscope is an STM where no external bias voltage is applied to the tunnel junction.[9] The mean-square noise voltage from the junction is measured in a broad bandwidth, and then maintained constant using a feedback loop which controls tip–sample spacing. Since the mean-square noise voltage is proportional to the gap resistance, this technique allows one to maintain a constant gap resistance as the tip is rastered across the sample. Beyond just being able to map the topography of the surface, this technique is also useful for providing an independent control of the gap while simultaneously making other measurements (for example, thermoelectric voltage) across the tunnel junction, in a band outside the bandwidth in which the noise is measured.

Another modification of the STM is the scanning tunneling potentiometer.[10] In this technique, a bridge method is used to measure the spatial variation in potential across the sample as the tip is controlled and scanned so as to track the surface topography. An AC voltage (typically a few kiloHertz frequency) is applied between tip and sample which generates an AC tunnel current. The amplitude of this current is then used to control the tip–sample spacing. An independent control loop, whose band (DC to 1 kHz) is outside the band of the gap control loop, is used to maintain zero DC tunnel current by con-

TABLE I. SXM Techniques and Capabilities

1.	Scanning Tunneling Microscope (1981) —G. Binnig, H. Rohrer —Atomic resolution images of conducting surfaces
2.	Scanning Near-Field Optical Microscope (1982) —D. W. Pohl —50 nm (lateral resolution) optical images
3.	Scanning Capacitance Microscope (1984) —J. R. Matey, J. Blanc —500 nm (lat. res.) images of capacitance variation
4.	Scanning Thermal Microscope (1985) —C. C. Williams, H. K. Wickramasinghe —50 nm (lat. res.) thermal images
5.	Atomic Force Microscope (1986) —G. Binning, C. F. Quate, Ch. Gerber —Atomic resolution on conducting/nonconducting surfaces
6.	Scanning Attractive Force Microscope (1987) —Y. Martin, C. C. Williams, H. K. Wickramasinghe —5 nm (lat. res.) non-contact images of surfaces
7.	Magnetic Force Microscope (1987) —Y. Martin, H. K. Wickramasinghe —100 nm (lat. res.) images of magnetic bits/heads
8.	"Frictional" Force Microscope (1987) —C. M. Mate, G. M. McClelland, S. Chiang —Atomic-scale images of lateral ("frictional") forces
9.	Electrostatic Force Microscope (1987) —Y. Martin, D. W. Abraham, H. K. Wickramasinghe —Detection of charge as small as single electron
10.	Inelastic Tunneling Spectroscopy STM (1987) —D. P. E. Smith, D. Kirk, C. F. Quate —Phonon spectra of molecules in STM
11.	Laser Driven STM (1987) —L. Arnold, W. Krieger, H. Walther —Imaging by non linear mixing of optical waves in STM
12.	Ballistic Electron Emission Microscope (1988) —W. J. Kaiser (1988) —Probing of Schottky barriers on nm scale
13.	Inverse Photoemission Force Microscope (1988) —J. H. Coombs, J. K. Gimzewski, b. Reihl, J. K. Sass, R. R. Schlittler —Luminescence spectra on nm scale
14.	Near Field Acoustic Microscope (1989) —K. Takata, T. Hasegawa, S. Hosaka, S. Hosoki, T. Komoda —Low frequency acoustic measurements on 10 nm scale
15.	Scanning Noise Microscope (1989) —R. Moller, A. Esslinger, B. Koslowski —Tunneling microscopy with zero tip–sample bias

TABLE 1. Continued.

16.	Scanning Spin-precession Microscope (1989) —Y. Manassen, R. Hamers, J. Demuth, A. Castellano —1 nm (lat. res.) images of paramagnetic spins
17.	Scanning Ion-Conductance Microscope (1989) —P. Hansma, B. Drake, O. Marti, S. Gould, C. Prater —500 nm (lat. res.) images in electrolyte
18.	Scanning Electrochemical Microscope (1989) —O. E. Husser, D. H. Craston, A. J. Bard
19.	Absorption Microscope/Spectroscope (1989) —J. Weaver, H. K. Wickramasinghe —1 nm (lat. res.) absorption images/spectroscopy
20.	Phonon Absorption Microscope (1989) —H. K. Wickramasinghe, J. M. R. Weaver, C. C. Williams —Phonon absorption images with nm resolution
21.	Scanning Chemical Potential Microscope (1990) —C. C. Williams, H. K. Wickramasinghe —Atomic scale images of chemical potential variation
22.	Photovoltage STM (1990) —R. J. Hamers, K. Markert —Photovoltage images on nm scale
23.	Kelvin Probe Force Microscope (1991) —M. Nonnenmacher, M. P. O'Boyle, H. K. Wickramasinghe —Contact potential measurements on 10 nm scale

tinuously causing the voltage on the tip to track the voltage on the sample as the tip is rastered across its surface. The tip voltage is then equal to the sample voltage at every point on the surface. This technique is useful for measuring nanometer scale potential variations on devices such as Schottky

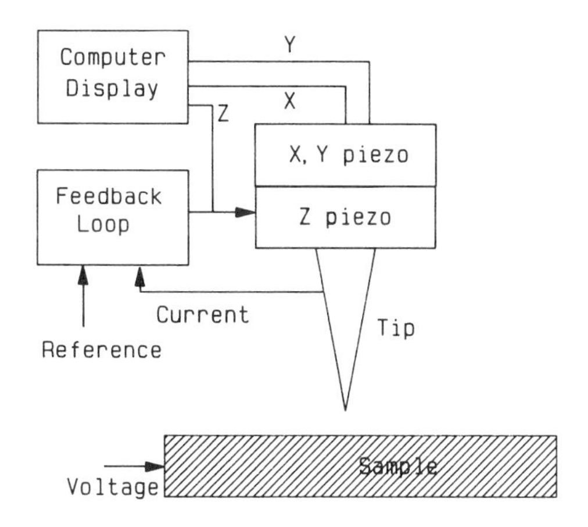

FIG. 1. Schematic of scanning tunneling microscope.

barriers, *pn*–junctions, and heterostructures. The voltage resolution is typically on the order of a few millivolts. As we shall see later, these techniques have now been extended to scanning force microscopy allowing one to measure potential distributions on insulating surfaces as well.

3.3.3 Optical Extensions

Important extensions of the STM have been demonstrated in the context of nonlinear optical mixing and inverse photoemission microscopy. Cat's whisker diodes were used as far back as 1901 to rectify RF signals.[11] Recently, the STM has been used for electrical rectification,[12] optical rectification,[13] and non linear optical frequency mixing. Unlike the Cat's whisker diode, the STM allows one to have a well defined and controllable junction.

In the experiments dealing with nonlinear mixing at the STM junction, the STM tip acts as a receiving antenna for the incident electromagnetic radiation. The generated current is rectified due to the nonlinearity of the STM *I–V* response. These nonlinearities arise from material, geometrical, or thermal asymmetrics between tip and sample. A power density of 10^8 W/cm^2 can generate typical fields as high as 10^8 V/cm at the tip apex. Typically, the rectified tunnel current generated is on the order of 1 nA for a focussed 10 μm wavelength CO_2 laser.

An important application of nonlinear mixing at the STM junction is to microscopy.[14] In the early experiments, a CO_2 laser was directed at the STM junction and the rectified tunnel current was used to control tip–sample spacing in order to record the image. Atomic resolution has been achieved on graphite surfaces. In later experiments, two frequencies of the CO_2 laser spaced by 9 GHz were directed at the STM junction and the mixed, re-radiated, signal at 9 GHz was used to control tip-sample spacing. In the latter experiments the microscope operated with no electrical connection to the tunnel junction.

The experiments just described suggest several new possibilities. First, the photon-driven STM, where all electrical connections to the junction are removed, opens up the intriguing possibility of imaging insulators with the STM. Second, the very high frequency response (10^{14} Hz) of the tunnel junction should allow one to perform optical spectroscopy on a nanometer scale; the rectified DC junction current being monitored as a function of incident wavelength to record the spectrum. Neither of these propositions have been realised to date.

The stimulation of photon emission by tunneling electrons in an STM was first observed in 1988.[15] This and further experiments[16] have served to demonstrate the potential of inverse photoemission microscopy. Early experiments recorded isochromat spectra, where the emission at a fixed wavelength was

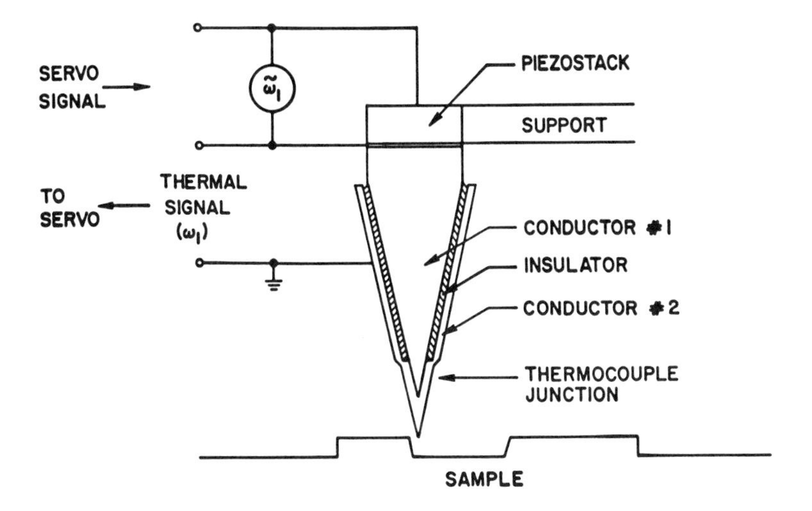

FIG. 2. Schematic of Scanning Thermal Probe.

monitored as a function of tip bias voltage. These spectra showed features which essentially followed the density-of-states features obtained from *I–V* spectra of the STM junction. In the case of metallic films, the emission wavelength has been shown to correspond to surface plasmon modes in the film that are inelastically excited by the tunneling electrons and then radiatively decay as light.[17] These plasmon modes are thought to originate as charge oscillations between the tip and sample in the STM junction and then scatter into light waves through the same junction.[18]

In further experiments, the tunneling parameters were maintained constant while the emission output was analysed in a spectrometer. Such experiments have successfully mapped the bandgap luminescence spectrum of semiconductors such as GaAs and deep level states in CdS.[19] As the STM experiments allow one to simultaneously record *I–V* spectra, it enables one to follow both the elastic and inelastic pathways of tunnel injection including the de-excitation processes.

3.4 Near-Field Thermal Microscopy and Extensions

The scanning thermal probe[20,21] was invented for profiling insulating surfaces. At that time, the atomic force microscope (AFM)-(26) which has since demonstrated the capability of profiling atomic features on both insulating and conducting crystals did not exist. The thermal probe consists of a thermal sensor (a thermocouple) built at the end of a fine tungsten tip (see Fig. 2). If a steady current is passed through this thermocouple junction,

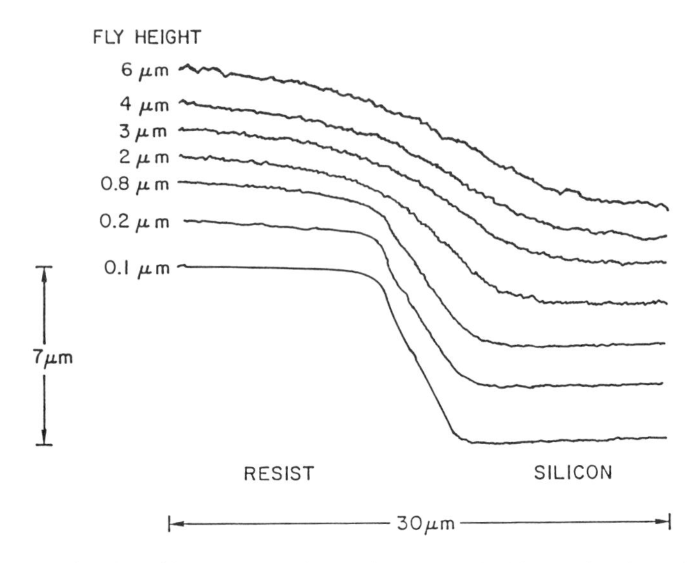

FIG. 3. Series of profiles over 7 μm photoresist step on Si with scanning thermal probe.

it heats up and comes to an equilibrium temperature above the ambient value. If the tip now approaches a sample surface (an insulator, conductor, or even a liquid), it cools down due to heat transfer from tip to sample. The tip temperature which is detected by the thermocouple can then be used to control tip–sample spacing (in much the same way as the tunnel current is used in an STM) as the tip is rastered across the surface. In operation, instead of measuring the DC thermoelectric voltage as described above, the tip is vibrated by a few tens of angstroms in the vertical direction, and the AC change in the thermoelectric voltage is used as a monitor of the tip–sample spacing. This renders the system immune to ambient temperature variations caused by room temperature fluctuations and air currents in the vicinity of the probe tip.

In our first experiments with the system, we took a series of scans over a step of photoresist on silicon and obtained profiles with the tip stabilised at varying distances over the surface. The results showed how the profile approached the true profile of the surface as the gap between tip and sample was progressively reduced (see Fig. 3).

The smallest detectable temperature change is determined by the thermoelectric coefficient of the thermocouple (typically a few microvolts per degree) and the Johnson noise from the junction. For a platinum–tungsten thermocouple with a junction resistance of 100 Ω the minimum detectable temperature change is 10^{-4} K in 1 Hz bandwidth.

By constructing smaller and smaller thermocouple junctions, the resol-

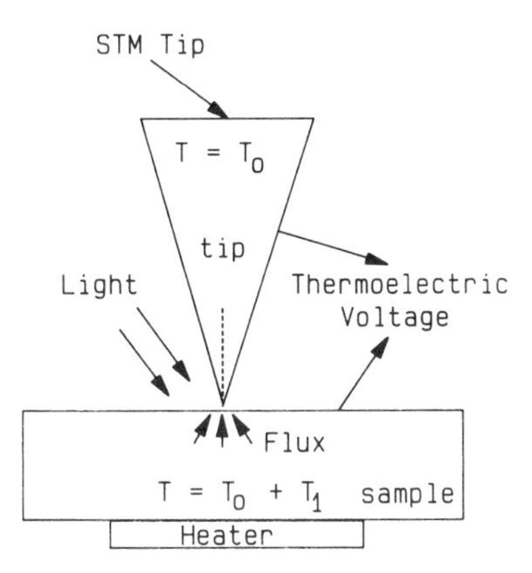

FIG. 4. Thermoelectric measurements with a "tunneling thermometer"; the temperature differential can be due to either the absorption of incident light or direct heating of the sample.

ution of the thermal profiler has been improved down to 350 Å.[21] However, as the tip–sample distance becomes less than the mean-free path of the air molecules (660 Å), classical mechanisms for heat conduction between tip and sample break down, and the AC modulation of the tip temperature (due to the tip vibration) diminishes markedly, making it difficult to stabilize the tip over the sample surface. Fortunately however, we observe a rapid change in thermal conduction from tip–sample as the tip approaches to distances below 100 Å from the sample. The conduction mechanism in this case is thought to be due to the near-field coupling of the optical phonon field between tip and sample.[22]

We have exploited this near-field coupling of the heat flux and extended the thermal measurements toward atomic resolution.[23,24] In these experiments, the thermocouple is formed by approaching a conducting tip toward a conducting surface of a different material using the tunnel current as the control parameter. The tunnel current is then periodically switched off (and simultaneously the z-control loop is put on hold) while the junction thermocouple voltage is measured. In this way, such a "tunneling thermocouple" can be rastered within tunneling range of the sample in order to measure atomic scale variations of the thermocouple voltage (see Fig. 4). The temperature sensitivity of the tunneling thermocouple, like the thermal probe, is limited by the Johnson noise in the tunnel resistance and the thermoelectric coefficient between tip and sample. For a junction resistance of 100 KΩ and a typical thermoelectric coefficient of 3 μV/K, the minimum detectable temperature change is 10^{-2}K in 1 Hz bandwidth.

Another way of controlling the gap in these experiments might be to use the scanning noise microscope described earlier. In this case, the thermocouple voltage and topography can be simultaneously measured on parallel channels.

The tunneling thermocouple maps the product of the local variations in the sample chemical potential gradient with temperature $\partial\mu_s/\partial T$ times ΔT—the temperature gradient across the top atomic layer being imaged. In principle one can therefore perform two separate experiments. In the first series of experiments, we measured the local variations in temperature ΔT of a gold surface due to the absorption of laser radiation—this allows one to do local spectroscopy. In the second series of experiments, we applied a constant temperature gradient ΔT normal to the top surface of the sample by heating it from the back and measured the local variations in the chemical potential gradient $\partial\mu_s/\partial T$.

In the optical absorption experiments,[23] we deposited a thin film of gold onto a cleaved mica surface and subsequently annealed it to form a crystal line gold layer with some mono-atomic steps. We then directed a tunable laser beam onto the surface and scanned a tip over it in order to simultaneously record a tunnel image and a local temperature image. In the tunnel image of the gold surface, some mono-atomic steps were clearly visible. The corresponding temperature image, with blue light focused onto the surface, also detected the atomic steps together with additional contrast, which we believe was due to local variations in the thermal properties of the surface. By stopping the scanner, and recording the thermoelectric signal versus wavelength, we were able to reproduce the absorption spectrum of the gold film on a nanometer scale.

In the chemical potential experiments,[24] we heated the back of a cleaved MoS_2 sample to 10 K above ambient and then mapped the variations in thermoelectric voltage. We were able to detect differences in the chemical potential signal $\partial\mu_s/\partial T$ between molybdenum and sulphur in this way. The chemical potential signal was high over the sulphur and low over the molybdenum in direct contrast to the STM response (see Fig. 5). Early theoretical considerations indicate that the variations in the thermoelectric coefficient can be related to variations in the logarithmic derivative of the local density-of-states.[25] Typical thermoelectric signals measured are in the range of mV/K for semiconductors and μV/K for metals.

3.5 Scanning Force Microscopy and Applications

The atomic force microscope[26] (AFM) was developed in order to study insulating surfaces. The first version worked in the repulsive mode; i.e., it measured the repulsive force between a diamond stylus and the sample with the stylus gently touching the surface. The force was detected by measuring the deflection of a cantilever (gold foil) attached to the stylus (see Fig. 6).

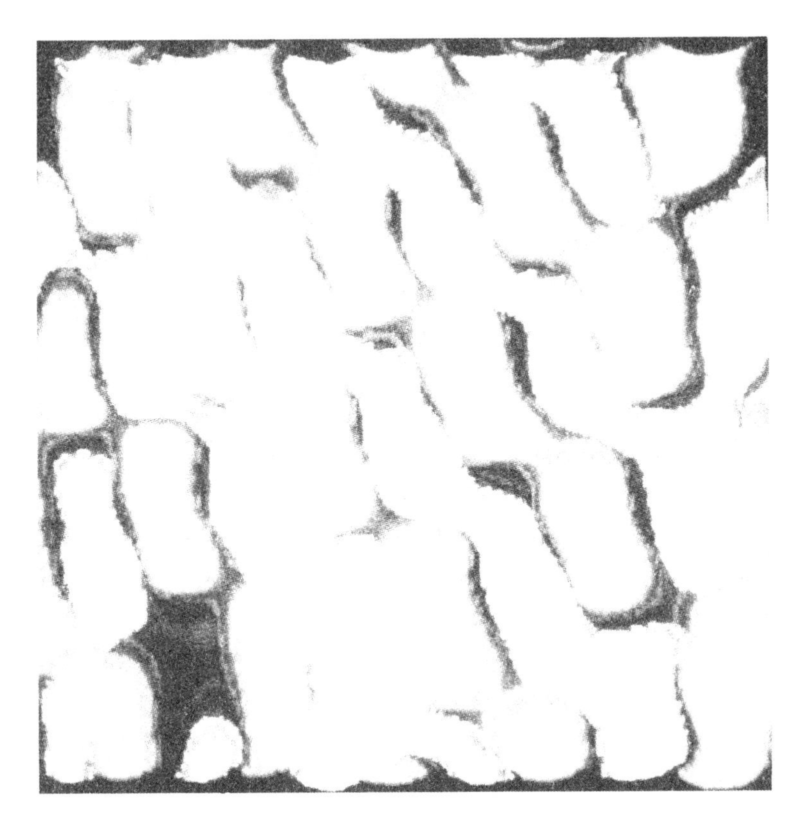

FIG. 5. Chemical potential image of a cleaved MoS_2 surface (red or bright spots) superimposed on a simultaneously recorded STM image (green or dark spots). The chemical potential signal is high over the sulphur atoms and low over the molybdenum atoms in direct contrast to the STM image. (See insert for color reproduction.)

Initial experiments used a tunneling sensor to detect this deflection. In these experiments, the tracking force was in the region of 10^{-7} N limited by the uncertainties in the force exerted by the tunneling tip on the gold foil/stylus and the relative motion of the tunneling atoms between tip and cantilever. Later experiments have replaced the tunneling sensor with optical sensors.[27] The AFM has demonstrated atomic resolution imaging on both conductors and insulators.[28] For a recent review of repulsive mode force microscopy see ref. 29.

In our group, we developed an AC version of the force microscope (see Fig. 7), which was capable of measuring van der Waal forces with a sensitivity down to 10^{-13} N and force gradients down to 10^{-6} N/m in the attractive mode.[30,31] This was important for us for the applications we had in mind in micro-electronics which required a non-destructive and non-contact

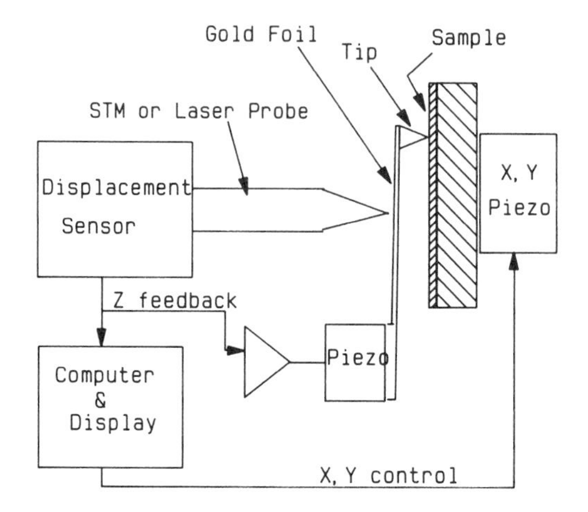

FIG. 6. Schematic of Atomic Force Microscope.

measurement. We excited a mechanical resonance of the cantilever vibration and detected a change in the resonance frequency due to the force interactions between tip and sample as the tip approached the sample. This signal was then used in a feedback loop to maintain a constant tip–sample spacing as the tip was rastered over the sample. In our experiments, we actually detected the resonance frequency change by keeping the excitation frequency constant and measuring a change in the vibration amplitude of the cantilever, using a sensitive laser heterodyne probe.[32–34] This probe is capable of detecting vibration amplitudes down to 5×10^{-5} Å in 1 Hz bandwidth. The laser heterodyne probe provides several advantages. It allows for a remote measurement of the tip vibration, thereby overcoming the difficulties presented by the tunneling sensor. It is immune to noise caused by microphonic and

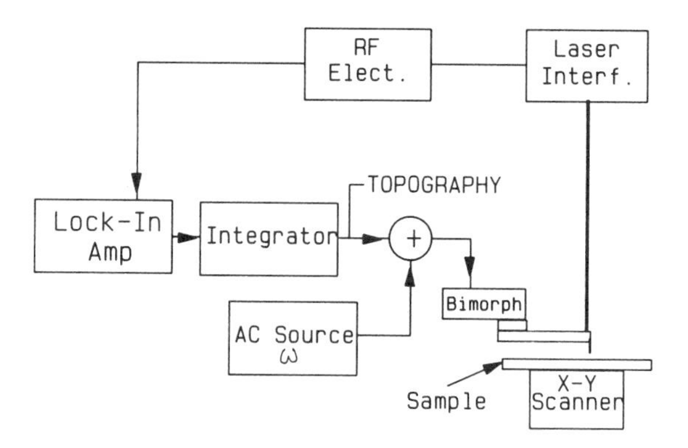

FIG. 7. Schematic of Scanning Force Microscope.

thermal fluctuations in the optical path. Finally, it provides a quantitative measurement of the tip vibration amplitude, related directly to the laser wavelength. However, the high sensitivity of the laser probe is not what presently determines the minimum detectable force gradient. Rather, it is limited by the thermally induced vibration of the cantilever.

We can obtain an expression for the minimum detectable value for the force gradient F' in the following way.[30] The force gradient F' between tip and sample causes a change in cantilever stiffness $\Delta k = F'$. We can relate the fractional change in cantilever resonance frequency ω to the fractional change in cantilever stiffness k by the expression

$$\frac{\Delta\omega}{\omega} = \frac{\Delta k}{2k} = \frac{F'}{2k}$$

The fractional change in resonance frequency is in turn related to the fractional change in cantilever resonance amplitude A through the Q factor.

$$\frac{\Delta A}{A} = \frac{4}{3\sqrt{3}} Q \frac{\Delta\omega}{\omega}$$

Relating these two expressions leads to an expression for ΔA in terms of F', Q, and k.

$$\Delta A = \frac{2QAF'}{3\sqrt{3}k}$$

The thermally excited noise amplitude N of the cantilever in a bandwidth B is

$$N = \sqrt{\frac{4KTQB}{k\omega}}$$

The minimum detectable force gradient F'_{m} is finally obtained by equating ΔA to N.

$$F'_{\mathrm{m}} = \frac{1}{A}\sqrt{\frac{27kKTB}{Q\omega}}$$

For $k = 10\,N/m$, $\omega/2\pi = 500\,\mathrm{KHz}$, $A = 1\,\mathrm{nm}$, and $Q = 500$, $F'_{\mathrm{m}} = 8.4 \times 10^{-5}\,N/m$, and $N = 5 \times 10^{-3}$ Å in 1 Hz bandwidth. The minimum vibration amplitude change that can be detected using the laser probe is 5×10^{-5} Å in 1 Hz bandwidth. This suggests that the full potential of the laser probe will only be realised by cooling the microscope to liquid helium temperatures thereby reducing the thermally excited vibration amplitude N towards its detection limit.

The attractive mode force microscope has been used to demonstrate a number of novel measurements ranging from non-contact profiling of surfaces to magnetic imaging [35–41] and electrostatic imaging.[42–45]

In magnetic force microscopy (MFM), the tip is replaced by a magnetic one—typically iron or nickel and a magnetic interaction force between tip and sample is detected. The "apparent" topography measured over a flat magnetic surface can be directly related to its magnetic features. Both hard and soft magnetic materials have been imaged and the highest resolution achieved to date is about 100 Å.[36] For a recent review of magnetic force microscopy, see Ref. 40.

In the electrostatic force microscope,[42] an AC voltage is applied between tip and sample and the induced force is measured. The force is proportional to the square of the applied voltage V^2 times the rate of change of tip–sample capacitance with spacing, $\partial C/\partial Z$. In our early experiments, we compared a regular surface profile and capacitive force image of a resist step on silicon. The capacitive force image showed a decrease in the force over the resist as compared with the bare silicon, as expected, because the electric field is largely dropped in the resist layer (as opposed to the gap) when the tip is over the resist. The effective change in capacitance that can be measured in this way is in the range of 10^{-21} F in a 1 Hz bandwidth.

The capacitance measurement technique can be used to map dopant profiles in semiconductors,[43] by measuring the change in tip–sample capacitance versus voltage at different sample bias voltages. We were able to map the dopant profile of the active regions of a cleaved MOSFET structure in this way, by scanning at a fixed bias voltage. The typical dopant densities varied from 10^{16}/cc in the gate region to 10^{20}/cc in the n^+ source and drain regions (see Fig. 8).

The electrostatic force microscope has also been applied to measure voltages on circuits.[42,44] In this case, the tip is usually connected to ground potential and a small dither voltage is superimposed onto the bias voltage. The induced vibration of the tip can then be used to determine the voltage on the circuit being probed. We were able to map the voltage distribution across circuit lines in this way.

In yet another experiment,[45] charge was deposited onto a PMMA substrate by applying a negative voltage pulse onto the tip, and its decay with time was imaged with the electrostatic force microscope. In this example, the deposited charge was estimated to be around 1200 electrons. With further refinements, it has been possible to observe discrete steps in the force versus time curve corresponding to the discharge of single charge carriers during the decay process.[46]

The UHV–STM has demonstrated the capability of measuring induced surface photovoltage on semiconductors with nanometer spatial resolution.[47] It is also possible to use the electrostatic force microscope in the poten-

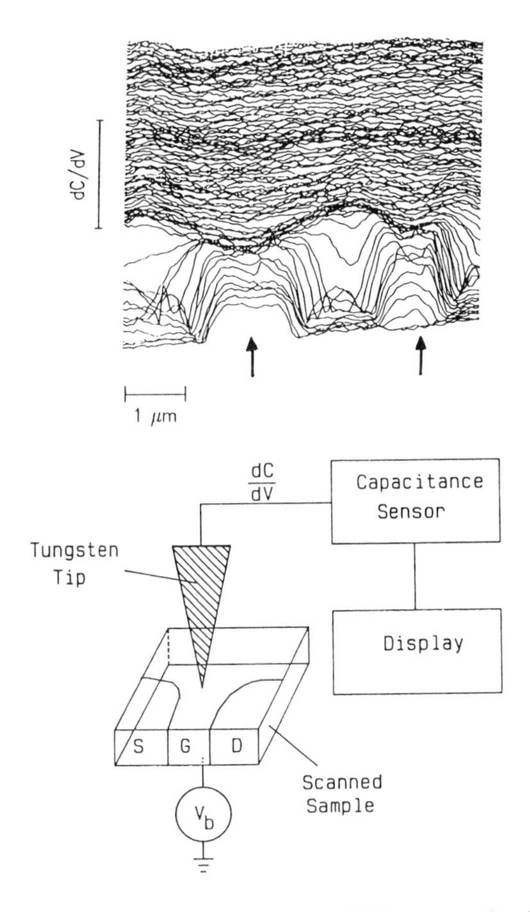

FIG. 8. Capacitance–Voltage image of a cleaved MOSFET structure showing the low doped (10^{16}/cc) p-channel regions (arrows) and the highly doped (10^{20}/cc)n^+ source/drain diffusions which extend 0.5 μm into the Si (© 1990 IEEE).

tiometric mode to make such measurements under ambient conditions.[48] In the force photovoltage experiments (see Fig. 9), in much the same way as in STM potentiometry, two independent control loops are used: one for topography and one for photovoltage. The cantilever is vibrated at frequency ω_t in the usual way to generate a control signal for topography. The voltage control loop, on the other hand, relies on the fact that the induced force on the tip varies as the square of the voltage.[49] Thus, if we apply an AC voltage at frequency ω_p across tip and sample, there will be zero induced force (and hence vibration) at this frequency when, and only when, the tip and sample are at the same DC potential. The vibration signal at ω_p can therefore be used in a feedback loop to continuously adjust the tip voltage to track the sample

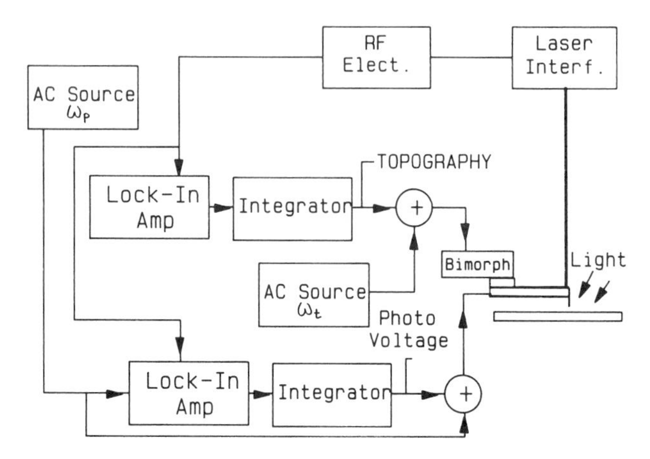

FIG. 9. Schematic for Scanning Force Photovoltage Microscopy.

voltage as the tip is rastered to form an image. Typical voltage resolution achieved with such a system is on the order of a few hundred μV. In our experiments, we scanned a Cr doped GaAs(100) surface illuminated with a HeNe laser. The topography images essentially showed little variation. In the surface photovoltage image, single dislocation loops were clearly visible. Typical signals were in the range of 200 mV.

Recent work has shown that even when the laser is switched off, there remains a finite voltage between tip and sample. One can map the spatial variations in this voltage over different conducting (metal or semiconductor) surfaces.[50] These variations, which are material dependent, are due to the differences in contact potential between the tip and sample regions being investigated. Figure 10 shows an example of a 5 μm palladium line on gold showing a difference in contact potential of 65 mV.

3.6 Conclusion

In this chapter, we briefly reviewed some of the history in the development of scanning tip microscopes. In discussing techniques, we have mainly focussed on work done within our group, but have also included the work of others where appropriate. These new microscopes allow one to obtain information about the physical and chemical properties of surfaces at an unprecedented resolution. Although many of these techniques are still only in their infancy, and in some cases the theory is not fully understood, they are already contributing to our understanding of the nano-world. We expect that with time, these techniques will play an important role in nanotechnology.

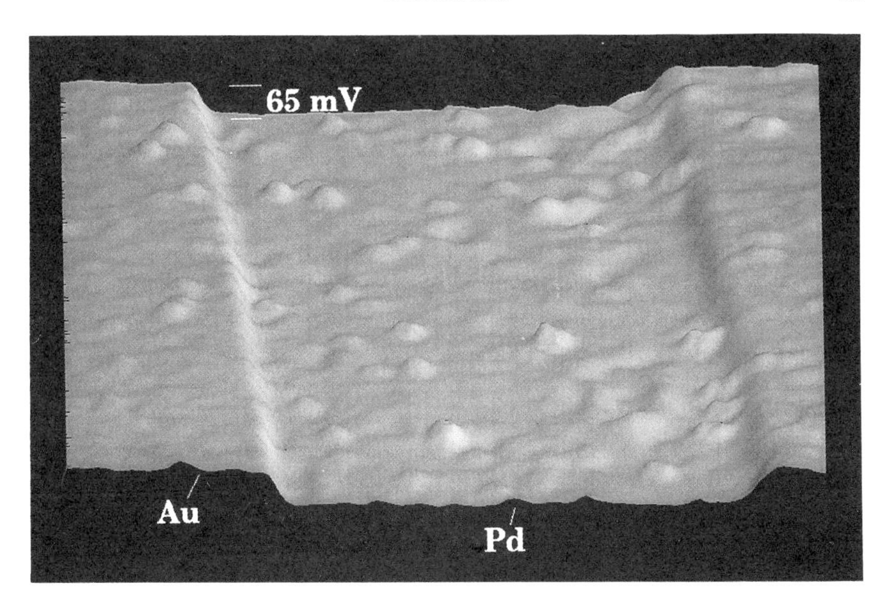

FIG. 10. Contact potential image of a palladium line on gold showing a 65 mV difference in contact potential between palladium and gold; field of view is 8 μm × 6 μm.

References

1. G. Binnig, H. Rohrer, Ch. Gerber, and E. Weibel, *Phys. Rev. Lett.*, **49**, 57 (1982).
2. E. Abbe, *Archiv. Microskopische Anat.*, **9**, 413 (1837).
3. E. H. Synge, *Phil. Mag.*, **6**, 356 (1928).
4. E. H. Synge, *Phil. Mag.*, **13**, 297 (1932).
5. J. A. O'Keefe, *J. Opt. Soc. Am.*, **46(5)**, 359 (1956).
6. A. V. Baez, *J. Opt. Soc. Am.*, **46(10)**, 901 (1956).
7. E. A. Ash and G. Nicholls, *Nature*, **237**, 510 (1972).
8. R. M. Feenstra and J. A. Stroscio, *J. Vac. Sci. Tech.* **B 5**, 923 (1987).
9. R. Moller, A. Esslinger, and B. Koslowski, *J. Vac. Sci. Tech A*, **8(1)**, 590 (1990).
10. P. Muralt and D. W. Pohl, *Appl. Phys. Lett.*, **48**, 514 (1986).
11. H. G. J. Aitken, p. 1900-1932, in *The Continuous Wave Technology and American Radio*, Princeton University Press, New York, 1985.
12. J. A. Strocio, R. M. Feenstra, and A. P. Fein, *Phys. Rev. Lett.*, **57**, 2579 (1986).
13. H. Q. Ngugen, P. H. Cutler, T. E. Feuchtwang, Z. Huang, Y. Kuk, P. J. Silverman, A. A. Lucas, and T. E. Sullivan, *IEEE Trans. Electron Dev.*, **36(11)**, 2671 (1989).
14. W. Krieger, T. Suzuki, M. Volcker, and H. Walther, *Phys. Rev.* **B 41**, 10229 (1990).
15. J. K. Gimzewski, J. K. Sass, R. R. Schlittler, and J. Schott, *Europhys. Lett.*, **8**, 435 (1989).
16. R. Berndt, A. Baratoff, and J. K. Gimzewski, "Basic Concepts and Applications of STM and Related Techniques". *Proc. NATO ASI, Erice Italy*, April 17-29, 1989, Kluwer Academic Publishers, 269 (1989).
17. B. N. J. Persson and A. Baratoff, *Bull. Am. Phys. Soc.*, **35**, 634 (1990).
18. P. Johansson, R. Monreal, and P. Apell, *Phys. Rev.* **B 42**, 9210 (1990).
19. R. Berndt, R. R. Schlittler, and J. K. Gimzewski, *Proc. Engineering Foundation Conference*

—*Scanned Probe Microscopy: STM and Beyond*, Santa Barbara, CA., Jan 6-11, 1991, AIP Conference Series, No. 241, (1992).
20. C. C. Williams and H. K. Wickramasinghe, *Appl. Phys. Lett.*, **49(23)**, 1587 (1986).
21. C. C. Williams and H. K. Wickramasinghe, *Proc. IEEE Ultrasonics Symposium*, IEEE Cat. No. 86CH2375-4, 393 (1986).
22. K. Dransfeld and J. Xu, *J. Microscopy*, **152(1)**, 35 (1988).
23. J. M. R. Weaver, L. M. Walpita, and H. K. Wickramasinghe, *Nature*, **342(6251)**, 783 (1989).
24. C. C. Williams and H. K. Wickramasinghe, *Nature*, **344(6264)**, 317 (1990).
25. J. A. Stovneng and P. Lipavsky, *Phys. Rev.*, **B 42 (14)**, 9214 (1990).
26. G. Binnig, C. F. Quate, and Ch. Gerber, *Phys. Rev. Lett.*, **56**, 930 (1986).
27. G. Meyer and N. M. Amer, *Appl. Phys. Lett.*, **53**, 1044 (1988).
28. T. R. Albrecht and C. F. Quate, *J. Vac. Sci and Tech.* **A 6**, 271 (1988).
29. D. Rugar and P. Hansma, *Phys. Today.*, **43(10)**, 23 (1990).
30. Y. Martin, C. C. Williams, and H. K. Wickramasinghe, *J. Appl. Phys.*, **61**, 4723 (1987).
31. G. M. McClelland, R. Erlandsson, and S. Chiang, p. 307, in *Review of Progress in Quantitative Nondestructive Evaluation*, Vol. 6B, D. O. Thompson and D. E. Chimenti, eds., Plenum, New York, 1987.
32. R. M. De La Rue, R. F. Humphryes, I. M. Mason, and E. A. Ash, *Proc. IEE.*, **119(2)**, 117(1972).
33. H. K. Wickramasinghe and E. A. Ash, *Proc. MRI Symposium on Optical and Acoustical Microelectronics*, Polytechnic Institute of Brooklyn, New York, 413 (1974).
34. D. Royer, E. Dieulesaint, and Y. Martin, *Proc. IEEE Ultrasonics Symposium*, IEEE Cat. No. 85CH2209-5, 432 (1985).
35. Y. Martin and H. K. Wickramasinghe, *Appl. Phys. Lett.*, **50**, 1455 (1987).
36. J. J. Saenz, N. Garcia, P. Grutter, E. Meyer, H. Heinzelmann, R. Wiesendanger, L. Rosenthaler, H. R. Hidber, and H. J. Guntherodt, *J. Appl. Phys.*, **62**, 4293 (1987).
37. Y. Martin, D. Rugar, and H. K. Wickramasinghe, *Appl. Phys. Lett.*, **52**, 244 (1988).
38. D. Sarid, D. Iams, and V. Weissenberger, *Optics Lett.*, **13**, 1057 (1988).
39. C. Schonenberger and S. F. Alvarado, *Z. Phys.* **B 80**, 373 (1990).
40. U. Hartmann, *J. Magn. Magn. Mat.*, in press, (1991).
41. H. J. Mamin, D. Rugar, J. E. Stern, B. D. Terris, and S. E. Lambert, *Appl. Phys. Lett.*, **53**, 1563 (1988).
42. Y. Martin, D. W. Abraham, and H. K. Wickramasinghe, *Appl. Phys. Lett.*, **52**, 1103 (1988).
43. J. A. Slinkman, C. C. Williams, D. W. Abraham, and H. K. Wickramasinghe, *Proc. IEEE IEDM Conference*, IEEE Cat. No. 90CH2865-4, 90-73 (1990).
44. D. W. Abraham, Y. Martin, and H. K. Wickramasinghe, *SPIE*, **897**, 191 (1988).
45. J. E. Stern, B. D. Terris, H. J. Mamin, and D. Rugar, *Appl. Phys. Lett.*, **53**, 2717 (1988).
46. C. Schonenberger and S. F. Alvarado, *Phys. Rev. Lett.*, **65(25)**, 3162 (1990).
47. R. J. Hamers and K. Markert, *Phys. Rev. Lett.*, **64**, 1051 (1990).
48. J. M. R. Weaver and H. K. Wickramasinghe, *J. Vac. Sci. and Tech.* **B 9**, 1562 (1991).
49. J. M. R. Weaver, and D. W. Abraham, *J. Vac. Sci. and Tech.* **B 9**, 1559 (1991).
50. M. Nonnenmacher, M. P. O'Boyle, and H. K. Wickramasinghe, *Appl. Phys. Lett.*, **58(25)**, 2921 (1991).

4. METHODS OF TUNNELING SPECTROSCOPY

Joseph A. Stroscio

Electron and Optical Physics Division, National Institute of Standards and Technology, Gaithersburg, Maryland

R. M. Feenstra

IBM Research Division, T. J. Watson Research Center, Yorktown Heights, New York

Inherent in the tunneling process is the potential difference between the two tunneling electrodes. By varying this potential difference, eigenstates of the tip and sample become available for tunneling and lend themselves to spectroscopic investigation. In this chapter we describe the application of tunneling spectroscopy to probe surface phenomena. Examples are drawn from a few of the earlier applications of STM which demonstrated the potential for spectroscopy. Further applications to semiconductors is found in Chapter 5, to metal surfaces in Chapter 6, and to interfaces in Chapter 7; applications to charge density waves and superconductors are described in Chapters 8 and 9, respectively.

Tunneling spectroscopy is not a new technique and has been used a great deal in examining properties of fixed tunneling junctions.[1-4] With the advent of the scanning tunneling microscope, two new variables come into play that were not accessible in fixed tunnel junctions. The first, and most important, is the scanning ability of the STM, which adds a great potential for atom-resolved probing of spectroscopic signals on a material. This ability has allowed investigators to probe the electronic properties ranging from individual adatoms on a surface to spatial properties of vortex states of superconductors. The second feature in STM spectroscopy is the variability of the tip–sample separation. This variability allows the probing of the potential barrier between the two electrodes and also allows the STM to function as an electron interferometer by changing the electron wavefunction pathlength between the sample and tip. Perhaps another advantage of tunneling spectroscopy with the STM is that perfect surfaces (or well prepared systems) can be examined in ultra-high vacuum environments; they have the advantage of known composition in contrast to the sometimes unknown composition and structure of tunneling barriers in fixed tunnel junctions. With all these advantages, tunneling spectroscopy still suffers from the

METHODS OF EXPERIMENTAL PHYSICS
Vol. 27

ISBN 0-12-475972-6

unknown contribution of the probe tip. This can lead to non-reproducibility in data resulting from tip instabilities with voltage change, tip composition, and/or structural dependencies. This brings to mind a quote from a well known text on tunneling, "Tunneling is an art, not a science."[5] To some degree this can be said of tunneling spectroscopy with the STM. Of course, one could counter this by saying, "one picture is worth a thousand words." Whether tunneling spectroscopy remains more an art than science will depend on progress with instrumentation, stability of the junction, reproducibility of results, and more information on probe contributions. To this end we hope this chapter will be useful to future investigators.

4.1 Instrumentation

The instrumentation needed for tunneling spectroscopy is simple and usually requires equal or more work on the software end of the data acquisition system than in hardware. For some applications, spectroscopic measurements, such as voltage-dependent imaging or conductance measurements with active feedback, can be made with a standard STM feedback loop. More sophisticated measurements, which demand a fixed tip–sample separation, require the addition of a sample-and-hold circuit to the feedback loop,[6] as shown in Chapter 2. The requirements of the sample-and-hold circuit are: 1) that it hold the tip position stable for a given time of the measurement, 2) it does not produce a glitch in the tip–sample separation that would be large enough to influence the measurement, and 3) its response time is fast enough for the application. Requirements 1 and 3 are easy enough to meet with most equipment. Requirement 2, however, can be a challenge. If we consider a standard feedback loop with a gain of about 400 Å per volt (this includes the piezo gain and high-voltage amplifier stage gain) after the sample-and-hold, then a 0.1 mV offset, common in most solid-state sample-and-hold circuits, would yield an unwanted tip movement of 0.04 Å; this translates into an 8% change in the tunneling current. To minimize these offsets some researchers have utilized reed relays in combination with a low pass filter for the sample-and-hold, which seem to give negligible offsets and droop, but suffer from limited time response, usually on the order of 0.5 ms.

The other hardware component necessary for spectroscopic measurements is a multi-channel data acquisition system for the measurements of the tip position and tunneling current. It is usually desirable to have an additional channel to measure the conductivity, dI/dV, and D–A channels for programming the tunneling voltage and tip position. The tip position can be programmed by adding a summing input to the high voltage z-piezo driver amplifiers. The data acquisition system used in the NIST laboratory is shown in Fig. 1.[7] In this system the data acquisition is programmed using a digital

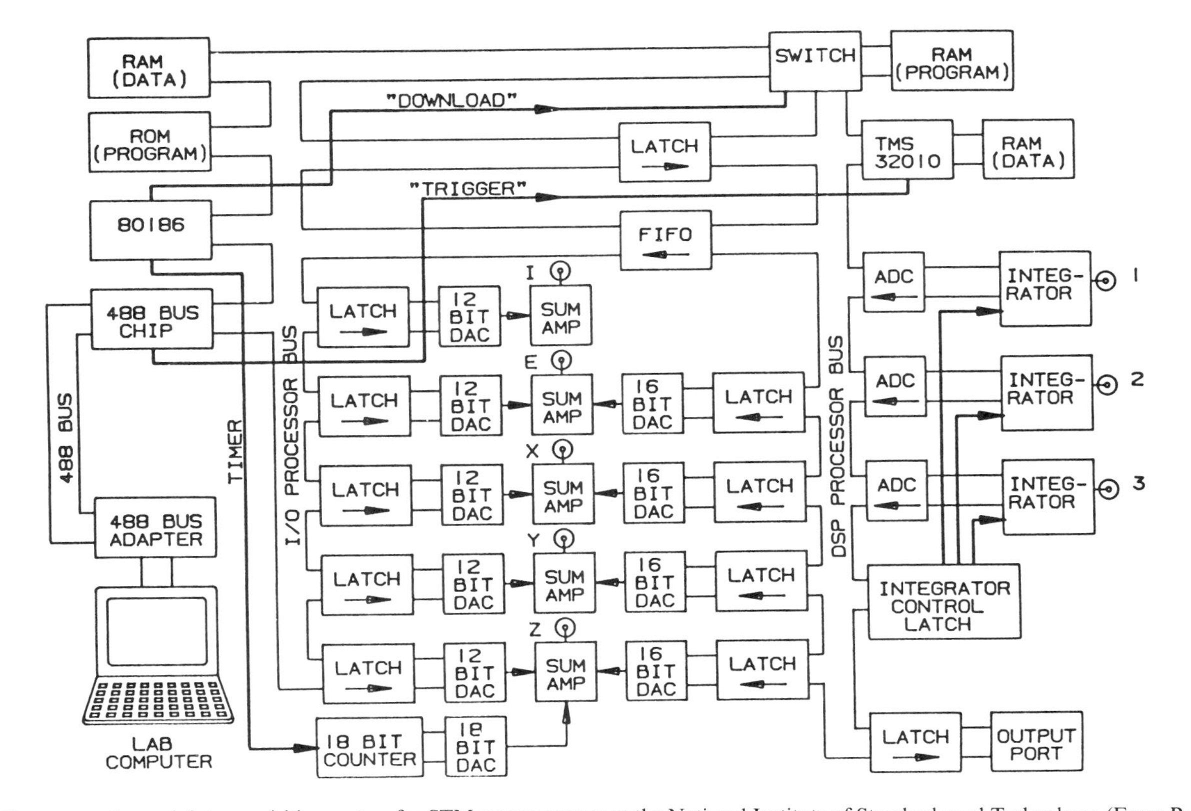

FIG. 1. Scan generator and data acquisition system for STM measurements at the National Institute of Standards and Technology. (From Ref. [7].)

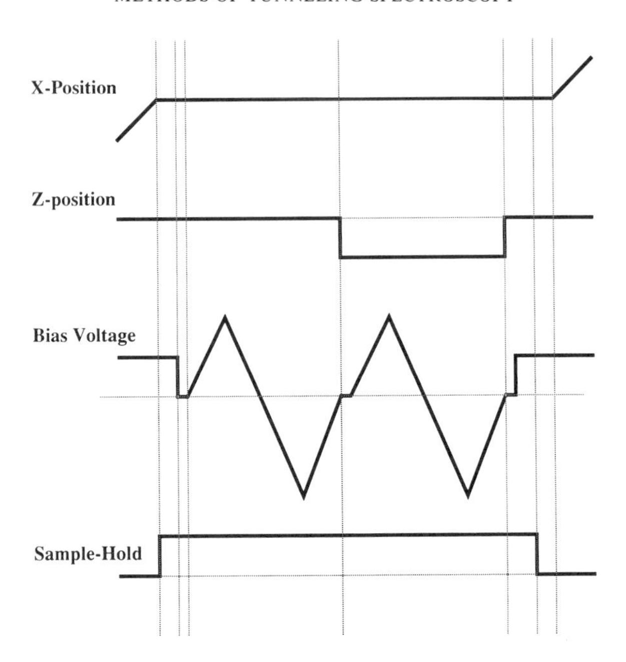

FIG. 2. Timing sequence used for current vs voltage measurement.

signal processor, which takes the load off the lab computer. In most spectroscopic applications, the need for fast acquisition necessitates the programming of part or all of the data acquisition in assembly language. In the system shown in Fig. 1, we use programmable integrators on the front end of the A–D converters, which are useful in enhancing the signal-to-noise in very sensitive measurements.

Figure 2 shows the timing sequence used in taking a fixed tip–sample separation current–voltage (I–V) measurement in the middle of an image acquisition. While an I–V measurement can be obtained in the absence of an image, it is beneficial to obtain the I–V measurement with an image so one knows the condition of the surface where the measurement was made. Figure 2 shows the sample-and-hold enabled while the tunneling voltage ramp is applied. Note a small delay is added after the sample-and-hold is disabled to allow the feedback loop to settle (this time delay is dependent upon the feedback loop time constant). After the first voltage ramp, the z tip-position can be changed for a second I–V measurement, as shown in Fig. 2. This allows a greater dynamic range to be obtained at the lower voltages, for example, if the initial starting tip–sample position was chosen with the tunneling voltage above 1 V (see Section 4.4). The speed of the voltage ramp is determined by the time constant of the tunneling unit and is influenced by

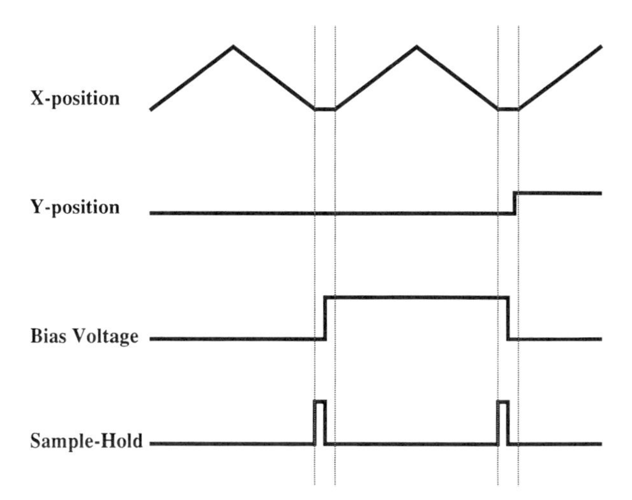

FIG. 3. Timing sequence used for voltage-dependent imaging.

the response of the current amplifier and the capacitance coupling between the tip and sample assembly. For this reason it is advantageous to shield the tip from the sample mount to reduce capacitive coupling. With the unit not tunneling (large tip–sample separation), the capacitive coupling can be measured by applying a sawtooth voltage waveform to the sample and measuring the induced displacement current on the tip. The response of the current preamp can be increased by splitting the gain in two stages; a good mix is to place a preamp on the tunneling unit with a fixed gain of $\sim 10^8\ \mathrm{V\,A^{-1}}$ with a variable-gain amplifier outside the tunneling unit to give another $\times$ 10–100 gain.

The sample-and-hold circuit is also useful in voltage-dependent imaging. To compare STM images at different voltages one wants to compare the registry of an image obtained at one voltage with another obtained at a different voltage. Taking images in sequence usually doesn't suffice since the images will be separated by some amount of thermal drift which is larger than the spatial separation of interest. To overcome the drift, multiple images can be built up simultaneously by interlacing line scans at different voltages.[8] Figure 3 shows the timing sequence used in voltage-dependent imaging. The sample-and-hold is used before switching to a new voltage; in this way the voltage can be switched very quickly without the servo active, until the new voltage is reached. Examples of voltage-dependent imaging are presented in Section 4.3.

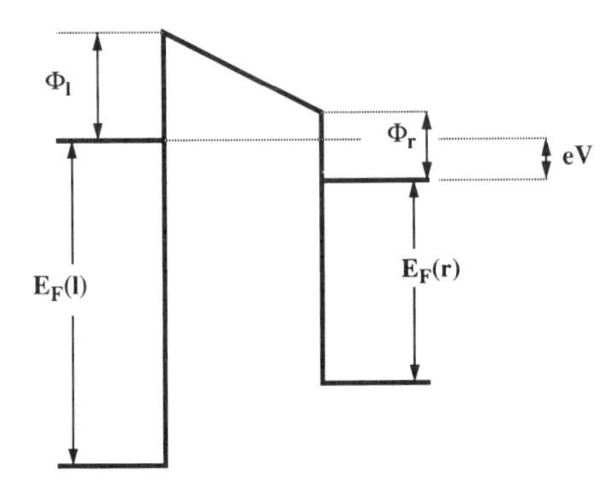

FIG. 4. Schematic diagram of trapezoidal potential barrier for tunneling. $E_F(l)$, $E_F(r)$ and ϕ_l, ϕ_r are the Fermi energies and workfunctions of the left and right electrodes, respectively.

4.2 General Current versus Voltage Characteristics

In examining tunneling spectroscopy data, it is useful to sort out characteristics that do not depend directly on the electron density-of-states (DOS) so that sample-specific structure in the sample DOS can be extracted. In this section we illustrate the general voltage and tip–sample separation dependence in metal–barrier–metal tunneling with some simple models of planar tunneling.

4.2.1 Metal–Vacuum–Metal Tunneling

To examine the main voltage and distance dependencies in tunneling, it is suitable to examine the non-interacting free-electron model for tunneling through a planar barrier.[1–4]

We begin by considering the energy barrier in Fig. 4. Taking the z direction normal to the electrodes, the tunneling current density is given by,

$$j = \frac{2e}{h} \int dE \, [\, f(E) - f(E + eV)] \int \frac{d^2k_{\|}}{(2\pi)^2} \, D(E, \mathbf{k}_{\|}), \tag{1}$$

where $f(E)$ is the Fermi–Dirac distribution function, D is the tunneling transmission probability, and $\mathbf{k}_{\|}$ is the wavevector component parallel to the junction interface. The free-electron metal approximation leads to a transmission probability, D, which only depends on the normal component of the energy, E_z, and therefore Eq. (1) can be simplified to an integral over the normal component of energy as,

$$j = \int_0^{\infty} dE_z D(E_z) N(E_z), \tag{2}$$

where $N(E_z)$ is normal energy distribution function given by,

$$N(E_z) = \frac{4\pi me}{h^3}\int_0^\infty [f(E) - f(E+eV)]\,dE_\parallel \tag{3}$$

$$= \frac{4\pi mek_B T}{h^3}\ln\left\{\frac{1+\exp[E_F - E_z)/k_B T]}{1+\exp[(E_F - E_z - eV)/k_B T]}\right\}, \tag{4}$$

where E_F is the Fermi energy.

Equations (2)–(4) can be used to calculate the current density given a form for the potential and a method to calculate the transmission probability, D. For a trapezoidal barrier, as in Fig. 4, D may be calculated using the WKB method, which yields for voltages less than ≈ 1 V,

$$D(E_z) \approx \exp\left[-s\left\{\frac{2m}{\hbar^2}(\bar{\phi} + E_F - \left|\frac{eV}{2}\right| - E_z)\right\}^{1/2}\right], \tag{5}$$

where $\bar{\phi}$ is the average workfunction of the two electrodes, and s is the tip–sample separation. For small voltages and energies at the Fermi level, $D \approx \exp(-2\kappa s)$, where κ is the vacuum decay constant, $(2m\bar{\phi}/\hbar^2)^{1/2}$.

Equation (5) shows the basic voltage and distance dependences in tunneling. Namely, the tunneling current is exponentially dependent on both voltage (for low voltages, j is ohmic) and tip–sample separation. The latter exponential dependence on separation is what makes the STM work by giving the extremely high sensitivity needed to servo the tip at constant distance above the surface. For a workfunction of 4.5 eV, $\kappa = 1.1$ Å^{-1}, which yields an order of magnitude change in tunneling transmission, and hence current, for a change in separation of about 1 Å. It is this exponential dependence of the current on separation which characterizes tunneling.[9,10]

The tunneling current dependences on distance and voltage can be calculated using Eq. (2) with D calculated directly by integrating Schrödinger's equation through the barrier to obtain the transmission probability. In this way more complicated forms of the vacuum barrier can be used to include image potential effects or finite probe tip radius effects.[11] A series of current–voltage characteristics calculated for various tip–sample separations is shown in Fig. 5. For small voltages, < 1 V, one sees a linear voltage-dependence characteristic of ohmic behavior. For larger voltages, the exponential dependence dominates the I–V characteristic. A strong exponential dependence, in addition, is observed on the tip–sample separation, with virtually no current density at low voltage for separations larger than 7 Å. The implications of this distance dependence on acquiring spectroscopic data is evident; for a fixed tip–sample separation, s, spectroscopic data can only be obtained over a limited voltage range, since the current detectors are linear amplifiers. To increase the dynamic range in voltage, the tip–sample separation must be varied. In practice, this is done by freezing the tip position using the sample-

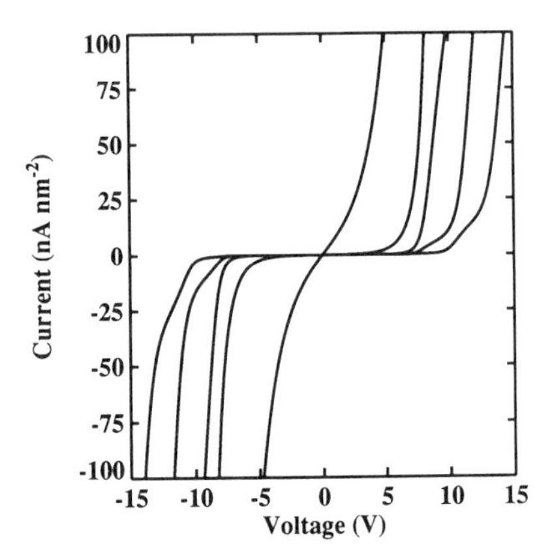

FIG. 5. Current–voltage characteristics obtained from Eq. (2) with electrical parameters for a Au–W junction. The curves correspond to tip–sample separations of 5, 7, 9, 11, and 13 Å.

and-hold technique described in Section 4.1, with different tip–sample separations obtained by adjusting the initial reference tunneling voltage while the feedback is active. Data acquired in this fashion are shown in Section 4.4.1 for the Si(111)2 × 1 surface.[8,11] Tip–sample separation can also be programmed, with sample-and-hold, for a series of decreasing separations from some initial value, but when the voltage is swept, some limiting value should be set to limit the current at the smaller separations. This current limiting can be done in software, for example, where the current is checked at each voltage during the ramp and the voltage ramp terminated at the current limit setpoint.

An alternate method for measuring spectroscopic data is to keep the feedback active while ramping the voltage at a fixed tunneling current. In this way the tip–sample separation is constantly adjusting itself and a large dynamic range in voltage can be obtained.[12] In this method the tip is following a separation versus voltage curve at constant current, such as simulated in Fig. 6. The disadvantage of this approach is that as the voltage is decreased a lower limit of about 0.1–1 V is encountered because of approaching tip–sample contact, and thus one cannot scan through zero bias with this method. The change in slope in the s–V data above 4 V in Fig. 6 is due to the transition from the vacuum tunneling regime to the Fowler–Nordheim or field emission regime, where the potential is raised above the electrode workfunction and the electron experiences a positive kinetic energy region. In this region, quantum interference effects can occur, leading to barrier resonances in the electron transmission at certain energies,[13–15] as described in the following section.

A third method of obtaining spectroscopic data is to combine the first two

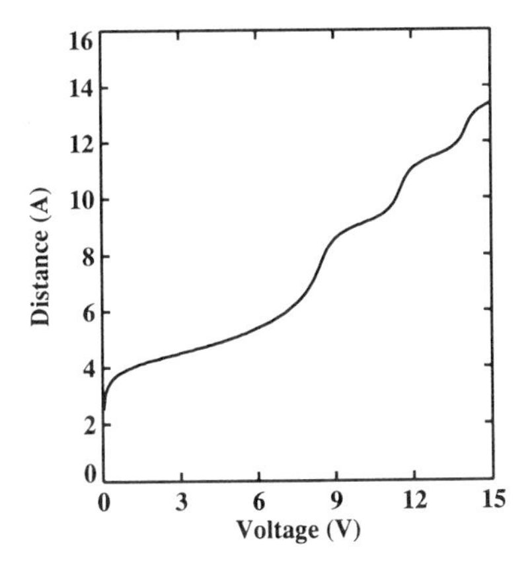

FIG. 6. Separation vs voltage at constant current obtained using Eq. (2) with parameters for a Au–W junction. The current is constant at a value of $100\,\text{nA}\,\text{nm}^{-2}$.

methods. Instead of following a separation versus voltage path determined by the tunneling junction, one can program a separation versus voltage path to simulate an approximate constant current path with the exception that the tip is now limited to some minimum tip–sample distance. With this method the conductivity, dI/dV, is measured along with the current, I, to get some measure of the sample density-of-states. Examples of this technique are described in Section 4.5.

4.2.2 Barrier Resonances

As indicated in Section 4.2.1, electron transmission resonances can occur in tunneling between two electrodes when the applied potential exceeds the work function. These transmission resonances show up as wiggles in the s–V data or as kinks in the constant distance I–V data, as seen in Figs. 5 and 6.[8,11] The best method of detection of the barrier resonances is to measure the tunneling conductivity, dI/dV, with the feedback loop active so that a large dynamic range can be obtained, as shown in Fig. 7.[15] The resonances are seen as a series of decaying oscillations in the tunneling conductivity as a function of voltage. Figure 8 illustrates the origin of the resonances; in the positive kinetic energy region, the quantum interference in the electron wave function occurs between the potential boundaries at the surface-vacuum interface and the region where the electron kinetic energy becomes positive. In this region standing wave type solutions are found which enhance the electron transmission at specific energies. In Fig. 8(a) the electron energy is such that two antinodes occur corresponding to the second resonance condition. As the electron energy is increased, further antinodes corresponding to a half wave-

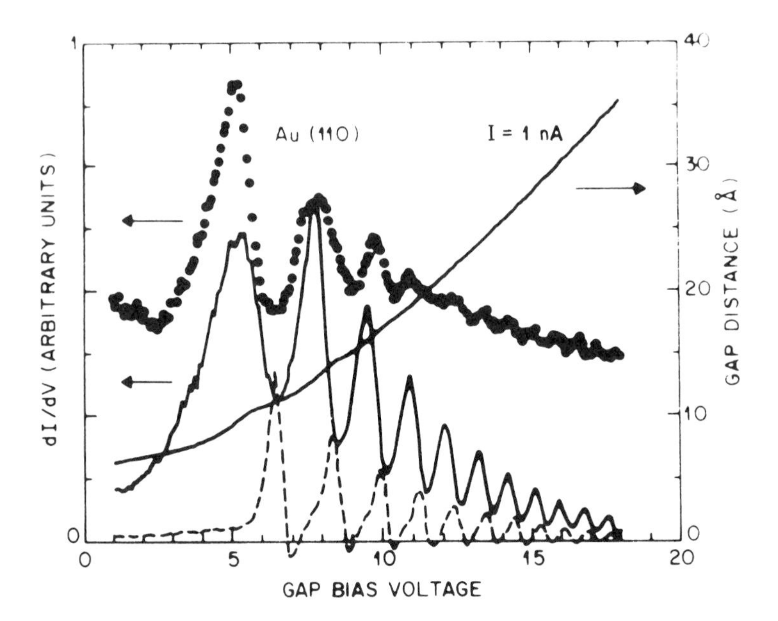

FIG. 7. Experimental curves of dI/dV (closed circles) and tip–sample separation vs bias voltage. The oscillatory solid curve is the theoretical barrier penetration factor. The dashed curve omits the image-potential contribution. (From Ref. [15].)

length appear in the barrier region resulting in a microscopic electron interferometer.

The position and strength of the resonances are sensitive to the details of the potential barrier.[16] In Fig. 7, inclusion of the image potential contribution to the barrier is seen to significantly affect the position of the transmission resonances.[15] Due to the difficulty in realistic calculations of the tunneling process, it is difficult to extract quantitative characteristics of the barrier potential from the barrier resonances. Attempts have been made using a free-electron model approximation.[16] Experimental data have also shown that the standing wave states can be used to probe the potential discontinuity at a buried interface.[17]

4.3 Voltage-Dependent Imaging Measurements

In the STM field one often thinks of the STM traces as corresponding to topographic features because one does, after all, end up with some sort of surface "map." To first order, the global features in the STM traces do correspond to surface topography as, for example, in observing atomic steps on surfaces. Finer details in the STM contours, such as surface corrugations or contours of atoms will undoubtedly display a combination of structural

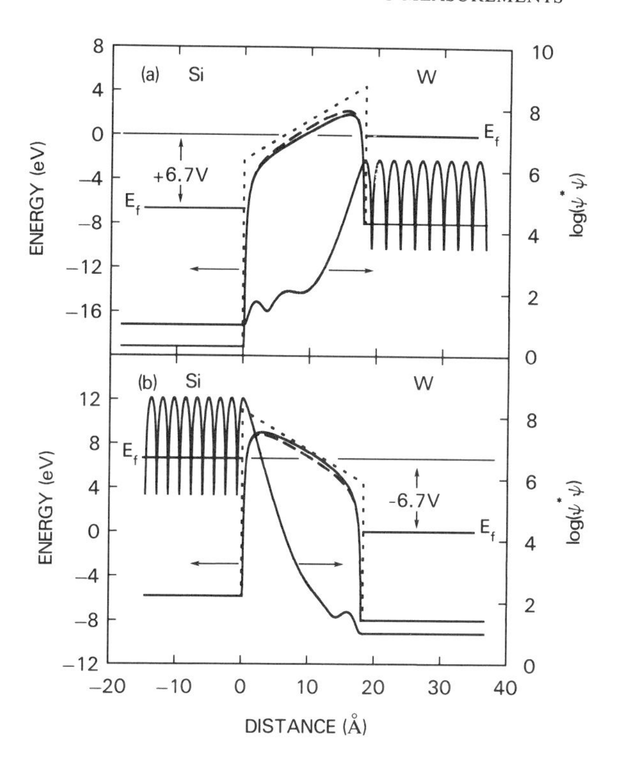

FIG. 8. Electronic potential and electron probability densities in a metal–vacuum–metal junction, with voltages of (a) +6.7 V and (b) −6.7 V applied to the sample (Si) relative to the probe-tip (W). The dotted line gives the trapezoidal barrier, the dashed line includes the effect of the image potential, and the solid line gives the potential including an inhomogeneous field arising from the probe-tip curvature. The probability density is shown for an electron with energy at the Fermi level of the negatively biased electrode. (From Ref. [11].)

and electronic properties. In particular, because of the localized states on semiconductor surfaces, it is more appropriate to think of the STM as measuring surface electron wavefunctions instead of atom core positions in these cases. Variation of the energy window of states contributing to the tunneling current allows spectroscopic information to be obtained. This information can be obtained by examining the voltage dependence of the constant current images described in this section. Energy resolved images may also be obtained from measuring the differential conductivity, or may be constructed from individual I–V measurements at each pixel in an image. These different measurements for obtaining energy-dependent information of the electronic properties of surfaces have advantages and disadvantages relative to each other that ultimately depend on the specific application. In this chapter a few of these methods are illustrated for the purpose of describing the measurement and the type of information that can be obtained.

4.3.1 Clean Surfaces: Si(111)2 × 1

The first direct example showing that the STM images surface wavefunctions rather than atoms occurred with voltage-dependent imaging of the Si(111)2 × 1 surface.[8,11,18,19] The Si(111)2 × 1 surface is a metastable surface formed by cleaving. The surface structure is characterized by quasi-one-dimensional zig-zag chains of Si atoms in the so-called π-bonded chain model,[20] shown superimposed on STM images in Fig. 9. The STM images are characterized by a single maximum in the unit cell of dimensions 6.65 × 3.84 Å. The observation of a single maximum in the 2 × 1 unit cell with a corrugation amplitude of ~ 0.6 Å is inconsistent with a purely geometric interpretation of the STM images; there are two atoms in the π-bonded chain model with a charge density corrugation < 0.05 Å, at the distances probed with the STM.[18]

The STM images make sense when one considers the spatial dependence of the *electronic* structure of the π-bonded chain model, as evidenced by the *voltage* dependence observed in the STM images in Fig. 9. The voltage-dependent images were obtained by recording a single trace at one bias, switching the tunneling bias to a new value (usually this is accomplished using a sample-and-hold delay before switching the new bias in), and then retracing the same measurement. In this way drift effects are minimized, since the time between a single trace is much smaller than that of an entire image acquisition.

The voltage-dependent images in Fig. 9 show that a maximum in the image taken at $+1$ V appears as a minimum at -1 V, as seen in the center of the crosshairs in the two images. This is more clearly seen in the line profiles in Fig. 9(c); a 180° phase shift is observed in the corrugation along the $[01\bar{1}]$ direction with respect to tunneling at positive and negative voltages.

One can understand these images by considering the electronic structure of the π-bonded chain model.[20,21] A simple starting point is to consider the π-bonded chain structure as a non-interacting, one-dimensional chain with a two-atom basis.[19] The two atoms in the unit cell are inequivalent due to the symmetry of the underlying lattice, and also possibly due to a buckling of the chain. Figure 10 shows a schematic of the dispersion of the energy band structure for a one-dimensional chain with parameters consistent for the Si(111)2 × 1 surface. Tunneling at positive voltages accesses the empty π^* band, while at negative voltage the tunneling current is probing the filled π band. At low voltages, the states primarily contributing to the tunneling process are from states near the surface Brillouin zone edge. It is instructive

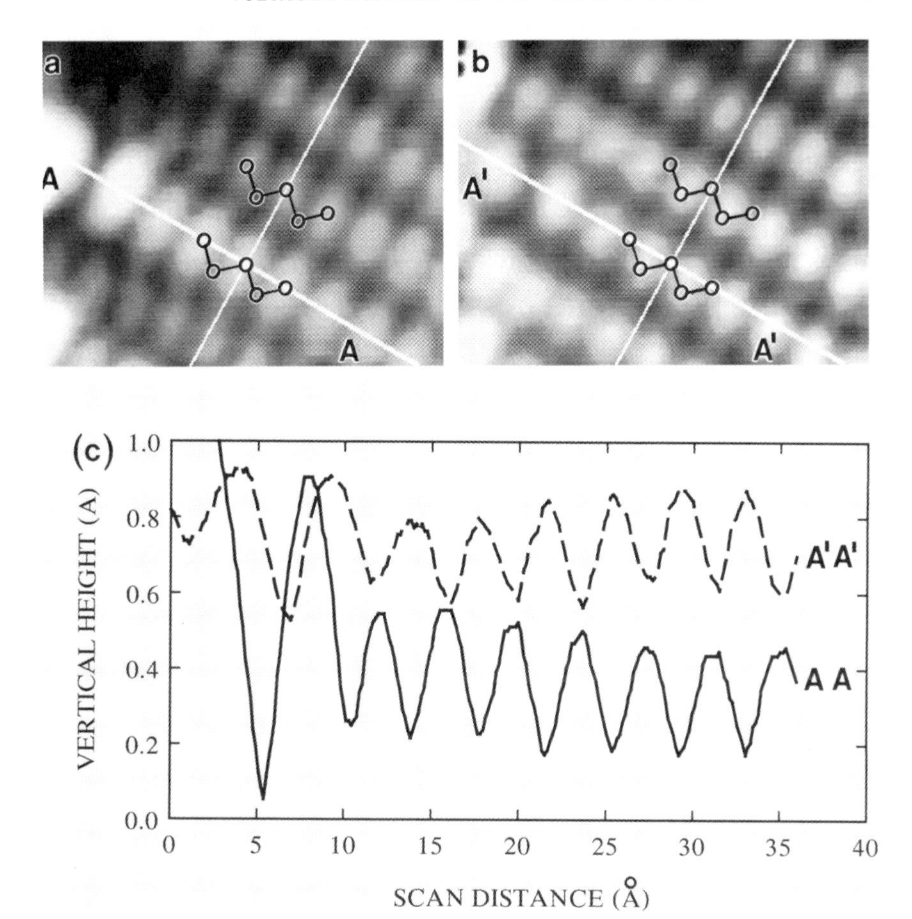

FIG. 9. 35 × 35 Å STM images of the Si(111)2 × 1 surface recorded simultaneously at (a) + 1 and (b) − 1 V. A top view of the π-bonded chain model is shown schematically in the center of both images. (c) Surface height along the [01$\bar{1}$] cross sections AA (solid line) and A′A′ (dashed line), which occur at identical positions in the two images. The A′A′ cross section is shifted up, with respect to AA, for clarity. (From Ref. [18].)

to consider the consequences for tunneling on either side of the gap at these voltages.

For simplicity, let's assume a two-plane wave expansion in a nearly free-electron model is adequate to describe the band structure near the zone edge in Fig. 10. In general, the surface wavefunction can be expanded in surface reciprocal lattice vectors. Since the higher Fourier components decay faster, a description based on a few Fourier components is adequate at a sufficiently large distance from the surface.[22] Keeping only two Fourier components the

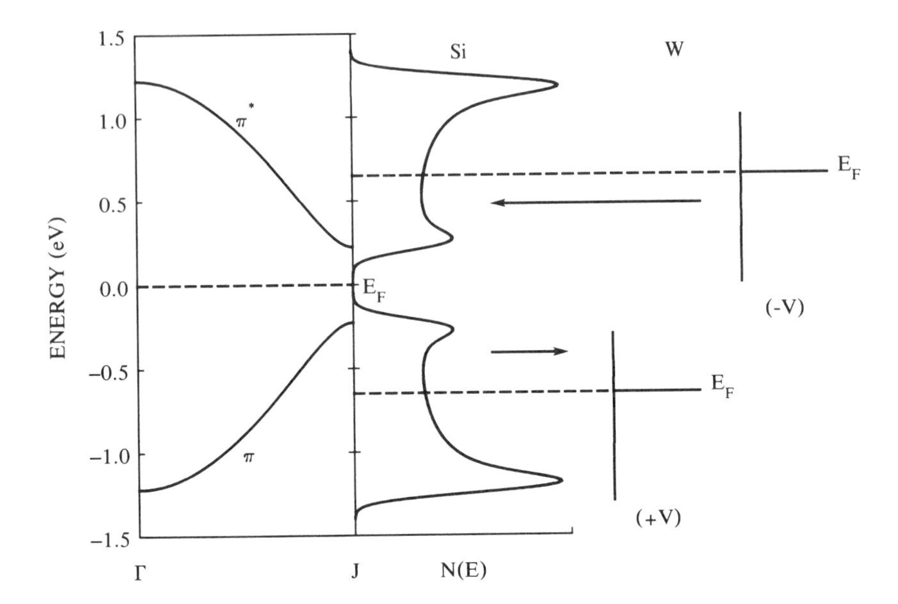

FIG. 10. Energy dispersion and density-of-states for a one-dimensional tight binding calculation for a two-atom basis chain model, simulating the π-bonded chain. The right-hand portion shows schematically that tunneling at positive and negative sample bias accesses the empty and filled Si surface state bands. (From Ref. [19].)

wavefunction can be expanded as,

$$\psi_k = C(k)e^{-z\alpha_k}e^{ikx} + C(k-G)e^{-z\alpha_{k-G}}e^{i(k-G)x}, \tag{6}$$

where $\alpha_k = [\kappa^2 + k^2]^{1/2}$ and $\kappa = \sqrt{2m\phi/\hbar^2}$ is the minimum decay constant. In the small bias limit, the tunneling current from Chapter 2 is,

$$I \propto \rho(\mathbf{r}_T, E_F) = |\psi_k|^2. \tag{7}$$

At the zone edge, $k = G/2$, the coefficient $C(G/2) = \pm C(-G/2)$, and the associated tunneling current becomes,

$$I^{\pm} = Ae^{-2z\alpha_{G/2}}[1 \pm \cos(Gx)], \tag{8}$$

where A is an experimentally dependent constant. In constant current operation, the tip will trace a contour $z(x)$ given by,

$$z^{\pm} = \frac{1}{2\alpha_{G/2}} \ln[1 \pm \cos(Gx)] - z_0, \tag{9}$$

where $z_0 = (1/2\alpha_{G/2}) \ln(I/A)$. The contours for the two bands as calculated from Eq. (9), are shown as dashed lines in Fig. 11.[19] The contours are characterized by an array of singular dips due to the nodes in the wavefunctions, which are standing waves at the zone boundary. One also observes that

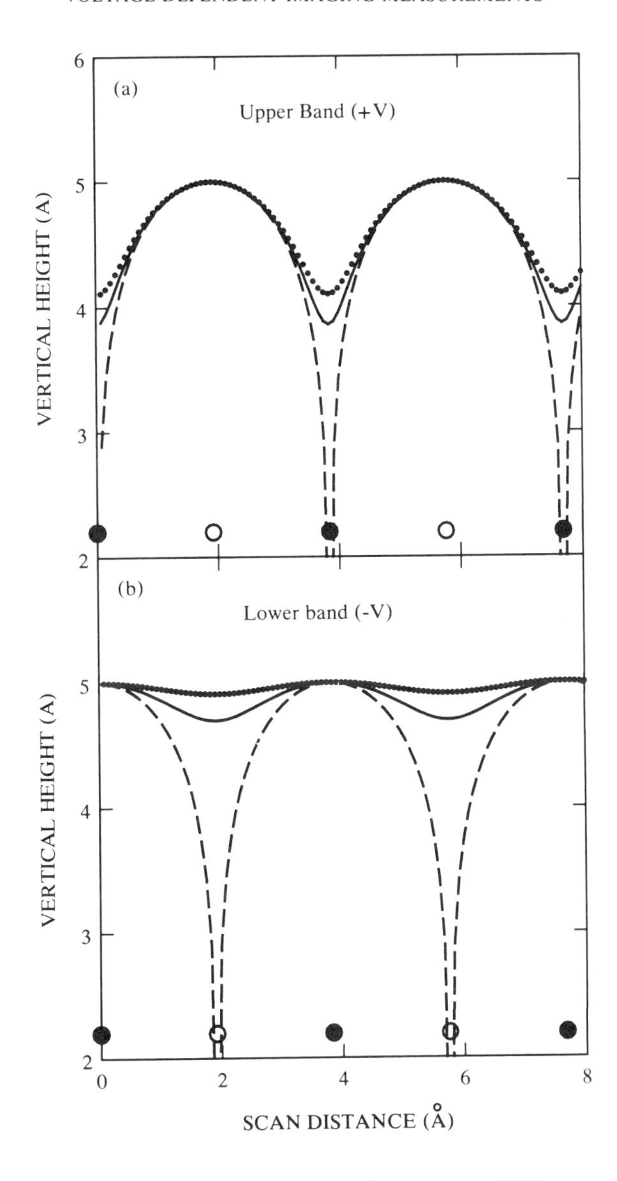

FIG. 11. Constant-current contours from the two-plane-wave model for a one-dimensional chain with a two-atom basis. Contours for the (a) upper, unoccupied and (b) lower, filled bands are shown. The contours corresponding to wave vectors $k = G/2$ (dashed line), $k = 0.75G/2$ (dotted line), and an integration over $k = (0.75 - 1.0)G/2$ (solid line). (From Ref. [19].)

the contours shift half a unit cell length in going from the upper to lower band. This results from the opposite parity of the wavefunctions in Eq. (6) at $k = G/2$. We see that this is precisely the same effect observed in the experimental images. The singularity observed in the contours at the zone edge is expected to be rounded out because of experimental limitations. The pathology is also removed if one integrates over a portion of the zone (which likely happens in practice), as shown by the solid line in Fig. 11, or if one moves away from the band edge as shown by the dotted line. One observes, however, that the phase shift between the two bands remains, which is consistent with experimental observations.

The simple analysis discussed earlier demonstrates that the STM images of the Si(111)2 $\times$ 1 surface represent the spatial dependence of the π-bonded surface state bands. It shows that the maxima observed in the images do not correspond simply to atomic locations, but are intimately related to the details of the bonding between the atoms, and hence, the surface band structure. At larger tunneling voltages, a tight-binding picture of the surface band structure would be more accurate to describe the wavefunctions away from the zone edge. Deviations from observing only one maximum in the images as one samples closer to the zone center $\bar{\Gamma}$ might be expected.[23] In practice, one usually observes the corrugation going to zero along the chain axis with increasing voltage. On rare occasions both atoms in the unit cell have been observed in one image.[24] This is probably very dependent on the tip resolution since the near-neighbor distance between the Si atoms is < 3 Å.

As a final note, we observe that shifts observed in the state density with respect to tunneling across the band gap are very analogous to optical absorption anisotropy.[25] This is not that surprising since the optical matrix elements also depend on the parity of the surface wavefunctions. In general, just as optical absorption can probe the existence of a band gap, voltage-dependent imaging with the STM will be expected to show phase shifts in the state density associated with any band gap, as the spatial separation of valence and conduction band states are intimately related to the existence of the energy gap. Shortly after the Si(111)2 $\times$ 1 discovery, voltage-dependent imaging for polar III–V materials was shown by the authors to lead to a selective imaging of the anion or cation species, which is discussed further in Section 5.3.

4.3.2 Adsorbate Covered Surfaces: O/GaAs(110)

Adsorbate covered surfaces offer a unique application in scanning tunneling microscopy, yielding an unprecedented atomic view in single atom imaging of individual adsorbates. STM measurements of adsorbate covered surfaces give a local view of the adsorption site, the surrounding geometry and its influence on the adsorbate, or local ordering within the adsorbate overlayer. An adsorbate is really a hybrid adsorbate–surface complex characterized by its interaction with the substrate.[26] As shown in Chapter 1 and previous calculations, this interaction can lead to energy-dependent structure in the local density-of-states probed by the STM.[27–29] Voltage-dependent imaging is one way to access this energy dependence with the consequence that the appearance of an adsorbate will depend on voltage.[19] The most extreme case is one where the adsorbate changes from a positive to a negative STM contour with a change in voltage; a characteristic of some strongly electronegative species.[26,29]

The first example of highly voltage-dependent STM contours occurred with the study of oxygen adsorption on the GaAs(110) surface.[30] At low coverages of 0.001 of a monolayer (ML), one can study the interaction of a single adsorbate on a surface. For oxygen on GaAs(110), these coverages are easily produced due to the low oxygen sticking probability. Figure 12 shows two STM images of oxygen exposed to the GaAs(110) surface. One sees protrusions appearing in the image that increase in density with exposure. The protrusions are slightly larger than expected for a single atom contour (see Chapter 1). The most striking effect is seen in the voltage-dependent images in Figs. 13 and 14. It is observed that the oxygen–GaAs complex displays a positive STM contour when sampling the filled electronic states, while a negative STM contour is observed when sampling the unoccupied states. This striking voltage dependence observed in the STM contours results from the energy dependence of the local density-of-states.[19] As discussed in Chapter 2, different elements will have characteristic variations in the energy spectrum of the state density. The oxygen data verifies the prediction that for a highly electronegative species, the state density can be reduced for energies above the Fermi level.[26,29] For this system and chemisorption on semiconductors in general, more complex effects due to band-bending and screening may influence the STM contours near adsorbates and defects.[30] Such effects can be measured with current versus voltage characteristics, as discussed in Section 4.4.

An example where voltage-dependent imaging of adsorbates has been very successful is in the application to the Si(111)7 $\times$ 7 surface, with studies of NH_3,[31] Cl,[32] and O adsorption,[33] just to name a few. The Si(111)7 $\times$ 7 surface is unique in that the large unit cell has three "varieties" of atoms to bond to: the Si rest atoms, center adatoms, and corner atoms. The reactivity of these different atoms can be seen in voltage-dependent imaging of NH_3 on Si(111)7 $\times$ 7.[31] The STM topographs at the different energies provide a unique ability to follow the spatial distribution of the surface reaction on an atomic scale. From such measurements it was concluded that Si rest atoms are more reactive than Si adatoms, and that center adatoms are more reactive

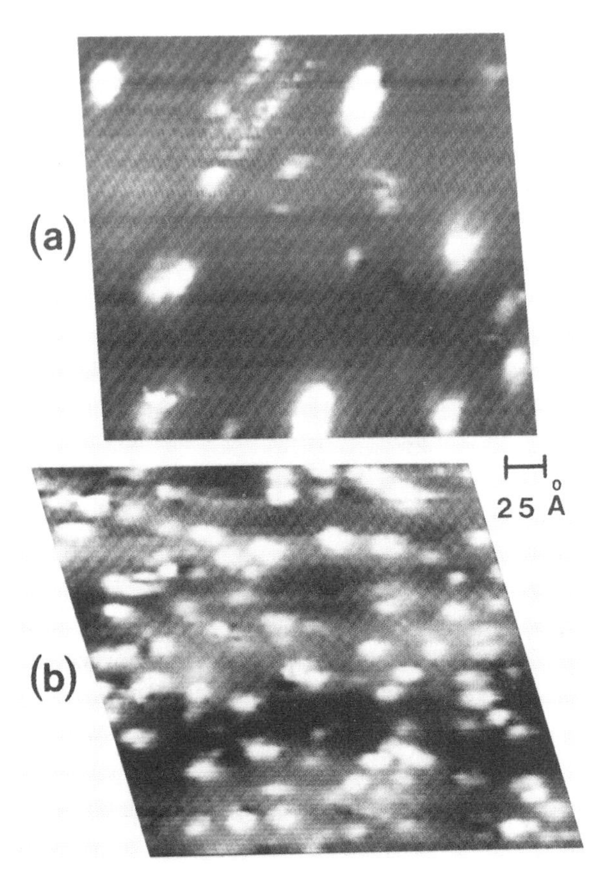

FIG. 12. STM images of the *n*-GaAs(110) surface exposed to oxygen. (a) Oxygen exposure 120 L (1 L = 10^{-6} Torr s). (b) Oxygen exposure 1480 L. Sample voltage was −2.5 V. (From Ref. [30].)

than corner adatoms for NH_3 adsorption (see Section 5.1 for more details). Further insight into bonding changes can be obtained from examining density–of–states rearrangements obtained from current versus voltage measurements, described in Section 4.4.

4.4 Fixed Separation *I*–*V* Measurements

4.4.1 Normalized Density of States: Si(111)2 × 1

In Section 4.3 we saw how voltage-dependent imaging of the Si(111)2 × 1 surface yielded images characteristic of the π and π^* surface bands. A more energy-resolved probe of these surface band features can be obtained with

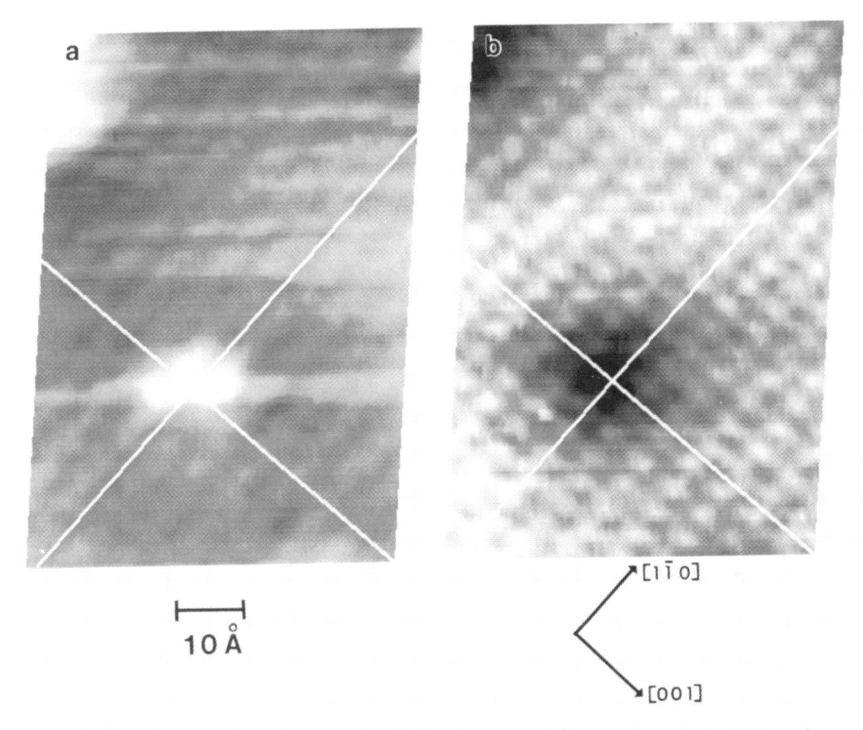

FIG. 13. Simultaneous STM images of an isolated oxygen defect on the *n*-GaAs(110) surface at (a) −2.5 and (b) 2.3 V. (From Ref. [19].)

I–V measurements. Fixed tip–sample separation *I–V* curves are obtained using the sample-and-hold technique with the timing scheme shown in Fig. 2. Figure 15 shows the *I–V* curves obtained at various fixed tip–sample separation for the Si(111)2 × 1 surface.[8,11] Also shown is the *s–V* data obtained at 1 nA tunneling current. Comparing the data in Fig. 15 with the simulated *I–V* characteristics for metal–vacuum–metal tunneling (Fig. 5), we observe similar characteristics such as a linear behavior at low voltages, becoming exponential for voltages > 1 V. Figure 15 also shows that only a limited voltage range is obtained at one gap resistance and a large dynamic range is obtained by sampling multiple tip–sample separations. One feature not seen in the simulated planar tunneling model curves (Fig. 5), is the rectification observed in the *I–V* characteristics at negative sample voltage. This rectification results from the enhanced electric field at the probe tip due to the finite radius. When this field is included in the planar tunneling model, good agreement can be obtained with data in Fig. 15.[11] In addition to the above characteristics, barrier resonances are seen in the *I–V* and *s–V* characteristics for voltages > 4 V, as predicted in the planar tunneling calculations.

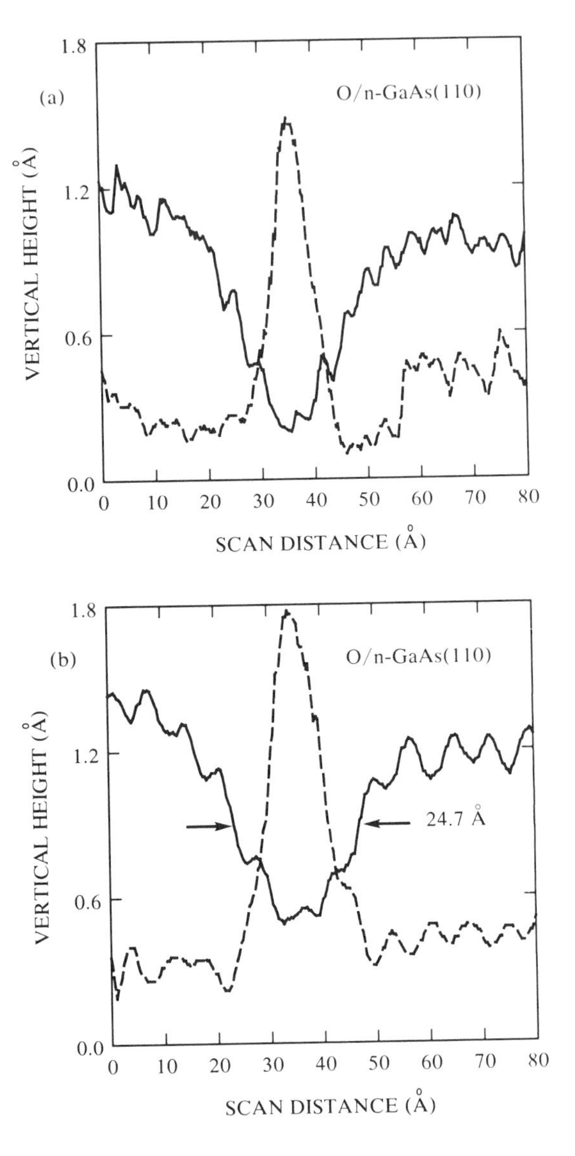

FIG. 14. Cross-sectional cuts in (a) the $[1\bar{1}0]$ direction and (b) the [001] direction in the two images of Fig. 13. Positive bias corresponding to Fig. 13(b) is shown by the solid line and negative bias corresponding to Fig. 13(a) is shown by the dashed line. (From Ref. [19].)

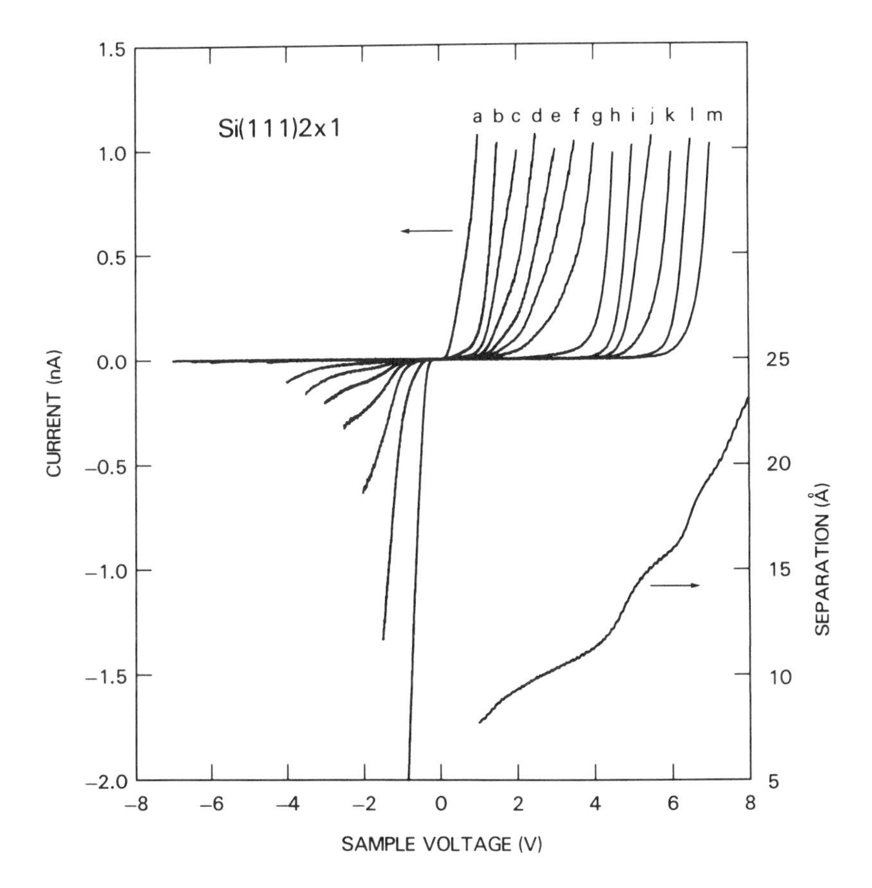

FIG. 15. Tunneling current vs voltage for a tungsten probe tip and Si(111)2 × 1 sample, at tip–sample separations of 7.8, 8.7, 9.3, 9.9, 10.3, 10.8, 11.3, 12.3, 14.1, 15.1, 16.0, 17.7, and 19.5 Å for the curves labeled a–m, respectively. These separations are obtained from a measurement of the separation vs voltage, at 1 nA constant current, shown in the lower part of the figure. (From Ref. [8].)

Structure in the *I–V* characteristics in Fig. 15, which are sample specific, are the kinks observed at voltages < 3 V, including the small band gap near zero bias. These kinks are due to density-of-states features contributing to the tunneling, and are more clearly seen in the derivative of the *I–V* data shown in Fig. 16.[11] A number of features are reproduced in the different tip–sample measurements. From Fig. 16 we see that the derivative highlights the density-of-states features, but the lack of dynamic range in a single measurement is evident. The large voltage and distance dependences result from the exponential dependence of the transmission probability discussed in Section 4.2 (see Eq. (5)). As first suggested by the authors, these exponential

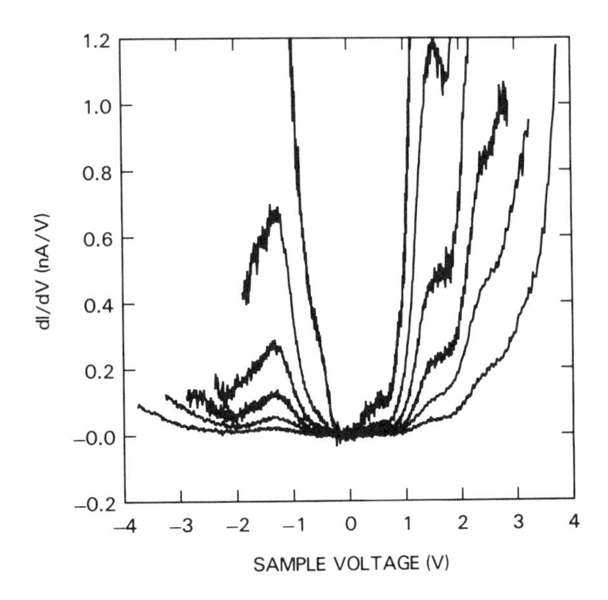

FIG. 16. Derivative of the current vs voltage data in Fig. 15. (From Ref. [11].)

dependences of separation and voltage can be removed by normalizing the differential conductance by the static conductance (I/V), producing a quantity which approximates a normalized density-of-states.[8,11]

Figure 17(b) shows the normalized conductance $(dI/dV)/(I/V)$ for the Si(111)2 × 1 surface computed from the data in Fig. 15. As observed in Fig. 17, all the dI/dV spectra in Fig. 16 collapse onto a *single* curve. Shown for comparison in Fig. 17(c) is a simple estimate of the surface density-of-state for the one-dimensional π-bonded chain model (see Section 4.3), and the Si bulk density-of-states (dashed line). As observed, the normalized conductance shows the main four-peaked structure expected for the surface state bands as well as higher-lying resonances, which may be derived from bulk-critical points.

Also shown in Fig. 17(a) is the inverse decay-length data, obtained from the distance dependence of the I–V measurements in Fig. 15, which is seen to support the conclusion that tunneling at the lower voltages on the Si(111)2 × 1 surface samples the states near the Brillouin zone edge (see Fig. 10).[11] The tunneling current is expected to decay with distance with an inverse decay length of $2\sqrt{\kappa^2 + k_\parallel^2}$, where κ is the vacuum decay constant (see Section 4.4.1). The large increase seen in the inverse decay length at $\pm 0.5\,\mathrm{V}$ implies a maximum value of $k_\parallel = 1.1\,\text{Å}^{-1}$, which is close to the value of the wavevector at the edge of the surface Brillouin zone of $0.94\,\text{Å}^{-1}$.

The removal of the exponential dependences of the tip–sample separation

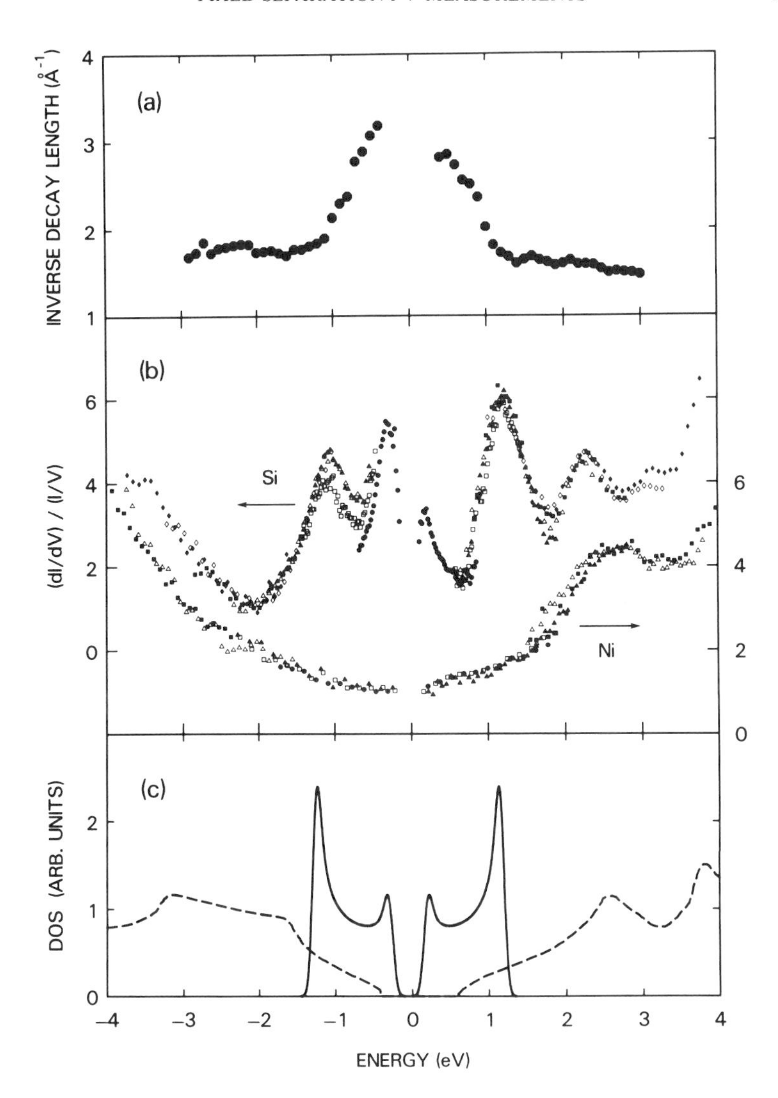

FIG. 17. (a) Inverse decay length of the tunneling current as a function of energy (relative to the Fermi level). (b) Ratio of differential to total conductivity for silicon and nickel. The different symbols refer to different tip–sample separations. For silicon, the circles, open squares, filled triangles, open triangles, filled squares, open lozenges, and filled lozenges refer to the curves a–g, from Fig. 15. (c) Theoretical DOS for the bulk valence band and conduction band of silicon (dashed line), and the DOS from a one-dimensional tight-binding model of the π-bonded chains (solid line). (From Ref. [8].)

and voltage in the normalized differential conductance can be demonstrated by looking at an approximate expression for the tunneling current (see Chapter 1),

$$I \propto \int_0^{eV} \rho_s(E) D(E)\, dE, \tag{10}$$

where ρ_s is the surface density-of-states, and D is the transmission probability in Eq. (5). The normalized conductance is then,

$$\frac{dI/dV}{I/V} = \frac{\rho_s(eV)D(eV)}{\frac{1}{eV}\int_0^{eV} \rho_s(E)D(E)\, dE} + \dots \tag{11}$$

Since the transmission probability appears both in the numerator and denominator, then, to first order, the normalized differential conductance is proportional to a normalized surface density-of-states,

$$\frac{dI/dV}{I/V} \approx \frac{\rho_s(eV)}{\frac{1}{eV}\int_0^{eV} \rho_s(E)\, dE}. \tag{12}$$

The second term neglected in Eq. (11) gives rise to a smoothly varying background. A closer examination of the cancellation of the transmission factors shows that the normalized differential conductance should favor unfilled states versus lower-lying filled states.[11]

The ability of the normalized conductance to show density-of-states features has been born out in theoretical calculations by Lang,[34] as shown in Fig. 14 in Chapter 1. The dashed line in that figure shows $(dI/dV)/(I/V)$ calculated using the expression in Eq. (10), which is seen to provide a good approximation to the surface density-of-states.

The normalization discussed here works fine for metals and small gap semiconductors. However, the normalized conductance is ill-defined at a band gap edge. This results from the fact that the current is going to zero at a gap edge faster than the differential conductance, which yields a divergence in the normalized conductivity. Experimentally, one sees large peaks on either side of a band gap which have nothing to do with density-of-states, but instead are an artifact of the normalization; usually the density-of-states goes to zero at gap edges. Therefore, for wide gap (> 0.5 eV) semiconductors, one needs to find an alternative solution. Basically, one needs to normalize to the transmission probability, which the current no longer adequately measures. Methods to achieve this are discussed separately in Section 4.5.

4.4.2 GaAs Current Versus Voltage Characteristics

Tunneling spectroscopy of semiconductor surfaces can reveal the semicon-

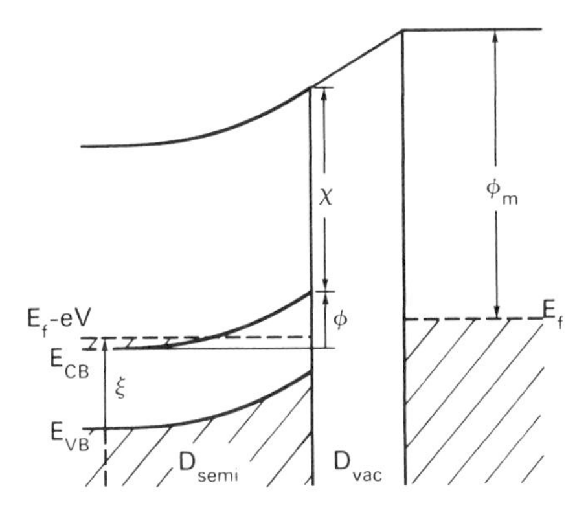

FIG. 18. Schematic view of the energy diagram for semiconductor–vacuum–metal tunneling. The semiconductor has its valence band maximum at E_{VB} and conduction band minimum at E_{CB}. The metal Fermi level is denoted E_F and the semiconductor Fermi level lies at E_F–eV. The metal work function is denoted by ϕ_m. The barrier height at the semiconductor surface, ϕ, is shown bent upward due to the electric field penetration into the semiconductor. (From Ref. [37].)

ductor properties such as surface band gaps, surface band bending, and dopant carrier densities. With the scanning ability of the STM, such measurements offer a wide-ranging potential to probe these properties on an atomic scale. In this section we illustrate some of the semiconductor tunneling characteristics with data from the GaAs(110) surfaces (see Section 5.3 for further information on GaAs surfaces). GaAs(110) is a convenient substrate as there are no surface states within the bulk band gap of 1.45 eV (the surface band gap, 2.4 eV, is in fact larger than the bulk band gap).[35,36] A schematic of the energy diagram for a semiconductor–vacuum–metal tunneling is shown in Fig. 18. In this section we will focus on the effects of the semiconductor band gap and surface band bending in the tunneling characteristics.

Figure 19 shows the *I–V* characteristics for *p*-GaAs(110) acquired at several fixed tip–sample separations.[37] The rectification observed at negative sample bias is similar to the Si *I–V* characteristics in Section 4.4.1. A significant feature of the GaAs(110) *I–V* characteristics is the absence of a band gap region where tunneling is not observed. In fact, curve (d) is virtually ohmic at zero bias. The experienced eye would note, however, a strong threshold behavior at 1.5 V in curve (c).

To see the *I–V* characteristics more clearly, we normalize the various curves to account for the different tip–sample separations, as shown in Fig. 20. This is accomplished by multiplying each successive *I–V* curve by the distance-dependent transmission term, $\exp(2\kappa\,\Delta s)$, where κ is the vacuum decay constant, $\sqrt{2m\phi}/\hbar$ (see Section 4.2). Thus the vertical scale in Fig. 20 is not the true measured current, but a normalized effective current. Note that

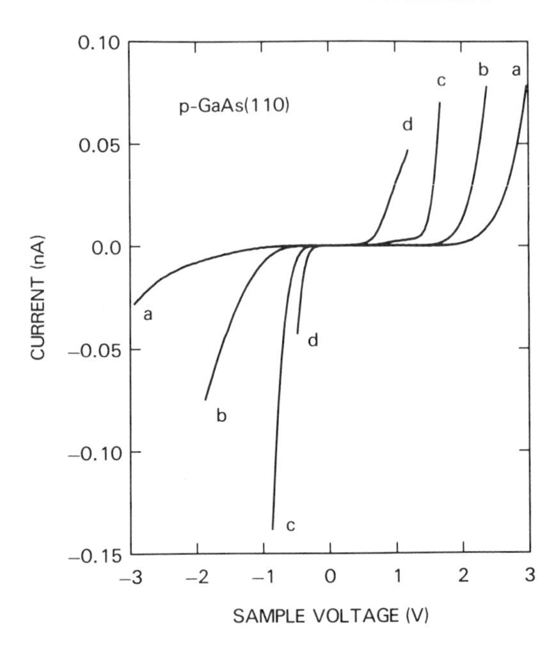

Fig. 19. Current vs voltage characteristics of p-GaAs(110). Each curve is acquired at constant tip–sample separation with the relative separations differing by (a) 0, (b) -1.2, (c) -3.2, and (d) -4.8 Å. (From Ref. [37].)

the various tip–sample distance measurements allow almost eight decades of current to be effectively measured.

The logarithmic plot of the I–V data in Fig. 20 shows the semiconductor details more clearly. In this figure, three regions of tunneling current are seen; a component beginning at 0 V and increasing with positive sample voltages, a second inflection point at $+1.5$ V, and a component beginning at 0 V which increases uniformly with negative sample voltages. The first component, labeled D in Fig. 19, obscures the presence of a band gap due to tunneling into unoccupied hole states at small positive voltages, due to the ionized dopant impurity concentration of $1 \times 10^{18}\,\text{cm}^{-3}$. The inflection at 1.5 V defines the conduction band edge and is due to tunneling into the unoccupied conduction band states (labeled C). Because of the p-type doping, the Fermi level at 0 V is at the top of the valence band and the valence band edge is seen by the turn-on of tunneling at negative voltage (labeled V). Tunneling into degenerate p-type semiconductors should give rise to a negative resistance region associated with tunneling into the hole states. This comes about because the tunneling current into the hole states decreases with increasing positive sample bias, since the barrier is increasing for these states while the number of hole states for tunneling remains fixed. This gives rise to a negative

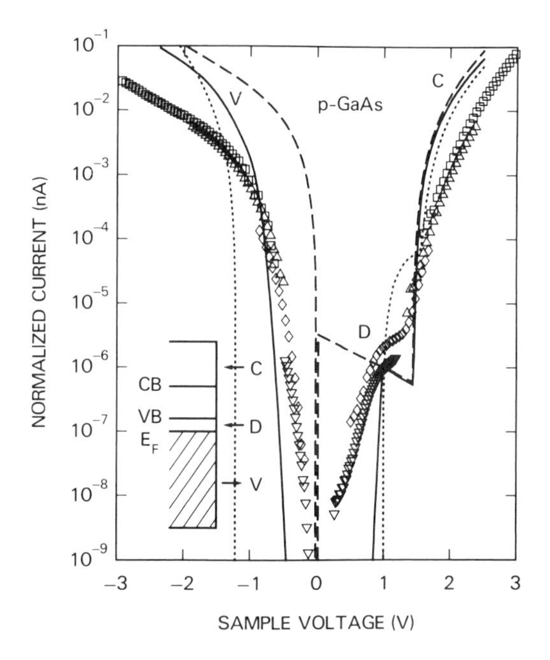

FIG. 20. Current vs voltage for p-GaAs(110). The open symbols show data computed from the data in Fig. 19 by multiplying the observed current by $\exp(2\kappa\,\Delta s)$, where $\kappa = (2m\phi)^{1/2}/\hbar = 1.09\,\text{Å}^{-1}$. Theoretical results are shown including band bending (dotted line), assuming flat band conditions (dashed line), and including both band bending and tunneling through the semiconductor space charge region (solid line). The various components of the current are shown in the inset: C, conduction band; V, valence band; and D, dopant induced. (From Ref. [37].)

resistance in tunneling calculations for this bias range,[38] which is usually observed only as a plateau in the STM I–V characteristics (see Fig. 20), as well as in fixed-tunneling junctions.[39]

Figure 21 shows the I–V characteristics for n-type GaAs(110).[37] In this case a single I–V curve is sufficient to span the displayed voltage range. Tunneling out of the occupied dopant impurity states now occurs in the voltage range 0 to -1.5 V. Below -1.5 V the current flows out of the valence band, and above 0 V the current flows into the conduction band. Both the p-type and n-type I–V curves are characteristic of tunneling into degenerate semiconductors, as seen in planar junctions as well.[40] The essential difference between the p-type and n-type curves is a shift in voltage by about 1.5 V, corresponding to the shift in the bulk Fermi level across the GaAs(110) band gap. In addition, the intensity of the D component differs between p-type and n-type, since in the former case the tunneling-barrier height for this component

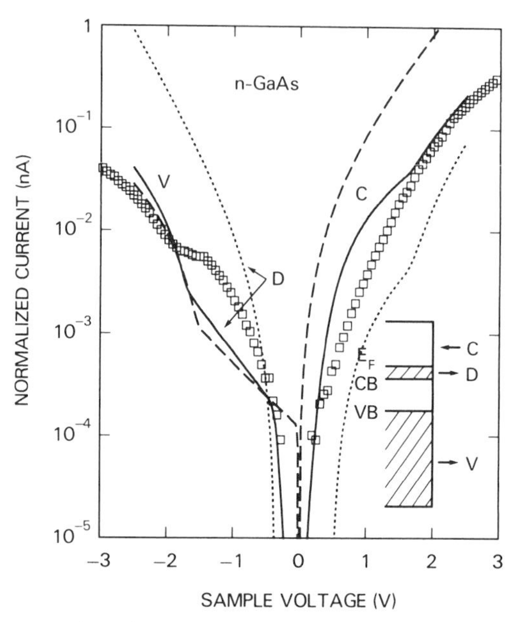

FIG. 21. Current vs voltage for n-GaAs(110). The open symbols show the data from a single current–voltage measurement. Theoretical results are shown for the same cases as in Fig. 20. The various components of the current are shown in the inset: C, conduction band; V, valence band; and D, dopant induced. (From Ref. [37].)

increases as the magnitude of the voltage increases, whereas in the latter case it decreases.

Both the n-type and p-type data were analyzed with the planar tunneling calculations,[37] as described in Section 4.2. Semiconductor characteristics were included within the effective mass model. In these calculations, two new effects were identified in the tunneling characteristics: first, the electrostatic potential in the semiconductor is effected by the junction electric field because of the relatively weak semiconductor screening. Second, a discernible tunneling transmission through the space charge layer at the semiconductor surface can be observed.

The field penetration into the semiconductor, illustrated in Fig. 18, can be calculated by integrating Poisson's equation through the semiconductor and matching the electric potential and electric displacement at the semiconductor—vacuum interface.[37] Figure 22 shows the resulting surface band bending as a function of applied sample voltage for n- and p-GaAs(110). The net effect is that the effective voltage, or potential, across the junction is not equal to the applied voltage, but reduced. This calculation neglects any screening by surface states, which will reduce this tip-induced band bending.[41] The importance of tunneling through the space charge region can be seen by estimating

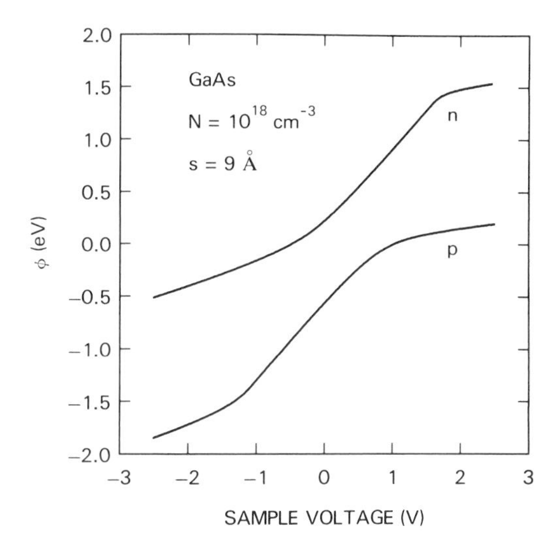

FIG. 22. Semiconductor surface band bending, ϕ (see Fig. 18), as a function of sample voltage for n- and p-type GaAs doped at $1 \times 10^{18}\,\mathrm{cm}^{-3}$. The quantity plotted equals the band bending at the surface of the semiconductor and is computed by numerical integration of Poisson's equation through the space charge region. The metal electrode is placed 9 Å from the semiconductor. (From Ref. [37].)

the transmission probability for space charge regions determined by the doping and surface potential. Zero-temperature calculations, including these various terms, are shown in Figs. 20 and 21. The best agreement with the experimental data is obtained when the calculation includes both the field penetration and tunneling through the space charge region.

4.4.3 Spatial Semiconductor Characteristics

The properties observed in the semiconductor I–V characteristics: band gaps, surface band bending, and carrier concentrations, may all be probed as a function of spatial position, including in the vicinity of adsorbates. In this section a few examples are presented illustrating these properties.

In the last section we saw that the bulk band gap can be observed in the clean surface I–V characteristics on GaAs(110). With the presence of an adsorbate, states can be introduced into the band gap with the effect of narrowing the observed tunneling gap. The extreme case of a metallic band will lead to the absence of a tunneling gap, since the Fermi level would intersect a partially filled band of states. An example of gap narrowing is observed in the case of Cs on GaAs(110), as illustrated in Fig. 23.[42] Corresponding structures for the 1- and 2-D Cs phases are shown in Fig. 24. Curve (a) shows the I–V characteristics of n-GaAs(110), with a 1.5 eV band gap. This differs from the curve in Fig. 19 in that the dopant contribution is not

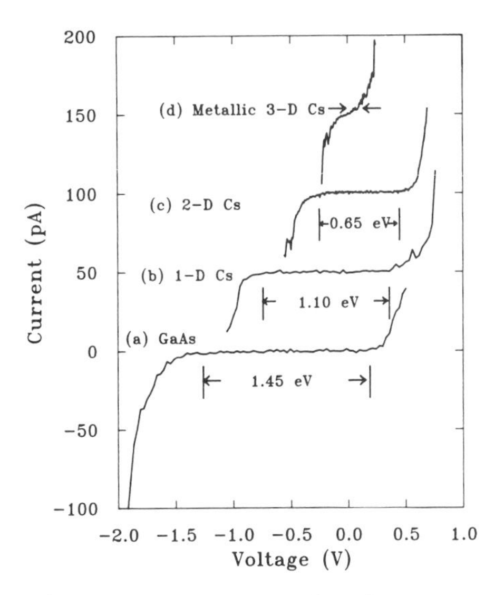

FIG. 23. Current vs voltage measurements over various Cs structures on GaAs(110): Curve a, region of clean *n*-GaAs(110); curve b, 1-D zig-zag chain on *n*-GaAs(110); curve c, 2-D Cs on *p*-GaAs(110); curve d, saturation 3-D bilayer on *p*-GaAs(110). The indicated band gaps were determined on a more sensitive logarithmic scale. Note that curves b–d are offset for clarity. (From Ref. [42].)

seen, since a small amount of band bending is present (due to the presence of nearby Cs), which pinches off any tunneling from the dopant induced carriers. *I–V* spectra recorded on the 1-D Cs chains, shown in Fig. 24(a), reveal a narrowing of the GaAs(110) band gap to 1.1 eV and an upward band bending shift of ~0.25 eV. Further narrowing to 0.65 eV is seen in the *I–V* characteristics of the 2-D phase. Fully metallic characteristics are seen with second layer nucleation, as seen in curve (d), which does not display a band gap. The narrowing of the GaAs(110) band gap indicates that states are introduced due to the presence of the adsorbates, but with the surprising result that 1- and 2-D phases of alkali metal atoms are insulating.[42] Mott-insulator type of behavior was also found in Cs on InSb(110) with the 2-D phase also displaying a 0.6 eV gap; however, in this case the 2-D alkali phase opened up a band gap larger than the InSb substrate band gap of 0.15 eV, which indicated the 0.6 eV band gap is a characteristic of the 2-D Cs phases.[43]

While the determination of the tunneling gap seems straightforward from the *I–V* measurements, care must be taken to ensure that sufficient dynamic range is present to measure the gap. This results from the small density-of-states near the gap edge, and as shown in the simulated *I–V* curves in Fig. 5 in Section 4.2, a false gap may be observed due to insufficient dynamic range

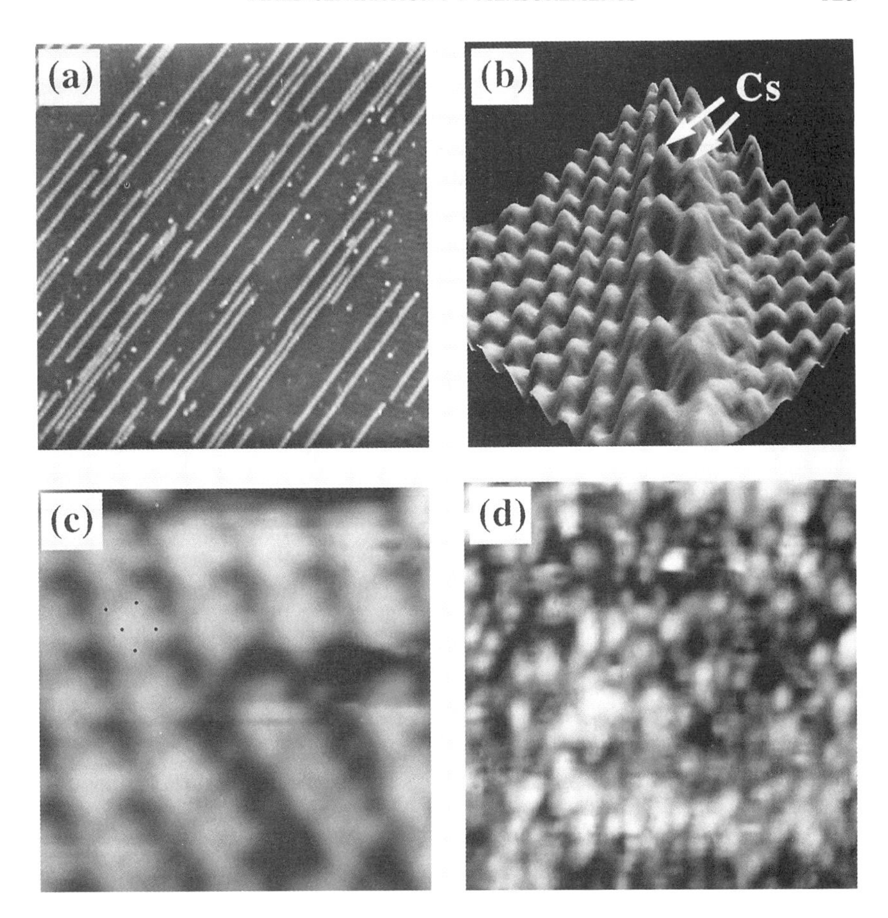

FIG. 24. STM topographic images of Cs on GaAs(110) corresponding to the I–V measurements in Fig. 23. (a) 1370 × 1370 Å image of Cs 1-D chains, (b) 70 × 70 Å, a single Cs zig-zag chain, (c) 50 × 50 Å, of the 2-D c(4 × 4) Cs phase. The small black dots denote the approximate positions of the five Cs atoms within one Cs polygon. (d) 400 × 400 Å, of a disordered 3-D Cs bilayer at saturation coverage. (From Ref. [42].)

to measure the small tunneling current if the tip–sample separation is too large. For the semiconductor gap measurements discussed earlier, we usually acquire a family of I–V curves at decreasing distances, as in Fig. 20, to ensure that the limiting gap is obtained. Usually the smallest tip–sample separation used in the measurement is just close enough that tip–sample contact is imminent. The variable separation methods described in Section 4.5 can be used to provide a more continuous measure of the entire I–V curve, thereby allowing subsequent normalization.

As mentioned above, the I–V characteristic can show a spatial dependence in the position of the semiconductor band edges with respect to the Fermi level, indicating a spatially varying potential at the surface. Potential variations in semiconductors are generally more long-range than in metals due to the lower carrier densities and hence longer screening lengths. For small potential disturbances the length scale over which the potential decays to an equilibrium value can be estimated by the Thomas–Fermi method which yields a screening length for n-type material as,[44]

$$\lambda = \left\{ \frac{\varepsilon\varepsilon_0 k_B T \mathscr{F}_{1/2}(\eta)}{ne^2 \mathscr{F}_{-1/2}(\eta)} \right\}^{1/2}, \tag{13}$$

where ε is the static dielectric constant (12.9 for GaAs), ε_0 is the permitivity of free space, n is the bulk electron concentration, $\mathscr{F}_k$ is the Fermi integral of order k, and $\eta = (E_F - E_c)/k_B T$ is the reduced Fermi energy. For non-degenerate carrier densities this simplifies to the so-called Debye length $L = (\varepsilon\varepsilon_0 k_B T/ne^2)$. For n-GaAs with an impurity concentration of $10^{18}\,\mathrm{cm}^{-3}$ the screening length is 56 Å. Long-range effects can be seen in the STM contours near adsorbates or defects which change the local charge balance.[30] Such defects occur naturally on cleaved III–V(110) surfaces, and display long-range contours which resemble shallow wells or raised mounds, depending on polarity and doping. These long-range contours indicate spatial variations in the surface band bending, as shown in Fig. 25. From the sign of the long-range contour (increasing or decreasing) and knowledge of the doping, the charge state of the defect can be inferred, although the exact defect origin (antisites, dopant atoms . . .) awaits further studies. Long-range contours can also be seen with adsorbates, such as observed in the contours of oxygen on GaAs(110) contours, shown in Fig. 14, and the raised As atoms near the 1-D Cs chain in Fig. 24.

The spatial variations in the surface band bending can also be probed with I–V characteristics. Two effects are usually seen. One is that the dopant-induced carrier contribution to the tunneling current decreases as the bands deviate from flat band conditions, as a result of the increasing difference between the Fermi level and a band edge. This affects the free carrier density through the Fermi distribution function as,

$$n(T) = N_c \mathscr{F}_{1/2}((E_F - E_c + e\phi)/k_B T) \tag{14}$$

$$p(T) = N_v \mathscr{F}_{1/2}((E_v - E_F + e\phi)/k_B T), \tag{15}$$

where n and p are the electron and hole concentrations, N_c and N_v are the effective conduction and valence band density-of-states, and ϕ is the spatially varying electrostatic potential.

The second effect observed in the I–V characteristics is that the positions of the valence and conduction band edges shift with respect to the Fermi level (0 V) with band bending. Figure 26 illustrates both of these effects with I–V

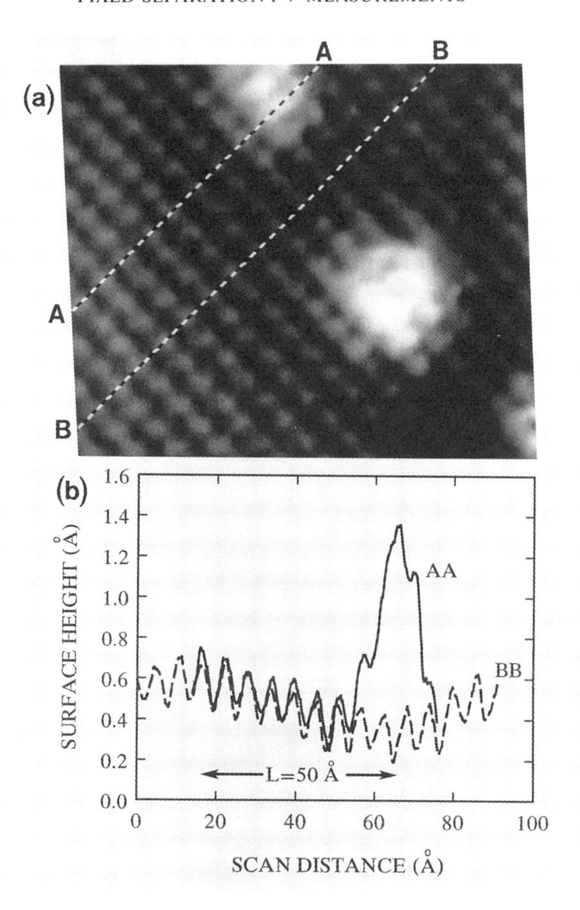

FIG. 25. (a) STM image of native defects on *n*-GaAs(110) at −2.5 V sample bias. (b) Surface height contours, in the [001] direction, along lines AA (solid line) and BB (dashed line) shown in (a). (From Ref. [30].)

characteristics acquired on the GaAs(110) surface in the vicinity of an Fe cluster.[45] In Fig. 26(a), the tunneling due to dopant-induced carriers for voltages between 0 and 1.4 V is evident for the *I–V* curve far from the Fe cluster, and is characteristic of flat bands at the surface. As we approach the cluster, this tunneling current decreases due to a downward shift of the semiconductor bands. The shift of the bands is also seen in the *I–V* curves at negative biases. Figure 26(b) quantifies these effects with the magnitude of the tunneling current at 0.8 eV versus spatial position. Since the tunneling barrier is changing very little with position ($< 2\%$), this tunneling current is directly proportional to the valence-band free-carrier hole density, as shown by the dashed line in Fig. 27, which is set to its bulk value at large distances. This

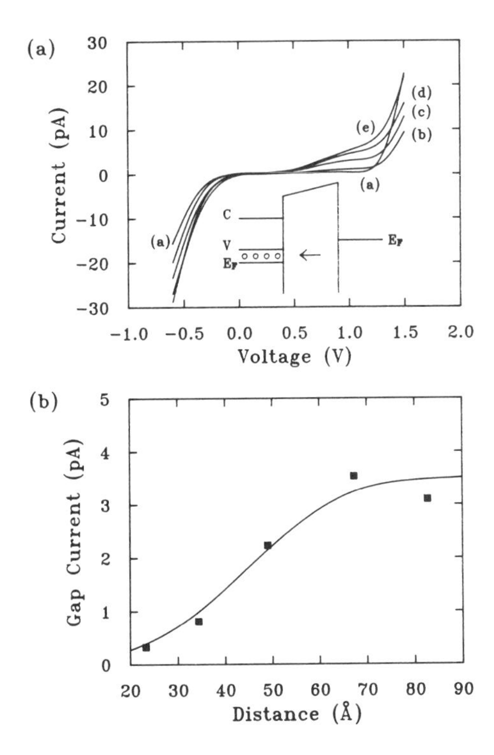

FIG. 26. (a) Tunneling current vs voltage measurements on p-GaAs(110) (impurity concentration $2 \times 10^{19}\,\mathrm{cm}^{-3}$) as a function of distance from an isolated Fe cluster on GaAs(110). Curves (a–e) correspond to distances from the cluster edge of 23.3, 34.3, 49.0, 67.3, and 82.7 Å, respectively. The inset shows the energy diagram (not to scale) for tunneling into holes at the top of the valence band with positive sample bias. (b) The tunneling current at V = 0.8 V as a function of distance from the Fe cluster (solid squares). (From Ref. [45].)

data is then fit (using Eq. (15)) to extract the spatial potential profile as shown by the solid line in Fig. 27. The screening potential is observed to encompass a range of about 0.1 eV. In the semi-classical approximation, this potential shifts the semiconductor bands uniformly with respect to the Fermi level, which we see qualitatively in the curves (a–e) in Fig. 26(a). The screened potential is expected to have an asymptotic limit of the zeroth-order modified Bessel function, for a solution of Poisson's equation with the surface boundary condition, of the form, $\phi \propto \exp(-\rho/\lambda)/\sqrt{\rho}$, where ρ is in the surface plane. A best fit to this functional form is shown in Fig. 27 with a screening length $\lambda = 21.7 \pm 0.4$ Å. This value can be compared to a bulk screening length of 18 Å estimated from Eq. (13), which is seen to be of similar magnitude. Note with this analysis the 0.1 eV variation of the potential in Fig. 27(b) is in agreement with the shifts in the I–V curves, but is

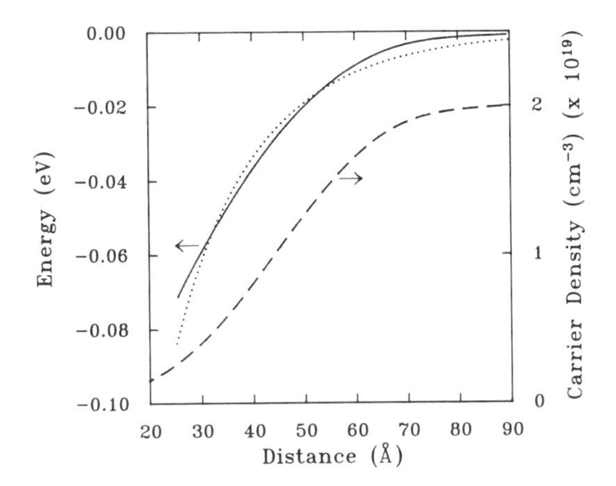

FIG. 27. Dashed line: Carrier density vs distance obtained from equating the spatial variation in the gap current (solid line in Fig. 26(b)) to the surface free carrier density. The plateau at 90 Å was scaled to the bulk carrier density of $2 \times 10^{19}\,\text{cm}^{-3}$. Solid line: Screening potential ϕ vs distance obtained from the carrier density shown by the dashed line. Dotted line: Best fit to the potential data shown by the solid line with the functional form $e^{-\rho/\lambda}/\sqrt{\rho}$, with $\lambda = 21.7 \pm 0.4$ Å. (From Ref. [45].)

measured (although indirectly) to better precision, since the band edges are not well-defined in Fig. 26(a).

4.4.4 Current Imaging Spectroscopy

Combining the scanning ability of the STM with the spectroscopic capabilities allows for high spatial resolution spectroscopy measurements, essentially on an atom by atom basis. Current imaging or current imaging tunneling spectroscopy (CITS) involves the measurement of the I–V characteristics at each pixel that the normal STM topograph is taken. One then has a normal STM image as well as current images (or derived quantities such as dI/dV), where the current images are obtained from the three-dimensional data structure of $I(V, x, y)$. The current image thus represents a slice, at a given voltage, of the current as a function of the lateral x, y coordinates. Variations on this theme can be performed whereby I–V characteristics are recorded at a subset of points in a grid or along a particular line in an image. The use of current imaging, in particular with conductance measurements, allows a more energy-resolved spectroscopy to be performed with spatial emphasis. One can see the location of states as a function of energy *and* position. Notable examples have been the surface states on the Si(111)7 × 7 surface by Hamers *et al.*[46] and the vortex states in superconductors by Hess *et al.*[47] (see Chapter 9). In this section we illustrate the technique with an example from

Fe/GaAs(110)[48] and show some of the problems in the interpretation of current imaging with data from the Si(111)2 × 1 surface.[19]

The timing sequence for current imaging is the same as shown in Fig. 2, with the adjustment that the sequence is repeated at each pixel sampled in the image. For shorter acquisition times, a smaller subset of pixels can be chosen for the I–V measurements, for example, in a coarser grid or along a line in the image (see Chapter 9). During data acquisition, the I–V measurements can be observed in real time with a monitor of the current signal on an oscilloscope. The conductance, dI/dV, can be measured directly with lock-in detection or the conductance can be computed numerically from the current measurements. Due to the lengthened time requirements of the I–V acquisition, it is desirable to have a high bandwidth preamplifier and very low tip–sample capacitance so that the I–V measurement can be made as fast as possible (assuming lock-in detection is not being used). If a measurement is made too fast compared to the system response (tip–sample junction), however, a large displacement current offset will appear giving rise to hysteresis in the I–V measurement. Signal-to-noise can be improved by signal averaging or lock-in detection at the expense of longer data acquisition times.

Figure 28 shows conductance images of Fe/GaAs(110) at two different voltages obtained from I–V measurements at each pixel in the STM topograph.[48] Fe clusters are located in the central black regions, where the conductance is large and off-scale in this image due to the large current from the metallic clusters. Current limiting, as suggested in Section 4.1, was used so that the conductance at these voltages was not even obtained over the clusters. Nearby the clusters on the GaAs(110) surface, finite conductance is observed at energies within the GaAs(110) band gap where there is usually no tunneling current observed. Note this finite conductance is not due to dopant carriers, discussed in the previous section, since this contribution is "pinched off" due to band bending (see Section 4.4.3). The conductance near the clusters is seen to decay exponentially with distance, reaching zero value far from the clusters. By analyzing the conductance images with energy, it is seen that this decay length depends on the energy within the GaAs (110) band gap, as seen by comparing Figs. 28(a) and (b), and is discussed in more detail in Section 5.3. The analysis of the Fe conductance images was relatively straightforward (the changes in the conductance occurred over regions where the topography was not changing appreciably). When large topographic features overlap the spectroscopic features, however, care must be given to other non-spectroscopic sources of contrast. One non-spectroscopic source of contrast can be a feedback loop which is simply proportional with low gain. The resulting finite error in z position over varying topography changes the tunneling transmission probability, thereby introducing false contrast in the spectroscopy images.

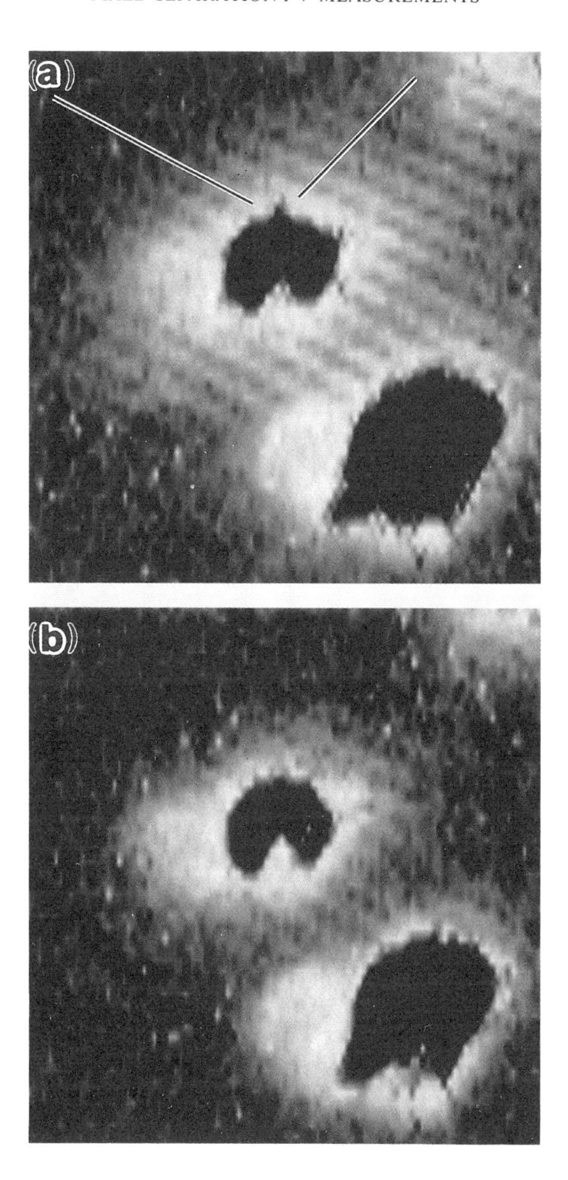

FIG. 28. Images of the logarithm of the differential conductance, dI/dV, obtained from the I–V characteristics of Fe/p-GaAs(110), recorded at each pixel in the normal topographic image. The images in the top (a) and bottom (b) panels correspond to voltages of 1.1 and 1.0 V, respectively. The images are shown with a grey scale ranging from 0 to 300 pA/V. The central black regions denote areas of large current over the metallic clusters. (From Ref. [48].)

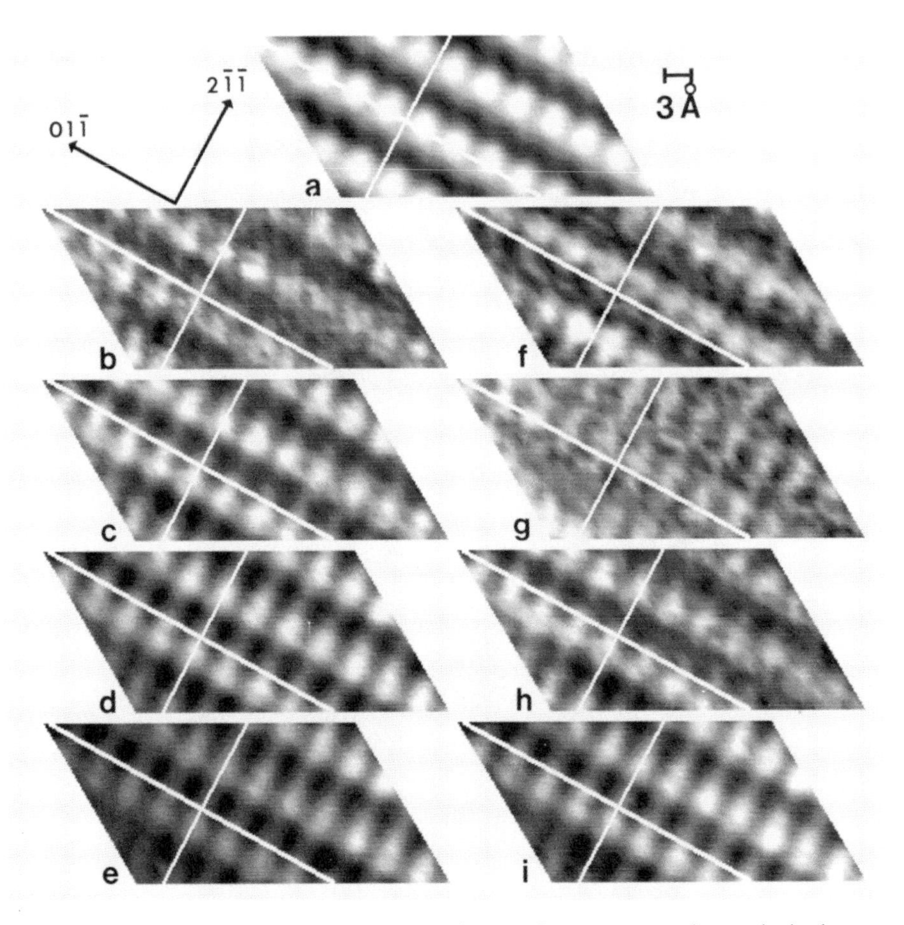

FIG. 29. STM images of the Si(111)2 × 1 surface. (a) Constant-current image obtained at a sample bias of 1.5 V. (b)–(e) Current images, obtained simultaneously with the constant current image (a), at sample voltages of 1.22, 0.93, 0.63, and 0.34 V, respectively. (f)–(i) Conductivity images, obtained by numerical differentiation, at sample voltages of 1.22, 0.93, 0.63, and 0.34 V, respectively. For all images, the magnitude of the appropriate quantity is plotted using a grey scale, with black representing small values and white representing large values. (From Ref. [19].)

Topographic-dependent contrast through the tunneling transmission factor can be inherent in the measurement, as seen in data from Si(111)2 × 1 surface, shown in Fig. 29.[19] Recall from Section 4.3 that the topographic features in the STM image [Fig. 29(a)] result from the localized surface states on *one* of the two atoms in the unit cell. As the voltage is lowered, going through the current images (b–e) and conductance images (f–i), one observes shifts in the position of the maxima at the center of the crosshairs, with a 180° phase change observed in [01$\bar{1}$] direction at the lowest voltage. Such a change

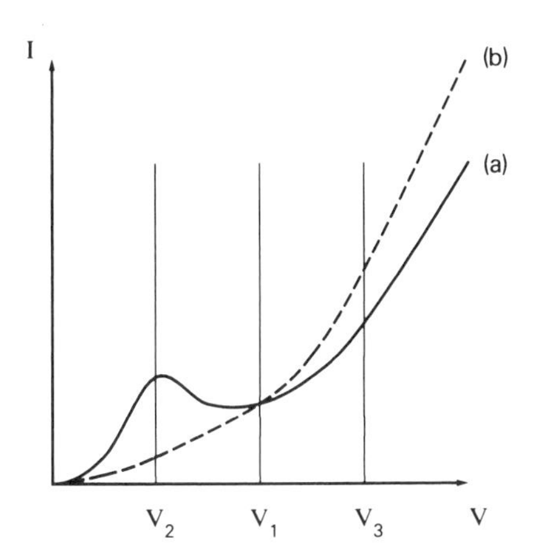

FIG. 30. Schematic illustration of current–voltage characteristics at two different spatial locations (a) and (b) on a surface. The currents are equal at the voltage V_1. At V_2, a surface state feature appears at the location (a), and it can be imaged at that voltage. An image formed at the voltage V_3 will show a maximum at spatial location (b), associated with the varying background level of the current. (From Ref. [49].) Reprinted by permission of Kluwer Academic Publishers.

of phase is *not* observed in the normal STM topograph (the phase shift is observed when the polarity is changed), and therefore this change in contrast is not derived from electronic density-of-states; rather the contrast results from changes in the tunneling transmission factor. This results from the fact that the STM constant current contours are not constant z contours, and when the tip is in the valley versus the peak of a maximum, the tunneling transmission factor (Eq. (5)) is different, since it depends on z. In analyzing normalized conductance images of $(dI/dV)/(I/V)$, one sees these effects minimized to some degree.[19] In general, however, one has to consider carefully the contrast observed in the spectroscopic images as many variables can be interwoven.

The general problem encountered in the interpretation of current (or conductivity) images, when topographic variation is encountered, is illustrated in Fig. 30.[49] We plot in that figure two possible I–V curves, assumed to exist at two different lateral positions (a) and (b) on the surface and at some particular values of tip–sample separation used in the measurement. If these curves were measured with the CITS method, then the currents would be constrained to be equal at the feedback voltage V_1, as shown in Fig. 30. In both of the curves, we show the current increasing rapidly with voltage. This increasing "background" on the current is always present in real data, and arises from the transmission term for electrons tunneling through the vacuum barrier, Eq. (5). In curve (a) of Fig. 30 we show a possible surface-state peak near the voltage V_2, which does not appear in curve (b). Now let us consider forming images of the tunneling current. At the voltage V_2 we will see a maximum in the image near the spatial position (a), and this maximum can correctly be interpreted as a surface-state related feature. However, at the voltage V_3, the images will show a maximum near the spatial position (b). In this case the maximum arises purely from varying background on the current, and is not related to a surface-state.

As illustrated in the previous example, one must exercise caution in interpreting the features seen in current images. Similar comments apply for differences between current images, and for conductivity images measured at constant current. In general, a complete spectrum of the surface states must first be examined before one can assign specific topographic features to individual surface states. These problems of the "background" are more severe for current images measured by the CITS method than for conventional constant-current images because the CITS images actually represent a current difference, and as such they are more sensitive to changes in the background level of the tunnel current.

4.5 Variable Separation Measurements

In this section we discuss methods for performing spectroscopy with the STM using a continuous variable separation between probe tip and sample. The aim of these methods is to achieve a wide dynamic range of tunneling current and conductivity. This goal is especially important for semiconductor surfaces, where, at low bias voltages of approximately $\leqslant 2$ V the presence of a band gap leads to tunneling conductivities which range over many orders of magnitude. Since STM measurements are limited to relatively low currents of approximately $\leqslant 1$ nA to prevent sample damage, it is difficult to measure, in a limited amount of time, such a large dynamic range of conductivity. Thus, one resorts to the use of variable tip–sample separation, in which the tunnel current at low bias voltage is amplified by reducing the tip–sample separation. Subsequent normalization of the spectrum is performed to remove the effects of varying the tip–sample separation. The most basic type of variable separation acquisition method has already been illustrated in Figs. 15 and 19, where different discrete values of the tip–sample separation are used to obtain the high dynamic range. The major problem with this type of measurement is that the spectrum ends up being acquired piecewise, and some effort is required to match together these pieces into the final complete spectrum. The solution to this problem is to vary the tip–sample separation in a *continuous* manner as the bias voltage changes. Two methods for acquiring data in this way are presented in Section 4.5.1.

The two most basic methods of spectroscopic normalization have already been illustrated above. The first method is shown in Figs. 15 and 17, where one computes a ratio of the differential conductivity dI/dV to the total conductivity I/V. This method has been succesfully used on a variety of metal and semiconductor surfaces with small surface-state band gaps, and the normalized conductivity $(dI/dV)/(I/V)$ is often said to be proportional to the surface state density. However, for the case of surfaces with large band gaps such as GaAs(110), the method completely breaks down; both the current and the conductivity approach zero at the band edges, and their ratio tends to diverge (or at least approach some large value).[50,51] These divergences are purely artifacts of the normalization method, and they do not reflect any real state density of the surface. Thus, we require some generalization of the I/V normalization method which can be applied to spectra which contain large band gap regions. One such method is described in Section 4.5.2.

The second basic type of normalization method, illustrated in Figs. 19–20, is to account for the differing values of tip–sample separation simply by using a lowest-order multiplicative correction of the form $\exp(2\kappa\Delta s)$, where Δs are the known values of the relative tip–sample separations, and κ is the inverse decay length of the tunnel current (times 1/2). This type of normalization is clearly also applicable to the case where the tip–sample separation varies continuously. The only parameter in this analysis is the inverse decay length κ, which in general depends on both the tip–sample separation and bias voltage, $\kappa = \kappa(s, V)$. These dependences are usually small, so that in most cases it is sufficient to assume a constant value, $\kappa = \kappa_0$, which can be easily obtained from a single measurement of tunnel current versus separation or simply by computing $\kappa_0 \equiv (2m\phi)^{1/2}/\hbar$. In some cases, however, one may wish to go further and include the complete form $\kappa(s, V)$ together with any possible variation in κ from point to point on the surface. It is then necessary to measure the decay length *simultaneously* with the measurement of the tunneling spectra. There exists a rigorous method for accomplishing this goal,[52,53] as described in Section 4.5.3.

4.5.1 Acquisition Methods

As discussed earlier, it is generally necessary to vary the tip–sample separation s during a spectroscopic measurement in order to amplify the tunnel current up to a conveniently measurable range. In this section we discuss methods in which the separation is varied continuously as a function of voltage, i.e., $s = s(V)$. One then measures the tunnel current, $I_m[s(V), V]$, where the subscript m refers to the measured quantity. In general, one also wants to determine the conductivity, but this cannot be obtained by numerical differentiation of the current because that calculation yields a *total* derivative which includes contributions from the varying $s(V)$ (see Eq. (19)). Thus, to obtain conductivity one must use a modulation technique, so that, at each voltage, the conductivity equals the derivative of the current with respect to voltage at *fixed* tip–sample separation,

$$\sigma_{\mathrm{m}}[s(V), V] \equiv \left.\frac{\partial I_{\mathrm{m}}}{\partial V}\right|_{s}. \tag{16}$$

Using a modulation frequency of about 1 kHz with a lock-in amplifier leads to a measurement time for the spectrum of a few seconds. The feedback loop of the STM must be disabled during this entire time period; thus, these methods are only possible when the residual thermal or piezoelectric drift rates of the STM are less than 0.1 Å/s.

The major decision which must be made in the variable gap methods is the choice of the tip–sample separation contour, $s(V)$ to use in the measurement. One of the simplest choices is to apply a linear ramp, moving the tip towards the surface as the magnitude of the voltage is decreased, and then, after the voltage crosses zero, retracting the tip as the magnitude of the voltage increases. Thus, one uses a contour of the form

$$\Delta s(V) = s(V) - s_0 = -a|V|, \tag{17}$$

where s_0 is the tip–sample separation at zero volts, and the parameter a determines the slope of the ramp. For general measurements on any surface, a value for a of 0.5–1 Å/V is suitable, although larger values can be used for systems in which higher sensitivity is required at low voltage. Figure 31 shows an example of this type of data, acquired on a monolayer of Sb on the GaAs(110) surface.[50] This system possesses a surface-state band gap of about 1.4 eV (nearly the same as the clean GaAs band gap). High dynamic range was desired in this case in order to probe the band gap region for the existence of states in the gap. Figure 31(a) shows the conductivity σ_{m}, (b) shows the current I_{m}, and (c) shows the applied variation in the tip–sample separation Δs which has a slope of 2 Å/V in this case. Figure 31(d) refers to the normalization of the spectrum, which we discuss in the following sections.

The choice of a V-shaped $s(V)$ contour is sufficient for a number of STM applications, but there are cases when this method fails to provide a suitable amplification of the current at the appropriate voltages. For example, consider the measurement of spectra from both *p*- and *n*-type GaAs surfaces, using the same measurement parameters for each spectrum. Such a situation would be encountered, for example, in spectroscopic measurements across a *pn*-junction. I–V curves for the *p*-type and *n*-type material are shown in Figs. 20 and 21, respectively. At a given voltage, the currents in the two materials can differ by more than a factor of 10, and this ratio varies widely as a function of voltage. The situation is even worse on material which is slightly depleted so that the D-components of the tunnel current seen in Figs. 20 and 21 are absent; then, the currents on *p*- and *n*-type materials can differ by many

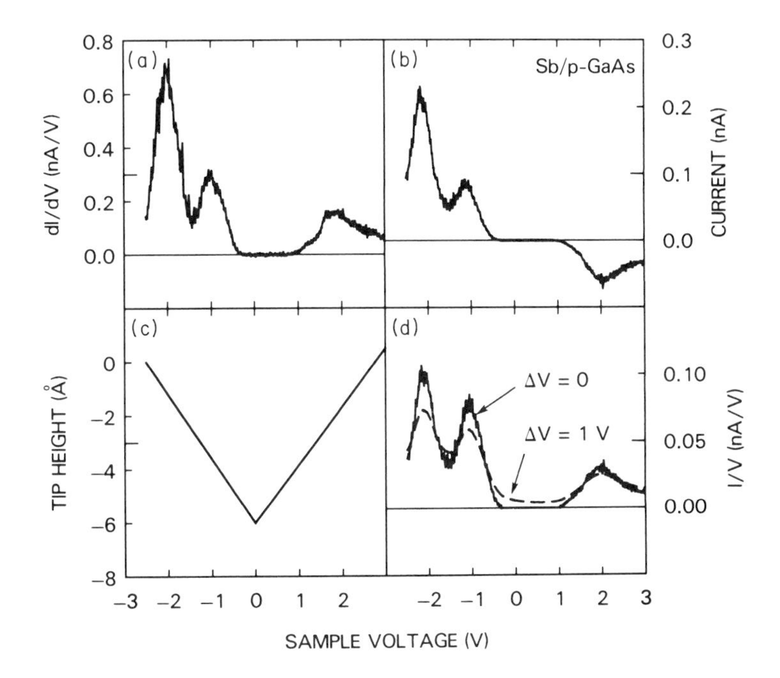

FIG. 31. Raw data for (a) the differential conductivity, and (b) the current, as a function of sample voltage, obtained from an ordered region of a monolayer Sb film on GaAs(110). The applied variation in tip–sample separation is shown in (c). The total conductivity I/V is shown in (d), with no broadening (solid line) and with a broadening of $\Delta V = 1$ V (dashed line). (From Ref. [50].)

orders of magnitude. Thus, for any choice of tip–sample separation which is specified *a priori*, it is not possible for both the *p*- and *n*-type material to have a current which is both large enough to be measurable and not too large to cause sample damage. In this case we require a more sophisticated choice of tip–sample separation, in particular, one that depends on the magnitude of the tunnel current at each voltage.

A general choice of $s(V)$ contour, suitable for spectroscopic measurement on any system, can be obtained by the following prescription: Starting from the initial high-voltage point of the spectrum, the current is maintained at a constant value (by leaving the feedback loop enabled), and the tip–sample separation follows the resulting constant-current contour. Then, when the measured tip displacement exceeds some preset value, the feedback loop is disabled and the tip–sample separation is held constant. After zero volts has been crossed and the voltage is increasing in magnitude, constant-current operation is re-established when the magnitude of the current starts to exceed its original constant value. An example of this type of data is shown in Fig.

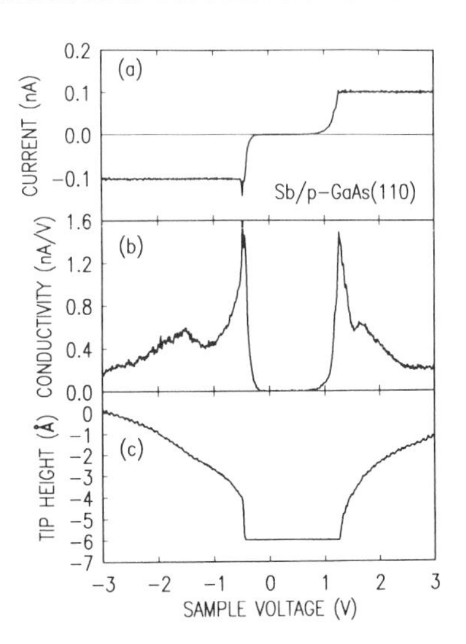

FIG. 32. Raw data for (a) the tunneling current, and (b) the conductivity, as a function of sample voltage, obtained from an ordered region of a monolayer Sb film on GaAs(110). The applied variation in tip–sample separation is shown in (c), illustrating a combination constant-current/constant-s method. The zero of tip height is arbitrary. (From Ref. [52].)

32, taken again from the Sb/GaAs(110) surface.[52] Figure 32(a) shows the current I_m, (b) shows the conductivity σ_m, and (c) shows the measured variation in tip–sample separation Δs. The preset value of the maximum tip displacement was 6 Å in this case. Comparing Figs. 31 and 32, it is clear that a much higher amplification of the current and conductivity has been achieved at low voltages in Fig. 32, because of the more sophisticated choice of tip–sample separation. On the other hand, it may be apparent to the reader that subsequent normalization of the data to eliminate the effects of the varying tip–sample separation is somewhat more difficult for the data of Fig. 32 compared with Fig. 31. This normalization is the subject of the following two sections.

4.5.2 Normalization to $\overline{I/V}$

In this section we consider normalization of spectroscopic data by taking the ratio of the differential conductivity σ_m to the total conductivity I_m/V. Figure 33(a) shows this ratio, computed for the raw data of Fig. 31. We see that near the band edges at -0.4 and 1.0 V, the normalized conductivity tends to diverge. This divergence is not surprising; it arises simply because, at a band edge, both the tunnel current and conductivity approach zero, and their ratio diverges. This breakdown of the analysis method occurs because I/V is no longer a valid estimator of the tunneling transmission term, which

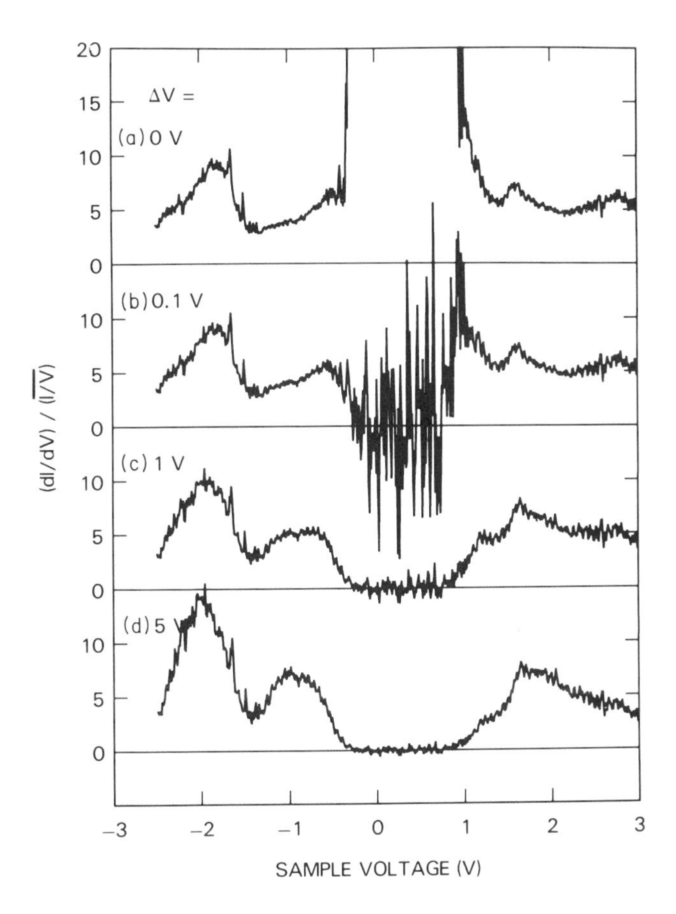

FIG. 33. Normalization of the conductivity spectrum (from Fig. 31), where I/V is broadened by various amounts, ΔV. In curve (a) the normalized data become extremely noisy near 0 V, and that section of the curve is not plotted. (From ref. [50].)

itself does *not* equal zero in a band gap. Thus, we must find some other means of estimating the transmission term in order to provide a suitable normalization quantity.

One method to eliminate the divergence in the normalized conductivity is to *broaden* the function I/V by convolution with a suitable function, thereby eliminating the zeros of I/V within the band gap.[50,51] An example of this is shown in Fig. 31(d), where we show I/V for the Sb/GaAs data with no broadening, and with a broadening width of $\Delta V = 1$ V. The broadening in this case was accomplished by convolution with an exponential function,

$$\overline{I_m/V} \equiv \int_{-\infty}^{\infty} (I_m/V') \exp\left\{\frac{|V' - V|}{\Delta V}\right\} dV'. \quad (18)$$

For ease of computation this convolution was Fourier transformed, in which case it corresponds to a low-pass filter of the form $1/(1 + f_V \Delta V)$ where f_V is the frequency component of the voltage.[50] The results for the normalized conductivity, $\sigma_m/(\overline{I_m/V})$, are shown in Figs. 33(b)–(d) for broadening widths of $\Delta V = 0.1$, 1, and 5 V. We see that, even with a small amount of broadening, the divergences at the band edges disappear, and the normalized conductivity within the band gap approaches the zero level. In general, one must use a ΔV value which is roughly equal to the band gap in order to eliminate all of the band edge divergences and noise within the gap. Away from the band edges the other features in the spectrum are only slightly affected by the broadening procedure.

The choice of an exponential convolution in Eq. (18) is made purely for convenience, and other functions (e.g., a Gaussian) could also be applied. Some authors choose simply to add some small constant value to I/V in order to eliminate the zeros within the band gap.[54–57] All of these procedures are, of course, only applied to the normalization quantity I/V, and do not affect the data values for the differential conductivity. We note that the use of such procedures in estimating a suitable normalization curve is actually much more widespread than might appear from a reading of the literature. Many, if not most, of the practitioners of STM spectroscopy have occasionally used these methods as an aid in computing the normalized conductivity.[50–57] Wherever a band gap can be seen in the normalized conductivity, i.e., $(dI/dV)/(I/V) \rightarrow 0$, one can be sure that some small amount of broadening has been included in the computation of I/V.

Although the procedure outlined above allows us to extend the I/V normalization procedure to cases where a band gap is present in the data, it suffers from the drawback that the broadening method will always involve one (or more) parameters and thus is somewhat more arbitrary than one might like. There is no solution to this problem, and we must accept this drawback of the method. When normalized spectra are compared with each other, one must check in each case that the details of the normalization are not affecting the spectral features in question. A more significant problem arises when we consider how to apply the $\overline{I/V}$ method, including broadening, to any arbitrary piece of data, i.e., acquired with arbitrary $s(V)$ contour. In the example shown in Figs. 31 and 33, the V-shaped contour provided a particularly favorable case for this method. However, if we consider, for example, data acquired with constant tip–sample separation, then the computation of $\overline{I/V}$ by Eq. (18) will, at each particular V, preferentially weight currents with $|V'| > |V|$ since those currents are exponentially greater in magnitude. Thus, the broadening procedure becomes rather asymmetric, which is a problem, especially for large band gap cases. An even more serious problem occurs if we consider data of the form shown in Fig. 32. There, the use of constant-current $s(V)$ contours up to the band edges, and constant-s within the band gap, leads to the sharply peaked features in the differential conductivity σ_m at the band edges. If we normalize σ_m to $\overline{I_m/V}$, then these peaks will remain in the ratio. Thus, the use of $\overline{I/V}$ according to Eq. (18) is applicable only to particular $s(V)$ contours, and some other method must be found for the general case. We describe such a method in the following section, where we first describe how to transform data acquired with arbitrary $s(V)$ into constant-s conductivity, and we then provide a generalization of the $\overline{I/V}$ method to this constant separation data. This more general method gives further justification for the $\overline{I/V}$ method, as both yield the same results for a given system.

4.5.3 Transformation to Constant Separation

In this section we describe a method by which spectroscopic data acquired with variable tip–sample separation can be transformed, in a parameter-free method, to constant separation.[52,53] To illustrate the feasibility of this procedure, consider the following total derivative:

$$\frac{dI(s, V)}{dV} = \left.\frac{\partial I(s, V)}{\partial V}\right|_s + \left.\frac{\partial I(s, V)}{\partial s}\right|_V \frac{ds(V)}{dV}. \tag{19}$$

The quantities on the far left- and far right-hand sides of this equation can be obtained by numerical differentiation of the measured $I_m[s(V), V]$ and $s(V)$, respectively, and the quantity on the right-hand side of the equals sign is identical to the measured conductivity. Thus, we can solve for $\partial I/\partial s|_V$, which, in principle, contains the information required to deduce the s-dependence of the current and conductivity. In practice, to avoid numerical differentiation, the analysis method proceeds using an integral transform approach as described next.

The analysis method is based on the use of the logarithmic derivative,

$$g_m[s(V), V] \equiv \frac{\sigma_m[s(V), V]}{I_m[s(V), V]}. \tag{20}$$

Since the logarithmic derivative equals the *ratio* of conductivity to current, it is approximately independent of tip–sample separation for small changes in separation. Denoting the logarithmic derivative at some constant-s value of $s = s'$ to be $g(s', V)$, we have

$$g(s', V) \simeq g_{\mathrm{m}}[s(V), V]. \tag{21}$$

This is the only approximation made in the analysis, and it has only relatively minor effects on the final results of the analysis procedure,[53] as further discussed later. Now, from the definition of $g(s', V)$ we have

$$g(s', V) \equiv \frac{\sigma(s', V)}{I(s', V)} = \frac{d}{dV} \log I(s', V), \tag{22}$$

(note that the total derivative on the right-hand side of this equation is valid only because the current is evaluated at constant $s = s'$). From eq. (22) it follows that

$$\frac{I(s', V)}{I(s', V')} = \exp\left\{\int_{V'}^{V} g(s', E)\, dE\right\}, \tag{23}$$

for any V and V' which have the same sign. Differentiating with respect to V then yields

$$\sigma(s', V) = I(s', V')g(s', V) \exp\left\{\int_{V'}^{V} g(s', E)\, dE\right\}. \tag{24}$$

This equation gives the conductivity at any value of s, in particular at $s = s'$. The value of V' is arbitrary; choice of a different V' simply introduces a scale factor multiplying the exponential term, which is canceled by the $I(s', V')$ multiplier as seen from Eq. (23).

Equation (24) provides the conductivity at constant tip–sample separation, which is the desired result of the analysis. To evaluate this equation in terms of the measured data, Eq. (21) is used to evaluate the logarithmic derivative. Furthermore, the term $I(s', V')$ is evaluated from the measured quantity $I_m[s(V'), V]$, in which case it is necessary to take $s' = s(V')$. The evaluation of Eq. (24) in terms of measured quantities is then given by

$$\sigma(s', V) \simeq I_m[s(V'), V']g_m[s(V), V] \exp\left\{\int_{V'}^{V} g_m[s(E), E]\, dE\right\}. \tag{25}$$

The evaluation of this equation must be performed separately for positive and negative voltages. The measured $s(V)$ contour need not be symmetric between positive and negative voltages (e.g., as seen in Fig. 32(c)); thus, to achieve consistent normalization between the two sides of the spectrum, one must choose the V' values on either side such that the $s(V')$ values are equal for positive and negative voltages.

Figure 34 shows the results for the conductivity at constant-s for the data of Fig. 32, evaluated as just described. We see that the conductivity at constant-s extends over six orders of magnitude, with clear definition of the band edges and other spectral features. Because we have transformed from variable-s to constant-s, the result implicitly contains values for the inverse decay length $\kappa(s, V)$. These can be explicitly obtained by considering the ratio of transformed to measured conductivity, as a function of $s(V)$, as

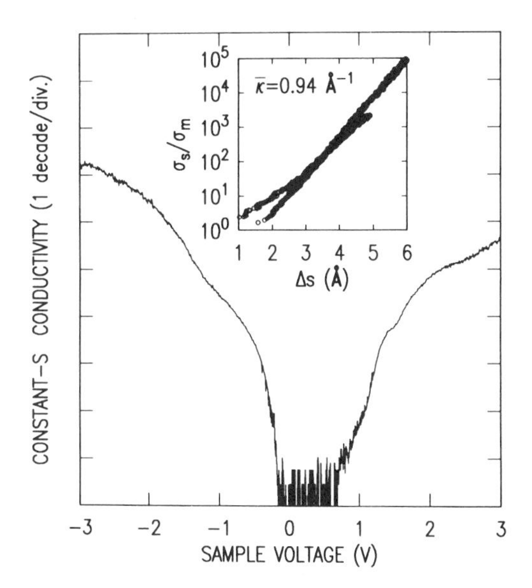

FIG. 34. Transformation of the conductivity spectrum from Fig. 32 to a constant value of tip–sample separation. The inset shows the ratio of transformed to measured conductivity, displaying an exponential dependence on the relative tip–sample separation, Δs. The two curves in the inset arise from the positive and negative voltage sides of the spectrum. Normalization points of $V' = -0.20$ and 0.71 V are used, corresponding to $s' = s(V') = -6.0$ Å on the scale of Fig. 32(c). (From Ref. [52].)

shown in the inset of Fig. 34. Values for $\kappa(s, V)$ can be obtained by measuring the slope of these curves, which, for Fig. 34, yields an average value of 0.94 Å^{-1}. A detailed analysis reveals that the values for $\kappa(s, V)$ obtained in this way are *identical* with those obtained by the conventional method of measuring the logarithmic derivative of the tunnel current with respect to separation.[53] In particular, this result holds true even when we include the effects of the approximate Eq. (21), and for any variation of κ as a function of (s, V). If we consider, rather than $\kappa(s, V)$, the evaluation of $\sigma(s', V)$ from Eq. (25), it turns out that the possible V-dependence of κ will affect the accuracy of this evaluation because of the use of Eq. (21). For a spectrum such as Fig. 34, which spans the range -3 to 3 V and extends over about six orders of magnitude, the approximations in the analysis can lead to an overestimate of the s-dependence of the conductivity by as much as one order of magnitude (i.e., the spectrum in Fig. 34 would, if actually measured at constant-s, extend over only about five orders of magnitude rather than six).[53] This overestimate will accumulate gradually over the spectrum, and will not introduce any significant features other than the overall uncertainty in the relative magnitudes. Thus, so long as one does not try to interpret the magnitude of the constant-s conductivity too literally, the analysis method provides a rigorous and reliable method for transforming the data to constant tip–sample separation.

We now return to the problem posed at the end of the previous section, namely, how to perform an $\overline{I/V}$ type normalization to data acquired with arbitrary $s(V)$ contour. The $\overline{I/V}$ method presented in the previous section worked well only for data acquired with a V-shaped $s(V)$ contour, whereas the analysis method discussed in this section provides a general method for transforming data to fixed tip–sample separation. A generalization of the $\overline{I/V}$ method, applicable to data acquired with arbitrary $s(V)$, can then be obtained by combining the elements of both these procedures. We perform an exponential scaling when computing $\overline{I/V}$, so that the broadening includes equal contributions from higher and lower voltages. Denoting the current at constant-s by I_S, we have

$$\overline{I_S/V} \equiv \exp(a|V|) \int_{-\infty}^{\infty} (I_S/V') \exp\left\{\frac{|V'-V|}{\Delta V}\right\} \exp(-a|V'|)\, dV'. \tag{26}$$

The current at constant-s is obtained either by direct measurement, or by integrating the transformed conductivity from Eq. (25),

$$I_S(V) \equiv \int_0^V \sigma_S(V')\, dV', \tag{27}$$

where $\sigma_S(V) \equiv \sigma(s', V)$. The parameter a in Eq. (26) can be chosen to provide a suitable scaling of the current such that it varies with voltage roughly linearly rather than exponentially. A typical value for a is about $1\,V^{-1}$, however, the precise value chosen has little effect on the final normalized conductivity. Given I_S and σ_S, the $\overline{I_S/V}$ normalization is performed by Eq. (26), and the normalized conductivity is given by $\sigma_S/(\overline{I_S/V})$.

Figures 35(a) and (b) show the results of the complete analysis procedure, applied to the data of Figs. 32 and 31, respectively. We see in the spectra very clear definition of the band gaps and various other spectral features. The quality of these spectra far exceeds any results which could be obtained by measurement at constant tip–sample separation (note, in particular, the very low noise level within the band gap of Fig. 35(a), which results from the small values of tip–sample separation used throughout that region). It is interesting to compare Fig. 35(b) with the normalized conductivity of Fig. 33(c). These results were obtained from the same raw data (Fig. 31) and using nominally the same analysis parameters, but with substantially different analysis methods. For Fig. 33(c), the normalization is done directly using Eq. (18), and then plotting $\sigma_m/(\overline{I_m/V})$. For Fig. 35(b), the analysis follows a more circuitous route, first transforming to constant-s through the use of Eqs. (25) and (27), then normalizing using Eq. (26) and finally plotting $\sigma_S/(\overline{I_S/V})$.

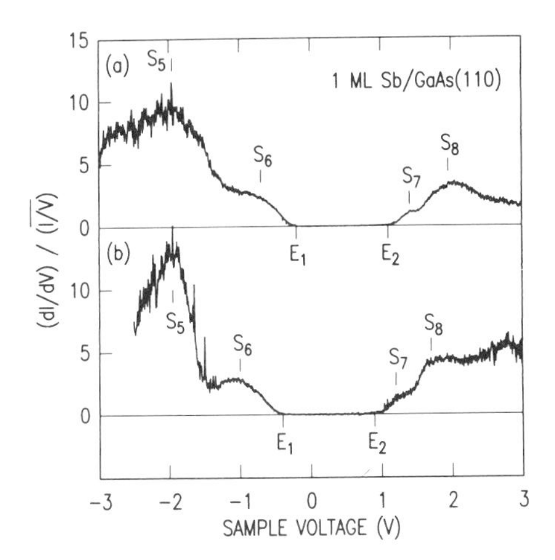

FIG. 35. Normalized conductivity vs sample voltage for two samples of monolayer Sb films on GaAs(110). Curves (a) and (b) come from the raw data of Figs. 32 and 31, respectively. The edges of the band-gap region are denoted by E_1 and E_2, with the slight shift in their positions between (a) and (b) arising from differing amounts of band bending. The various spectral peaks are labeled S_5–S_8. The sample voltage corresponds to the energy of a state relative to the Fermi level (0 V). (From Ref. [52].)

Comparing Figs. 35(b) and 33(c), we see that the two schemes give almost the same results, with only some relatively minor differences in the shapes of spectral features. Thus, for data acquired with a V-shaped $s(V)$ ramp one can use either the "short-cut" method of Section 4.5.2, or the longer method of the present section. For data acquired with arbitrary $s(V)$ contour, one must use the methods described in the present section. The choice of final data display depends on the particular application: One can simply plot the constant-s data on a logarithmic scale, thus displaying the full range of the data without any arbitrary parameters. Alternatively, one can perform the $\overline{I/V}$ normalization, thereby enhancing small features of the spectrum and allowing display on a linear scale. For most applications, both types of display are useful.

References

1. C. B. Duke, "Tunneling in Solids," Suppl. 10 of *Solid State Physics*, (F. Seitz and D. Turnbull, eds.), Academic Press, New York, 1969.
2. *Tunneling Phenomena in Solids*, (E. Burstein and S. Lundqvist, eds.), Plenum, New York, 1969.
3. *Inelastic Electron Tunneling Spectroscopy*, (T. Wolfram, ed.), Springer-Verlag, New York, 1978.
4. E. L. Wolf, *Principles of Electron Tunneling Spectroscopy*, Oxford Univ., New York, 1989.

5. C. B. Duke, *op. cit.*, p. vii.
6. A. P. Fein, J. R. Kirtley, and R. M. Feenstra, *Rev. Sci. Instrum.* **58**, 1806 (1987).
7. R. D. Cutkosky, *Rev. Sci. Instrum.* **61**, 960 (1990).
8. J. A. Stroscio, R. M. Feenstra, and A. P. Fein, *Phys. Rev. Lett.* **57**, 2579 (1986).
9. R. Young, J. Ward, and F. Scire, *Phys. Rev. Lett.* **27**, 922 (1971).
10. E. C. Teague, Ph.D. Thesis, North Texas State University, 1978, reprinted in *J. Res. National Bureau of Standards* **91**, 171 (1986).
11. R. M. Feenstra, J. A. Stroscio, and A. P. Fein, *Surf. Sci.* **181**, 295 (1987).
12. R. S. Becker, J. A. Golovchenko, D. R. Hamann, and B. S. Swartzentruber, *Phys. Rev. Lett.* **55**, 2032 (1985).
13. G. Binnig and H. Rohrer, *Helvetica Physica Acta* **55**, 726 (1982).
14. G. Binnig, K. H. Frank, H. Fuchs, N. Garcia, B. Reihl, H. Rohrer, F. Salvan, and A. R. Williams, *Phys. Rev. Lett.* **55**, 991 (1985).
15. R. S. Becker, J. A. Golovchenko, and B. S. Swartzentruber, *Phys. Rev. Lett.* **55**, 987 (1985).
16. J. Bono and R. H. Good, *Surf. Sci.* **188**, 153 (1987).
17. J. A. Kubby, Y. R. Wang, and W. J. Greene, *Phys. Rev. Lett.* **65**, 2165 (1990).
18. J. A. Stroscio, R. M. Feenstra, and A. P. Fein, *J. Vac. Sci. Technol. A* **5**, 838 (1987).
19. J. A. Stroscio, R. M. Feenstra, D. M. Newns, and A. P. Fein, *J. Vac. Sci. Technol. A* **6**, 499 (1988).
20. K. C. Pandey, *Phys. Rev. Lett.* **47**, 1913 (1981).
21. K. C. Pandey, *Phys. Rev. Lett.* **49**, 223 (1982).
22. J. Tersoff, *Phys. Rev. Lett.* **57**, 440 (1986).
23. R. M. Feenstra and J. A. Stroscio, *Physica Scripta T* **19**, 55 (1987).
24. R. M. Feenstra and J. A. Stroscio, *Phys. Rev. Lett.* **59**, 2173 (1987).
25. P. Chiaradia, A. Cricenti, S. Selci, and G. Chiarotti, *Phys. Rev. Lett.* **52**, 1145 (1984).
26. N. D. Lang and A. R. Williams, *Phys. Rev. B* **118**, 616 (1978).
27. N. D. Lang, *Phys. Rev. Lett.* **55**, 230 (1985).
28. N. D. Lang, *Phys. Rev. Lett.* **56**, 1164 (1986).
29. N. D. Lang, *Phys. Rev. Lett.* **58**, 45 (1987).
30. J. A. Stroscio, R. M. Feenstra, and A. P. Fein, *Phys. Rev. Lett.* **58**, 1668 (1987).
31. R. Wolkow and Ph. Avouris, *Phys. Rev. Lett.* **60**, 1049 (1988).
32. J. S. Villarrubia and J. J. Boland, *Phys. Rev. Lett.* **63**, 306 (1989).
33. Ph. Avouris, In-Whan Lyo, and F. Bozso, *J. Vac. Sci. Technol. B* **9**, 424 (1991).
34. N. D. Lang, *Phys. Rev. B* **34**, 5947 (1986).
35. X. Zhu, S. G. Louie, and M. L. Cohen, *Phys. Rev. Lett.* **63**, 2112 (1989).
36. H. Carstensen, R. Claessen, R. Manzke, and M. Skibowski, *Phys. Rev. B* **41**, 9880 (1990).
37. R. M. Feenstra and J. A. Stroscio, *J. Vac. Sci. Technol. B* **5**, 923 (1987).
38. C. B. Duke, *op. cit.*, p. 68.
39. C. B. Duke, *op. cit.*, p. 101.
40. C. B. Duke, *op. cit.*, p. 97.
41. J. A. Stroscio and R. M. Feenstra, *J. Vac. Sci. Technol. B* **6**, 1472 (1988).
42. L. J. Whitman, J. A. Stroscio, R. A. Dragoset, and R. J. Celotta, *Phys. Rev. Lett.* **66**, 1338 (1991).
43. L. J. Whitman, J. A. Stroscio, R. A. Dragoset, and R. J. Celotta, *Phys. Rev. B* **44**, 5951 (1991).
44. R. B. Dingle, *Philos. Mag.* **46**, 831 (1955).
45. J. A. Stroscio, P. N. First, R. A. Dragoset, and R. J. Celotta, unpublished.
46. R. J. Hamers, R. M. Tromp, and J. E. Demuth, *Phys. Rev. Lett.* **56**, 1972 (1986).
47. H. F. Hess, R. B. Robinson, and J. V. Waszczak, *Phys. Rev. Lett.* **64**, 2711 (1990).

48. P. N. First, J. A. Stroscio, R. A. Dragoset, D. T. Pierce, and R. J. Celotta, *Phys. Rev. Lett.* **63**, 1416 (1989).
49. R. M. Feenstra, p. 211 in *Scanning Tunneling Microscopy and Related Methods*, (R. J. Behm, N. Garcia, and H. Rohrer, eds.), Kluwer, Dordrecht, 1990.
50. P. Mårtensson and R. M. Feenstra, *Phys. Rev. B* **39**, 7744 (1989).
51. R. M. Feenstra, *J. Vac. Sci. Technol. B* **7**, 925 (1989).
52. C. K. Shih, R. M. Feenstra, and P. Mårtensson, *J. Vac. Sci. Technol. A* **8**, 3379 (1990).
53. C. K. Shih, R. M. Feenstra, and G. V. Chandrashekhar, *Phys. Rev. B* **43**, 7913 (1991).
54. M. Prietsch, A. Samsavar, and R. Ludeke, *Phys. Rev. B* **43**, 11850 (1991).
55. J. J. Boland, *Surf. Sci.* **244**, 1 (1991); J. J. Boland, private communication.
56. Ph. Avouris, I.-W. Lyo, and F. Bozso, *J. Vac. Sci. Technol. B* **9**, 424 (1991); Ph. Avouris, private communication.
57. R. S. Becker, G. S. Higashi, and A. J. Becker, *Phys. Rev. Lett.* **65**, 1917 (1990); R. S. Becker, private communication.

5. SEMICONDUCTOR SURFACES

5.1. Silicon

Russell Becker and Robert Wolkow

AT&T Bell Laboratories, Murray Hill, New Jersey

5.1.1 Introduction to Silicon

The semiconductor industry, along with a large segment of the condensed matter physics community, devotes the lion's share of its resources to element 14 in the periodic table, silicon. Silicon lies just below carbon in Group IV, and, together with germanium, is an intrinsic semiconductor with an indirect bandgap energy of 1.17 eV. Silicon principally forms tetrahedral sp3 bonds, taking the diamond crystal structure displayed by carbon. The stability of silicon oxide with respect to attack by common processing reagents such as water and methanol, coupled with the abundance of this element in the earth's crust, has made it pervasive in advanced technology—from semiconductor microelectronics to power rectifiers and solar cells. The vast effort devoted for more than a quarter century to silicon technology has resulted in its continued dominance over other materials, notably the III–Vs, in all areas of solid state electronics save optical devices.

The first semiconductor surface imaged with the scanning tunneling microscope (STM) was the 7×7 reconstruction of Si(111).[1] One of the properties of the elemental semiconductors is that their clean surfaces typically undergo a process termed "reconstruction," whereby the fundamental periodicity of the structure taken by the surface atoms is different from that of the underlying bulk material. This process is due to the covalent nature of their bonds; a simple bulk termination at the surface leaves a large number of unsatisfied (dangling) bonds that result in a large free energy. In order to mitigate the energy associated with these dangling bonds, the surface atoms rearrange themselves to diminish the dangling body density, reducing the free energy, generally at some cost in increasing the component of the free energy derived from surface stress.[2] The terminology "$m \times n$" refers to the two-dimensional Miller–indices needed to describe the surface unit cell (unit mesh) in terms of bulk lattice vectors. The surface superlattice, since it is larger than the bulk lattice, will produce fractional order beams observed in diffraction studies of these reconstructions. The experimental and theoretical basis for surface reconstructions is a field that has occupied many surface scientists for the last

METHODS OF EXPERIMENTAL PHYSICS
Vol. 27

ISBN 0-12-475972-6

30 years. Since the initial observations, using low energy electron diffraction (LEED), of the 7×7 on Si(111) surfaces in 1959 by Schlier and Farnsworth,[3] a large number of models have been tendered in explanation of the unusual symmetry displayed by this surface. The inability of the various studies to settle on a particular structure reflects the difficulty of solving the inverse scattering problem in three dimensions; it is not an easy task to determine the scattering potential given the diffraction intensities. This problem is exacerbated in LEED with the large contribution of multiple scattering to the diffracted beams, making detailed analysis of LEED I–V data formidable for even the simplest structures. While the tunneling images of the 7×7 shown by Binnig *et al.* did not unequivocally determine the detailed structure of this phase, they did reject a number of models and cleared the way for the dimer-adatom-stacking fault (DAS) model proposed by Takayanagi *et al.*[4] Indeed, it can be argued that this single observation by Rohrer and Binnig "made" the new field of scanning tunneling microscopy, bringing it forcibly to the attention of mainstream surface scientists world-wide. Even more significant is the effect the STM images have had on the outlook of surface studies, turning them from a collective, band structure-like perspective to a more atomistic viewpoint. As has been stated many times, "A picture is worth at least a thousand words."

The pioneering studies by the Zurich group were followed by further investigations with the STM from other laboratories, many of which were carried out on silicon surfaces under ultra-high vacuum (UHV) conditions. The first images of the fundamental defect at a surface, the atomic step, on Si(111)–7×7 surface were shown by Becker *et al.*[5] The strong correlation between the position of the step riser with respect to the 7×7 unit mesh evident in the tunneling images suggested that a close connection existed between these features. These workers further demonstrated that the partial stacking fault inherent in the 7×7 DAS structure may be directly imaged by measuring the spatial variation in the differential conductivity (dI/dV) at a bias condition near an unoccupied stacking fault state,[6] demonstrating that at least some of the electronic features of this reconstruction were spatially distinct. The 7×7 is not the only stable phase that may be found on Si(111). A reconstruction of 2×1 symmetry exists on the cleaved surface; the 7×7, by contrast, is found on surfaces prepared by *in situ* annealing to temperatures in excess of 875°C. The first tunneling images of the Si(111)–2×1 reconstruction were shown by Feenstra *et al.*,[7] where the surface features demonstrated consistency with the π-bonded chain model proposed by Pandey.[8] In this study, the initial STM measurement of a tunnel junction I–V characteristic disclosed a surface state energy gap of ~ 0.5 eV. Tromp *et al.*[9] then showed high-resolution tunneling images of the Si(001)–2×1 surface, confirming that the dimerization of neighboring surface atoms clearly accounted for the principal features of this crystal face. This work fueled, rather than quenched, the controversy over the detailed nature of Si–Si dimers on this surface, for the tunneling images showed both dimers that were symmetric and dimers that were asymmetric. Further exploration of the spatial configuration of electron states on the Si(111)–7 × 7 was carried out by Hamers *et al.*,[10] using a variation on the interrupted feedback method in Ref. 7, where selected parts of the junction I–V characteristic were displayed in surface plan view. These current images clearly suggested that some of the surface states were associated with the adatom features, while others were more characteristic of the partial stacking fault, in accordance with earlier work.[6] Plainly, the interest in characterizing all aspects of silicon surface features had driven the STM from an instrument with which only interesting pictures could be taken to a powerful tool for the determination of both surface geometry and surface electronic features.

Since this promising start, a large range of surface physics experiments have been performed on silicon surfaces. Metals such as Ag, Au, Al, Ga, B, K, Li, As, Sb, Sn, Pd, Cu, Ni, and In have been deposited and their surface properties examined. Gases such as H, O, Cl, NH_3, and NO have been reacted with clean silicon surfaces under a variety of conditions. Transitions involving both surface phases and step phases have been explored under a range of temperatures, and the effects of varying the surface stress has been examined on Si(001). Experiments examining the atomic details of homo- and hetero-epitaxy and surface diffusion on silicon are in the literature, as well as measurements on the relative formation energies of fundamental surface defects such as steps and kinks. The remainder of this chapter devotes itself to these briefly mentioned results.

5.1.2 Clean Surfaces

A fundamental understanding of the nature of the clean surfaces of silicon is necessary before examining more complex issues such as adsorbate interactions, diffusion, and epitaxy. Each of the low-index faces of silicon will be examined in turn, starting with the (111).

5.1.2.1 Si(111). As mentioned in passing in the introduction, the clean (111) face of silicon exhibits two common reconstructions, the 2 × 1 and 7 × 7, although others related to the 7 × 7 may exist under some sample preparation conditions.[11] The 7 × 7 is stable up to ~875°C whereupon it makes a transition to a "1 × 1" phase.[12,13] The 2 × 1 is formed by cleaving at or below room temperature, and is metastable with respect to the 7 × 7, reverting to the latter upon annealing at or beyond 380°C.[14] Both have been studied extensively, with a number of models advanced in support of the available data.

5.1.2.1.1 Si(111)–7 × 7. The history of the study of the clean Si(111)–7 × 7 surface is almost a history of surface science itself, starting in 1959 with the first LEED observations of this and other low-index semiconductor surfaces[3] and continuing nearly to the present day. In fact, the currently accepted structural model for this surface was not proposed until 1985 by Takayanagi *et al.*[4] Much of the reason for the difficulty in arriving at a satisfactory model for the 7 × 7 is due to the large size of the unit mesh; with 49 surface atoms the structure is almost inaccessible to detailed theoretical analysis, rendering exhaustive comparison of experimental data with calculations untenable. Over the years, as more surface science techniques were brought to bear on this problem, a large number of structural models were proposed to explain the experimental results.

Geometrical models for the Si(111)–7 × 7 may be simply divided into two categories: those that consist of a periodic distortion or buckling of an otherwise 1 × 1 surface layer, and those that involve placing an array of entities of some kind on a silicon substrate. One buckling distortion model was advanced by Chadi *et al.*,[15] with a theoretical comparison to LEED I–V data giving support.[16,17,18]. In contrast to the buckling models, which do not involve a large amount of mass transport, are the various entity models. In his early studies, Lander *et al.*[19] proposed an array of 13 vacancies of local 2 × 2 order in the 49 surface-site 7 × 7 unit mesh. Later, Harrison[20] proposed an array of adatoms (silicon atoms bonded to three surface atoms) in the vacancy sites of Lander's model. This was expanded to consider adatom clusters ("milk stools") in these same vacancy sites.[21] While all of these proposals had merits, none was satisfactory in explaining all of the experimental results on the 7 × 7, leaving this fundamental surface phase unsolved as surface science entered the 1980s. This was the state of the Si(111)–7 × 7 surface structure when Binnig and Rohrer published the first real-space images of this enigmatic surface in 1983.[1]

Figure 1 shows a tunneling image of the Si(111)–7 × 7.[1] In this figure the STM topographic data is rendered as a gray scale, with lighter shades representing higher points, and darker shades lower points. The total relief, from light to dark, is approximately 1 Å. A number of remarkable features are immediately apparent without regard to theoretical models. First, the surface is tiled by a network of entities, apparently protrusions, with a nominal spacing of two lattice vectors (2 × 2). Second, a further network of deep holes connect the corners of the 7 × 7 unit mesh. Third, one-half of the unit mesh is apparently slightly higher (0.2 Å) than the other, breaking the symmetry of the 7 × 7 along the short diagonal. From these observations a pure buckling structure may be ruled out, while the presence of the 12 protrusions in each 7 × 7 mesh lend strong support to some sort of entity model. The existence of the deep depressions at the corners of the unit mesh

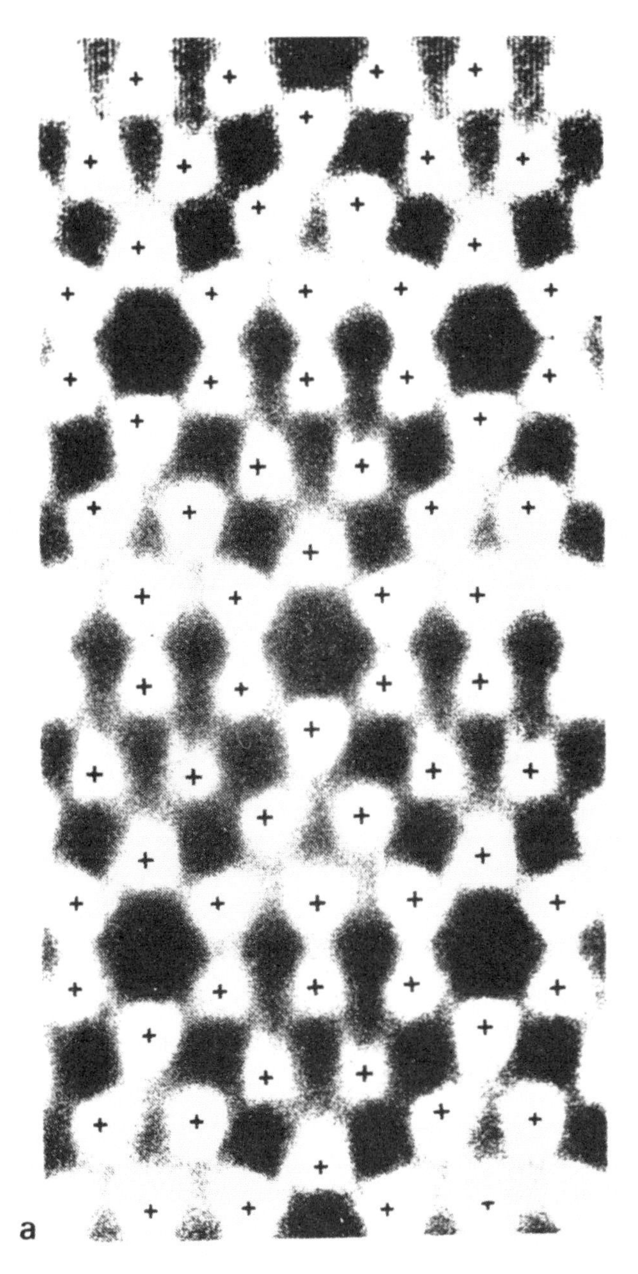

FIG. 1. Plan view tunneling image of the Si(111)–7 × 7 surface. The data is rendered as a gray scale, with light representing high areas and dark keyed to areas of lower height, with a total range of ~ 1Å. Taken from Figure 3 of Ref. 1.

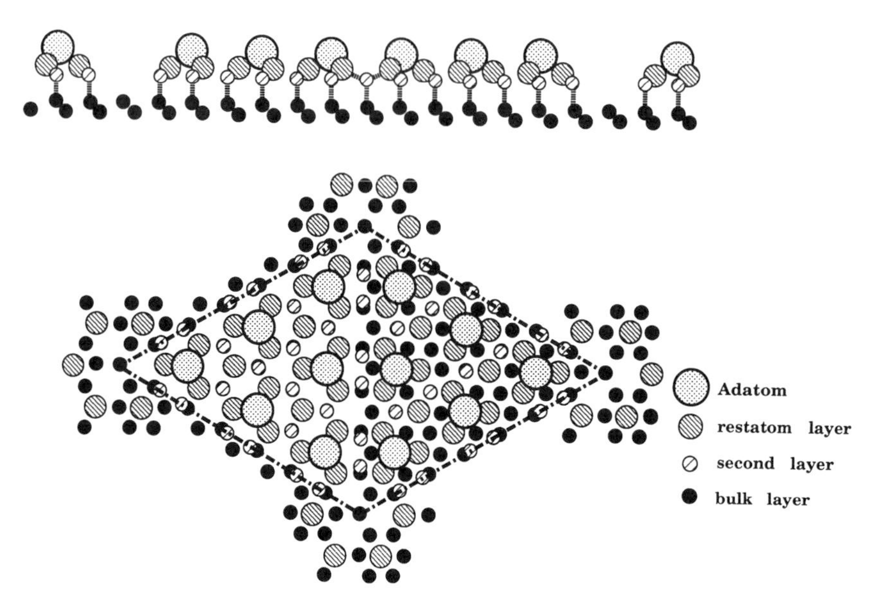

FIG. 2. Schematic of the Si(111)–7 × 7 DAS structure. A unit mesh is outlined.

suggest that the entitites may be a combination of vacancies *and* adatoms, together numbering 13, as in the early suggestion by Lander *et al.*[11] The observation that the two halves of the unit mesh have somewhat different apparent heights introduces a new feature into 7 × 7 structural models, which the previous proposals failed to address.

A number of new models for the 7 × 7 structure, taking into account the STM observations, were quickly introduced, combining ideas involving stacking faults with some of the earlier entity models to explain both these observations, and new data from ion scattering experiments[22] and transmission electron diffraction.[4] The outcome of all this was the dimer-adatom-stacking fault (DAS) model of Takayanagi *et al.*[4] Figure 2 shows a schematic view of the DAS structure in the 7 × 7 symmetry. One of the features of the DAS model is that it involves considerable rearrangement of both the surface and top double-layer silicon atoms. The protrusions readily apparent in the STM image in Fig. 1 are the 12 top-layer adatoms in the DAS structure. These are each bonded to three atoms in the surface double-layer at what are referred to as T_4 sites. A double-layer on the Si(111) surface has two kinds of independent three-fold sites; those characterized by a hollow with no atom directly below (H_3), and the neighboring three-fold site possessing an atom directly below (giving effectively a fourth atom to coordinate with, hence the subscript four). The local $\sqrt{3}$ symmetry across the short diagonal is accommodated by incorporating a partial stacking fault into the top double-layer. The presence of this partial stacking fault also accounts for the asymmetric appearance of the mesh halves in the STM images because the apparently higher side incorporates the stacking fault. The edges of the two mesh halves are then zipped together with three pairs of dimers in the top double-layer on each of the mesh sides giving 18 pairs of dimers in all.

The images published by Binnig *et al.* were supplemented in 1985 by tunneling images of Si(111)–7 × 7 atomic steps,[5] and are shown in Fig. 3 in plan view as a gray scale. Studying this image reveals additional information about the 7 × 7 structure. First, the step risers are naturally incorporated into the reconstruction by placing the large depressions (corner holes) at the corners of the unit mesh at the step edge, giving a 7th-order periodicity to the step riser. This feature has been confirmed in further studies of atomic steps on equilibrium Si(111)–7 × 7 surfaces.[23] The edge of the lower terrace also terminates at the short diagonal of the 7 × 7 mesh (although this may be violated under certain annealing/growing conditions, notably fast cooling[24]). Together, these observations imply that parallel steps on the equilibrium surface are separated by an integral number of 7 × 7 meshes. Further, the persistence of the 7 × 7 right up to, and including the step riser, indicates that the reconstruction is dependent on short-range, local interactions. In the context of the DAS structural model, this position for the step riser minimizes the number of dangling bonds exposed by the presence of the step. The asymmetry between the heights of the two halves of the unit mesh is also evident, although in the opposite sense from that reported in the earlier work by Binnig *et al.* In retrospect, this apparent disagreement is a consequence of the fact that the images obtained by the Zurich group were taken in occupied sample states by tunneling from the sample to the tip, while the Bell Laboratories group imaged the 7 × 7 in unoccupied sample states by tunneling from the tip to the sample. As became evident in later work, the apparent height is a stronger function of the spatial distribution of electronic states than the actual geometric position of the atoms; the two sets of observations were simply accessing electronic states with opposing phase.

Spatial mapping of the distribution of electronic states with the STM was demonstrated by Becker *et al.*[6] using the procedure of measuring the differential conductivity, dI/dV, at the applied tunneling bias, V_{bias}, as a function of position. In the context of the basic tunneling theory, the current I_t when tunneling from the sample to the tip is a function of both the tunneling barrier T and the local density-of-states ρ:

$$I_t(V_{\text{bias}}, \boldsymbol{r}) = \int_0^{V_{\text{bias}}} T(V)\rho(\text{eV}, \boldsymbol{r})\, dV .$$

For small changes in V_{bias} the tunneling barrier T is nearly a constant in this

FIG. 3. Tunneling image of atomic steps on the Si(111)–7 × 7 surface. The gray scale is keyed to apparent height. A unit mesh is outlined. Taken from Figure 2 of Ref. 5.

approximation, so that

$$\frac{dI}{dV}(V_{\text{bias}}, \boldsymbol{r}) \propto \rho(\text{eV}_{\text{bias}}, \boldsymbol{r}).$$

The spatial map of dI/dV may then be compared to the simultaneously acquired tunneling image in order to locate density-of-states (DOS) features in the unit mesh. In practice, the measurement of dI/dV is accomplished by applying a small (0.1 V) dither signal to the junction bias at a frequency

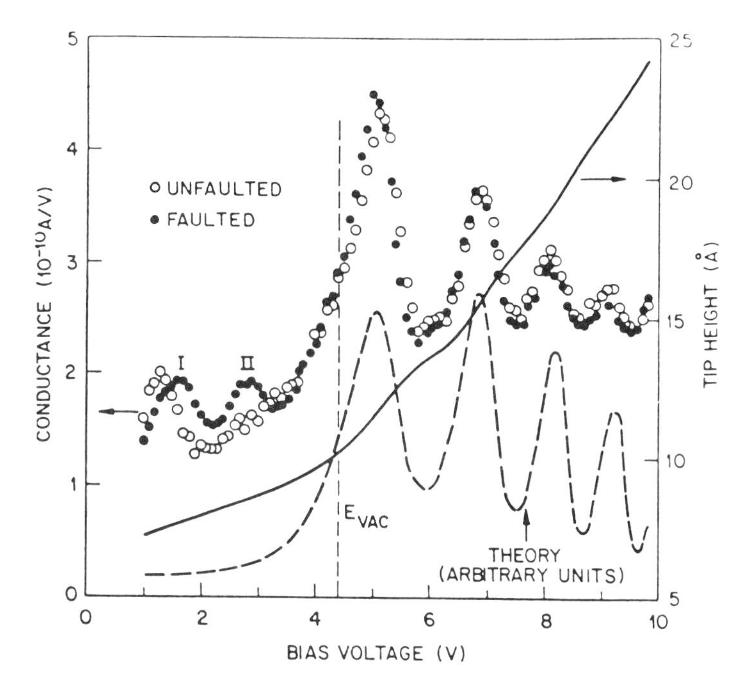

FIG. 4. Differential conductivity (dI/dV) spectra of the Si(111)-7 × 7 surface. Filled circles represent data taken over the faulted half of the unit mesh. The vacuum level is indicated: peaks above this point are due to barrier resonances in the vacuum region between the tip and sample, while those below are DOS features. Taken from Figure 2 of Ref. 6.

(~4 KHz) which is beyond the response bandwidth of the STM feedback loop. The differential conductivity is therefore measured at constant demanded tunneling current. In this regard, the height of the tip above the sample will vary during measurement as the feedback loop servos the tip in order to maintain the demanded current. Figure 4 shows differential conductance spectra for the two halves of the 7 × 7 mesh measured at 1 nA tunneling current.[6] The large oscillations above the vacuum level are due to barrier resonances (electron standing waves) formed in the vacuum region between the tip and the sample,[25] and are nearly identical in both data sets. The features below the vacuum level are due to the density of unoccupied electron states in the sample, and are not identical in both halves of the unit mesh; in particular the peak labeled II at 2.8 V is present on the faulted half of the mesh and nonexistent on the unfaulted half. The peak labeled I has been shifted in energy position by nearly 0.3 eV between the two halves of the unit mesh. When the STM bias is set to −2.0 V, just at the position where peak I shifts, a map of the spatial localization of this feature may be recorded and is shown in Fig. 5.[6] Here we clearly see that this DOS feature varies spatially

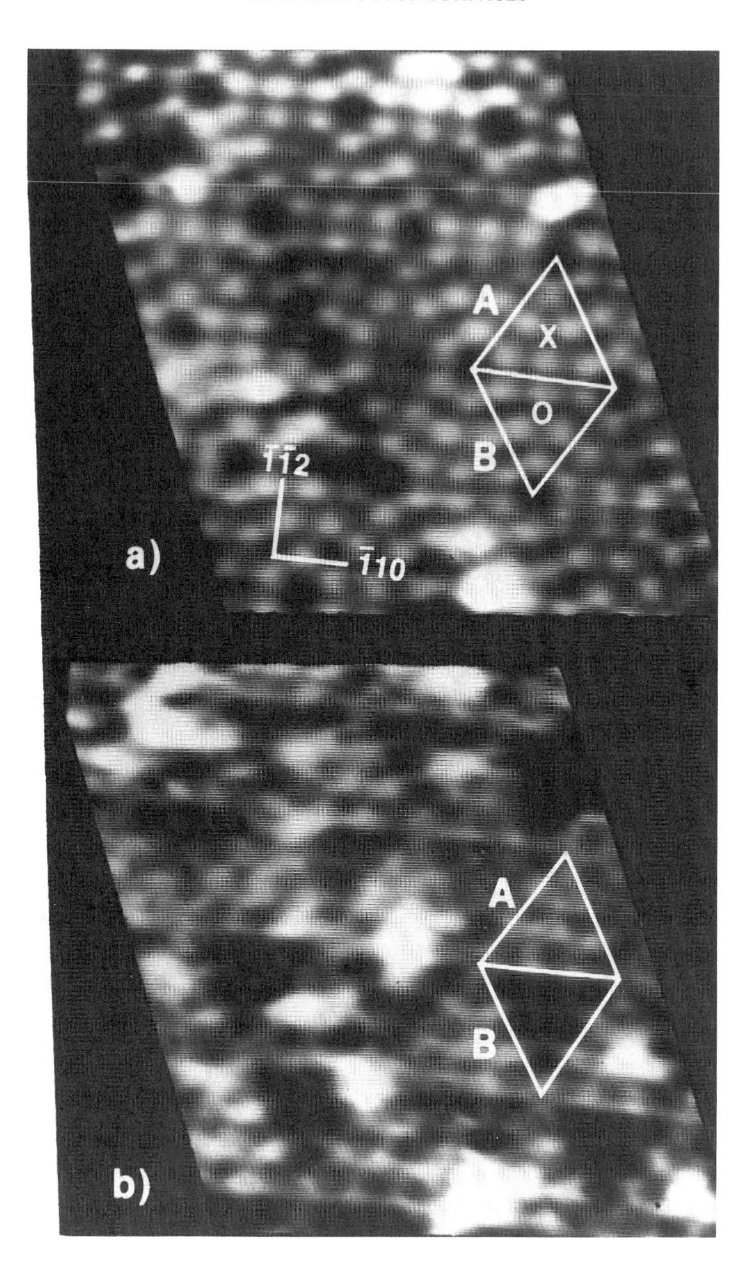

FIG. 5. Topographic (a) and differential conductivity (b) images of the Si(111)–7 × 7 surface. A unit mesh is shown in the topograph, with the same area outlined in the differential conductivity image. [A] indicates the unfaulted half of the unit mesh. Taken from Figure 1 of Ref. 6.

over the 7 × 7 surface with the presence and absence of the stacking fault on one side of the unit mesh that is inherent in the DAS structure. These authors presented a slab calculation demonstrating that the presence of a stacking fault could conceivably result in such features in the DOS. More recently, Kubby *et al.*[26] have extended these spatial measurements of dI/dV on the Si(111)–7 × 7 to a wider range of bias conditions and observe a sequence of phase reversals in the orientation of the contrast in the conductance images. They conclude that the DOS feature due to the partial stacking fault may be described as a layer resonance between the sample surface, taken to be the adatom layer, and the last bulk-like layer of the crystal. The presence of the partial stacking fault modulates this depth, Z_0, with a 7th order periodicity, giving rise to the contrast in the conductance images. Figure 6 shows this sequence of phase reversals between mesh halves, and the interplay between these and the appearance of asymmetry in the associated topographic tunneling images.[26] Here, I represents the tunneling images, while II denotes the simultaneous conductance image. Figures (a) through (h) are data sets representing V_{bias} ranging from 0.25 V up to 5.0 V. One may see that the orientation of the bright features in the conductance image reverses phase as a resonance condition is passed. Figure 7 shows a schematic of the potential energy and resonance conditions for this set of measurements.[26] These authors conclude that the state imaged by Becker *et al.*[6] arose from the $n = 2$ layer resonance.

Another technique for spatially imaging electronic features with the STM was introduced by Hamers *et al.*[10] in 1986 (see Chapter 4 for a detailed discussion of the various tunneling spectroscopic methods). This method consists of interupting the feedback loop of the STM, quickly ramping the applied bias, and recording the resulting tunneling current as a function of both V_{bias} and $\boldsymbol{r}$. The total conductance, I_t/V_{bias}, at various positions on the surface may be plotted and compared to other energy sensitive probes such as valance-band photoemission. Alternatively, spatial maps of the *current differences* bracketing regions of interest in the conductance may be displayed and compared with the simultaneously acquired tunneling image. For this reason, Hamers *et al.* christened this method current imaging tunneling spectroscopy (CITS). Figure 8 shows a series of CITS images of a 7 × 7 unit mesh.[10] In the upper panel, centered around −0.35 V bias, 12 maxima are visible with more intensity arising from the faulted half of the unit mesh. These 12 maxima are associated with the dangling bonds on the 12 adatoms of the DAS structure. In the center panel, at −0.8 V bias, six maxima are visible, three on each mesh half, and an additional maxima located in the center of each corner hole. These are associated with the rest atom dangling bonds located between each of the adatoms sitting on the top double-layer, and the dangling bond in the deep corner holes of the DAS model. The bottom panel shows the tunneling current distribution at −1.75 V bias, and

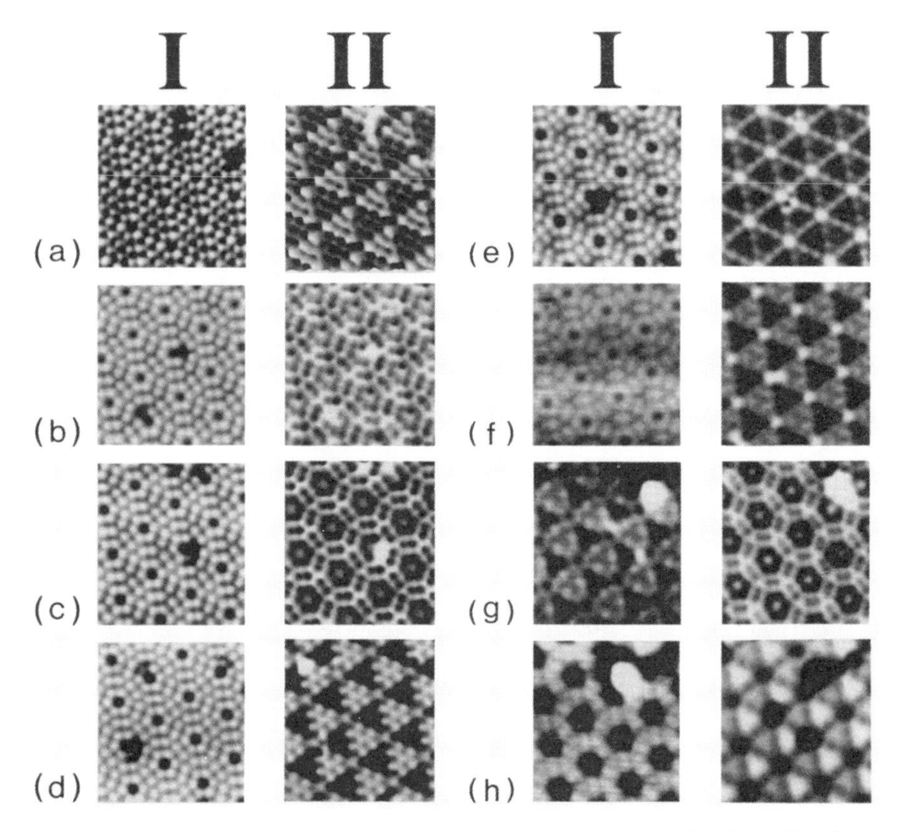

FIG. 6. Topograph (I) and differential conductivity (II) images of the Si(111)–7 × 7 surface. A range of bias conditions is shown, with (a) corresponding to 0.25 V, (b) 1.0 V, (c) 1.5 V, (d) 2.0 V, (e) 2.5 V, (f) 3.5 V, (g) 4.5 V, and (h) 5.0 V, all tunneling from the tip to the sample. Taken from Figure 1 of Ref. 26.

has been suggested as arising from the backbond electron states in the surface double-layer. These authors go on to point out that the states seen in this set of current images may be analogous to the S_1, S_2, and S_3 states identified in ultraviolet photoemission spectroscopy (UPS) studies of the 7 × 7 surface.[27]

While the current images shown in Fig. 8 are quite remarkable, and the agreement between these STM results, the Takayanagi DAS structure, and the UPS data is good, this agreement is somewhat fortuitous due to the nature of the electron states that are imaged in this work. Angle-resolved UPS studies[28] show that the bands corresponding to the S_1 and S_2 states are quite flat, indicating that these features are highly localized and atomic-like (as would be expected for isolated dangling bonds). Under these conditions, coupled with the preponderance for the STM to access electron states with

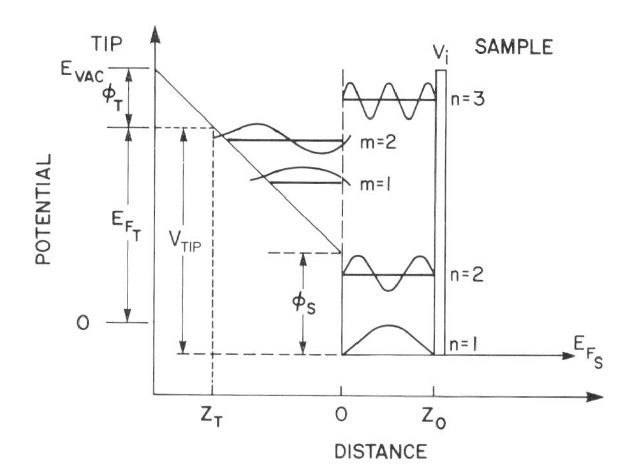

FIG. 7. Schematic showing the potential energy diagram for both the vacuum barrier resonances and the layer resonances in Figure 6. Taken from Figure 2 of Ref. 26.

$\mathbf{k}\| = 0$, current difference imaging may be expected to produce quite accurate results. However, it has been demonstrated[29,30] that a variety of conditions can bring about erroneous results since the magnitude of the tunneling current can be more sensitive to the tip–sample geometry than to changes in the local DOS.

A more general purpose technique for visualizing the relative spatial relationship between various surface electron states, especially those on either side of the Fermi level, is to employ voltage-dependent imaging.[30,31] In this method, multiple junction bias conditions are used at each point in the tunneling image, either by switching between a set of biases at each point, allowing adequate time for the feedback loop to reach equilibrium, or to employ one bias when scanning from left to right, and another when scanning from right to left, with any residual hysterisis between the image pairs removed by later display processors. *Tunneling images*, since they integrate all of the current between the source Fermi level and the drain Fermi level, are less susceptible to uncontrolled variations in the tip–sample geometry and tip electronic structure than the analogous *current difference images*, which derive contrast from only a small portion of the tunneling current.

Figure 9 shows a pair of multiple-bias images for the Si(111)–7 × 7 surface.[30] Here, the upper image (a) is taken at −2.0 V tip bias (tunneling into unoccupied sample states), while the lower image (b) is taken at +2.0 V bias (tunneling out of filled sample states). At this bias, the empty state image shows the 12 adatoms of the 7 × 7 unit mesh at essentially equal apparent height, with no features to distinguish one side of the unit mesh from the other. The filled state image appears much as the −0.35 V CITS image of

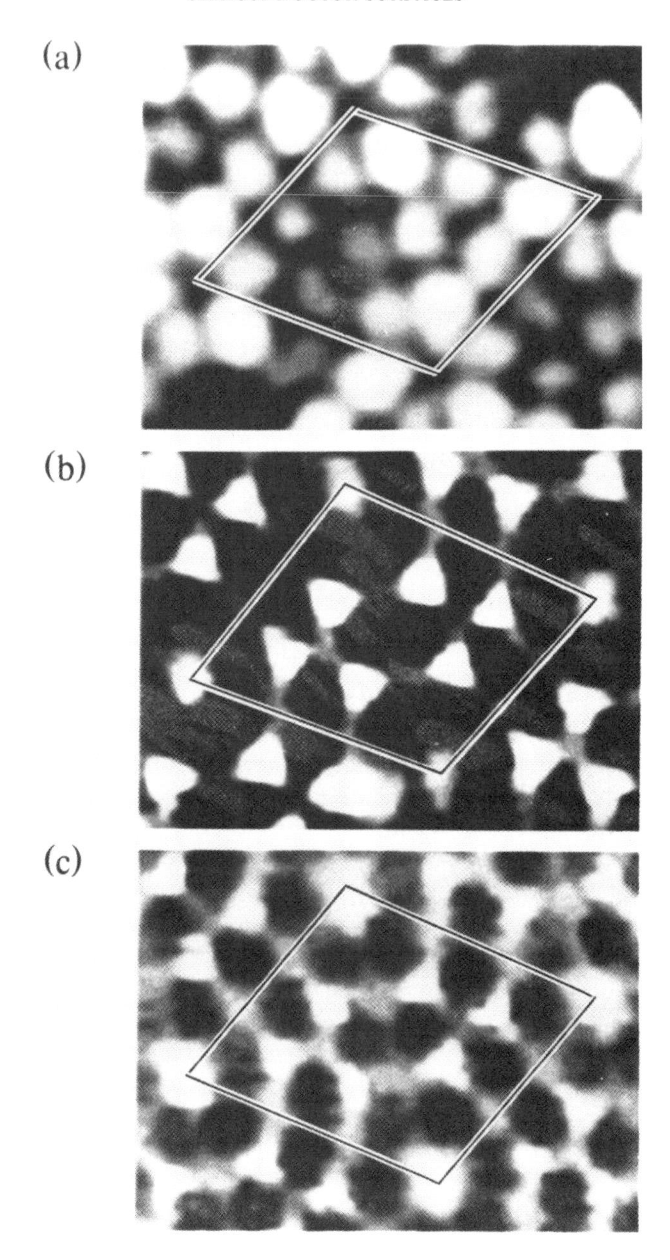

FIG. 8. CITS images of the Si(111)–7 × 7 surface. (a) corresponds to −0.35 V bias, (b) −0.8 V bias, and (c) −1.75 V bias, all tunneling from occupied sample states to the tip. Bright indicates more current, while dark corresponds to less tunneling current. Taken from Figure 3 of Ref. 10.

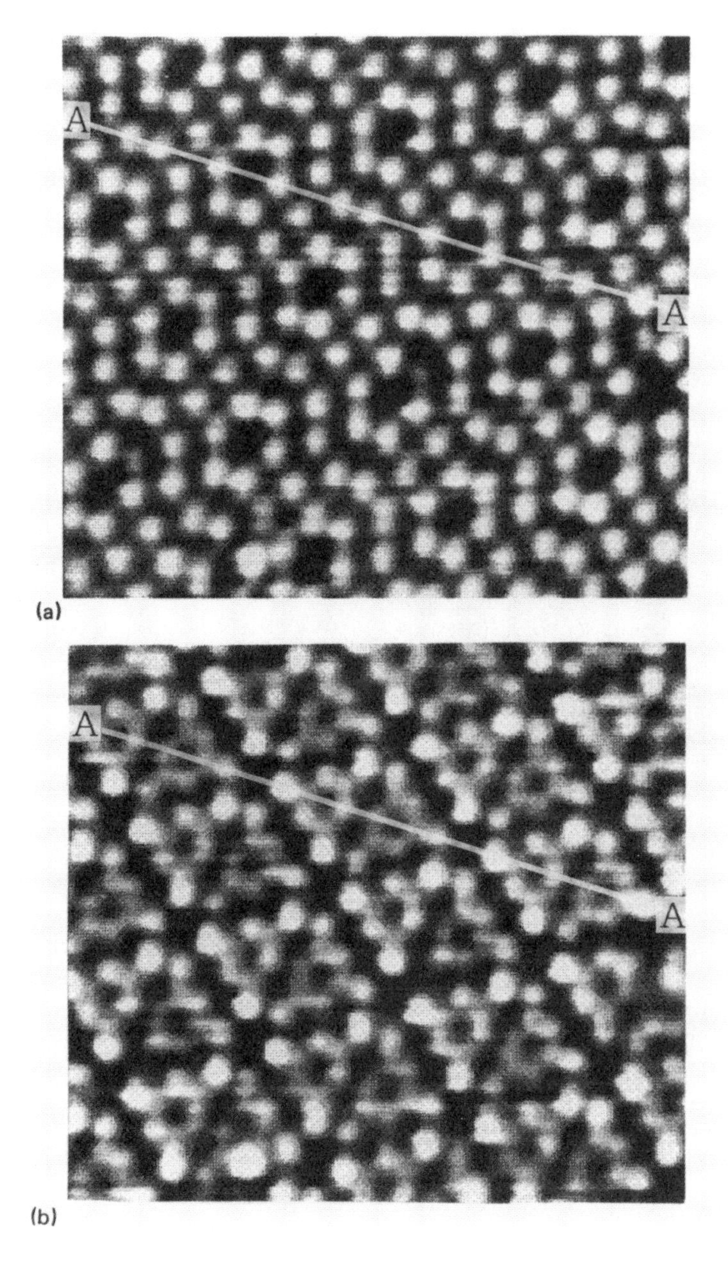

FIG. 9. Dual polarity tunneling images of the Si(111)–7 × 7 surface. The upper image (a) is taken at −2.0 V bias tunneling from the tip to the unoccupied sample states, while the lower image (b) is at ±2.0 V bias tunneling from occupied sample states to the tip. Taken from Figure 1 of Ref. 30.

Fig. 8, with the six atadoms in the faulted half of the mesh much higher than the six in the unfaulted half of the mesh. Further, the adatoms located at the corner positions are higher than those at the middle positions, a feature which is common to adatoms in the DAS family of structures (5×5, 7×7, 9×9) as reported in Ref. 31. In fact, the 9×9 structure is the first to contain an adatom not adjacent to the dimer-domain walls of the partial stacking fault. Images of this structure[31] show this adatom much lower in apparent height (~ 0.5 Å) than the adatoms at the edge of the corner hole. This similarity of the filled state STM image taken at -2.0 V sample bias and the -0.35 V CITS image is due to the fact that when tunneling proceeds from the sample to the tip, the sample states lying closest to the sample Fermi level, E_F, constitute the majority of the tunneling image. This unequal contribution is simply because the tunneling barrier is greater for states further removed from E_F, increasing by nearly an order of magnitude for each electron volt. Since the dangling bond adatom states lie closest to E_F, in fact forming a partially occupied band pinning the surface Fermi level, they are quite prominent in the tunneling image. Careful inspection also reveals the presence of saddle points in the filled state images at the position expected for the rest atoms in the DAS structure, a feature not seen in the unoccupied state image. This hints at the presence of a state at these spatial positions lying within 2 eV of E_F, likely the UPS S_2 state at -0.8 eV. The relative lack of height difference between the adatom species in the unoccupied state images indicates that there is little difference in unoccupied DOS between these sites, consistent with a fractionally occupied band with considerably less than 50% filling (this renders the relative variations in unoccupation between adatoms less discernible). The lack of a saddle-point feature similar to those in the occupied state images suggests that this feature is completely filled, and not partially filled as the adatom band.

These features are easily interpreted in the context of first-principles pseudo-potential calculations on related Si(111) adatom structures.[32] Here, the complicated 7×7 DAS structure is simplified to a set of adatoms on T_4 sites with adjacent rest atom dangling bonds. These calculations find a partially occupied band due to adatom dangling bonds lying at, and pinning, the Fermi level, and a deeper band at 1.5—2.0 eV below E_F, corresponding to substrate dangling bonds coupling to adatom px and py orbitals. The band at -0.8 eV is due to a completely filled restatom dangling bond band filled by charge transfer from the higher-lying adatom dangling bond band. Since the corner adatoms are adjacent to only one rest atom, they are less efficient at transfer than the middle adatoms, which are adjacent to two. In the 9×9 geometry, the interior adatom is adjacent to three restatoms, giving almost complete charge transfer, producing the markedly lowered apparent height. In effect, the interior adatom is behaving as Ge adatoms do in the $c2 \times 8$

structure found on the Ge(111) surface (see Section 5.2 for further details on germanium). Both theoretically and experimentally, the surface bands of the 7×7 are found to have small dispersion, with highly localized atomic-like states responsible for the large contrast seen in the STM images. It is clear that the presence of the partial stacking fault in the DAS structure breaks the surface adatom termination into small, relatively independent 2×2 islands, resulting in the localization of the various dangling bond states.

One conclusion that may be drawn from the various STM studies of the 7×7 is that the admixture of geometric and electronic features in the tunneling images complicates their interpretation. In fact, more than one study has concluded that electronic features dominate the composition of the images for elemental semiconductors. In general, the high points in the STM images may be traced to dangling bonds on the surface structure for both occupied and unoccupied state images.

5.1.2.1.2 Si(111)-2 × 1. The native reconstruction on the cleaved Si(111) surface, the 2×1, has nearly as long and checkered a history as its more famous cousin, the 7×7. Nearly as many models have been proposed for this structure as for the 7×7, with most of them involving some form of buckling or other minor rearrangement of the atomic positions. The original buckling model, consisting of alternating rows of high and low silicon atoms, was proposed by Haneman in 1961[33] after the ground-breaking LEED work by Schlier and Farnsworth.[3] After more thorough study of the cleaved and annealed (111) surfaces of both silicon and germanium, Lander *et al.* proposed a double-bond model.[11] Seiwatz suggested a conjugated chain model,[34] Selloni and Tosatti a pairing model,[35] and Harrison proposed an adatom model.[36]

The difficulty with all of these models was their inability to adequately account for the experimental data. Calculations on the various buckling models indicate a rather large amount of electron transfer between the raised and lowered atoms, giving a higher degree of ionicity than was observed using core level photoemission. Even more, the large separation on the surface between the buckled atoms precluded the strong optical absorption at 0.45 eV seen on this surface, and failed to account for its anisotropy. These considerations led Pandey to propose, in 1981, a new structure for the 2×1.[8] This geometry, referred to as the π-bonded chain, consists of a rearrangement of the bonds between the top double-layer and the substrate, giving a succession of seven- and five-membered rings along the surface $[\bar{2}11]$, resulting in alternating chains of surface silicon atoms, with the raised chains characterized by nearest neighbor π-bonds. Figure 10 shows a schematic of this model, with the seven- and five-membered rings, surface chains, and π-bonds denoted, along with a 1×1 surface for comparison. When viewed from the side along a $[0\bar{1}1]$, it is clear that the combination of seven and five member

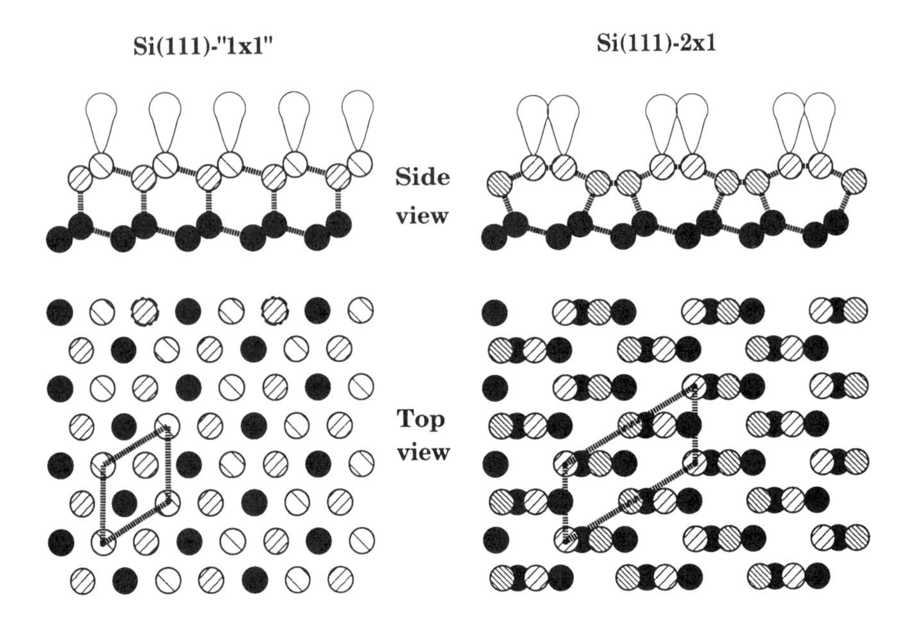

FIG. 10. Atomic models of both the unreconstructed Si(111)–1 × 1 surface and the Si(111)–2 × 1 π-bonded chain surface. Bulk atoms are shown as filled circles with the dangling bonds indicated.

rings may be considered a rebonding of the surface double-layer to the substrate. In this manner, the symmetry of the surface is lowered, the dangling bonds are brought together and aligned, lowering the free energy through the formation of the π-bonds between neighboring atoms on the raised chain.

The first STM images of the 2 × 1 structure were published by Feenstra *et al.*[7] In this work they reported the observation of row-like features that were consistent with the Pandey model, but incompatible with many of the other suggested structures, notably the buckling models. Also, no evidence for adatom-like features, such as those evident on the Si(111)–7 × 7, was seen. A tunnel junction I–V characteristic was presented with a visible energy gap of ~ 0.5 eV between filled and empty state tails. The authors further demonstrated that disorder features found on the surface were quite prominent when the tunneling bias was reduced to the edges of the energy gap, providing strong evidence that disorder results in the propagation of midgap states (most likely by the introduction of unsatisfied dangling bonds).

The details of the electronic structure of the Si(111)–2 × 1 were analyzed in a series of STM studies by these same authors, leaving little doubt that the π-bonded chain structure accounted for the 2 × 1 symmetry.[37,38] Figures 9(a) and (b) in Chapter 4 show a set of dual-polarity tunneling images of the 2 × 1 surface.[37] Several notable features are quite apparent after studying the data in this figure. First, the row-like features reported in the earlier study[7] are now resolved into individual high points, separated by one lattice spacing (3.84 Å) along $[0\bar{1}1]$, *in both polarities*. Second, a phase shift of 180° along $[0\bar{1}1]$ exists between the image taken tunneling from the tip into unoccupied sample states, and the simultaneously acquired image taken tunneling from the occupied sample states back to the tip. Third, a small phase shift of $\sim 35°$ along $\langle\bar{2}11\rangle$ occurs between the two images. Under the auspices of the Pandey model, the dangling bonds along the raised chain are distorted to a nearly parallel orientation, forming the π-bonds. The degeneracy of neighboring atoms is broken by the inequivalence of their bonding to the substrate layer; further, a 180° phase shift exists between the π states at the bottom of the surface energy gap and the π^* states at the top. Together, these result in the bonding states strongly localized to one set of atoms in a chain, with the antibonding states localized to the alternate set. This accounts for the misregistration along $[0\bar{1}1]$, apparent in the STM image pair. In effect, between the two images, every atom in the raised chain is visible, while the lowered chains are not imaged at all. The small phase shift perpendicular to the chains, along $[\bar{2}11]$ is consistent with the zig-zag nature of the raised chain, seen in the schematic in Fig. 10. One of the puzzles about the 2×1 surface, the large optical absorption at 0.45 eV, is answered by the large phase shift along $[\bar{1}10]$. This nearly complete separation between filled and empty surface states, together with the close proximity (one lattice constant), account for the efficiency of the process, while the anisotropy in the phase shifts in the tunneling images duplicate the anisotropy in the optical activity.

Interrupted feedback loop tunneling spectroscopy data provides additional information on the nature of the 2×1 surface phase. Figure 17(b), in Chapter 4 shows several sets of tunnel junction I–V characteristics plotted as the normalized conductivity.[37] The different symbols in the figure denote sets of data acquired at several tip–sample separations; the very good superposition of them is a measure of the quality of the normalization process. An energy gap of ~ 0.5 eV is visible between filled and empty surface states, with two other features apparent at higher values of the tunneling bias. These agree quite well with a simple calculation of the DOS resulting from the π and π^* bands, which is depicted at the bottom of the figure.

As was the case for the 7×7, imaging of atomic steps on the vacuum-cleaved Si(111) surface is useful both as a further verification of the structural model for the flat terraces, and to explore the existence of various step geometries that may be displayed. High resolution STM images of double-layer steps on the Si(111)-2×1 have been obtained by Feenstra and Stroscio.[39] For the case where the step riser is parallel to the chain direction on both upper and lower terraces, these authors find a simple reconstruction

Si(111)-2x1 step structure

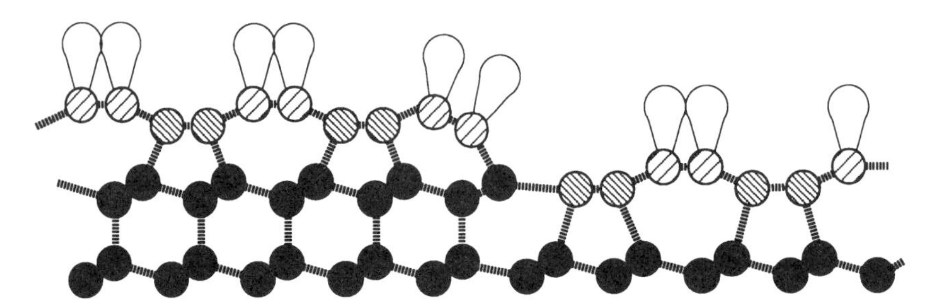

FIG. 11. Atomic model of reconstructed steps on the Si(111)–2 × 1 surface. Taken from Ref. 39.

which is a natural evolution of the basic Pandey model. Figure 11 shows a schematic section view of this configuration viewed along a surface $[\bar{1}10]$. In this model, the agreement between the high points in the tunneling image and the position of the dangling bonds is quite good, allowing the authors to reject other basic configurations for the step riser.

5.1.2.2 Si(001). The Si(001) surface displays a reconstruction that is a relatively simple modification of the bulk terminated structure, in which each surface atom has two dangling bonds and is bonded to two subsurface atoms. A small displacement, involving no bond breaking, allows surface atoms to pair-up to form dimers, thereby attaining a stabilized structure in which the dangling bonds are reduced from two per atom to one. In their pioneering LEED studies, Schlier and Farnsworth detected half-integral beams which they understood could not arise from surface atoms in a bulk-like configuration.[3] They proposed that the observed 2 × 1 surface mesh was consistent with a structure created when adjacent rows of surface atoms moved together in a bonding interaction. A diagram of the Si(001) surface is shown in Fig. 12.

Surface atom pairs, or *dimers*, were not widely accepted for many years because studies subsequent to those of Schlier and Farnsworth revealed higher order diffraction spots, of an intensity and sharpness that was highly dependent on sample treatment.[40,41,42] It became apparent that a simple symmetric dimer structure could not account for all observations. Various vacancy[41,43] and conjugated chain models[34] were proposed. When electronic structure calculations by Appelbaum *et al.* on the dimer and vacancy models,[44] and by Kerker *et al.* on the conjugated chain model,[45] became available for comparison with the photoemission data of Rowe,[46] the surface dimer model was found to give the best agreement. The chain model was

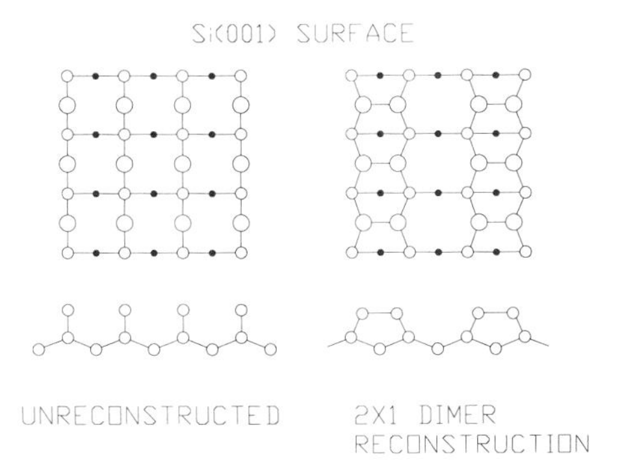

FIG. 12. Schematic diagram the Si(001) reconstructed surface.

supported by early LEED analyses that only considered displacements in one or two surface layers.[47,48] However, after Appelbaum and Hamann showed that substantial atomic displacements were expected in the first five layers at least,[49] more elaborate LEED analyses were performed and the dimer model was then found to provide the best fit.[50]

Two decades after the first LEED study, dimers were generally recognized to be the principal feature of the reconstructed Si(001) surface, yet some dissatisfaction with the dimer model lingered because of two key inconsistencies. These were: (1) the dimer model could only account for integral and half integral diffraction, yet 1/4 order beams were sometimes observed, and (2) the half-filled band derived from the remaining dangling bonds implies a metallic surface, but, angle-resolved photoemission measurements by Himpsel and Eastman indicated that the surface was not metallic.[51]

Because of these difficulties the idea of asymmetric dimers was pursued. It was recognized that symmetric dimers should be subject to a Peierls distortion which would result in a stabilized asymmetric structure. The buckling of dimers could induce the degenerate dangling bonds of a symmetric dimer to split, causing a gap to open in the band associated with those states, in correspondence with the experimental observation that the surface is non-metallic. In addition, if arranged in a 4×2 mesh, buckled dimers could account for the observation of 1/4 order diffraction beams. Chadi explored these possibilities with an empirical tight-binding total energy calculation of the Si(001) reconstruction.[52] He found that the symmetric dimer was indeed unstable with respect to the buckled structure. Further, it was found that the geometric buckling is coupled to an electronic asymmetry, which results in a relatively full dangling bond on the up atom, and depleted occupation of the dangling bond on the low atom. The calculated band structure showed a gap of 0.6 eV. These results, though very compelling, did have a weak point: the intra-atomic Coulomb repulsion (U) was not explicitly accounted for but only estimated, and since a larger value of U could lead to a preference for symmetric dimers, debate continued. Yin and Cohen performed more thorough self-consistent pseudo-potential calculations and also found in favor of the buckled dimer,[53] but Pandey, also using the psuedo-potential method, subsequently pointed out that, because of sensitivity to computational parameters, the uncertainty in the method did not allow either the symmetric or the asymmetric structure to be identified with certainty as more stable.[54] Chadi's results were followed by new experimental efforts. Medium and low energy ion scattering measurements were suggestive of buckled dimers.[55,56,57] Complex dynamical LEED analyses were performed but did not result in compelling evidence of the buckled dimer model.[58,59,60,61]

5.1.2.2.1 Room Temperature Imaging of Si(001)–2 × 1. The emergence of the STM led to a turning point in the long effort to understand the reconstructed Si(001) surface. Tromp, Hamers, and Demuth,[9,62,63] reported occupied-state images that revealed the dimers as oblong protrusions with the expected periodicity, and with corrugations along and between dimer rows of approximately 0.5 Å and 0.1 Å respectively, as shown in Fig. 13. These images, while leaving a number of issues unresolved, clearly established a number of important points. Foremost among these was verification of the dimer model, other models such as the conjugated chain and missing row simply did not match the observed topographic features. Interestingly, both symmetric and buckled dimers were observed, as were very numerous defects. The buckled dimers displayed local 2 × 2 and 4 × 2 symmetry. Further, it appeared that the buckling of dimers was associated with defects. These observations led to a plausible explanation for diffraction measurements that showed predominant 2 × 1 symmetry, and, depending on sample preparation, variable degrees of 4 × 2 order. Higher-order diffraction should be associated with increased contamination.

Occupied state images of Si(001) fortuitously correspond to one's imagined view of the reconstructed surface. Unoccupied state images, however, are distinctly different than those of the filled states, and are more difficult to interpret because the topographic features do not simply conform to the geometric structure of the underlying nuclei. Maxima are not observed over dimers, but between them. This can be seen when filled and empty state images of a step edge are juxtaposed as in Fig. 14. Focusing on the upper terrace at the left side of the image, the occupied state image shows rows, of a regular width, up to the edge of the terrace. In the unoccupied state image, by contrast, the last row is clearly more narrow than the others. If the unoccupied maxima are interpreted as dimers, we are forced to conclude that atoms in the last row on the terrace are unpaired. It is clear that this is not the case, for energetic reasons, and also since the resulting two dangling bonds per atom would be very distinct in both occupied and unoccupied state

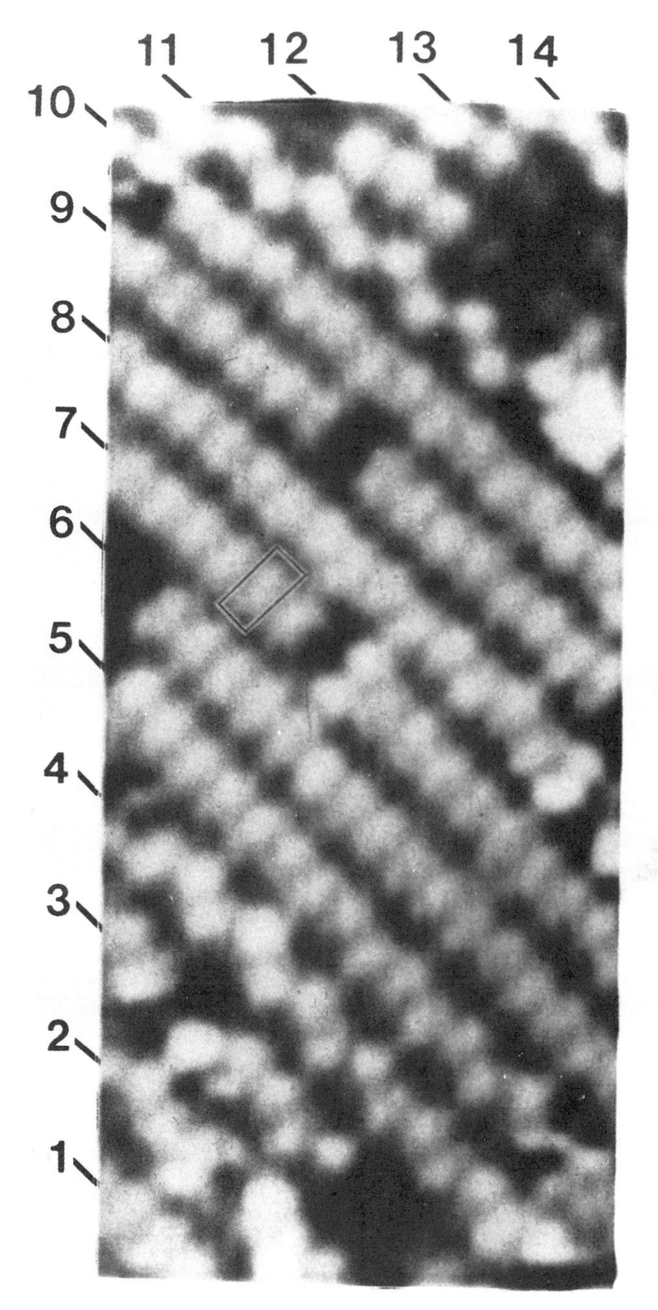

FIG. 13. STM image of Si(001) showing dimers, both symmetric and buckled, and numerous defects. From Ref. 62.

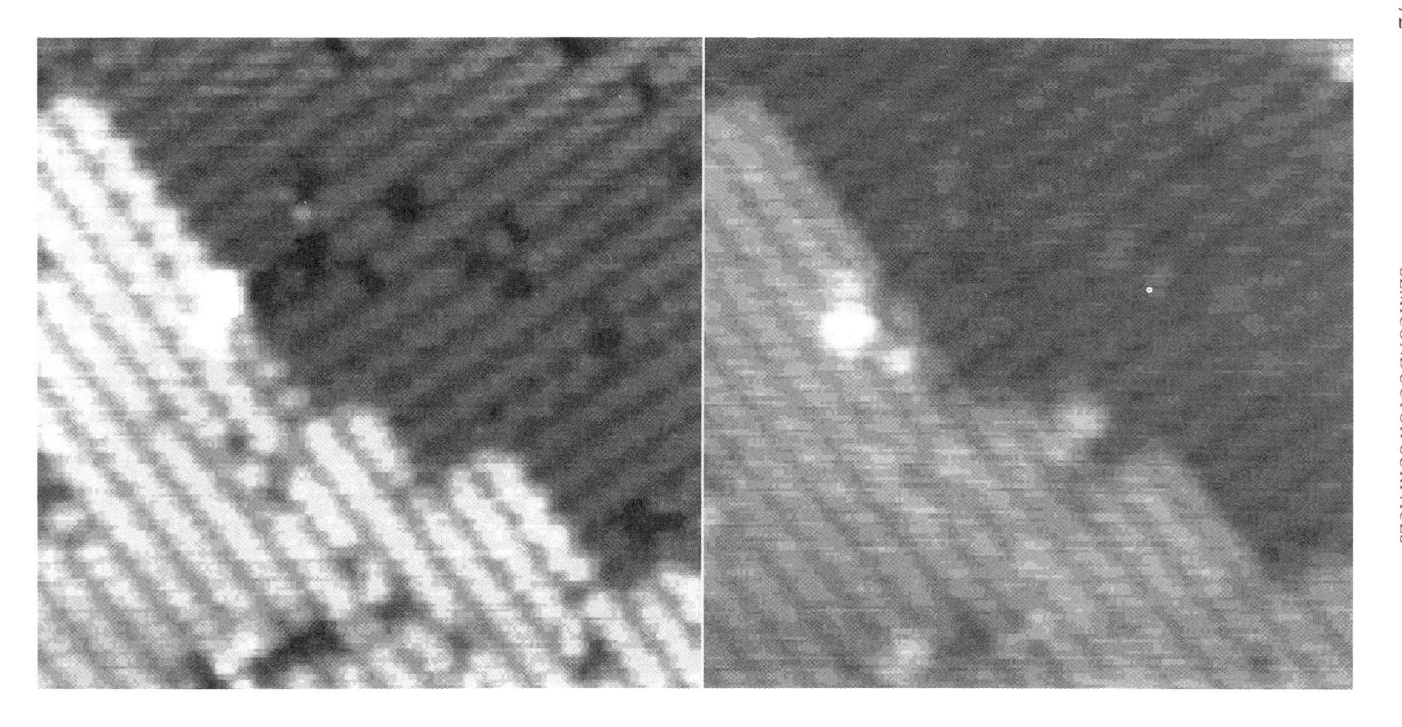

FIG. 14. STM images of a step on the Si(001) surface. At the left is an occupied state image, and at the right is an unoccupied state image of the same area. From R. A. Wolkow, previously unpublished.

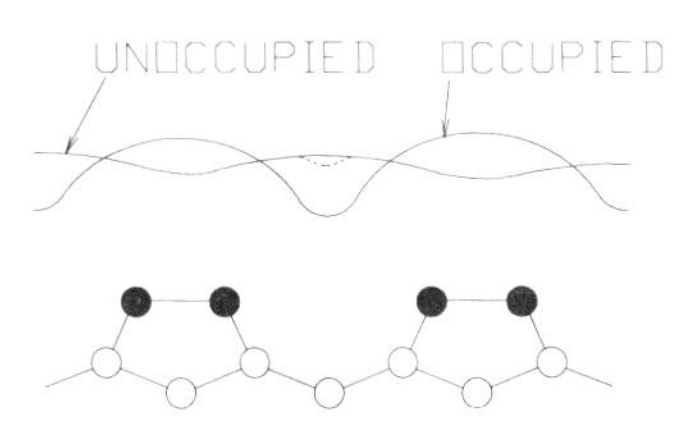

FIG. 15. Schematic STM occupied and unoccupied state contours and their relation to the underlying dimers. The dotted portion of the unoccupied state contour shows a secondary minimum seen occasionally with extraordinary tips.

images. It can therefore be concluded that the unoccupied state electronic structure leads to maxima between the dimers, while occupied state images show maxima which conform to the underlying dimers. In Fig. 15, schematic STM contours for both occupied and unoccupied states are shown. The measured corrugation is typically 0.1 to 0.2 Å. Occasionally, with an uncommonly sharp tip[64,65] a more shallow topographic minimum (shown as a dotted line) is observed between dimers.

STM images of the Si(001) surface led to universal acceptance of the dimer model while also explaining the confusing higher-order diffraction which had often been observed. The observation of both symmetric and buckled dimers, however, fueled rather than settled the controversy involving the dimer configuration. While it was recognized that the symmetric appearance of dimers might result from averaging by the STM as the dimer rapidly flipped between buckled configurations, this could not be confirmed.

Other investigations directed at the symmetric versus asymmetric dimer question were carried out. In one recent calculation spin effects were incorporated and it was concluded that symmetric dimers are most stable.[66] High-resolution photoemission data, on the other hand, indicated two inequivalent types of surface silicon atoms, which were interpreted to be the up and down atoms of buckled dimers.[67] Thorough pseudo-potential calculations by Pandey,[54] Payne, *et al.*,[68] and Roberts and Needs[69] show that if there is an energy lowering associated with buckled dimers, it is a subtle effect, perhaps 10 to 60 meV, and is too small to predict with certainty.

5.1.2.2.2 Low Temperature Imaging of Si(001). While the majority of studies, both experimental and theoretical, concluded that dimers have an asymmetric nature, significant uncertainty remained. Clear evidence of asymmetric dimers at last became available with the advent of a variable temperature, UHV STM capable of operating at temperatures down to approximately 100 K.[70] Wolkow showed that on cooling to 120°K, the number of buckled dimers increased at the expense of symmetric-appearing dimers, and concluded that only bistable dimers could account for this observation.[71] Since the Si(001) surface is always observed to have numerous defects, the

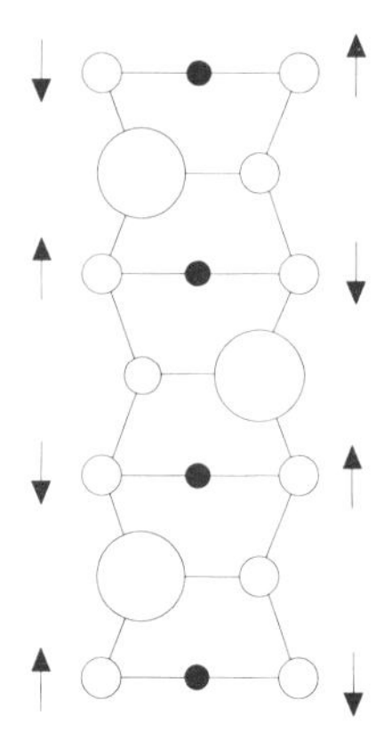

FIG. 16. A simple schematic view of the lattice strain which acts to couple adjacent dimers in an anticorrelated manner. From Ref. 71.

study concentrated on extracting the innate dimer characteristics from observations of dimers under the influence of defects.

Wolkow began by enumerating the effects of defects observed at room temperature. Symmetric-appearing defects, such as the "missing dimer" type, do not induce buckling, while defects which are themselves asymmetric in appearance cause buckling in adjacent dimers. As had been observed in the original STM images of Si(001),[9] the magnitude of buckling decayed along a row with increasing distance from a defect, with the sense of buckling alternating from one dimer to the next. A simple strain argument was presented to account for the anti-correlated buckling pattern.[72] As shown in Fig. 16, the up end of a buckled dimer causes second-layer atoms to come together while the down end of the dimer pushes second-layer atoms apart. Adjacent dimers most naturally accommodate this distortion by buckling in the opposite direction. Electrostatic effects, if a factor in coupling dimers, would also lead to this alternating pattern. By examining dimers situated between two buckle-inducing defects in the same row, the effect of supportive or conflicting influence of defects was observed. Defects placed, by chance, such that an in-phase influence is imposed on intervening dimers, cause longer-range buckling to occur, compared to a situation where buckled dimers were under the influence of only one defect. In addition, the out-of-phase effect of defects was observed to cause a relatively sharp decay to buckling.

In sharp contrast to room temperature observations, images recorded at 120 K show extended regions of buckled dimers (see Fig. 17). While room temperature buckling shows a characteristic decay length of six to eight dimers, at 120°K buckling extends along a row, in most cases, with no

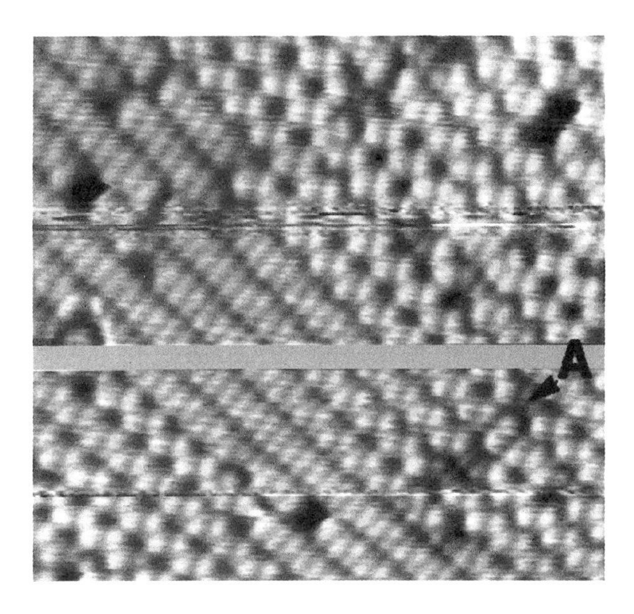

FIG. 17. 110 × 110 Å topograph of Si(001) taken at a temperature of 120 °K at −2 V sample bias and 0.4 nA. (The horizontal band resulted from an equipment malfunction during acquisition.) Taken from Ref. 71.

apparent decay in magnitude. The row containing the defect labeled A, for example, shows 27 buckled dimers running from the defect to the edge of the image. Adjacent buckled rows can interact in two ways. The more common c(4 × 2) arrangement results when the zig-zag pattern of buckled rows is out of phase. The in-phase p(2 × 2) arrangement is also present and can be seen in the lower-right corner of the image in Fig. 17. Given that defects are randomly scattered on the surface, it might be expected that adjacent buckled rows would take on p(2 × 2) order as often as c(4 × 2). Since at low temperature c(4 × 2) dominates, it is concluded that row–row coupling, with a preference for c(4 × 2), plays a significant role in forming extended two-dimensional ordered domains. Examination of larger area 120 K scans shows that approximately 80% of the surface is buckled and that c(4 × 2) ordering between rows is more than five times more common than p(2 × 2). Antiphase boundaries (i.e., p(2 × 2) ordering) can be observed between c(4 × 2) domains.

Asymmetric dimers are consistent with the observation of increased buckling since they would require thermal activation to switch orientation. To decisively establish this point however, the effect of defects (which cannot be eliminated since step edges also lead to buckling) must be clarified. To help visualize the effect of dimer–dimer and dimer–defect coupling, schematic

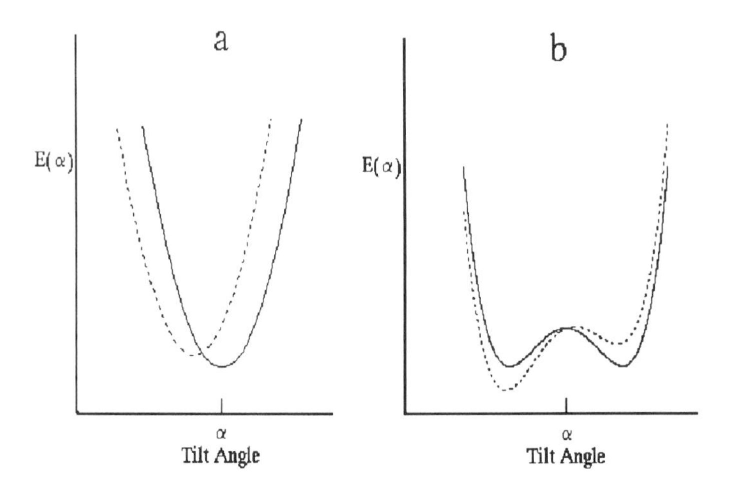

FIG. 18. Schematic potential energy versus tilt angle curves. (a) Solid curve represents a symmetric dimer, dotted curve represents a normally symmetric dimer which is forced to buckle. (b) Solid curve depicts a bistable asymmetric dimer. Dotted curve depicts a strain induced asymmetry.

potential energy curves are presented in Fig. 18. The solid curve in Fig. 18a represents an unperturbed symmetric dimer; the dotted curve depicts a normally symmetric dimer that is forced to buckle. The bistable potential shown in Fig. 18b describes the asymmetric dimer model. The effect of coupling is represented by giving a preference to one direction of tilt, as shown by the dotted line. Qualitatively the distortion of a dimer is the same whether induced by a neighboring buckled dimer or a defect. Defects will induce asymmetry that is more pronounced, and since defects do not switch orientation, the asymmetry is static. With increasing distance from the defect, strain dissipates and the asymmetry forced upon the dimers is reduced. Dimers that are far from defects will experience a relatively small asymmetry as a result of coupling to neighboring dimers. This asymmetry is not static, indeed any potential curve one might sketch to depict this interaction should be viewed as describing a transient configuration since the single dimer potential actually changes continually as nearby dimers fluctuate.

Order does not extend as a result of defect-induced strain changing with temperature, the strain field is essentially constant with temperature (certainly over the narrow range studied). Some dimers may experience a defect-induced asymmetry (represented by the dotted curve in Fig. 18b) yet appear symmetric at room temperature. This asymmetry on cooling might cause the dimer to appear buckled. A pessimistic view holds that all dimers, at the observed defect density, may be subject to some defect induced asymmetry, and that it is this effect that causes buckling to extent at low temperature.

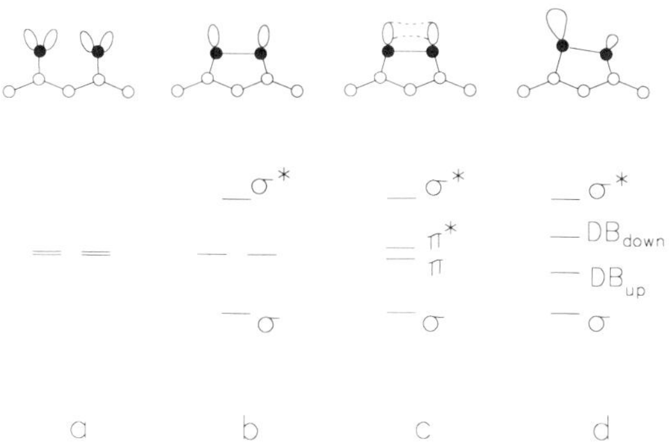

FIG. 19. Energy level diagram representing (a) four degenerate dangling bonds (DB) from two bulk-like surface atoms, (b) dimer bond formation depicted by a σ bonding level and a σ^* antibonding level, two dangling bonds remain with one on each atom, (c) splitting of the DB levels as a result of π-π^* interaction, (d) buckled dimer with sharply split DB states associated with the up and down dimer atoms.

Indeed this may be true but, in addition, the dimers must have an innate bistable asymmetric character. This is clearly seen if one considers the alternative view, that dimers are symmetric. A symmetric dimer cannot "hide" a distortion through thermal activation. A buckled monostable dimer (as in Fig. 18(a)) would appear to the STM to have a constant mean angle of tilt whether in the ground state or vibrationally excited. (Assuming it is not highly excited, and it clearly is not, the dimer rocking mode is associated with phonon structure at 25 meV,[72] kT at room temperature is ~25 meV.) From these arguments we can extract the native character of the dimer, and it is asymmetric.

Defects act to smear out the transition from disorder (symmetric appearance) to order (ordered buckled domains). An abrupt order–disorder transition[73,74,75] does not occur; rather, as the temperature is lowered, order extends gradually from defects.

5.1.2.2.3 Dimers and STM Imaging. The essential features of the dimer electronic structure are depicted in the simple energy level diagram shown in Fig. 19. At the left extreme of the diagram are four degenerate levels representing the dangling bonds of two bulk-terminated surface atoms. Proceeding to the right, dimer bond formation is represented by the σ bonding level and the σ^* antibonding level. The remaining dangling bonds, one on each atom, remain noninteracting. This simple description of the electronic structure is consistent with the ball-and-stick model of the dimer, and like the ball-and-stick model it describes gross features correctly. Two of the four available electrons fill the dimer bonding level and the remaining two electrons half-fill the dangling bonds. Since the STM is most sensitive to

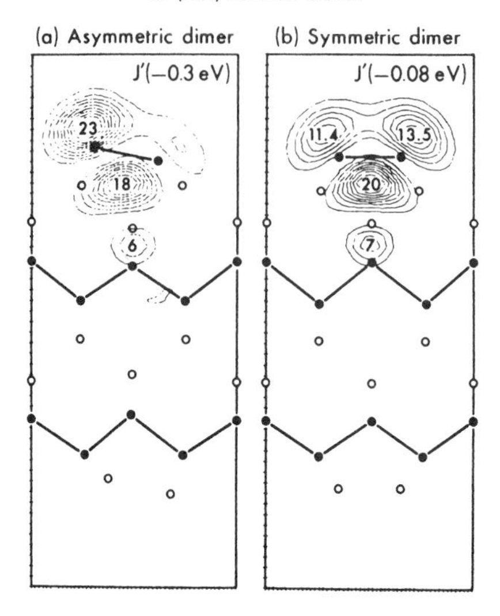

FIG. 20. Charge density plots contrasting the character of occupied states near the Fermi level for the (a) asymmetric and the (b) symmetric dimers. From Ref. 76.

electron states nearest the Fermi level, dimer images are expected to be derived primarily from the dangling bonds. STM images of the Si(001) surface tend, however, to be described as due to the dimer bond,[63] or the dimer π bond.[64,65] The term "dimer bond" is unfortunate as it suggests, quite incorrectly, that electrons are derived from the dimer σ bond. The σ bond is largely inaccessible to the STM tip both energetically and spatially; the electron density does not protrude into the vacuum but is concentrated between the atoms of the dimer pair. The use of dimer π-bond is a better choice as it refers to a state derived from the π-bonding interaction of the dangling bonds. In the original theoretical description of the (symmetric) dimer reconstruction Appelbaum, Baraff, and Hamann[44] showed that the dangling bonds do interact to form a π-bonding, and a higher energy π-antibonding level, but that the interaction is very weak. The splitting due to the π–π^* interaction is represented in Fig. 19(c).

Since the true dimer configuration is asymmetric, not symmetric as assumed in the calculation by Appelbaum *et al.*, and since we associate distinctly different electronic structure with the buckled dimer [Fig. 19(d)], we must re-examine the assignment of STM topographic features to the dimer π-bond. Theoretical calculations which contrast the electronic structure of the symmetric and asymmetric dimers show that the π–π^* interaction induces a small splitting compared to that created by buckling.[76,77,78] Figure 20 shows the spatial character of the π-bonding interaction associated with the symmetric dimer, and for comparison, the distinctly different buckled dimer. Buckling of the dimer leads to states which are like dangling bonds in spatial character, but which are energetically split such that the state at the dimer up atom becomes relatively full at the expense of the state localized at the down atom. Dimers which are forced by a nearby defect to buckle, or which naturally settle into a buckled configuration at low temperature, appear in occupied state images to be strongly tilted (the up atom is approximately 0.5 Å higher than the down atom). This observation matches very well the picture of an asymmetric dimer with modified dangling bonds as described earlier. At room temperature the barrier to switching direction of tilt is readily overcome and the STM captures an average picture of the dimers fluctuating between buckled extremes, and an elongated, symmetric protrusion, centred over the dimer results. The symmetric-appearing dimer has routinely been described as due to the dimer π-bond. This is at least partially correct since the dimer will pass through a symmetric intermediate state while switching between extremes. The dangling bond states associated with the up and down ends of the buckled dimer also contribute. The weighting of these various contributions is unclear at this time.

Similar to the occupied state images, the states associated with the up atom and the down atom of the buckled dimer must figure into the room temperature unoccupied state image, however, the π^* description seems compelling. This state has a node at the center of the dimer bond, matching the observed minimum in STM images, and maximum density at the dimer ends where the topographic maximum is observed. It may be, however, that the unoccupied state image has contributions from the dimer antibonding level (which has maximum state density at the ends of the dimer) in addition to the dangling bonds. The calculations of Kruger *et al.*[77,78] suggest that the dimer σ antibonding level is very near the bottom of the conduction band and is therefore energetically accessible. While it does not have the spatial characteristics of a dangling bond, the dimer antibonding state does not decay with distance from the surface as rapidly as does a back bond.[78]

In addition to producing maxima which are not centered over the dimer units, unoccupied state images show another interesting feature. In contrast to occupied state images, unoccupied state images do not show buckling, even in dimers adjacent to buckle inducing defects. It has been suggested that the field imposed on a dimer during scanning with the STM could tend to flatten buckled dimers.[75,79] Kochanski *et al.*[75] argued that when imaging unoccupied states, the electric field set up by the negative tip may repel the atom, causing the dimer to appear symmetric, while a positive tip would enhance buckling. A dimer held flat would not show contributions characteristic of the strongly split buckled dimer. If this mechanism were acting, it

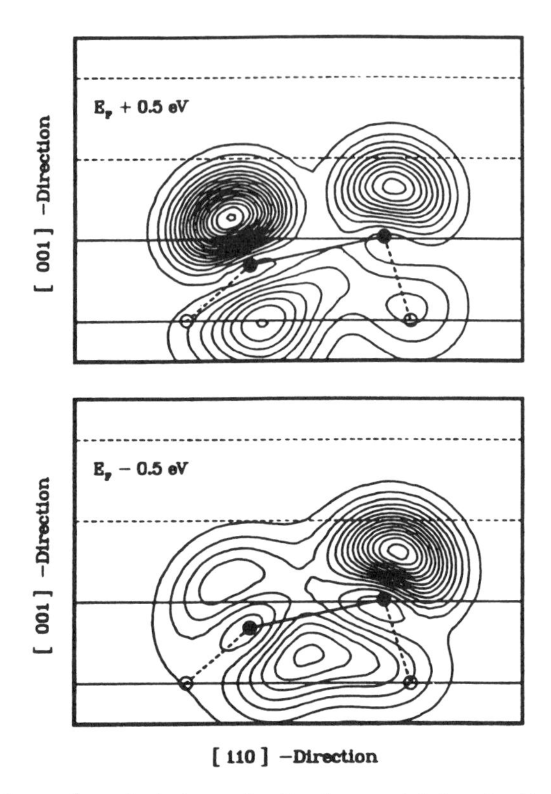

FIG. 21. Contours of constant charge density above and below the Fermi level for the asymmetric dimer. (From Ref. 78.)

might be expected that different degrees of asymmetry could be induced as a function of applied field, while no such dependence has as yet been observed. A simpler explanation for the absence of buckling in unoccupied state images arises from the asymmetry in the electronic structure of the buckled dimer. In contrast to the occupied state situation, where the up atom has the most pronounced state density, tending to amplify the buckling effect, when imaging unoccupied states the down atom has the greatest state density and tends to cancel the effect of the actual displacement of the nuclear centers. Figure 21 contrasts charge density plots above and below the Fermi level.

5.1.2.2.4 Stepped Surfaces. Steps on Si(001) display complex behaviour known to depend on a number of factors including the angle of miscut, annealing and growth conditions, contaminants, and surface stress. Recent atomically resolved STM images, together with theoretical modeling, have led to considerable improvement in our understanding of the fundamental

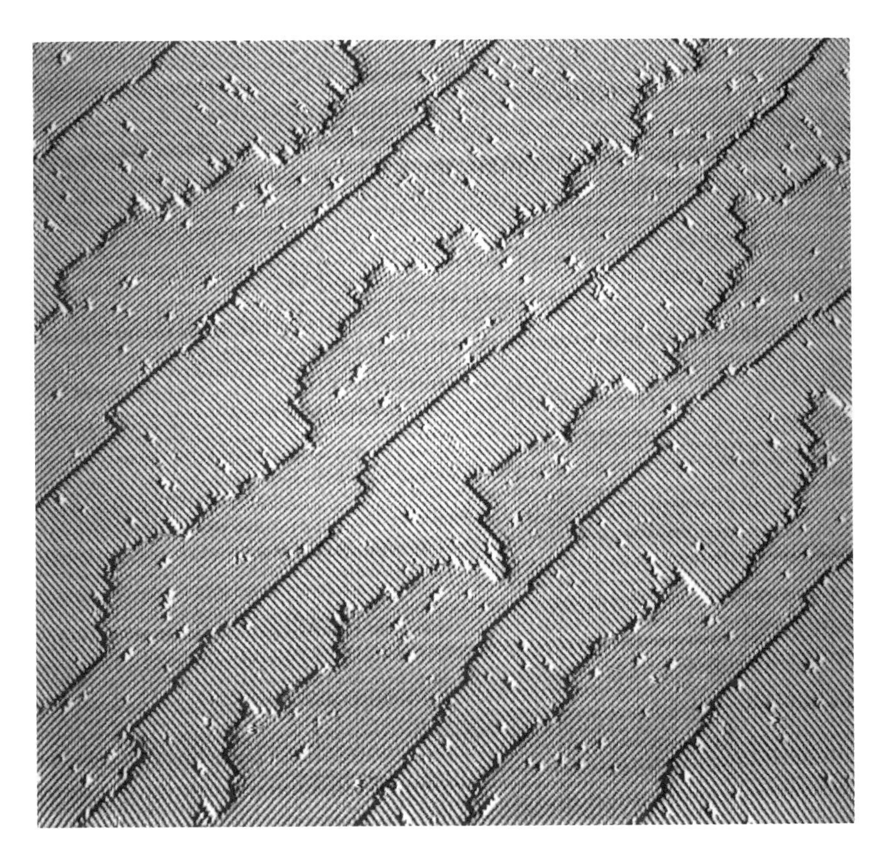

FIG. 22. Tunneling image of a stepped Si(001) surface showing the type *A* (smooth) and type *B* (rough) step risers. Taken from Ref. 88.

structures and mechanisms underlying these factors. In this section we begin with a discussion of the basic step structure; we next consider kinks, and conclude with a discussion of step interactions as effected by tilt angle, strain, and annealing.

Step Structure. The single step height in the (001) direction for the diamond lattice is $a/4 = 1.36$ Å, where $a = 5.431$ Å is the bulk silicon lattice constant. On traversing a single atomic step the dimer bond orientation rotates by 90°. Surfaces miscut from the (001) direction, known as "vicinal" surfaces, display a mean distance between single steps given by $a/4 \tan \alpha$ where α is the miscut angle. Surfaces cut toward the $[1\bar{1}0]$ or the [110] direction display two distinct types of single step, as seen in Fig. 22. When dimer rows on the upper terrace run parallel to the step edge, the step is referred to as type *A*. The other possibility, referred to as a type *B* step, has

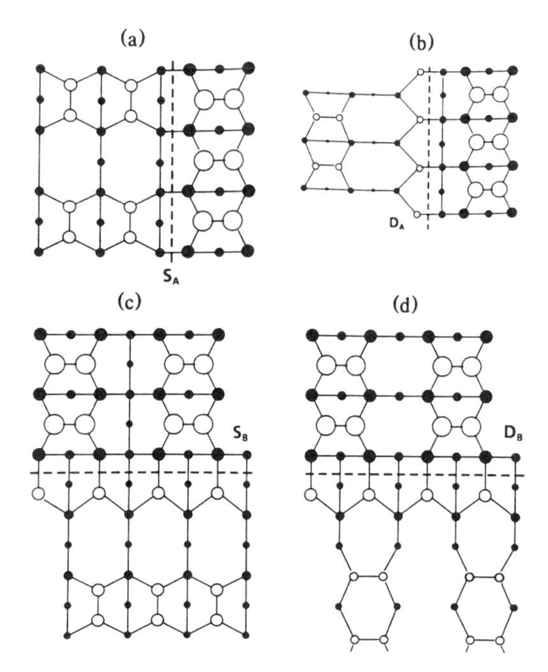

FIG. 23. Top views of single and double steps of type *A*, labelled S_A and D_A, single and double steps of type *B*, labelled S_B and S_B. Dimer bonds are aligned in [$\bar{1}$10] direction. Dashed lines that run parallel to [$\bar{1}$10] or [110] axes indicate the step positions. Open circles denote atoms with dangling bonds. Edge atoms (shaded circles) (b)–(d) are rebonded, i.e., they form dimerlike bonds with lower terrace atoms. Larger circles are used for upper terrace atoms. Only some sublayer atoms are shown for the sake of clarity. From Ref. 80.

rows on the upper terrace perpendicular to the step edge. When the surface is tilted toward a direction intermediate between [1$\bar{1}$0] and [110], steps of mixed type *A* and type *B* character result. Type *A* steps are found to be very smooth, while type *B* steps show a marked tendency to form kinks. A typical single stepped vicinal surface is shown in Fig. 22. Surfaces with a miscut angle larger than a few degrees form type *B* double steps. Since the dimer bond orientation is the same on terraces separated by a double step, surfaces prepared to have exclusively double steps are referred to as single domain or primitive.

Chadi has proposed models for single and double steps of types *A* and *B*.[80] These structures, which are found to correspond well with STM images, are shown in Fig. 23. Formation energies of type *A* and *B* single steps were calculated to be 0.01 eV and 0.15 eV respectively, while type *A* and *B* double step formation energies were calculated to be 0.54 eV and 0.05 eV respectively. Chadi found that type *B* step structures with "rebonded" atoms, as shown in Fig. 23, were most stable. A type *B* single step without rebonded

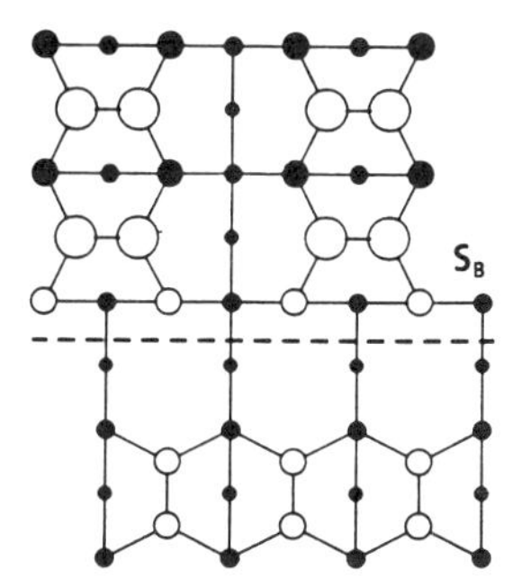

FIG. 24. Top view of a simple "nonbonded" edge geometry for a type *B* single step. Open circles denote atoms with dangling bonds. This structure is characterized by having a dangling bond on each second-layer edge atom and is not as energetically favorable as the rebonded atomic configuration of 23(c).

atoms was also considered and found to have a formation energy of 0.16 eV. The corresponding structure is shown in Fig. 24. Both kinds of type *B* single steps have been observed.[62] The double type *A* step which Chadi predicts to have a relatively high formation energy is not observed. Figure 25 shows an image of a type *B* double step. An earlier model of the double step due to Aspnes and Ihm[81] invoked a pi-bonded chain structure like that seen on Si(111)–(2 × 1) surface.[39] STM images do not support this model, however, since a periodic structure is observed at the step edge with twice the spacing expected for the π-bonded chain. While the spacing is also twice that expected for rebonded atoms, it is felt that Chadi's model correctly describes the double step. It appears that the rebonded atoms buckle, like the dimers on the (001) terrace, and only every other rebonded atom is visible.[82] The buckling of the rebonded dimers is caused by the asymmetric strain imparted from the asymmetric dimers on the upper terrace. Careful examination of both occupied and unoccupied state images reveals that all rebonded atoms predicted by Chadi are present.

Vicinality versus Single Double Steps. For a number of years it had been thought that vicinal Si(001) surfaces had only one equilibrium structure in which only double layer steps are present.[83] Calculated step energies substantiated this view by showing the double layer type *B* step to be energetically preferred over single layer steps.[80,81] It has been observed, however, that small vicinal angles, on the order of 1° or less, lead to stable single stepped surfaces. Recently Alerhand *et al.* have provided a compelling explanation for these observations. They demonstrated that, at vicinal angles exceeding approximately 2°, the equilibrium surface is double stepped, but for small misorientations the single step is at equilibrium.[84] Further, Alerhand *et al.* established that, in addition to step energies, two other factors

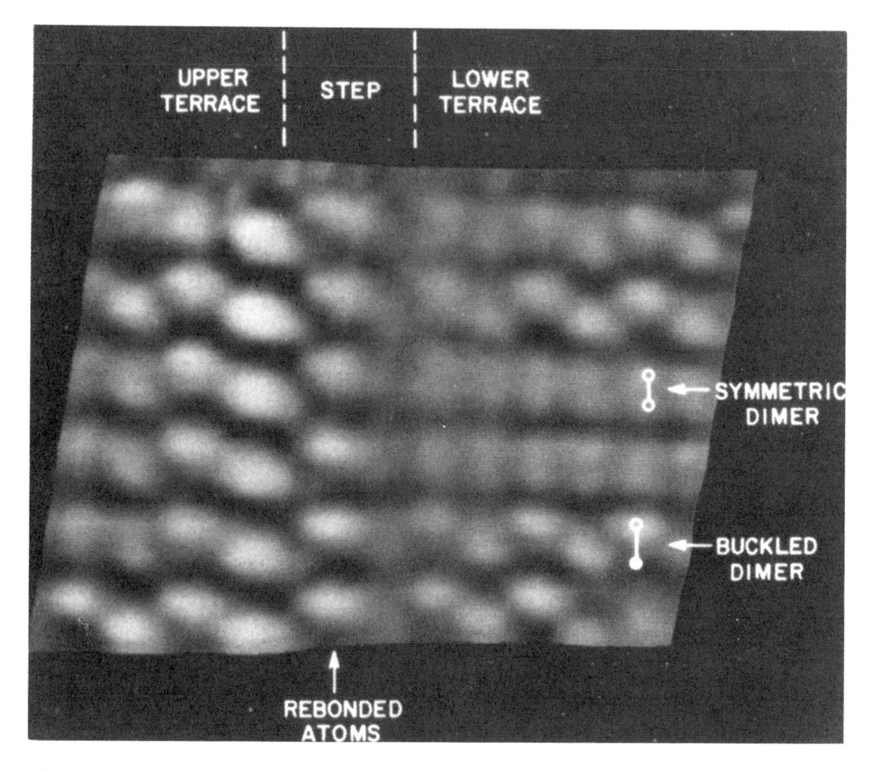

FIG. 25. Occupied state STM image of a type B double step. The region covered is 46 × 50 Å. Six dimer rows run laterally across the field of view while the step runs from bottom to top. Curvature keyed shading is used. The simple bias was −1.6 V. From Ref. 82.

play a role in determining the relative stability of single and double steps. The first is the long range effect of the anisotropic stress associated with a dimerized terrace, and the second is the effect of thermal fluctuations (i.e., kinks) in step edges. At zero temperature the latter does not come into play. At angles of a few degrees or larger, the dominant factor is that double steps are energetically favored over single steps by approximately 0.1 eV/atom. Since the dimerized surface has a compressive stress perpendicular to the dimer bond and a tensile stress parallel to the dimer, crystal strain can be relieved by forming a single stepped surface since the dimers on adjacent terraces are then perpendicular. The strain per unit terrace width becomes more significant at small miscut angles and eventually dominates at a calculated angle of 0.05 at 0 K. The entropic contribution of thermally induced kinks in step edges are found to significantly shift this transition angle to larger angles at higher temperatures. Diffusion leading to step fluctuations is frozen out at temperatures well above room temperature but below typical anneal temperatures. The temperature at which mass transport stops during cooling of a crystal is not clear, if we take a reasonable value of 500 K[84] we can read the experimentally relevant transition angle from the calculated phase diagram to be approximately 2°. This value is in very good agreement with experimentally determined upper and lower bounds of approximately 3.5° and 1° respectively.[84]

Strain Effects on Si(001) Steps. The LEED investigations of Saloner *et al.*, established that steps on vicinal surfaces were evenly spaced, clearly suggesting a step–step repulsive interaction.[85] Menn *et al.* investigated the effect of applying a uni-axial stress to the Si(001) surface and provided striking evidence that the repulsion between steps was related to strain in the crystal.[86] A surface, which previous to the application of stress, showed an even distribution of 1 × 2 and 2 × 1 terraces, could, by bending the crystal, be transformed to a nearly primitive surface. Alerhand *et al.* successfully accounted for this behavior, and in so doing, showed that the bulk elastic strain is the primary cause of interaction between steps.[84] It is demonstrated that the dimerized surface has a compressive stress perpendicular to the dimer bond and a tensile stress along the dimer direction, and as a result, domains for which an applied compressive stress is directed along the dimer bond, grow at the expense of the other domain. Figure 26 compares the experimental data of Menn *et al.* with the calculations of Alerhand *et al.* The experimental data clearly shows the coordinated shift in 1 × 2 and 2 × 1 LEED intensities as a function of applied strain. The light lines result from a model which assumes that the mean number of steps changes as a function of stress. The heavy lines, which fit the data very well, result when the number of steps is conserved.

Kinks and Step-Kink Interactions. Step and kink energies have been the subject of many investigations.[87] Recently Swartzentruber *et al.* have analyzed equilibrium distributions of kinks and steps to determine kink step separations and lengths, and from these extract measures of step and kink energies.[88] They recorded STM images of thoroughly annealed Si(001) surfaces miscut 0.3° toward [110]. Kink lengths and the separations between kinks were tabulated for type A and type B steps, and the probability of finding kinks separated by a given distance was plotted as a function of separation between kinks. The data were fit very well by a function that assumed the creation of each kink was a statistically independent event. This is somewhat surprising, since the long-range strain that leads to step–step repulsion[89] must be manifest in kink formation probabilities, otherwise steps would not tend to lie at the midpoint between adjacent steps. It is possible that the large separation between steps on the surfaces studied masked this effect by producing too small a gradient in the local strain field. In any case the assumption of independent kinks allows the number of kinks as a

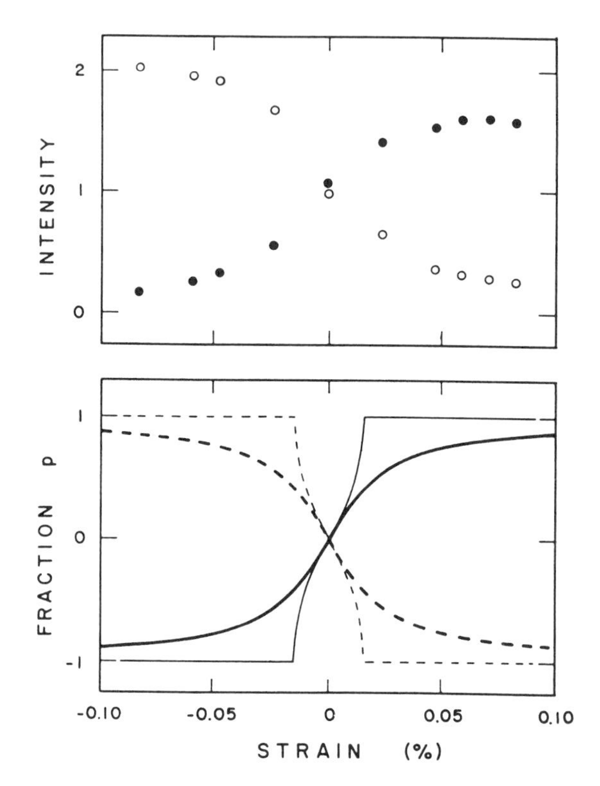

FIG. 26. Top: Experimental intensity of (1 × 2) (open circles) and (2 × 1) domains on the Si(001) surface as a function of applied external strain. Bottom: Fit by theory, thick (thin) lines correspond to quasi (global) equilibrium. From Ref. 89.

function of length to be written $N(n) \propto e^{-E(n)/kT}$, where $E(n)$ is the energy of a kink of length n. Plots of $E(n) = -\ln(N(n)/2N(0))$ versus n were fit according to $E(n) = n\varepsilon_S + C$, where ε_S is the unit single step energy and C is described as an additional energy due to the corner structure at kinks. Values of $\varepsilon_{S_A} = 0.028 \pm 0.002$ eV/atom, $\varepsilon_{S_B} = 0.09 \pm 0.01$ eV/atom, and $C = 0.08 \pm 0.02$ eV were found.

A more detailed discussion, particularly of double steps and of earlier developments which provided the background to work on steps described here, may be found in the review by Griffith and Kochanski.[87]

5.1.2.3 Si(110). While a vast quantity of effort and expertise has been devoted to the study of the (111) and (001) faces of silicon, by comparison little is known about the other low-index face, the (110). A perusal of the literature shows that even the elemental symmetry displayed by this face was not agreed upon as little as ten years ago. Partly as a consequence of the study of this surface with the STM, it has become apparent that trace amounts of

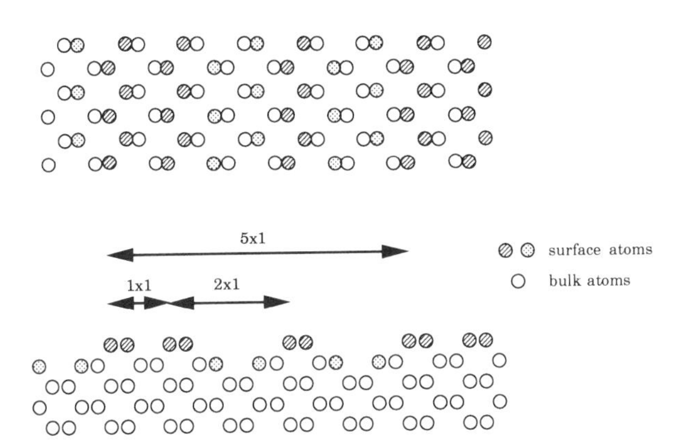

FIG. 27. Schematic model of Si(110) "missing row" surface structure. Both a plan view and side view along a surface [1$\bar{1}$0] are shown. Surface atoms are denoted by shaded circles.

metallic impurities has a profound effect on the fundamental reconstruciton of this crystal face, more so than the (111) or (001) surfaces.

The first LEED investigation of the Si(110) surface structure[90] reported a wide range of surface structures that depended on the preparation history (heating, sputtering, annealing, etc.) of the sample. Upon heating to ~1200°C a 5 × 1 diffraction pattern was observed, with a narrowing around the 2/5, 3/5 beams suggesting these were in reality a splitting of a 1/2 order beam due to the presence of steps or facets. This suggestion is strengthened by the weakness of the 1/5 and 4/5 order spots. The existence of similar higher-order reconstructions (7 × 1, 9 × 1) was derived from an observed narrowing of the 1/2 ⇒ 2/5, 3/5 splitting, suggesting these were now 3/7, 4/7 or 4/9, 5/9 beams. Upon annealing at a lower temperature (700°C to 800°C) a complicated diffraction pattern, referred to as the X pattern, was seen and assigned to twinning or faceting of one or more of the lower symmetry structures.

Olshanetsky and Shklyaev further investigated the Si(110) surface phases using LEED and reported that the various structures seen earlier were related phases, and further that the X structure was a faceted series of (15 17 1) planes.[91] They designated this phase the Si(15 17 1)–2 × 1. Figure 27 shows a schematic of the missing row category of structures that had been widely proposed as the basis for the multiplicity of phases on the Si(110) surface. A natural bulk truncation at the (110) surface would lead to a surface consisting of [$\bar{1}$10] "tubes" consisting of six-membered rings, somewhat reminiscent of the five- and seven-membered rings in the Pandey model of the Si(111)–2 × 1

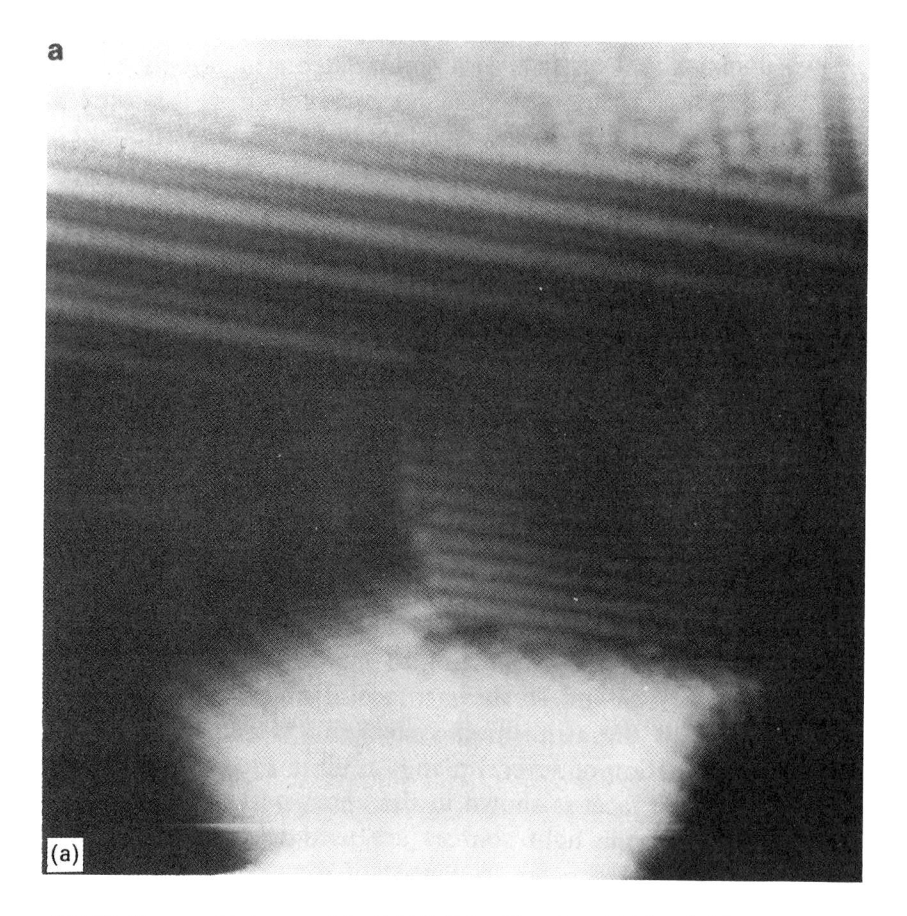

FIG. 28. Topographic images of the Si(110) "16 × 2" surface. (a) Maps height to gray scale, while (b) has the contrast enhanced to bring out subtle details. Image scale is 800 × 910 Å with a vertical relief of 20 Å. "A" denotes a "2 × 16" region oriented in the [1$\bar{1}$2]. "D" designates a disordered area. "E" a "2 × 16" domain oriented in the [$\bar{1}$12] direction. The other letters denote various crystal planes exposed to the STM. From Ref. 95.

surface. Using this bulk truncation as a basis, a 2 × 1 structure may be achieved by removing every other raised row of zig-zag atomic chains as shown in Fig. 27. One can see that 5 × 1, 7 × 1, and 9 × 1 structures may be easily formed from the basic 2 × 1 missing row structure. More evidence for the simple missing row phase came from ARUPS measurements by Martensson *et al.*,[92] which showed a filled surface state at -1.0 eV below E_F with positive dispersion, similar to that reported for the Si(111)–2 × 1. The similarity in bonding between the π-bonded chains and the zig-zag atoms at the

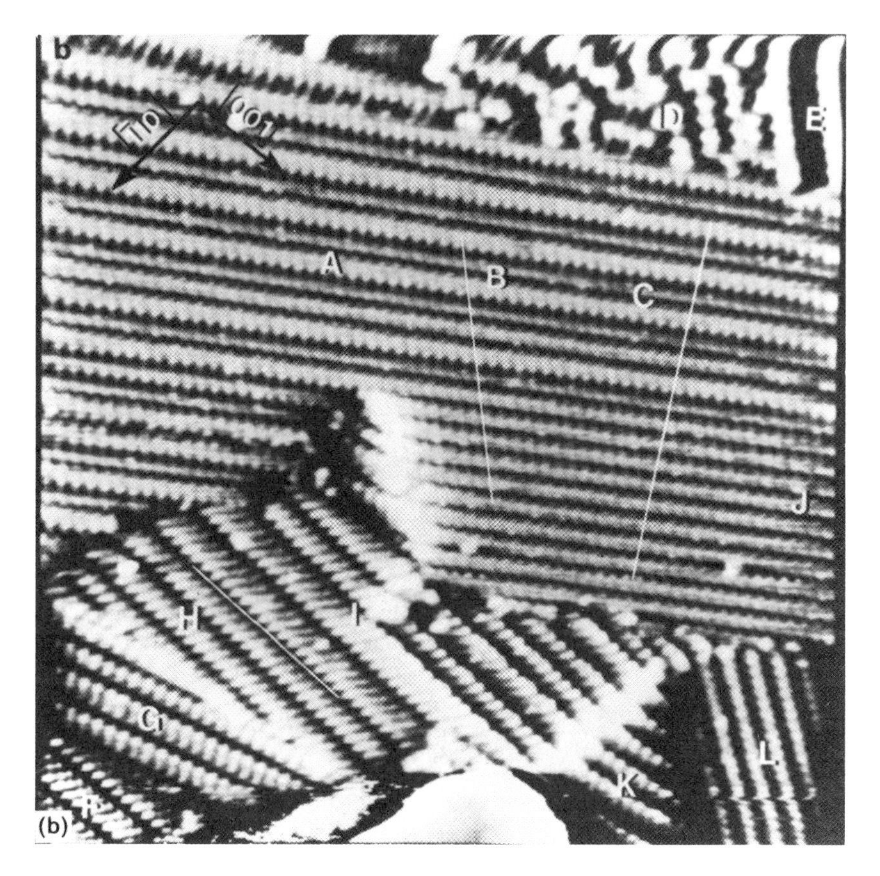

FIG. 28. Continued.

surface of a [$\bar{1}$10] tube seemingly strengthened the missing row models for the Si(110)–$n \times 1$ surface phases.

The first STM images of the Si(110) showed what appeared as an octet of protrusions arranged with a spacing 27.5 Å along ⟨001⟩.[93] This accounts for the $5a$ periodicity seen in the diffraction studies, while the individual quartets in the octet appeared to account for the underlying $2a$ periodicity reported earlier. Indeed, these authors suggested a model consisting of adatom decoration on the array of missing rows consistent with the earlier suggestions from the LEED measurements. Little ordering was noted along the surface ⟨$\bar{1}$10⟩, accounting for the streakiness of the diffraction beams in this direction and the apparent 1x symmetry. This work also reported on tunneling spectroscopy measurements on the Si(110), reporting an unoccupied state at +2.0 V

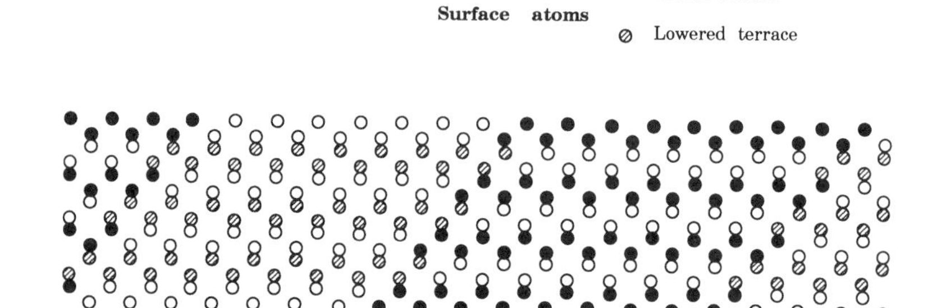

FIG. 29. Schematic model of the "16 × 2" structure. Raised terraces are solid circles, while shaded circles represent the lowered terraces.

sample bias, and an occupied state at − 1.0 V sample bias, in agreement with the ARUPS measurements.

Later work with the STM by Neddermeyer and Tosch[94] using short time annealing at 800°C reported a 4 × 5 phase composed of similar 2 × 5 subunits, and claimed that the $n \times 1$ phases reported earlier were likely the result of faceting, bringing about a number of local reconstructions.

All of this changed with the publication by van Loenen *et al.*[95] of tunneling images of the Si(110) surface containing varying amounts of metal contamination. These images, shown in Fig. 28, demonstrated conclusively that the clean surface phase was of "2 × 16" symmetry, constructed of the same quartet units already imaged in the 5 × 1 phase. By combining Rutherford Backscattering (RBS) measurements of surface nickel contamination with STM images, they demonstrated that small amounts of metal contamination vastly altered the symmetry of the surface phase. This was in accordance with work by Ichinokawa *et al.*[96] and Ampo *et al.*[97] which suggested that the complex phases seen in earlier studies were the result of trace metal contamination. The complex 4 × 5 and 2 × 5 phases reported by Neddermeyer *et al.*[94] were found to result from 0.1 ML of surface nickel, while the 5 × 1 seen both by LEED and the earlier STM work[93] resulted from ~0.01 ML surface nickel contamination. By eliminating all measurable nickel, a "2 × 16" symmetry phase, shown in Fig. 29,[96] results, with two twinned orientations. The surface structure consists of alternately raised and lowered rows running along $[1\bar{1}2]$, with entities in the rows resembling the quartets of the 5 × 1. The surface reconstruction and the steps run along $[1\bar{1}2]$ and $[\bar{1}12]$ directions, while the 5 × 1 is oriented along the [001] as reported earlier. The distance measured between pair of upper and lower rows is ~ 51.4 Å, corresponding to the nominal 16*a* spacing. Examining the tunneling image in Fig. 28, the terraced "2 × 16" structure proposed in Ref. 96 may be present, but is likely decorated with some sort of adatom structure to account for the quartets. The authors of Ref. 95 declined to propose a detailed structural model for the 2 × 16, leaving the determination of the atomic structure unknown at present.

5.1.3 Adsorbates on Silicon

While a good knowledge of the structure and properties of the clean surfaces is essential to understanding more complex surface systems, a large body of information may be gathered from probing simple adsorbate reactions, comparing and contrasting them to the behaviour of the generally less complicated clean surface phases. Despite the short history of vacuum tunneling microscopy in the surface science arena, an already wide variety of elemental and molecular adsorbates have been deposited on silicon surfaces and scrutinized with the STM.

By far, the largest body of experiment has been carried out on the Si(111)–7 × 7 surface, due both to the relative ease of its production, and the equally large array of adsorbate studies performed with more established surface probes such as low energy electron diffraction, valence band photoemission, surface extended x-ray absorption, surface x-ray diffraction, transmission electron diffraction, x-ray standing wave spectroscopy, inverse photoemission spectroscopy, core level photoemission, etc., on this very famous surface phase. These experiments may be divided into two broad categories: gas adsorption experiments and solid phase deposition experiments. When the solid phase material is a semiconductor, the deposition experiments are usually referred to as epitaxial growth studies. Many groups worldwide have plunged into experimental studies in these areas; this review will highlight some selected investigations on the Si(111)–7 × 7 surface.

5.1.3.1 Gases. A number of gases have been adsorbed onto the Si(111)–7 × 7. A partial list includes H, O_2, and NH_3. The adsorption of simple atomic and molecular gases on both metals and semiconductors has long been an established procedure for distinguishing surface from bulk features for the spatially averaging spectroscopic probes, such as valence band photoemission and its companion, inverse photoemission. The fundamental idea is simple: resonances due to surface-derived features will be affected or quenched, while those due to bulk features will not. Largely because of the semiconductor device industry, possibly the greatest amount of interest is in the reactions of hydrogen and oxygen with the elemental semiconductor surfaces, since these are the prototypical reactions that are believed to govern most of the interface chemistry during device processing.

NH_3 Absorption. In the first study clearly demonstrating the power of the STM to probe surface chemical reactions, Wolkow *et al.* monitored the reaction of ammonia (NH_3) on the Si(111)–(7 × 7) surface.[98] The very same area of the surface could be scanned before and after reaction, and unreacted and reacted sites could be unambiguously identified. The Si(111)–(7 × 7) surface is an interesting surface with which to explore the possibility of monitoring surface reactions with STM because it offers a variety of chemically different surface sites. The 12 adatoms per unit cell may be divided into two groups, according to the faulted and unfaulted sides of the unit cell. These may be further divided into "corner" and "middle" as previously discussed in the section on the Si(111)–7 × 7 clean surface. There are six rest atoms, and one atom in the corner hole. Figure 30 shows occupied state images of the unreacted (top left), and reacted surface, with associated tunneling *I–V* curves (displayed as normalized conductivity). The spectra of the clean surface show a distinct occupied state feature associated with the rest atom, and occupied and unoccupied portions of the adatom dangling bond states. The rest atoms, which are difficult to see in topographs, were followed by recording large arrays of site-specific *I–V* curves in coordination with topographs. Spectrum A, for the reacted surface, shows the complete absence of the strong occupied state feature associated with the rest atom. This dangling bond (or "lone-pair" since it is fully occupied) has been replaced by a chemical bond to an ammonia fragment, with an associated binding energy well outside the STM energy window. Reacted spectrum B, dotted curve, is typical of either the middle or corner type of adatom after reaction. When the dangling bond state is replaced with a chemical bond, the tip must come closer to the surface (~ 1 Å closer) to get the demanded current, with the result that the topograph shows a darkened area rather than a prominent adatom. Reacted surface spectra B and C, solid lines, point to unreacted adatoms, which have reacted neighboring rest atoms. The interesting aspect of these spectra is that they are very similar. On the unreacted surface, corner and middle adatoms show different levels of occupation. The middle adatoms, which have two rest atom neighbors, donate more charge toward filling the rest atom state than do corner adatoms, which have only one rest atom nearby. Upon reaction at the rest atom sites, this charge transfer is partly undone, with the result that unreacted corner and middle adatoms adjacent to the reacted rest atoms appear electronically very similar. Rest atoms are found to be most reactive, middle adatoms are next most likely to react, and corner adatoms are least reactive.

The preference for reaction at rest atoms is a surprise. It may be that adatoms have a significant bonding interaction with the atoms directly below them, and as a result are not as reactive as expected. Different degrees of lattice strain associated with reaction at corner and middle adatoms presum-

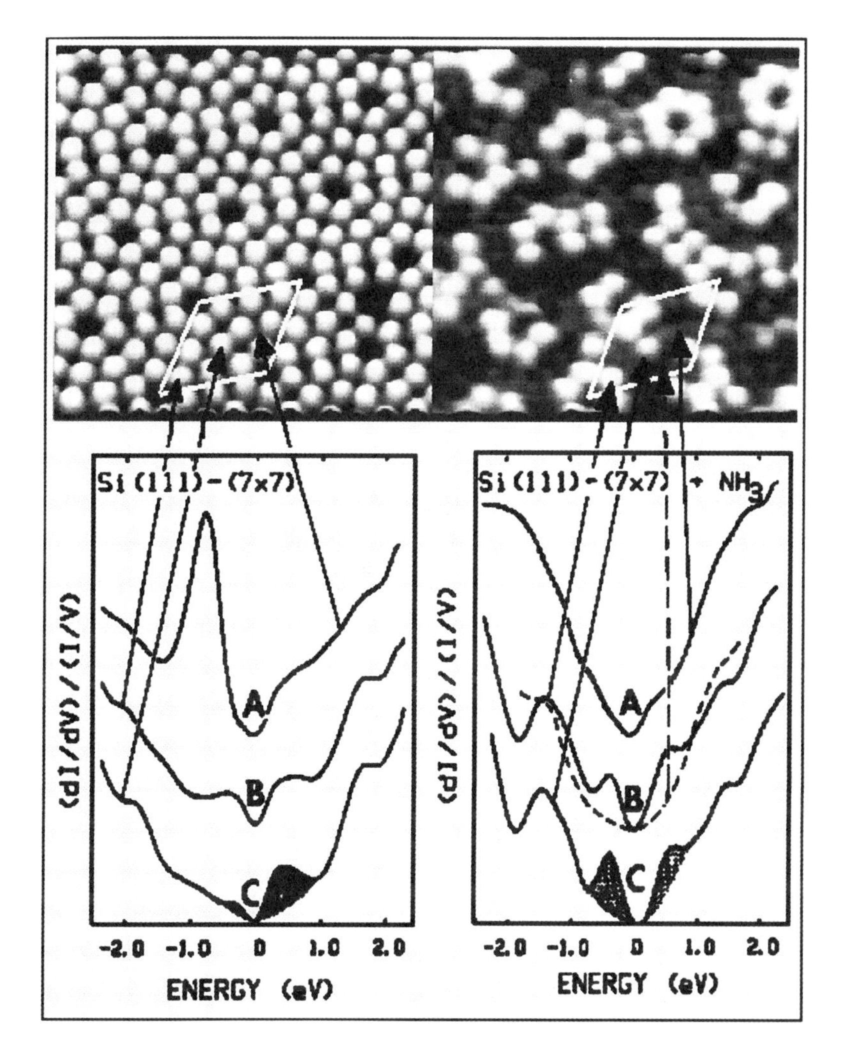

FIG. 30. (a) Topograph of the unoccupied states of the clean 7 × 7 surface (above) andatom resolved tunneling spectra (below). Curve A is the spectrum over a restatom site, curve B the spectrum over a corner adatom site, and curve C is the spectrum over a middle adatom site. Negative energies correspond to occupied states, while positive energies indicate empty states. (b) Topograph of the unoccupied states (above) and atom resolved tunneling spectra (below) for the NH_3 reacted surface. Curve A is the spectrum over a *reacted* restatom site, curve B (dashed line) the spectrum over a *reacted* corner adatom, while curves B (solid line) and C are the spectra over unreacted corner and middle adatoms respectively. Taken from Ref. 98.

ably leads to differentiation between these two sites. No specificity toward the faulted or unfaulted side of the unit cell, as occurs for room temperature Pd deposition and O_2 adsorption, was observed. Rather featureless *I–V* curves resulted over the corner atoms both before and after reaction, and no comment could be made regarding corner atom reactivity.

H Absorption. The reaction of hydrogen with the surface of Si(111) has been studied extensively with a wide number of surface probes. Despite this concentrated effort, the detailed nature of the reaction of H with the Si(111)-7×7 is yet unknown, principally due to the difficulty that area averaging probes have in isolating reactions which can, of course, be mediated at multiple sites. *In situ* hydrogen reactions are brought about generally by the exposure of clean Si surfaces to H_2 gas, some fraction of which is cracked into the atomic species using a hot filament within line-of-sight of the sample. After a measured exposure the H-reacted surface is then examined under UHV conditions.

Boland[99] has carried out extensive work on hydrogen gas reactions with silicon surfaces. For the 7×7 DAS surface he postulates that much of the confusion regarding the saturation coverage, ~ 1.25 ML, is due to the fact that the large number of strained bonds inherent in the reconstruction allow the H atoms a greater number of potential reaction sites. When these bonds are relaxed by hydride formation, this opens up new bonds for further attack, continuing, ideally, until the entire rest atom layer is 1×1 termined with H atoms, with the former adatoms herded into new 1×1 H-terminated terraces. The presence of the partial stacking fault tends to inhibit the process at lower temperatures (< 500°C), as is shown in Fig. 31.[99] The H-terminated areas may be increased upon further annealing under a hydrogen background pressure, but as the sample temperature is raised, competition between hydrogen adsorption and desorption occurs, resulting in a 7×7 surface as temperature is increased beyond 750°C.[100] Under no set of annealing conditions has the ideal H-terminated surface been observed. Neither of these STM investigations find any evidence for the etching of the silicon surface by atomic hydrogen; adatoms which appear to be missing at lower bias conditions (< 2.0 V) reappear as the applied bias is increased. This is simply a manifestation of the rearrangement of the DOS around the Fermi level when the hydrogen begins to adsorb on individual adatoms, similar to that seen for oxygen adsorption discussed next.

O Absorption. The initial stages of oxidation of silicon surfaces is an area of great interest to the semiconductor device industry, both because of the need to produce clean surfaces and to grow high quality oxide films for devices. A great deal remains unknown about the oxidation process on the atomic scale, including such fundamental questions as the geometric structure, kinetics, electronic structure, and site reactivity. These questions make

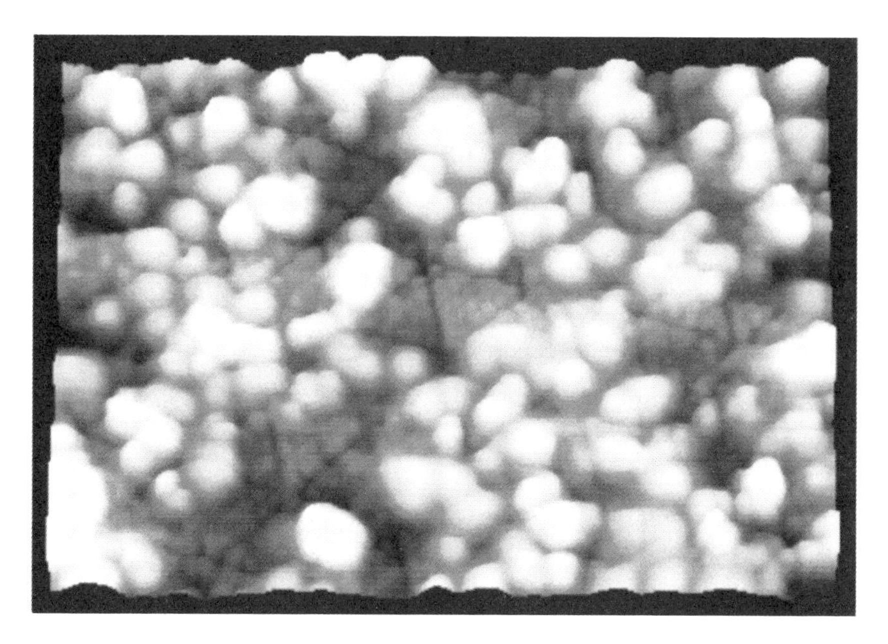

FIG. 31. STM topograph of the Si(111)–7 × 7:H surface recorded after a room temperature saturation of H atoms. The sample bias is +2.0 V and the image scale is 230 × 130 Å. Taken from Ref. 99.

the vacuum STM a natural choice for probing the low-coverage, initial stages of the growth of SiO_x on silicon at the atomic level.

The STM was first used by two different groups to study the *in situ* oxidation of Si(111)–7 × 7.[101,102] Both exposed the sample to molecular oxygen while scanning, making determination of the exposure and therefore the sticking coefficient problematic. Despite these drawbacks the tunneling images showed a reacted Si adatom site which was generally found near defective areas of the silicon sample, suggesting that these may serve as nucleation sites.

Shortly thereafter, Pelz and Koch[103] examined the same system with the STM but exposed the samples to measured amounts of molecular oxygen with the tip removed from the surface so as not to influence the dosing process. Figure 32 shows the appearance of one area of the 7 × 7 surface during successive oxygen exposures.[103] Noting the changes in adatom configuration and appearance that occur as the exposure in increased, Pelz and Koch determined that at least two reaction states exist, S_1 and S_2, which they suggest are related by at least a two-stage reaction process. The S_1 states are manifest as "bright" or higher adatoms, while the later S_2 states are characterized by "darker" or lower adatoms. This is generally true for all bias

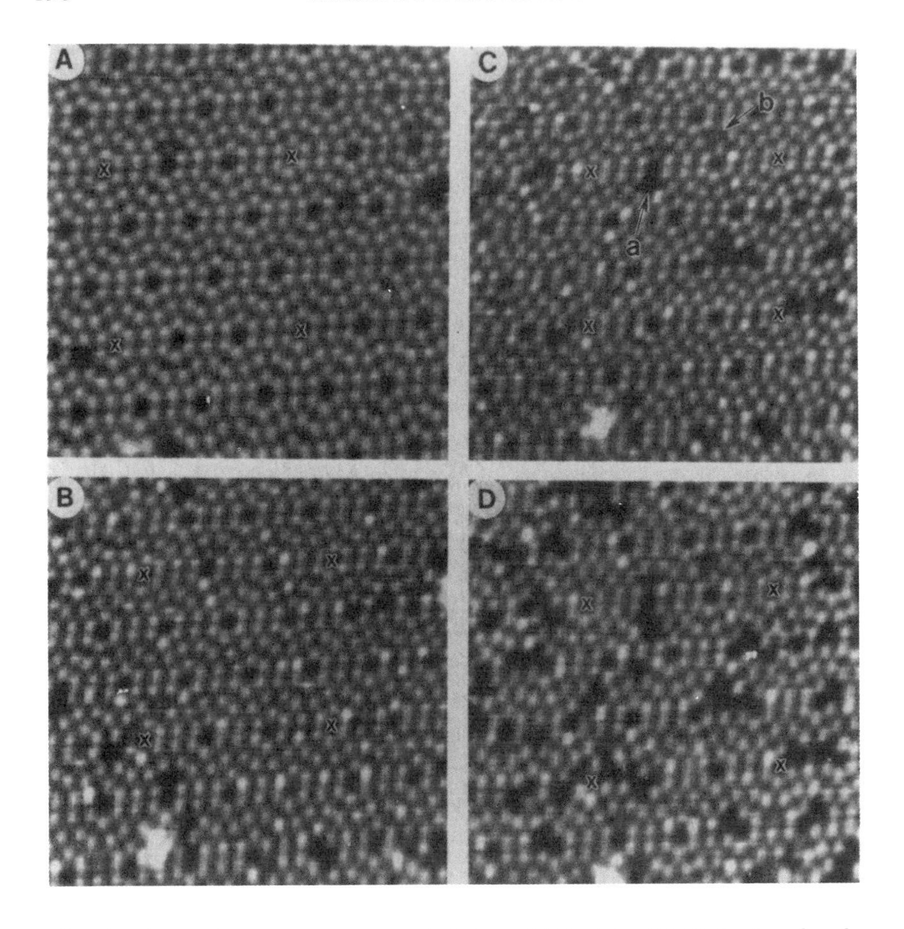

FIG. 32. Gray scale topographs measured at $V_{bias} \cong 2.0\,V$ of the Si(111)–7 × 7 surface for oxygen exposures of (a) 0 L, (b) 0.3 L, (c) 0.4 L, and (d) 0.6 L total O_2 dose. Sites marked "a" and "b" in (c) refer to S_2 adatom sites with different contrast levels. Image scales are ~200 × 200 Å. From Ref. 103.

conditions between $-3.0\,V$ and $+3.0\,V$, indicating that the S_1 state has somewhat enhanced state density near E_F, while the S_2 state has decreased state density relative to unreacted adatoms. As the oxygen exposure is increased, more S_2 adatoms are noted, and the bulk of both are located on the faulted half of the 7 × 7 unit mesh, suggesting increased reactivity at this point, similar to that found in experiments where Pd was deposited at room temperature. Statistics obtained from the images also indicate some preference for corner adatoms over middle adatoms, as was found in studies on NH_3 absorption on Si(111)–7 × 7. These authors also point out that the spatial distribution for the S_2 final state is more uniform than for the S_1, indicating that S_2 may be reached by alternate paths. Without exposing to further amounts of oxygen, this conversion from S_1 to S_2 may be accelerated by gently annealing the sample to $\sim 650°C$ for short periods of time. The authors suggest a model of successive binding of oxygen to different bonds in the adatom structure, but point out that none of the models considered are adequate to explain all of their data. Similar STM observations of this process have been published by another set of investigators.[104]

H Adsorption by Wet Chemical HF Methods. While the previous discussion has concerned the adsorption of simple gases on silicon surfaces *in situ*, it is also possible to absorb atomic hydrogen on Si(111) surfaces by using wet chemical methods. The interest here is the simplicity of the procedure when compared to UHV methods, while a flawless termination of the semiconductor surface with hydrogen would result in complete passivation. Both of these qualities make this procedure attractive from a device manufacturing viewpoint.The cleanliness and surface morphology of Si(111) prepared by wet chemical HF etching methods varies a great deal depending on the process used. Higashi *et al.*[105] have developed an etching recipe which results in smooth, well-defined defined Si(111)–H surfaces as measured using IR vibrational spectroscopy. In order to check the surface models and roughness measurements, Becker *et al.*[106] used The STM under UHV conditions to image these H-terminated surfaces. The Si(111)–H surfaces were prepared outside the UHV chamber containing the STM, then transferred in using conventional load-locking methods. Without any further surface preparation, the tunneling images revealed a well-ordered 1×1 surface with contamination in the form of "white balls" covering ~ 0.01 of the surface. Previous STM studies of HF-etched Si(111) had failed to detect any atomic features,[107] most likely due to the small corrugation amplitude of 0.07 Å. Spectroscopic measurements showed a 1.2 eV energy gap with no prominent features out to 2.0 eV on either side of E_F, indicating that surface-derived states had been removed by the H-termination (all dangling bonds were eliminated). Since the surfaces had been prepared in ambient conditions, the relative reactivity vis-à-vis conventional, clean semiconductor surfaces was many orders of magnitude attenuated. Even more interesting were *in situ* desorption experiments performed on the H-terminated surface using the STM tip as a local electron source. These investigators showed that the hydrogen could be selectively desorbed, from areas as small as ~ 40 Å in extent, by exposing the sample to a short pulse of electrons with kinetic energies of 2–10 eV. The resulting areas were transformed irreversibly to the 2×1 π-bonded chain upon desorption of the hydrogen, as determined both by examining the tunneling images and performing spectroscopy on both H-terminated and desorbed areas. A clue to the reaction mechanism may be gleaned from the data in Fig. 33, which shows the normalized $1 \times 1 \Rightarrow 2 \times 1$

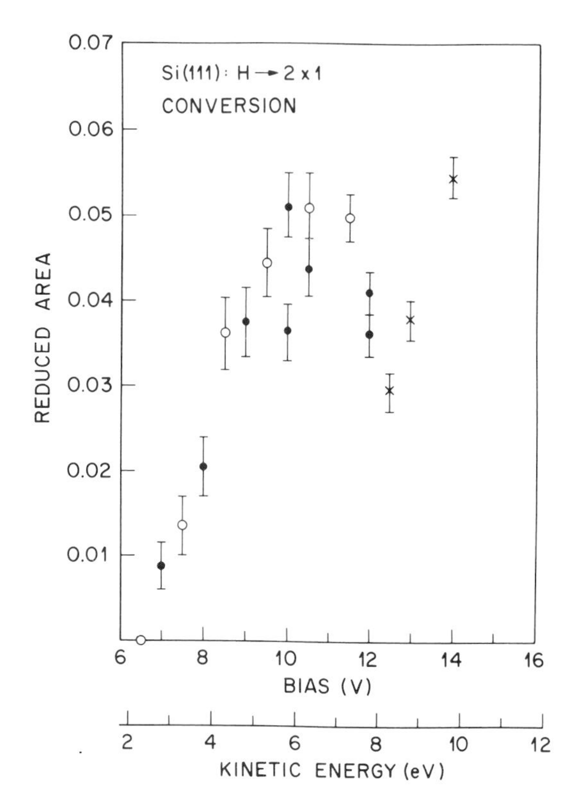

FIG. 33. Reduced 2 × 1 conversion area vs. end-point bias for the electron stimulated desorption of atomic hydrogen from the Si(111)–1 × 1:H surface. The different symbols refer to different data runs with the error bars derived from the statistics of the number of 2 × 1 sites. The lower abscissa indicates the bombarding electron kinetic energy. Taken from Ref. 106.

conversion area as a function of incident electron kinetic energy.[106] A peak is visible at ~6 eV, with an upturn in efficiency as the kinetic energy is increased beyond 10 eV. The peak is resonant with the indirect gap between Si–H bonding and antibonding bands, indicating that the reaction path is one of promoting an electron from the bonding to the antibonding band. Of course, the cross-section for this is quite low; approximately 10^8 electrons are required for each surface site converted. This work demonstrated that the STM could be used as a tool to instigate local chemical reactions.

Morita *et al.*[108] have also used the STM under UHV conditions to examine the H-terminated surface of Si(111) using dilute (1%) HF etches. They have found regions with triangular symmetry similar to those reported by Becker *et al.*, but with a spacing a between surface sites of 2.2 Å, rather than the 3.9 Å observed in the earlier work. They speculate that these features are due not to a Si–H monohydride species but to a trihydride termination of the surface atoms; each of the bright points in the STM images correspond to the terminating H atoms. This conclusion is not confirmed in IR spectroscopy measurements carried out on this system.

5.1.3.2 Metals. A large number of metals have been deposited on Si(111)–7 × 7 including the noble metals Ag, Au, Pd, the group III elements B, Al, Ga, and In, group V materials As and Sb, transition metals Cu and Ni, Group IV metals Sn, and alkali metals such as Cs, Li, and K. The results of some of these experiments are now described.

Ag. Silver-induced reconstructions on Si(111) have been studied extensively with many surface probes, with no consensus reached as to the details of the atomic structure. Much of this is related to the fact that the introduction of a small amount of Ag on the clean Si(111) surface at elevated temperatures causes a significant amount of subsurface rebonding. It is therefore quite likely that the Ag atoms themselves are not isolated to sit upon the surface of the silicon, but are incorporated into a more complex structure in the near surface region. This has led to the proposal of a number of structural models variously described as honeycomb, hexagonal, and trimer[109] to explain the $\sqrt{3} \times \sqrt{3}$ symmetry typically displayed by Ag coverages of $\sim$1 ML deposited at temperatures of 200 to 500°C. Other surface structures exist at various combinations of coverage and temperature, but the Si(111)–Ag$\sqrt{3}$ is the most widely investigated.

The first STM studies of the Si(111)–Ag system were by Wilson and Chiang[110] and Van Loenen *et al.*[111] Both studies reported nearly identical images of the $\sqrt{3} \times \sqrt{3}$ reconstruction, but concluded that the STM data supported different models for the surface structure, namely Ag–honeycomb and embedded trimer respectively. Wilson and Chiang initially prepared their samples by depositing $\sim$2 ML of Ag at room temperature, then subsequently annealing at 480°C for a few minutes to produce the $\sqrt{3}$ structure. Upon annealing at even higher temperatures ($\sim$600°C), regions of local 1 × 3 symmetry were imaged, but not at atomic resolution, thus precluding a detailed structural model. Tunneling images of the $\sqrt{3} \times \sqrt{3}$ areas showed them as six-membered rings of protrustions, suggesting a honeycomb structure for the surface. In contrast, Van Loenen *et al.* held a clean Si(111)–7 × 7 sample at 460°C and deposited Ag at 0.5 ML/min until LEED showed a sharp $\sqrt{3} \times \sqrt{3}$ diffraction pattern. Using the previously discussed CITS technique, they acquired a number of current difference images and used these to argue for the embedded trimer structure capped by a honeycomb of Si atoms. This configuration was based on arguments pertaining to charge transfer and chemical bonding. In essence, both the Almaden and Yorktown Heights groups agreed that the STM images closely resembled a honeycomb, with the former arguing that these were surface Ag atoms and the latter surface Si atoms. Unfortunately, the STM has great difficulty in distinguishing elemental species.

Not content to let matters rest at this stage, Wilson and Chiang carried out further experiments[112] on the Si(111)–Ag $\sqrt{3}$ system. In this later work, they held a clean Si(111)–7 × 7 sample at 500°C while depositing silver. Limiting the coverage to ~0.1 ML, they were able to produce a number of regions where the 7 × 7 and $\sqrt{3} \times \sqrt{3}$ phases coexisted. By imaging the phase boundary at high resolution, and then projecting the known surface lattice positions as determined by the DAS 7 × 7 structure out over the $\sqrt{3}$–Ag regions, they established that the tunneling images failed to support the embedded trimer model proposed by Van Loenen *et al.* but were consistent with either the simple honeycomb or at missing-top-layer honeycomb model proposed by Kono *et al.*[113] Some doubt has been cast over the validity of this projection method, since one of the figures in Ref. 112 extends the Si surface lattice across terraces separated by one atomic step in height. Yet the lattice may be seen to fit the protrusion network everywhere in the image, despite the expected phase shift in position encountered in vertical displacements of one double-layer.

Au. Gold has long been an attractive material for deposition on silicon surfaces both for the ease of carrying out the evaporation and the proliferation of gold contacts in the electronics industry, along with the wide variety of surface phases displayed as a function of both coverage and thermal treatment. Measurements carried out with traditional surface probes have reported structures with symmetries of 5 × 1, 5 × 2[114,115,116,117], $\sqrt{3} \times \sqrt{3}$, and 6 × 6,[118] depending on the coverage and annealing process. For the 5 × 1 and 5 × 2 phases, the reconstruction is described as rows of Au atoms oriented along a $\langle\bar{1}01\rangle$ with a $5a$ spacing, while the $\sqrt{3} \times \sqrt{3}$ and 6 × 6 structures have had a number of models invoked to explain their symmetry. These are related to the similar three-fold phases seen for Ag on Si(111) and consist of the honeycomb structure,[119] the hexagonal structure,[120] and the trimer structure.[121]

The first STM studies of Au-induced reconstructions on Si(111) were reported by Salvan *et al.*[122] of the $\sqrt{3} \times \sqrt{3}$ phase. In this study the authors concluded that the weight of the data favored the simple hexagonal model. In a later study by Dumas *et al.*,[123] 0.6 ML of Au was deposited on a Si(111)–7 × 7 surface while the sample temperature was held at 700°C, resulting in sharp $\sqrt{3}$ LEED patterns. Tunneling images of this structure, taken tunneling from occupied sample states at 0.45 V bias, shows a triangular array of entities with ~0.7 Å relief rotated at 30° to the orientation of the starting Si(111)–7 × 7 surface. The authors noted a somewhat triangular shape to the entities, which led them to conclude that a trimer arrangement of Au atoms was responsible for the STM images.

Subsequent to these experiments, Hasegawa *et al.*[124] used the STM to examine this system under a wider variety of conditions. As in the earlier measurements by Dumas, these workers started with a clean 7×7 surface, but had no way of determining the Au coverage on the silicon sample after deposition, so the experiments simply proceeded from lowest to highest coverage. At the lowest coverages, a 5×5 arrangement of protrustions was seen between terraces of nearly clean 7×7. As the coverage was increased, a 5×2 structure was observed, with the $5a$ direction running along $\langle\bar{1}10\rangle$. These consisted of two somewhat resolved rows with bright protrusions arranged approximately periodically along these rows. At still higher coverages a $\sqrt{3} \times \sqrt{3}$ was seen, similar in appearance to that reported by Dumas *et al.*[123] In this study, no triangular shape was detected for the protrusions, leading the authors to equivocate as to whether the data supported any of the previously suggested structural models. In a further study,[125] these same investigators used the STM to examine the initial transition from the clean 7×7 surface to the Au-induced 5×1 or 5×2, where they concluded that only a small amount of Au (<0.1 ML) was required to cause the Si(111) surface to reorder to 5×2 symmetry. In a similar manner, Baski *et al.*[126] examined the transition at higher coverages between the 5×1 and the $\sqrt{3} \times \sqrt{3}$ surface structures. As has been the case for the Ag-induced phases, the STM is not able to distinguish metal atoms from silicon in the tunneling images, making detailed conclusions difficult at best.

Pd. Palladium, another noble metal, has also been deposited on Si(111)–7×7 and the resulting sequence of growth imaged with the STM. Unlike the Ag and Au systems, where the introduction of the metal adsorbate at elevated temperatures induced a radically different surface phase, the deposition of Pd at room temperature tends to promote the growth of Pd_2Si material with the electrical characteristics of a silicide. Kohler *et al.*[127] have examined this system at successive stages of evolution and find a number of interesting features. First, the Pd atoms selectively adsorb to the faulted portion of the Si(111)–7×7 unit mesh, with nuclei containing ~ 13 Pd atoms. Figure 34 is an STM image showing these Pd nuclei on the 7×7 DAS structure, with a schematic depicting the approximate heights in the STM images.[127] This selection of the faulted side of the unit mesh strongly suggests a greater chemical reactivity, which is related to the small electronic differences between the two halves of the DAS structure. This tendency was also observed in the O_2 adsorption experiments of Pelz and Koch.[103] Further deposition does not lead to an increase in the number of nucleation sites, but rather, a lateral growth of the existing nuclei in an essentially two-dimensional

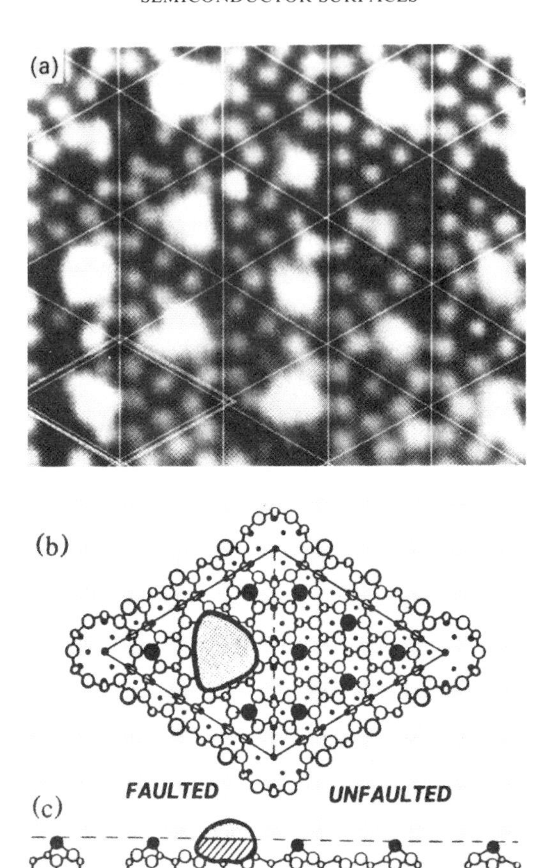

FIG. 34. STM topograph of a 115 × 92 Å are of Si(111)–7 × 7 with 0.25 ML Pd deposited at room temperature (upper panel). The grid indicates the 7 × 7 superlattice (center panel). Schematic of the DAS 7 × 7 with the location of the $PdSi_x$ nuclei indicated (bottom panel). Cut through the long diagonal of the 7 × 7 structure showing the vertical extension of the silicide nuclei. Taken from Ref. 127.

manner. The authors also report that surface defects such as atomic steps do not appear to affect this adsorption process.

B, Al, Ga, In. The first four elements in Group III of the periodic table have all been deposited on the Si(111) surface under various conditions. These materials have the common trait that all exhibit a structure with $\sqrt{3} \times \sqrt{3}$ symmetry, along with other phases peculiar to each of them. Boron exhibits only a $\sqrt{3}$ surface phase, while aluminium takes up both a $\sqrt{3}$ and a $\sqrt{7}$ surface phase. Gallium shows the $\sqrt{3}$ symmetry, a less understood "6.3 × 6.3" phase, and others that are even more poorly understood. Indium displays a number of phases, including one of $\sqrt{3}$ symmetry and one of 4×1 symmetry. All of these have been examined with the vacuum STM. The fact that a phase with $\sqrt{3}$ symmetry is formed by all of these elements suggests that a common structure may account for this phase. Both from theoretical considerations, and experimental evidence, the $\sqrt{3}$ phases are suspected as simple adatom phases, with the metal atoms arranged on an otherwise 1×1 Si double-layer with each adatom bonding to three surface Si atoms. This geometry is quite similar to that displayed by the adatoms on the Si(111)–7×7 surface, where the presence of the additional restatoms gives a 2×2 unit mesh with four surface atoms against a $\sqrt{3} \times \sqrt{3}$ mesh with three surface atoms.

Hamers and Demuth investigated the Si(111)–Al$\sqrt{3}$ surface with STM, and also reported on a related phase having $\sqrt{7} \times \sqrt{3}$ symmetry.[128] They found that the tunneling images were consistent with a simple adatom phase where the Al atoms are arranged in $\sqrt{3}$ symmetry on an otherwise 1×1 silicon substrate. They noticed that two distinct species of adatoms existed, with the minority appearing brighter (higher) than the remainder at one polarity, and conversely darker (lower) in the other, denoting a shift in surface state occupation at these sites. This minority species was inferred as substitutional Si adatoms in an otherwise Al adatom matrix. All of the adatoms were placed on T_4 sites in accordance with total energy calculations.[32] Based on spectroscopic measurements over both types of adatoms, they suggested that the defective adatoms were consistent with creation of a localized donor–acceptor set of states, rather than a simple shift in the DOS due to the presence of the additional bond for the Si adatoms. While appealing in some aspects, this explanation is not that well supported by the available data, and is not offered in explanation of similar substitutional features seen in the Si(111)–Ga and Si(111)–In systems. In further work, adatom-like areas with local $\sqrt{7} \times \sqrt{3}$ symmetry were observed and ascribed to essentially the same model.

The Si(111)–Ga system has been examined by at least two different groups. Nogami *et al.*[129] deposited 0.3 ML of Ga on Si(111)–7×7 at room temperature, then annealed to 475°C to get a well-ordered $\sqrt{3} \times \sqrt{3}$ structure. Similar to the previous work by Hamers and Demuth using aluminium, the Si(111)–Ga surface appeared as a simple adatom phase on a 1×1 substrate. Left open was whether or not the predominantly Ga adatoms were located on H_3 or T_4 sites. This question was answered by reducing the amount of surface Ga until mixed phase regions of 7×7 and $\sqrt{3} \times \sqrt{3}$ were found, then using the relative registry of the DAS 7×7 to determine the simple adatom sites, which were consistent with the T_4. Another phase in the Si(111)–Ga system, the so-called "6.3×6.3" phase[130] was studied by Chen *et al.*[131] with the vacuum STM. This odd phase is achieved by depositing additional Ga beyond the 1/3 ML required for the $\sqrt{3} \times \sqrt{3}$ structure and annealing at 300–500°C. The tunneling images revealed a surface consisting of some regions of the aforementioned Ga adatom phase interspersed with areas of hexagonally close-packed "supercells." These supercells are ~24 Å in lateral extent, and consist of a triangular array of protrusions with 4.1 Å nearest-neighbor spacing. The lateral extent of the supercells accounts for the 6.3*a* periodicity, while the boundaries appear as misfit dislocations. The authors suggest that the supercells are a "graphitic" phase, made up of co-planar interpenetrating Si and Ga lattices, that is weakly bonded to the substrate below. They point out that x-ray standing wave measurements of this structure indicate a contraction of the outer crystal layer by ~0.5 Å, which may allow for the lateral expansion of ~0.3 Å from the bulk Si lattice spacing.

Moving one row down in the periodic table but staying in Group III we have the Si(111)–In system. Like the Al and Ga systems previous, this metal exhibits a $\sqrt{3} \times \sqrt{3}$ symmetry when the surface coverage is ~0.3 ML. Nogami *et al.*[132] have examined this phase, and a related 4 × 1 phase with the STM. Like the metals before it, the $\sqrt{3}$ phase is consistent with a T_4 adatom phase. The best surfaces were obtained with In coverages of 0.3–0.5 ML and an anneal at 450°C after deposition at room temperature. By increasing the coverage 0.7–1.1 ML with the same annealing conditions, a 4 × 1 phase is produced. This consists of double-ridged rows spaced 4*a* apart in the lateral direction, with a 1*a* periodicity along the rows. Although the STM data rules out constructing the 4 × 1 from pairs of 2 × 1 subunits, no atomic model was suggested by the authors for the 4 × 1. In this study, epitaxial growth of In islands on the 4 × 1 was observed upon continued deposition of In metal.

At the top of the Group III column is boron, an element which serves as the chief dopant for p-type silicon, since it possesses a very high solid solubility. While the other elements in this group, usually at a coverage of ~1/3 ML, show a simple adatom $\sqrt{3}$ phase, boron was not expected to follow suit because of the much smaller size of the atom than the metals Al, Ga, and In. In the T_4 binding sites characteristic of the simple adatom phases, free energy is reduced by trading three dangling bonds on the surface layer for one on the adatom, but at some strain energy cost for distorting the surface bonds from their normal p_z character. The effect of this is to pull in the three atoms the adatom bonds with, and rotate the adatom bonds from tetrahedral 108° internal angles to something closer to 90°. Given the small size of boron, the stress introduced by "pulling in" the substrate top-layer atoms makes the simple adatom reconstruction energetically unstable. For these reasons, the existence of a Si(111)–B$\sqrt{3}$ phase was unexpected. This structure was examined in detail using x-ray diffraction methods by Headrick *et al.*,[133] who concluded that an adatom structure was responsible for the phase, but the adatoms were silicon rather than boron. These workers instead

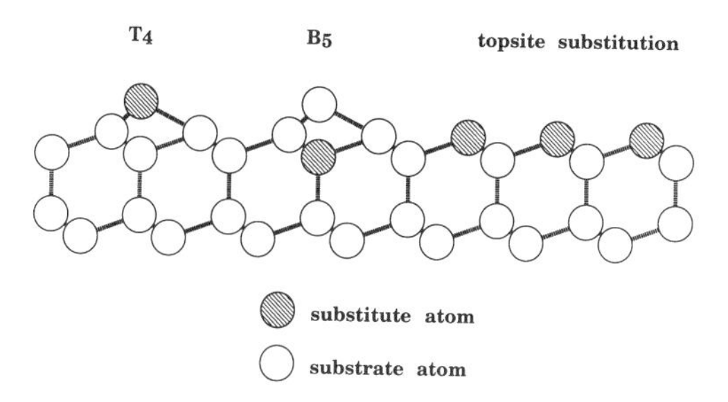

FIG. 35. Substitutional sites on Si(111) surfaces. Shown are (left to right) top T_4 site, bottom B_5 site, and the top double-layer substitutional site.

suggested that the boron atoms occupied the site directly beneath the T_4 silicon adatoms, and referred to this site as B_5 (bottom, five fold coordinated). Figure 35 shows a side view schematic of typical (111) surface sites, with T_4, B_5, and substitutional indicated. STM images of the Si(111)–B$\sqrt{3}$ surface[134,135] were consistent with this structure since they showed a uniform $\sqrt{3}$ lattice bearing strong resemblance to the previous simple adatom surface phases. Both groups of investigators published theoretical total energy calculations which supported the B_5 boron site over the conventional T_4 site. In addition, the presence of dark adatoms in the tunneling images was noted and explained as Si adatoms with a Si atom in the B_5 site rather than the majority boron atoms. Lyo *et al.*,[135] by steadily increasing the substrate Si temperature as the boron was deposited, showed images which suggest that the boron initially occupies the top T_4 site, moving to the energetically favorable B_5 site at higher temperatures. Thibaudau *et al.*[136] also presented tunneling images showing the Si(111)–B$\sqrt{3}$ structure but suggest instead the boron atoms occupy a subsurface site adjacent to and in between, the T_4 silicon adatoms. This configuration is neither supported by theoretical work, nor ion scattering data, on this system.

As. One of the more challenging problems in surface science is the creation of an ideal surface. The principal requirement of the ideal surface is that it have extremely low, or infinitesimal, reactivity. Elsewhere in this chapter we have demonstrated that the truncation of the bulk structure necessarily generates a large number of unsatisfied bonds (essentially one per surface site) which are diminished by reconstruction of the surface atoms. As we have seen, even this process does not eliminate all dangling bonds—the (111) and (001) surfaces of silicon are yet quite reactive with lifetimes measured in hours even under UHV conditions. A number of methods have been introduced to further reduce the reactivity of these surfaces. One is to terminate the surface bonds with some suitable chemical species, such as hydrogen, as was previously discussed. Another is to substitute the surface silicon with another element, so that the resulting electronic configuration leads to reduced reactivity.

Bringans *et al.*[137] showed that a nearly ideal passivation of the Si(111) surface may be brought about by replacement of the outermost Si atoms with As. These As atoms bond to three equivalent Si substrate atoms, put their remaining two electrons into a lone pair state, creating a filled band and eliminating the Si dangling bond surface states. This substitutional site is shown in Fig. 35. This substitution results in a 1×1 Si(111)–As surface with reactivity several orders of magnitude less than that of silicon itself. These surfaces are prepared by exposing clean Si(111)–7×7 surfaces to As_4 flux at a substrate temperature of 700°C. STM images of this surface phase[138] clearly show an array of protrusions with apparent height of ~ 0.1Å spaced ~ 3.9Å apart, i.e. 1×1. There were no discontinuities evident in the tunneling images, which may have signaled the continued existence of a partial stacking fault left over from the starting 7×7 DAS structure. Tunneling spectroscopy performed on this surface showed the introduction of a surface state energy gap of ~ 2.0 eV compared with the small (< 0.5 eV) or nonexistent energy gap displayed by the starting Si(111)–7×7 surface. These authors also presented a theoretical calculation of the surface band energies including the quasiparticle corrections, which are required to achieve a realistic energy gap. The calculations predicted an indirect surface state energy gap of 2.2 ± 0.2 eV, in good agreement with the STM measurements, confirming the substitutional Si(111)–As model.

Ni, Cu. The transition metals, nickel and copper, are responsible for inducing reconstructions on Si(111) that have not yet been resolved. Small amounts of elemental nickel (< 0.1 ML) cause Si(111) to form a structure with $\sqrt{19} \times \sqrt{19}$ symmetry. Given the radical effect trace amounts of this material have on the Si(110) surface, or even the Si(001)–2×1, this is not particularly surprising. The Si(111)–Cu system is even more interesting, having been the object of several STM investigations with very different model structures.

Wilson and Chiang have examined the effects of small amounts of elemental nickel on the Si(111)–7×7 with the tunneling microscope.[139] Their tunneling images show a surface phase that strongly resembles the regular array of ring-like 10 Å structures packed in a hexagonal manner. Every other three-fold "doughnut junction" has a bridge-like structure; these tend to be missing in defective areas. In support of the STM images, and the relatively small amount of Ni required to induce the reconstruction, they propose a model whereby an embedded six-fold coordinated Ni atom causes six Si adatoms to cluster about it in a fashion that intermixes metallic and covalent bonding. The bridge sites are then conventional T_4 Si adatoms located between every other ring feature junction.

The Si(111) surface annealed in the presence of ~ 1ML of copper displays a LEED pattern that is nominally "5 × 5." Close examination of the diffraction pattern reveals the presence of both 1/5 and 1/6 order spots; a number of measurements have settled on a value for the principal surface lattice vector of ~ 5.55*a*. Several STM investigations of the Si(111)–Cu system have been carried out. Wilson *et al.*[140] showed tunneling images which depicted this surface as a disordered array of discrete 5 × 5 meshes arranged loosely with 5–7*a* lateral spacing. These images led the authors to suggest the Si(111)-5 × 5 Cu structure as a set of incommensurate 5 × 5 cells separated by domain walls. Demuth *et al.*[141] subsequently examined this system with the STM, using Fourier filtering methods to decompose the complex tunneling and CITS images into their component features. They report that the 5 × 5 is incommensurate with the underlying Si lattice, with at least three distinct surface phases present in the reconstruction. These consisted of two electronically distinct "1 × 1" CuSi subcells separated by, and coordinated with, a network of slightly depressed regions ("craters"). Demuth *et al.* argued that these features were due to the existence of two distinct silicide phases with markedly different strain fields. The network resulted from arranging the various regions in a manner that reduced the overall strain energy of the surface. This proposal is similar to, but somewhat more complicated than, the model proposed by Chen *et al.*[131] for the Si(111)-"6.3 × 6.3" Ga system. More recently, Mortensen[142] has studied the Si–Cu "5 × 5" using the STM. At low bias conditions (< 0.5 V) tunneling out of occupied sample states, the surface appears as a regular array of ~ 12 Å "balls"; these then appear to have a random arrangement in a network at higher bias conditions (~ 2.0 V). Not only are the bright features random in location, but they undergo movement at the room temperature tunneling conditions employed in this study. Based on junction *I–V* characteristics, Mortensen suggests these features are purely electronic and not associated with distinct geometric features on the surface. Further, he argues that the overall 5 × 5 appearance is a Moiré effect of a nominally CuSi graphite-like lattice incommensurate with the underlying Si(111)-1 × 1 bulk. This argument is bolstered by x-ray standing wave and diffraction measurements which suggest a contraction of the outermost surface layer (containing the Cu atoms) toward the bulk, together with a lateral expansion of ~ 10% and a rotation of 3.3°. These features collectively give a lateral periodicity of ~ 5.55*a*. The 3.3° rotation gives rise to two distinct domains which, in turn, create domain boundaries. He interprets the bright balls as electronic features of the domain boundaries which move because of collective relaxation fluctuations at the boundaries.

In essence, the CuSi domains are "frustrated" in their attempt to form the lowest free energy configuration due to the ground state degeneracy.

5.1.4 Epitaxy, Diffusion, and High Temperature Work

The unrivaled ability of the vacuum STM for observation of surface features at atomic length scales has provided an unprecendented method of glimpsing the mechanisms of processes that are fundamental to much of technology. Chief among these is epitaxial growth, which is just emerging as an arena of experiment for the STM. Other significant areas are studies of surface diffusion (a difficult to measure quantity) and high temperature phase transitions, which are at the forefront of investigation.

5.1.4.1 Homoepitaxy of Silicon

Si(111). The epitaxial growth of Si on Si(111)–7 × 7 substrates has been explored with the STM by Kohler *et al.*[143] They find a variety of features at submonolayer coverage as a function of substrate temperature during the deposition process. At low temperatures, ~20°C, STM images show the first fraction of a monolayer grows in an amorphous manner, with nucleation appearing at the restatom sites of the unit mesh. As the substrate temperature is raised to ~250°C, ordering begins to appear, in that the islands gain lateral extent, with some of them showing adatom-like features characteristic of the starting 7 × 7 surface. As the temperature is increased to ~350°C, growth becomes predominantly epitaxial, with irregularly shaped quasi-7 × 7 islands forming. As the growth temperature is further raised to ~500°C, these islands become triangular and display a tendency to form $\langle\bar{1}\bar{1}2\rangle$ step risers with near perfect registry of the 7 × 7 meshes between the growth islands and the substrate. At these elevated temperatures, the diffusion is sufficient to create denuded zones around substrate steps, demonstrating that these are sinks for deposited Si atoms. The authors also noted a significant tendency for the growth islands to nucleate at substrate defects, such as 7 × 7 domain boundaries, and suggested that this was due to the increased dangling bond density at these sites. The observed tendency to begin second-layer growth before the first layer was complete is also consistent with such an explanation, for the growth islands have a much greater density of domain boundary defects than the substrate surface.

Si(001). Factors affecting the growth of epitaxial silicon have been studied most productively with the STM. Examination of island shapes and the effects of deposition temperature and annealing have provided insight into the roles of anisotropic diffusion and inequivalent step edges.

Individual silicon atoms diffuse freely on the Si(001) surface at room temperature and are never observed in STM images; the smallest unit seen is the dimer. Islands formed at deposition temperatures (i.e., the temperature of the crystal) of up to 570°K are found to be needle-like in appearance, with aspect ratios of 30 to 1 or more.[144,145,146,147] Figure 36 shows typical elongated islands formed at low deposition temperatures. Since the type *A* step is less energetic than the type *B* step,[80,88] it might be assumed that the observed island shapes are simply a manifestation of this fact and represent the equilibrium configuration. However, the aspect ratio of the shapes depends on the deposition temperature, and, elongated islands may be annealed to a more nearly round shape with aspect ratios near 2 to 1 if held at 600°K for several minutes.[148] Islands after annealing are shown in Fig. 37. The long islands arise as a result of strongly anisotropic growth conditions and are not equilibrium shapes. Anisotropic diffusion across terraces and an anisotropic probability of sticking to growing islands, play a role in determining the shapes.

Anisotropic sticking has been clearly demonstrated in a series of experiments by Hoeven *et al.*[149,150,151,152] Figure 38 shows a two domain surface (i.e., single atomic stepped with alternating 2×1 and 1×2 terraces) with a 0.5° vicinal angle after a 0.2 monolayer dose of silicon at 750 K deposition temperature. As is typical in these experiments, the crystal was cooled to room temperature before imaging. Before deposition the terraces were of roughly equal area, but after deposition the type *B* terrace (the upper terrace of a type *B* step) has grown at the expense of the type *A* terrace. It is apparent that the majority of incoming atoms are incorporated at the type *B* step edge. Note at this deposition temperature, islands, which are unstable with respect to step edge incorporation, are not commonly observed. On increasing the coverage to 0.5 monolayer, the surface becomes essentially single domain. This nonequilibrium surface, grown under kinetic control, reverts to the single step structure upon annealing.

Diffusion coefficients are notoriously difficult quantities to determine, for the self-diffusion of silicon, estimates of activation energies have varied widely.[153] Recently, Brocks, Kelly, and Car have used first-principles total-energy calculations to study the diffusion of silicon atoms on the dimer reconstructed Si(001) surface.[154] The calculated total energy surface covering the 2×1 unit cell is shown in Fig. 39(a). The solid circles are the dimer atoms, dimer bonds run in the X direction. The absolute minimum point on the energy surface, *M*, is the preferred binding point of an adatom. The corresponding geometry is shown in Fig. 39(b). The adatom is bound to two atoms, of two different dimers, and the dimers are in the same row. The adatom is asymmetrically located between dimer rows and does not interact with dimers in the adjacent row. Point *H* on the energy surface is a local minimum with a total energy 0.25 eV higher than the absolute minimum. Points *B* and *D* are saddle points with total energies of 1.0 eV and 0.6 eV higher than the absolute maximum. Diffusion along a row is predicted to

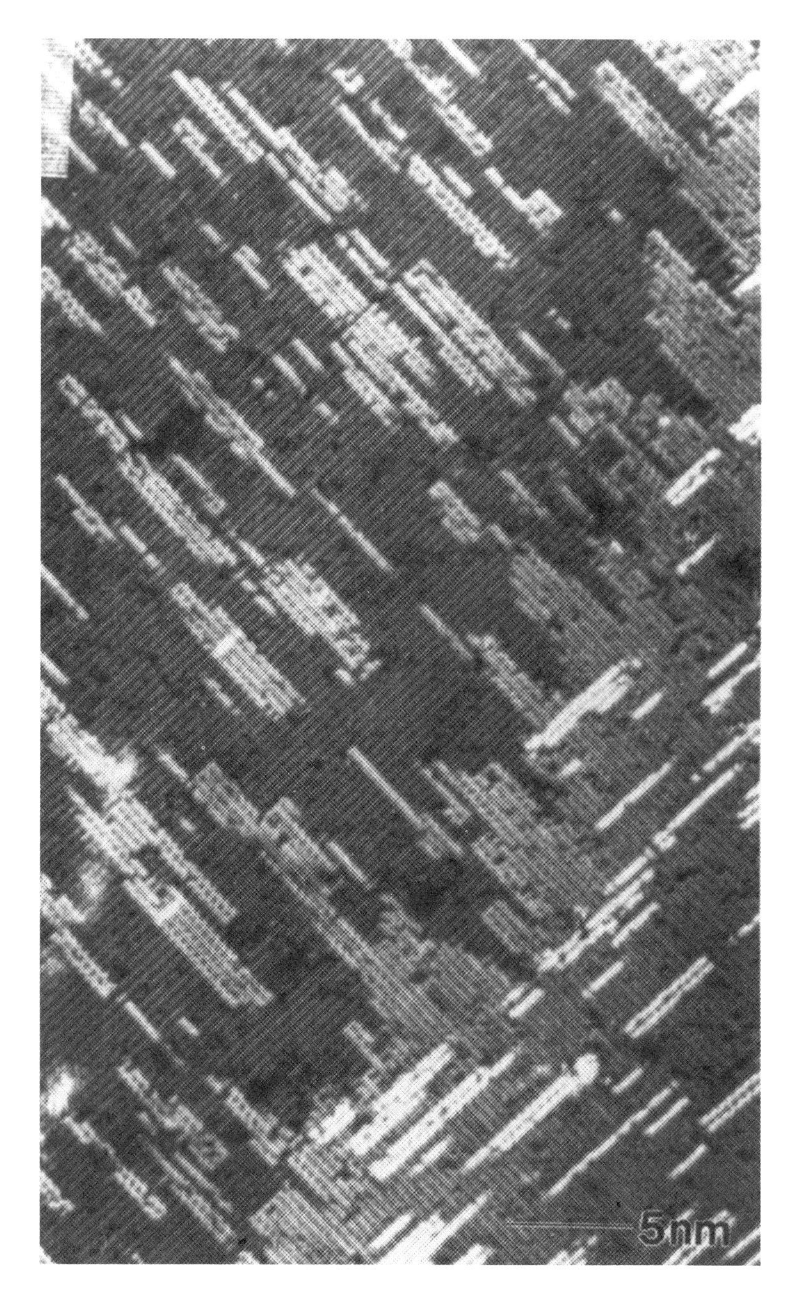

FIG. 36. STM image of the Si(001) surface after deposition of 0.1 monolayer silicon at a deposition temperature of 580 K. From Ref. 144.

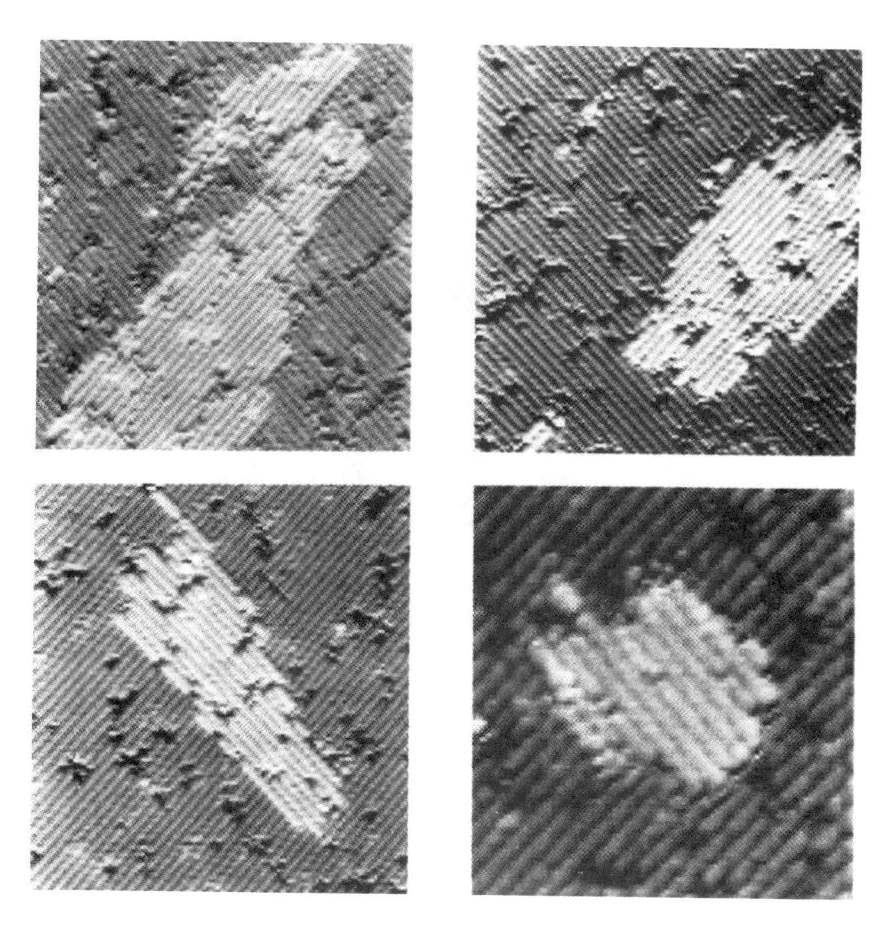

FIG. 37. STM images of island shapes after annealing structures such as those shown in Fig. 36 for two minutes at 600 K. From Ref. 148.

proceed in a C-shaped path, starting at point *M*, the adatom moves to point *H*, then to the half-way point *D*, and then on to *H* and finally *M* in the next unit cell. The effective activation energy is 0.6 eV. Proceeding more directly in the row direction (the Y-axis) leads to a barrier of 0.8 eV. To hop across rows, an adatom must overcome a barrier of 1.0 eV on passing through point *B*, predicting that diffusion is highly anisotropic.

Mo *et al.* argued, based on silicon island shapes, that diffusional anisotropy is small.[148,149] Mo *et al.* later studied denuded zones formed after Si deposition and concluded, from this more direct evidence, that diffusion anisotropy is quite considerable.[155] Most recently Mo *et al.* have attempted to measure the activation barrier to diffusion through an analysis of the

FIG. 38. Vicinal Si(001) with a micut of 0.5° after deposition of 0.2 monolayers of silicon at 750°K. From Ref. 149.

number density of islands formed during deposition.[156] The activation energy and pre-exponential for diffusion along rows were determined to be 0.67 ± 0.08 eV and approximately $10^{-3}\,cm^2/s$ respectively.

5.1.4.2 Phase Transitions at Elevated Temperature. From the beginning of vacuum STM observations in 1982, up until very recently, tunneling images were acquired at or below room temperature in all investigations. However, since 1990, a number of investigators have begun to explore the fertile ground that may be accessed by examining samples at elevated temperatures, enabling direct access to a number of important processes ranging from diffusion to chemical reactions. One must bear in mind that the STM is a relatively slow mechamism; the events that are observable will be those that take place on time scales greater than or equal to the framing time of the

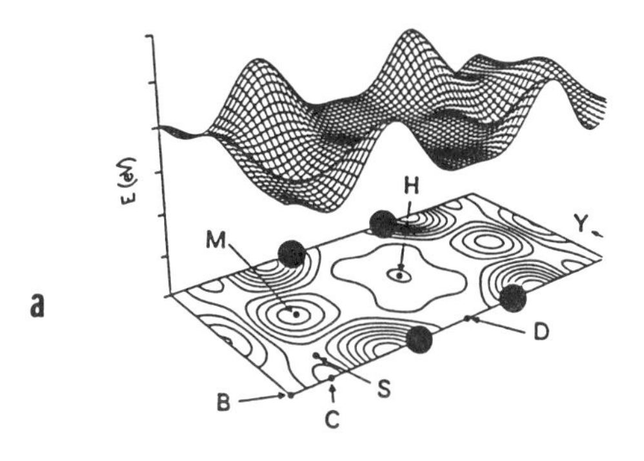

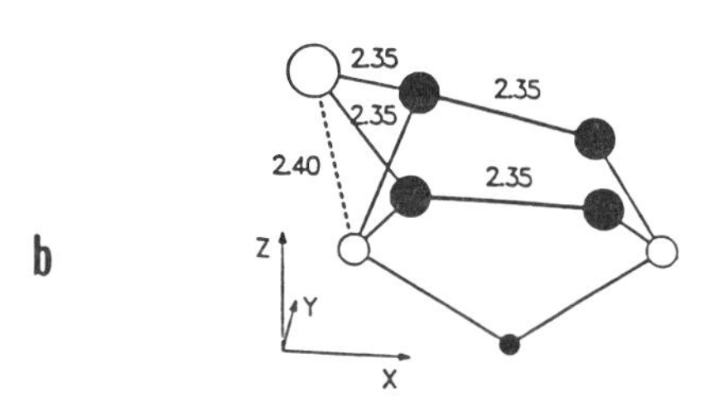

FIG. 39. (a) Perspective view and contour plot of the total energy surface covering a 2 × 1 unit cell. The first contour is at 0.1 eV with respect to the absolute minimum (M) and the contour spacing is 0.2 eV. The interval on the vertical energy axis is 1 eV. The X axis is chosen to pass through a dimer. (b) Perspective view of the equilibrium binding configuration showing the first, second, and third layer substrate atoms. Bond lengths are shown in Angstroms, the open circle represents the adatom. From Ref. 154.

instrument. Since most interesting processes at elevated temperatures occur quite rapidly, the development of faster scanning devices is crucial to this area of investigation.

Si(111)–7 × 7 to 1 × 1. The fundamental phase transition on the clean Si(111)–7 × 7 surface is the transformation of the stable, DAS 7 × 7 reversibly to a "1 × 1" structure at temperatures of ~875°C. As with most other surface phenomena, this process has been studied using other techniques, with no consensus as to the structure of the 1 × 1 phase. Previous STM observations were carried out below this transition temperature, leaving the study of the 7 × 7 ⇔ 1 × 1 outside the realm of atomic scale real-space imaging. Using reflection electron microscopy (REM), Osakabe *et al.*[157]

demonstrated that upon cooling from the 1 × 1 phase, the 7 × 7 nucleated at the upper edges of step risers and proceeded inwards along the upper terrace. Low energy electron microscopy (LEEM) measurements showed triangular regions of 7 × 7 existing on samples quenched from high temperature,[158] but made no determination of when this structure appeared.

Recently, Miki *et al.*[159] published tunneling images taken *at temperature* showing the 7 × 7 ⇔ 1 × 1 transition upon slowly cooling from the 1 × 1 phase. They find that, as in the REM data, the 7 × 7 nucleates at the upper terrace at a step riser, straightening the step riser through incorporation into the reconstruction as previously reported,[5] and is never seen in domains smaller than six unit meshes. Figure 40 shows a tunneling image of the Si(111)–1 × 1, 7 × 7 surface at temperature, along with a schematic depicting the features in the data.[159] The tunneling images show triangular regions with acute top angles ($<60°$), and the authors speculate that this is related to the slow scanning speed of the STM, compared to the fluctuation rate of the domains. Fluctuations occurring more rapidly than the acquisition time of the instrument will tend to draw out the triangular shapes when the fast scanning direction is perpendicular to the general direction of the step riser. While this explanation accounts for the ragged nature of the step risers adjacent to the 7 × 7 domains, it is not an entirely satisfactory explanation for the acute top angles. The minimum domain size of six mesh units is likely due to an interaction between the free energy gained by forming the DAS structure, against that lost by creating a domain boundary between the 1 × 1 and the 7 × 7. By carefully observing the fluctuation of 7 × 7 domains with changing temperature, the authors determine a critical temperature of 868°C for the transition.

Si(111)–2 × 1 to 5 × 5, 7 × 7. Another interesting temperature-dependent phase transition on Si(111) is the conversion of the 2 × 1 π-bonded chain, found on the cleavage surface, to the DAS 7 × 7 phases characteristic of samples treated by thermal processing under UHV conditions. Over the years, this process has been extensively studied using LEED. A related DAS phase, the 5 × 5, may also be found on this surface under certain conditions.[11] The 5 × 5 has been imaged with the STM,[31] and established as a smaller example of the well-studied 7 × 7. Examination of the DAS 7 × 7 schematic (Fig. 2) shows that it may be expanded/contracted to form a series of structures of $(2n+1) \times (2n+1)$ symmetry, where $n = 1, 2, 3, \ldots$. Examples of this structure having a 9 × 9 configuration were also reported in Ref. 31. Since it is well known that the 2 × 1 cleavage phase transforms irreversibly to the 7 × 7 upon annealing, an investigation of the mechanism of this at the atomic scale may prove instrumental in understanding both this process, and the stability of the 2 × 1 at room temperature.

Recently, Feenstra and Lutz[160] employed the STM in a study of the

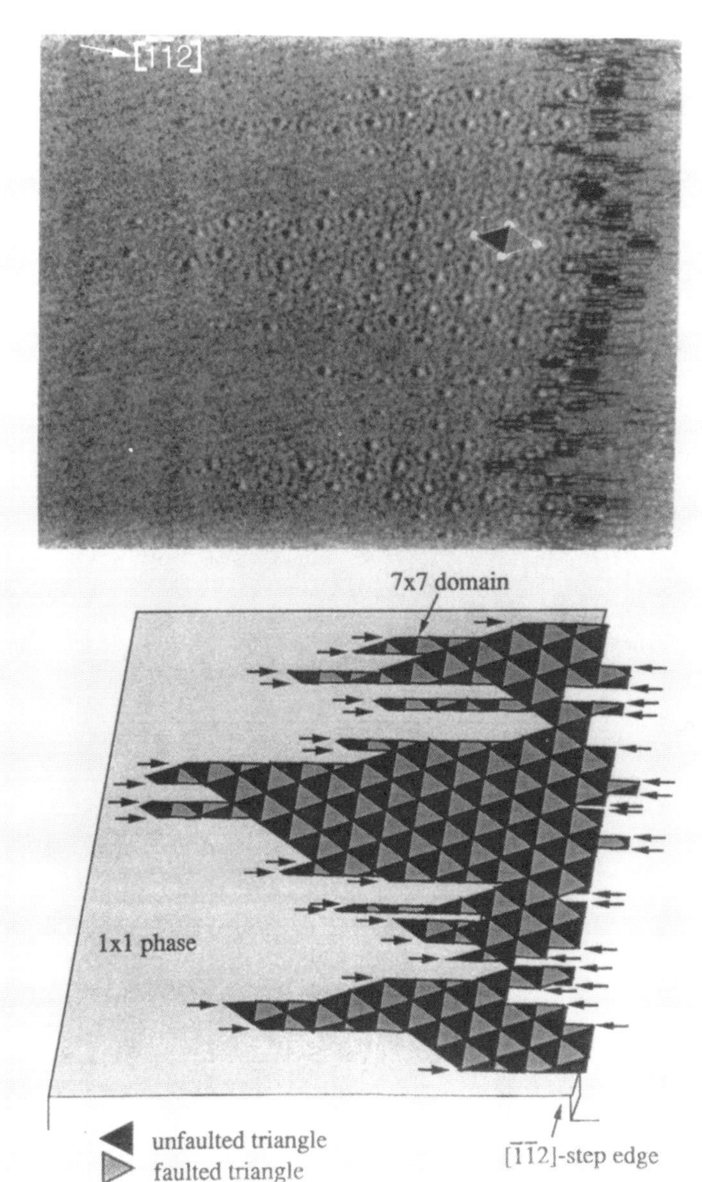

FIG. 40. Tunneling image taken at temperature of the Si(111) surface during the "1×1" $\Rightarrow 7 \times 7$ phase transition. The upper panel shows an atomic step and several domains of 7×7 superstructures. The bottom panel is a schematic of the data in the top panel indicating the 7×7 meshes and partial meshes imaged. Taken from Ref. 159.

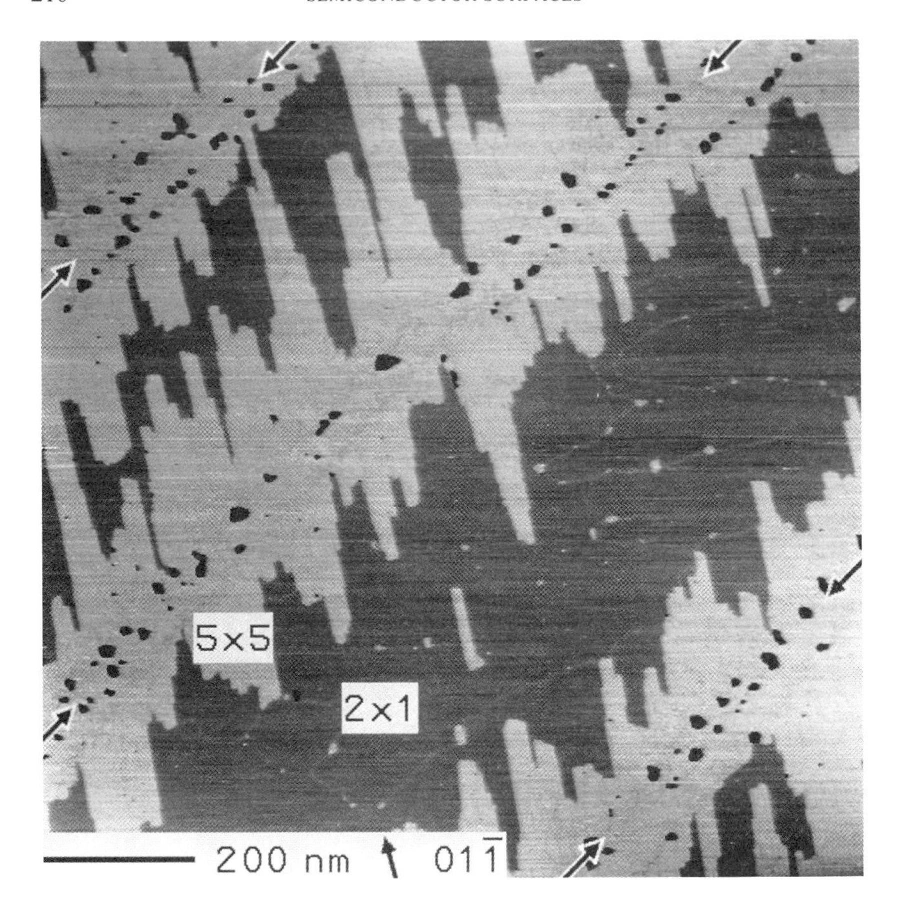

FIG. 41. Large-scale STM image obtained from a cleaved Si(111) sample, annealed at 425°C for 20 s. A series of domain boundaries in the original 2×1 structure are indicated by tick marks at the edge of the image, with the transformed regions growing out from those boundaries. Taken from Ref. 160.

$2 \times 1 \Rightarrow 7 \times 7$ phase transition. They found that two different reaction paths existed, depending on the relative density of nucleation sites (steps or domain boundaries) on the surface. For a sufficiently large concentration of nucleation sites, a path of $2 \times 1 \overset{300\,C}{\Rightarrow} 1 \times 1, 5 \times 5, 7 \times 7 \overset{400\,C}{\Rightarrow} 7 \times 7$ dominates, whereas in the absence of nucleation sites (the centers of large terraces, etc.) a second path $2 \times 1 \overset{350\,C}{\Rightarrow} 5 \times 5 \overset{600\,C}{\Rightarrow} 7 \times 7$ prevails. The existence of two distinct paths, depending on the roughness of the starting surface, is consistent with the past variable observation of the 5×5 phase. Figure 41 shows a tunneling image of the partially transformed Si(111)–2×1 surface, with 2×1 and 5×5 regions indicated, along with tick marks designating 2×1 grain boundaries.

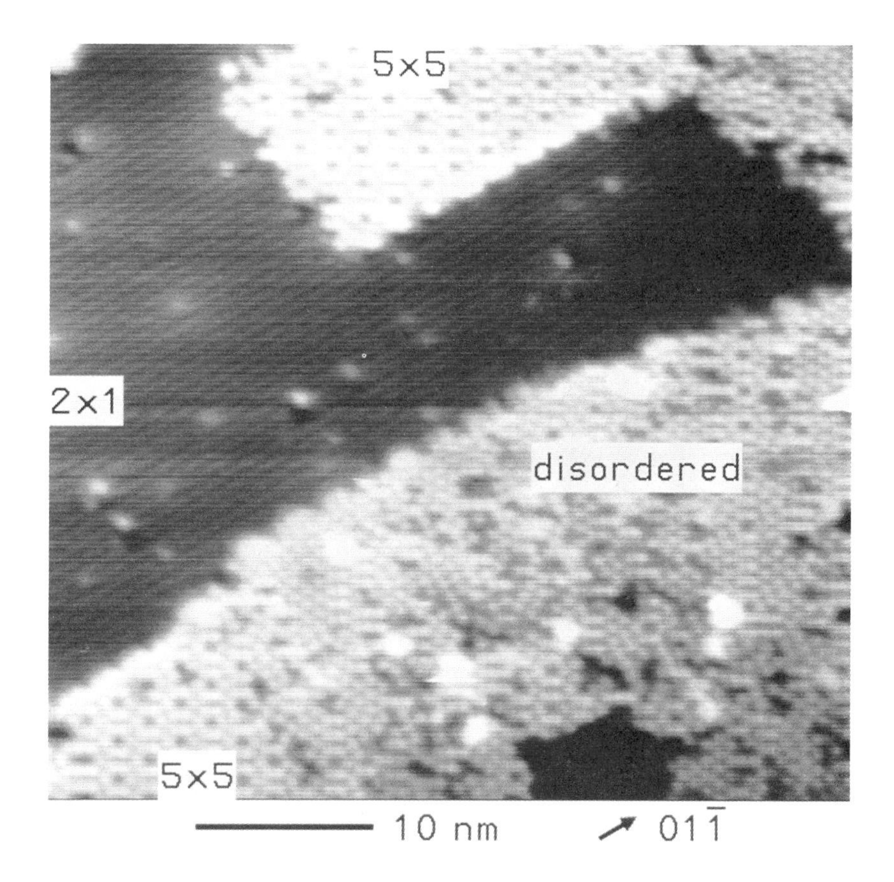

FIG. 42. STM image obtained from a cleaved Si(111) surface annealed at 300°C for 900 s. The image shows a typical example of the structures formed during annealing, and regions of differing structure are indicated. Taken from Ref. 160.

It is immediately clear that the 5 × 5 areas (of lighter shade) are growing out from the domain boundaries, which serve as nucleation sites. Figure 42 is a smaller scale image showing the details of the 2 × 1, 5 × 5, and disordered areas referred to as amorphous 1 × 1. Careful study of a number of images similar to those in the figure led the authors to conclude that the number density of additional surface atoms were crucial to the phase transition, i.e., the 2 × 1 and DAS 5 × 5 have the same number density of surface atoms, but the DAS 7 × 7 has a 4.1% higher density, while a simple adatom structure such as that comprising the 1 × 1 areas possesses a 12.5% higher density. The structures observed from point to point are driven by a trade-off between surface diffusion and free energy. The 7 × 7 has a slightly lower free energy than the 5 × 5, but requires more surface atoms to form. Without a con-

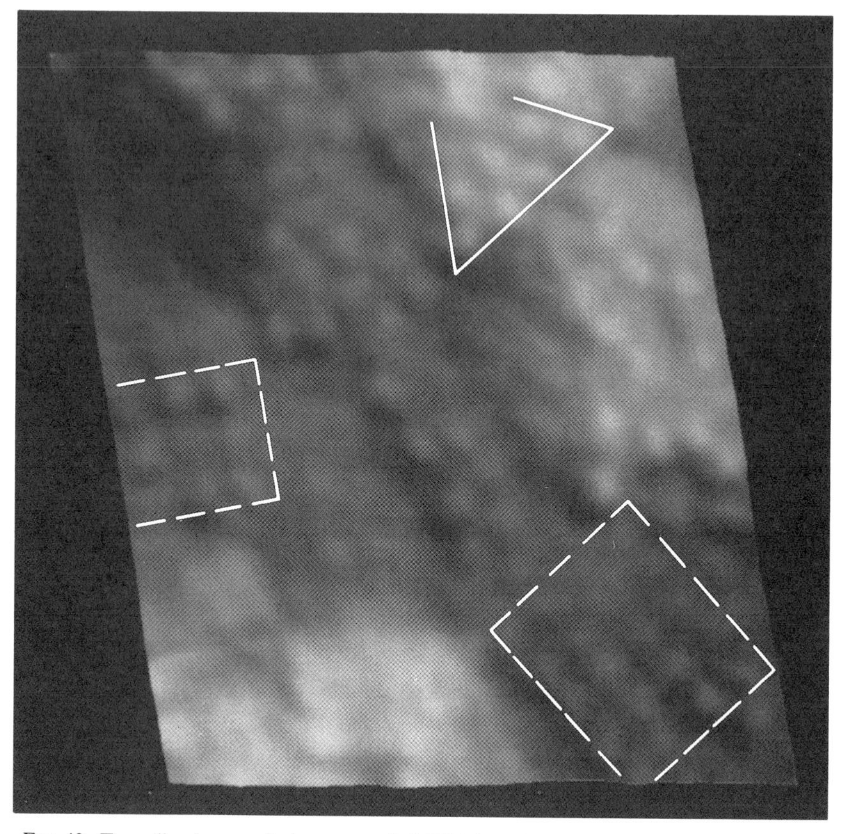

FIG. 43. Tunneling image of a laser-annealed Si(111) surface. Dashed lines enclose regions of c4 × 2 reconstruction and solid line delineate 2 × 2 regions. The gray scale is keyed to height, with a range of ~4 Å. The two lighter gray terraces are one double-layer up from the central region containing the c4 × 2. The image scale is 95 × 105 Å. This image was taken at a tip bias of −0.5 V at 1.0 nA. Taken from Ref. 164.

venient source for the atoms (step, grain boundary), formation is impaired, and the 5 × 5 is created instead. In both cases the 5 × 5 is an intermediate stage in the transition with the number density of surface atoms serving as a kinetic barrier in the process.

Si(111)–7 × 7 to Laser Annealed 1 × 1. A long-standing source of controversy concerned the nature of the surface formed on Si(111) upon exposure to Q-switched laser pulses.[161] One suggestion was that the high temperatures created during laser illumination in conjunction with the subsequent high cooling rate may have allowed the previously mentioned high temperature 1 × 1 phase to remain "frozen in" at room temperature.[162] Further studies by other researchers led to different conclusions regarding

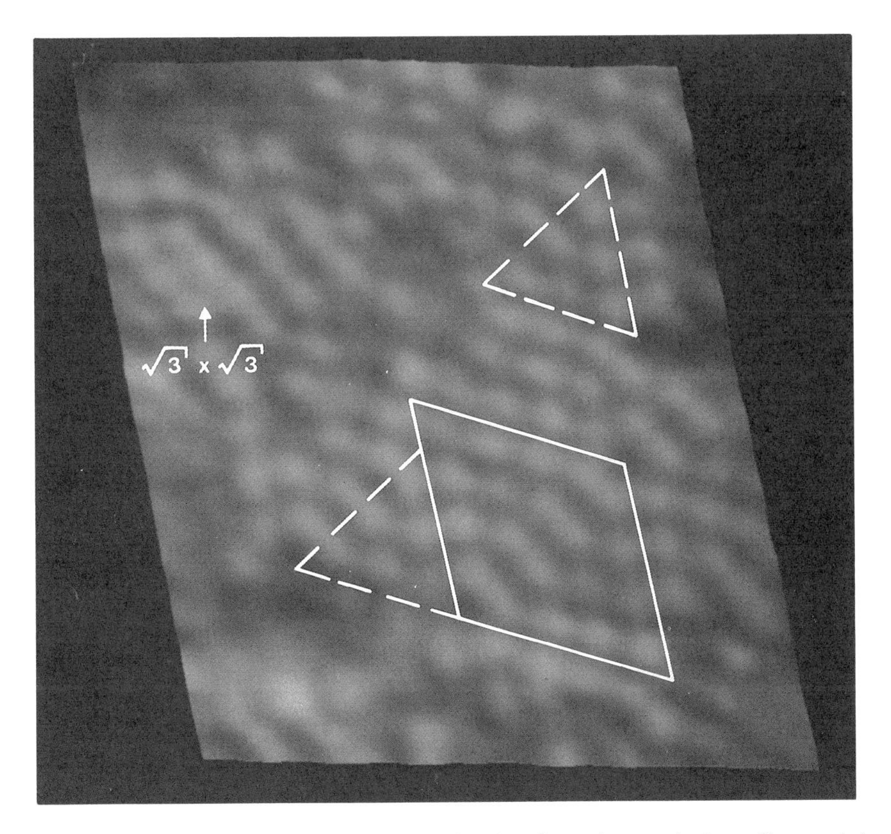

FIG. 44. Tunneling image of laser-annealed Si(111) surface subsequently thermally annealed at 600°C. The dashed lines indicated partial 7×7 meshes while the solid line surrounds a 9×9 mesh. A small $\sqrt{3}$ region is visible at the upper left. Image scale is $\sim 95 \times 105$ Å with the gray scale keyed to height. Taken from Ref. 164.

this surface, the consensus settling on a "disordered 7×7" structure.[163] The high specificity and direct space imaging of the STM made this question a natural candidate for study.

Becker *et al.*[164] showed tunneling images demonstrating that the laser-annealed Si(111) surface consists of a random 2×2 adatom array, exhibiting surface phases that are the building blocks of the Ge(111)–c2×8. A typical region of the laser-annealed surface is shown in Fig. 43 with 2×2 and c4×2 regions outlined.[164] These are distributed over the various twinned configurations possible on the (111) surface. When the laser-annealed sample is later gently annealed for ~ 1 minute at 600°C, the STM reveals that a number of intermediate DAS structures are formed, with 5×5, 7×7, and 9×9 meshes observed. These features are evident in Fig. 44. Here a 9×9 DAS mesh is outlined. This is consistent with the aforementioned results of Feenstra and Lutz on the $2 \times 1 \Rightarrow 7 \times 7$ phase transition, bearing in mind that the laser-stabilized surface has many more sources for adatoms to create the various DAS structures. Upon further annealing to 800°C for several minutes, the pristine 7×7 surface is recovered, albeit with an order of magnitude less surface relief than the starting sputtered and annealed surface. The great attenuation of the surface relief strongly argues for the complete melting of the surface during the laser pulse. These authors further suggest that the room temperature structure of the laser-stabilized surface is representative of the phase seen at elevated temperatures, with the quench rate exceeding the ability of the high-temperature phase to achieve long-range order.

This particular work underlines a tremendous advantage of the STM over other area-averaging surface probes: the ability to image and draw conclusions from structures as small as a single unit mesh.

5.1.5 Conclusion

As has been demonstrated throughout this chapter, the atomistic viewpoint of the vacuum tunneling microscope is a powerful tool for the examination of structures and processes on the scale of atoms. In the barely ten years of existence allocated to the STM, several long-standing questions in silicon surface physics have been answered, and equally important questions raised in many other areas of research.

References

1. G. Binnig, H. Rohrer, C. Gerber, and E. Weibel, *Phys. Rev. Lett.* **50**, 120 (1983).
2. R. D. Meade and D. V. Vanderbilt, *Phys. Rev. B* **40**, 3905 (1989).
3. R. E. Schlier and H. E. Farnsworth, *J. Chem. Phys.* **30**, 917 (1959).
4. K. Takayanagi, Y. Tanishiro, M. Takahashi, and S. Takahashi, *J. Vac. Sci. Tech. A3*, 1502 (1985).
5. R. S. Becker, J. A. Golovchenko, E. G. McRae, and B. S. Swartzentruber, *Phys. Rev. Lett.* **55**, 2028 (1985).
6. R. S. Becker, J. A. Golovchenko, D. R. Hamann, and B. S. Swartzentruber, *Phys. Rev. Lett.* **55**, 2032 (1985).
7. R. N. Feenstra, A. P. Fein, and W. A. Thompson, *Phys. Rev. Lett.* **56**, 608 (1986).
8. K. C. Pandey, *Phys. Rev. Lett.* **47**, 1913 (1981); *Phys. Rev. Lett.* **49**, 223 (1982).
9. R. M. Tromp, R. J. Hamers, and J. E. Demuth, *Phys. Rev. Lett.* **55**, 1303 (1985).
10. R. J. Hamers, J. E. Demuth, and R. M. Tromp, *Phys. Rev. Lett.* **56**, 1972 (1986).
11. J. J. Lander, F. Gobeli, and J. Morrison, *J. Appl. Phys.* **34**, 2298 (1963).
12. J. J. Lander, *Surf. Sci.* **1**, 125 (1964).
13. A. Bennet and M. B. Webb, *Surf. Sci.* **104**, 74 (1980).
14. P. P. Auer and W. Monch, *Surf. Sci.* **80**, 45 (1979).

15. D. J. Chadi, R. S. Bauer, R. H. Williams, G. V. Hansson, R. Z. Bachrach, and J. C. Mikkilson, Jr., *Phys. Rev. Lett.* **44**, 799 (1980).
16. J. D. Levine, P. Mark, and S. W. MacFarlane, *J. Vac. Sci. Tech.* **14**, 878 (1977).
17. D. J. Miller and D. Haneman, *J. Vac. Sci. Tech.* **16**, 1270 (1979).
18. D. J. Miller and D. Haneman, *Surf. Sci. Lett.* **104**, L237 (1980).
19. J. J. Lander and J. Morrison, *J. Appl. Phys.* **34**, 1403 (1963).
20. W. A. Harrison, *Surf. Sci.* **55**, 1 (1976).
21. L. C. Snyder and J. W. Moskowitz, *J. Vac. Sci. Tech.* **16**, 1266 (1979).
22. R. M. Tromp and E. J. van Loenen, *Surf. Sci.* **155**, 441 (1985).
23. X. S. Wang, J. L. Goldberg, N. C. Bartelt, T. L. Einstein, and E. D. Williams, *Phys. Rev. Lett.* **65**, 2430 (1990); J. L. Goldberg, X. S. Wang, J. Wei, N. C. Bartelt, and E. D. Williams, *J. Vac. Sci. Tech.* **A9**, 1868 (1991).
24. If the Si(111) wafer is cooled too rapidly from a temperature above the $1 \times 1 \Leftrightarrow 7 \times 7$ phase transition, kinetic limitations may result in step terminations other than at the corner holes. Under equilibrium conditions, Si(111)–7×7 steps are terminated at the partial stacking fault.
25. R. S. Becker, J. A. Golovchenko and B. S. Swarzentruber, *Phys. Rev. Lett.* **55**, 987 (1985).
26. J. A. Kubby, Y. R. Wang, and W. J. Greene, *Phys. Rev. B* **43**, 9346 (1991).
27. P. Martensson, A. Cricenti, L. S. O. Johansson, and G. V. Hansson, *Phys. Rev. B* **34**, 3015 (1986).
28. Martensson, W. X. Ni, G. v. Hansson, J. M. Nicholls, and B. Riehl, *Phys. Rev. B* **36**, 5974 (1987).
29. J. A. Stroscio, R. M. Feenstra, and A. P. Fein, *J. Vac. Sci. Tech.* **A5**, 838 (1987).
30. O. Nishikawa, M. Tomitori, F. Iwawaki, and N. Hirano, *J. Vac. Sci. Tech.* **A8**, 421 (1990).
31. R. S. Becker, B. S. Swartzentruber, J. S. Vickers and T. Klitsner, *Phys. Rev. B* **39**, 1633 (1989).
32. J. E. Northrup, *Phys. Rev. Lett.* **57**, 154 (1986).
33. D. Haneman, *Phys. Rev.* **121**, 1093 (1961).
34. R. Seiwatz, *Surf. Sci.* **2**, 473 (1964).
35. A. Selloni and E. Tosatti, *Sol. St. Comm.* **17**, 387 (1975).
36. W. A. Harrison, *J. Vac. Sci. Tech.* **14**, 883 (1977).
37. J. A. Stroscio, R. M. Feenstra, and A. P. Fein, *Phys. Rev. Lett.* **57**, 2579 (1986).
38. J. A. Stroscio, R. M. Feenstra, and A. P. Fein, *J. Vac. Sci. Tech* **A5**, 838 (1987).
39. R. M. Feenstra and J. A. Stroscio, *Phys. Rev. Lett.* **59**, 2173 (1987).
40. J. J. Lander and J. Morrison, *J. Chem. Phys.* **37**, 729 (1962).
41. T. D. Poppendieck, T. C. Ngoc and M. B. Webb, *Surf. Sci.* **75**, 287 (1978).
42. M. J. Cardillo and G. E. Becker, *Phys. Rev. Lett.* **40**, 1148 (1978); *Phys. Rev. B* **21**, 1497 (1980).
43. W. A. Harrison, *Surf. Sci.* **55**, 1 (1976).
44. J. A. Appelbaum, G. A. Baraff, and D. R. Hamann, *Phys. Rev. Lett.* **35**, 729 (1975); *Phys. Rev. B* **11**, 3822 (1975); *ibid B* **12**, 5749 (1975); *ibid B* **14**, 588 (1976); *ibid B* **15**, 2408 (1977).
45. G. P. Kerker, S. G. Louie, and M. L. Cohen, *Phys. Rev. B* **17**, 706 (1978).
46. J. E. Rowe, *Phys. Lett. A.* **46**, 400 (1974).
47. F. Jona, H. D. Shih, A. Ignatiev, D. W. Jepson, and P. M. Marcus, *J. Phys. C* **10**, L67 (1977).
48. S. J. White and D. P. Woodruff, *Surf. Sci.* **64**, 131 (1977).
49. J. A. Appelbaum and D. R. Hamann, *Phys. Rev. Lett.* **74**, 21 (1978).
50. S. Y. Tong and A. L. Maldondo, *Surf. Sci.* **78**, 459 (1978).
51. F. J. Himpsel and D. E. Eastman, *J. Vac. Sci. Technol.* **16**, 1297 (1979).
52. D. J. Chadi, *Phys. Rev. Lett.* **43**, 43 (1979).

53. M. T. Yin, and M. L. Cohen, *Phys. Rev. B* **24**, 2303 (1981).
54. K. C. Pandey, p.55, *Proc. 17th Int. Conf. Phys. Semicond.* in D. J. Chadi and W. A. Harrison (eds.), Springer, New York, 1985.
55. M. Aono, Y. Hou, C. Oshima, and Y. Ishizawa, *Phys. Rev. Lett.* **49**, 567 (1982).
56. I. Stensgaard, L. C. Feldman, and P. J. Silverman, *Surf. Sci.* **102**, 1 (1981).
57. R. M. Tromp, R. G. Smeenk, and F. W. Saris, *Phys. Rev. Lett.* **46**, 9392 (1981).
58. Y. Chabal, *Surf. Sci.* **168**, 594 (1986).
59. S. J. White, D. C. Frost, and K. A. R. Mitchell, *Solid State Commun.* **42**, 763 (1982).
60. W. S. Yang, F. Jona, and P. M. Marcus, *Phys. Rev. B* **28**, 2049 (1983).
61. B. W. Holland, C. B. Duke, and A. Paton, *Surf. Sci.* **140**, 1269 (1984).
62. R. J. Hamers, R. M. Tromp, and J. E. Demuth, *Phys. Rev. B* **34**, 5343 (1986).
63. R. J. Hamers, R. M. Tromp, and J. E. Demuth, *Surf. Sci.* **181**, 346 (1987).
64. R. J. Hamers, Ph. Avouris, and F. Bozso, *Phys. Rev. Lett.* **59**, 2071 (1987).
65. J. J. Boland, *Phys. Rev. Lett.* **67**, 1539 (1991).
66. E. Artacho and F. Yndurain, *Phys. Rev. Lett.* **62**, 2491 (1989).
67. G. K. Wertheim, D. M. Riffe, J. E. Rowe, and P. H. Citrin, unpublished.
68. M. C. Payne, N. Roberts, R. J. Needs, M. Needels, and J. D. Joannopoulos, *Surf. Sci.* **211**, 1 (1989).
69. N. Roberts and R. J. Needs, *Surf. Sci.* **236**, 112 (1990).
70. R. A. Wolkow, submitted to *Rev. Sci. Instrum.*
71. R. A. Wolkow, *Phys. Rev. Lett.* **68**, 2636 (1992).
72. O. L. Alerhand and E. J. Mele, *Phys. Rev. B,* **35**, 5533 (1987).
73. T. Tabata, T. Aruga, and Y. Murata, *Surf. Sci.* **179**, L63 (1987).
74. J. Ihm, D. H. Lee, J. D. Joannopoulos, and J. J. Xiong, *Phys. Rev. Lett.* **51**, 1872 (1982).
75. G. P. Kochanski and J. E. Griffith, *Surf. Sci.* **249**, L293 (1991).
76. J. Ihm, M. L. Cohen and D. J. Chadi, *Phys. Rev. B* **21**, 4592 (1980).
77. P. Kruger, A. Mazur, J. Pollmann, and G. Wollgarten, *Phys. Rev. Lett.* **57**, 1468 (1986).
78. J. Pollmann, P. Kruger, and A. Mazur, *J. Vac. Sci. Technol. B* **54**, 945 (1987).
79. P. Badziag, W. S. Verwoerd, and M. A. Van Hove, *Phys. Rev. B.* **43**, 2058 (1991).
80. D. J. Chadi, *Phys. Rev. Lett.,* **59**, 1691 (1987).
81. D. E. Aspnes and J. Ihm, *Phys. Rev. Lett.* **57**, 3054 (1986).
82. J. E. Griffith, G. P. Kochanski, J. A. Kubby, and P. E. Wierenga, *J. Vac. Sci. Technol.* **A7**, 1914 (1989).
83. M Henzler and J. Clabes, "Structural and electronic properties of stepped semiconductor surfaces", in *Proceedings of the 2nd Int. Conf. on Solid Surfaces, Jpn. J. Appl. Phys., Suppl. 2* 389 (1974).
84. O. L. Alerhand, A. N. Berker, J. D. Joannopoulos, D. Vanderbilt, R. J. Hamers, and J. E. Demuth, *Phys. Rev. Lett.*, **64**, 2406 (1990).
85. D. Saloner, J. A. Martin, M. C. Tringides, D. E. Savage. C. E. Aumann, and M. G. Lagally, *J. Appl. Phys.*, **61**, 2884 (1987).
86. F. K. Menn, W. E. Packard, and M. B. Webb, *Phys. Rev. Lett.*, **61**, 2469 (1988).
87. J. E. Griffith and G. P. Kochanski, *Critical Reviews in Solid State and Materials Science*, **16**, 255 (1990).
88. B. S. Swartzentruber, Y. W. Mo, R. Kariotis, M. G. Lagally, and M. B. Webb, *Phys. Rev. Lett.*, **65**, 1913 (1990).
89. O. L. Alerhand, D. Vanderbilt, R. D. Meade, and J. D. Joannopoulos, *Phys. Rev. Lett.*, **61**, 1973 (1988).
90. F. Jona, *IBM J. Res. Dev.* **9**, 375 (1965).
91. B. Z. Olshanetsky and A. A. Shklyaev, *Surf. Sci.* **67**, 581 (1977).
92. P. Martensson, G. V. Hansson, and P. Chiaradia, *Phys. Rev. B* **31**, 2581 (1985).

93. R. S. Becker, B. S. Swartzentruber, and J. S. Vickers, *J. Vac. Sci. Tech. A* **6**, 472 (1988).
94. H. Neddermeyer and St. Tosch, *Phys. Rev. B* **38**, 5784 (1988).
95. E. J. van Loenen, D. Dijkkamp, and A. J. Hoeven, *J. Micros.* **152**, 487 (1988); A. J. Hoeven, D. Dijkkamp. E. J. van Loenen, and P. J. G. M van Hooft, *Surf. Sci* **211/212**, 165 (1989).
96. T. Ichinokawa, H. Ampo. S. Miura, and A. Tamura, *Phys. Rev. B* **31**, 5183 (1985).
97. H. Ampo, S. Miura, K. Kato, Y. Ohkawa, and A. Tamura, *Phys. Rev. B* **34**, 2329 (1986).
98. R. A. Wolkow and Ph. Avouris, *Phys. Rev. Lett.*, **60**, 1049 (1988); Ph. Avouris and R. A. Wolkow, *Phys. Rev. B*, **39**, 5091 (1989).
99. J. J. Boland, *Surf. Sci.* **244**, 1 (1991).
100. H. E. Hessel, A. Feltz, U. Memmert, and R. J. Behm, p.391, *Am. Vac. Soc. 38th National Symposium*, (1991).
101. F. M. Leibsle, A. Samsavar, and T. Chiang, *Phys. Rev. B* **38**, 5780 (1988).
102. H. Tokumoto, K. Miki, H. Murakami, H. Bando, M. Ono, and K. Kajimura, *J. Vac. Sci. Tech. A* **8**, 255 (1990).
103. J. P. Pelz and R. H. Koch, *J. Vac. Sci. Tech. B* **9**, 775 (1991).
104. I. Lyo, P. Avouris, B. Schubert, and R. Hoffmann, *J. Phys. Chem.* **94**, 4400 (1990).
105. G. S. Higashi, Y. J. Chabal, G. W. Trucks and K. Raghavachari, *Appl. Phys. Lett.* **56**, 656 (1990).
106. R. S. Becker, Y. J. Chabal, G. S. Higashi, and A. J. Becker, *Phys. Rev. Lett.* **65**, 1917 (1990).
107. W. J. Kaiser, L. D. Bell, M. Hecht, and F. J. Grunthaner, *J. Vac. Sci. Tech.* **A6**, 519 (1988).
108. Y. Morita, K. Miki, and H. Tokumoto, *App. Phys. Lett.* (1991).
109. G. LeLay, *Surf. Sci.* **132**, 169 (1983).
110. R. J. Wilson and S. Chiang, *Phys. Rev. Lett.* **58**, 369 (1987).
111. E. J. van Loenen, J. E. Demuth, R. J. Hamers, and R. M. Tromp, *Phys. Rev. Lett.* **58** 373 (1987).
112. R. J. Wilson and S. Chiang, *Phys. Rev. Lett.* **59**, 2329 (1987).
113. S. Kono, K. Higashiyama, T. Kinoshita, T. Miyahara, H. Kato, H. Oshawa, Y. Enta, F. Meda, and Y. Yaegashi, *Phys. Rev. Lett* **58**, 1555 (1987).
114. H. Lipson and K. E. Singer, *J. Phys. C* **7**, 12 (1974).
115. J. H. Huang and R. S. Williams, *Surf. Sci.* **204**, 445 (1988).
116. L. E. Bermann and B. W. Batterman, *Phys. Rev. B.* **38**, 5397 (1988).
117. M. Ichikawa, T. Doi, and K. Hayakawa, *Surf. Sci.* **159**, 133 (1985).
118. S. Ino, Jap. *J. Appl. Phys.* **1**, 891 (1977).
119. J. H. Huang and R. S. Williams, *Phys. Rev. B* **38**, 4022 (1988).
120. K. Oura, M. Katayama, F. Shoji, and T. Hanawa *Phys. Rev. Lett.* **55**, 1486 (1985).
121. K. Hiagshiyama, S. Kono, and T. Sagawa, *Jpn. J. Appl. Phys.* **25**, L117 (1986).
122. F. Salvan, H. Fuchs, A. Baratoff, and G. Binnig, *Surf. Sci,* **162**, 634 (1985).
123. P. Dumas, A. Humbert, G. Mathieu, P. Mathiez, C. Mouttet, R. Rolland, F. Salvan, and F. Thibaudau, *J. Vac. Sci. Tech.* **A6** 517 (1988).
124. T. Hasegawa, K. Takata, S. Hosaka, and S. Hosoki, *J. Vac. Sci. Tech.* **A8**, 241 (1990).
125. T. Hasegawa, K. Takata, S. Hosaka, and S. Hosoki, *J. Vac. Sci. Tech.* **B9**, 758 (1991).
126. J. Nogami, A. A. Baski, and C. F. Quate, *Phys. Rev. Lett.* **65**, 1611 (1990).
127. U. K. Kohler, J. E. Demuth, and R. J. Hamers, *Phys. Rev. Lett.* **60**, 2499 (1988).
128. R. J. Hamers and J. E. Demuth, *Phys. Rev. Lett.* **60**, 2527 (1988); R. J. Hamers, *Phys. Rev. B* **40**, 1657 (1989).
129. J. Nogami, S. Park, and C. F. Quate, *Surf. Sci.* **203**, L631 (1988).
130. M. Otsuka, and T. Ichikawa, *Jpn. J. Appl. Phys.* **24**, 1103 (1985).

131. D. M. Chen, J. A. Golovchenko, P. B. Bedrossian, and K. Mortenson, *Phys. Rev. Lett.* **61**, 2867 (1988).
132. J. Nogami, S. Park, and C. F. Quate, *Phys. Rev. B* **36**, 6221 (1987).
133. R. L. Headrick, L. C. Feldman, I. K. Robinson, and E. Vlieg, *Phys. Rev. Lett.* **63**, 1253 (1989).
134. P. Bedrossian, D. M. Chen, D. Vanderbilt, J. A. Golovchenko, K. Mortenson, and R. D. Meade, *Phys. Rev. Lett.* **63**, 1257 (1989).
135. I. W. Lyo, P. Avouris, and E. Kaxiras, *Phys. Rev. Lett.* **63**, 1261 (1989).
136. F. Thibaudau, P. Dumas, P. Mathiez, A. Humbert, D. Satti, and F. Salvan, *Surf. Sci.* **211**, 148 (1989).
137. M. A. Olmstead, R. D. Bringans, R. I. G. Uhrberg, and R. Z. Bachrach, *Phys. Rev. B* **34**, 6401 (1986).
138. R. S. Becker, B. S. Swartzentruber, J. S. Vickers, M. S. Hybertsen, and S. G. Louie, *Phys. Rev. Lett.* **60**, 116 (1990).
139. R. J. Wilson and S. Chiang, *Phys. Rev. Lett.* **58**, 2575 (1987).
140. R. J. Wilson, S. Chiang, and F. Salvan, *Phys. Rev. B* **38**, 12696 (1988).
141. J. E. Demuth, P. Kaplan, U. K. Koehler, and R. J. Hamers, *Phys. Rev. Lett.* **62**, 641 (1989).
142. K. Mortensen, *Phys. Rev. Lett.* **66**, 461 (1991).
143. U. K. Kohler, J. E. Demuth, and R. J. Hamers, *J. Vac. Sci. Tech.* **A7**, 2860 (1989).
144. R. J. Hamers, U. K. Kohler, and J. E. Demuth, *J. Vac. Sci. Tech.* **A8**, 195 (1990).
145. Y. W. Mo, R. Kariotis, D. E. Savage, and M. G. Lagally, *Surf. Sci.* **219**, L551 (1989).
146. M. G. Lagally, R. Kariotis, B. S. Swartzentruber, and Y. W. Mo, *Ultramicoscopy* **31** 87 (1989).
147. Y. W. Mo, R. Kariotis, B. S. Swartzentruber, M. B. Webb, and M. G. Lagally, *J. Vac. Sci. Technol. A* **8**, 201 (1990).
148. Y. W. Mo, B. S. Swartzentruber, R. Kariotis, M. B. Webb, and M. G. Lagally, *Phys. Rev. Lett.* **63**, 2393 (1989).
149. A. J. Hoeven, J. M. Lenssinck, D. Dijkkamp, E. J. van Loenen, and J. Dieleman, *Phys. Rev. Lett.* **63**, 1830 (1989).
150. A. J. Hoeven, E. J. van Loenen, D. Dijkkamp, J. M. Lenssinck, and J. Dieleman, *Thin Solid Films* **183**, 263 (1989).
151. A. J. Hoeven, D. Dijkkamp, E. J. van Loenen, J. M. Lenssinck, and J. Dieleman, *J. Vac. Sci. Technol* **8** 207 (1990).
152. A. J. Hoeven, D. Dijkkamp, J. M. Lessinck, E. J. van Loenen, and J. Dieleman, *J. Vac. Sci. Technol* **8**, 3657 (1990).
153. F. Allen and E. Kasper, in *Silicon Molecular Beam Epitaxy*, Vol. I, p.65, (E. Kasper and J. C. Bean, eds.), CRC Press Boca Raton, FL, 1988.
154. G. Brocks, P. J. Kelly, and R. Car, *Phys. Rev. Lett.* **66**, 1729 (1991).
155. Y. W. Mo and M. G. Lagally, *Surf. Sci.*, **248**, 313 (1991).
156. Y. W. Mo, J. Kleiner, M. B. Webb, and M. G. Lagally, *Phys. Rev. Lett.* **66**, 1998 (1991).
157. N. Osakabe, Y. Tanishiro, K. Yagi, and G. Honjo, *Surf. Sci.* **109**, 133 (1981).
158. M. Mundschau, E. Bauer, W. Telieps, and W. Swiech, *Phil. Mag.* **61**, 257 (1990).
159. K. Miki, Y. Morita, H. Tokumoto, T. Sato, M. Iwatsuki, M. Suzuki, and T. Fukuda, to appear in *Ultramicroscopy*, May 1992.
160. R. M. Feenstra and A. M. Lutz, *Phys. Rev. B* **42**, 5391 (1990).
161. D. M. Zehner, C. W. White, and G. W. Owenby, *App. Phys. Lett.* **36**, 56 (1980).
162. P. L. Cowand-and J. A. Golovchenko, *J. Vac. Sci. Tech.* **17**, 1197 (1980).
163. R. M. Tromp, E. J. van Loenen, M. Iwami, and F. Saris, *Sol. St. Comm.* **44**, 971 (1982).
164. R. S. Becker, J. A. Golovchenko, G. S. Higashi, and B. S. Swartzentruber, *Phys. Rev. Lett.* **57**, 1020 (1986).

5.2. Germanium

Russell Becker

AT&T Bell Laboratories, Murray Hill, New Jersey

5.2.1. Introduction

While silicon has dominated the solid state electronics industry as the semiconductor of choice for device applications for more than 30 years, it is interesting to recall that the first transistor was constructed of germanium. Located two levels below carbon, and just below silicon in the fourth column of the periodic table, germanium has a lattice constant five percent larger than that of silicon, with a consequently smaller bandgap of 0.7 eV compared to the 1.17 eV of silicon. Both germanium and silicon are indirect gap materials with otherwise remarkably similar band structure. Germanium forms tetrahedrally coordinated covalent bonds and takes the diamond crystal structure in its solid phase, like both carbon and silicon. Germanium has a lower melting point than silicon, 937°C versus 1410°C. However, in contrast to silicon, the oxide of germanium is water soluble, rendering it much less useful from a technological standpoint, to the point that its availability from semiconductor vendors was imperiled as little as ten years ago. In the past few years, scientific and commercial interest in germanium has been increasing as heteroepitaxial growth methods improve[1] resulting in attractive Ge–Si alloy structures with applications in bandgap engineering for optoelectronic devices. There are some indications that a suitably ordered alloy of germanium and silicon may prove to be a direct gap semiconductor with enormous implications for the integration of optical and electronic devices, given the relative experience and familiarity of the commercial electronics industry with the manufacture and engineering of silicon and germanium compared to III–V materials.

Of no less importance than silicon in the understanding of chemistry and physics at semiconductor interfaces are the equivalent surfaces of germanium. Considerable insight into the nature of processes at silicon surfaces may be obtained by studying those same processes on the equivalent germanium surfaces. Unfortunately, while a quick perusal of the literature reveals a considerable body of work on silicon surfaces, there exists significantly less on germanium surfaces. Nonetheless, a thorough understanding of the native reconstructions of the low index germanium surfaces, together with their simple surface chemical reactions is of fundamental importance in

METHODS OF EXPERIMENTAL PHYSICS
Vol. 27

ISBN 0-12-475972-6

understanding the role of atomic geometries, steps, point defects, and related phenomena in semiconductor surface physics. In the short time that the STM has been available as an accepted tool of the surface scientist, the majority of studies on both metal and semiconductor surfaces has been limited to imaging their surface structure, although some relatively recent studies have been carried out in fields such as atomic-scale lithography, electrochemistry, surface dynamics of epitaxy, and surface chemistry. Most of the existing work on germanium surfaces is concerned with the details of the fundamental surface atomic geometry.

Chronologically, the Ge(111) surface with its c2 × 8 reconstruction was the second semiconductor surface imaged by the STM,[2] three years after Rohrer and Binnig presented real-space images of the Si(111)–7 × 7 surface.[3] This early work showed that the Ge(111)–c2 × 8 was less ordered than the 7 × 7, and was composed of simpler subunits, of 2 × 2 and c4 × 2 symmetry, that, when appropriately arranged, described one of the three equivalent twin domains of the c2 × 8. While the existence of three twinned domains had been derived from earlier low energy electron diffraction (LEED) studies, the presence of smaller building blocks was unsuspected, as the subunits' diffraction spots were superimposed on those of the c2 × 8. The next STM study carried out on germanium was by Kubby and co-workers[4] who used the microscope to determine both the spatial symmetry and electronic characteristics of the Ge(001) surface. Structurally, the (001) face is most similar to Si(001), with both surfaces exhibiting a 2 × 1 reconstruction described by dimerization of neighboring surface atoms, decreasing the surface energy principally by reducing the number of dangling bonds by half. A comprehensive study of Ge(111)–c2 × 8 was carried out by Becker *et al.*[5] where the various surface features found on germanium were compared to the dimer–adatom–stacking fault[6] (DAS) 5 × 5, 7 × 7, and 9 × 9 reconstructions of Si(111), and the simpler adatom structures imaged on the laser-annealed Si(111) surface. They concluded that the c2 × 8 is a pure adatom structure, with germanium adatoms occupying the top T_4 sites on a nominally 1 × 1 substrate, rather than the hollow H_3 sites. These adatom sites are the same as those occupied on the Si(111)–7 × 7. This model was confirmed in a recent study by Feenstra and Slavin who started from the cleaved Ge(111)–2 × 1 surface and annealed to get the c2 × 8.[7] Subsequently, the surfaces of germanium have been involved in a variety of STM studies and experiments. Both the (111) and (001) germanium surfaces terminated by a monolayer of arsenic have been studied with the STM, and contrasted to related Si(111) : As and Si(001) : As systems.[8] Heteroepitaxy of germanium on Si(001) has been studied by Mo *et al.*[9] Their STM images suggest four distinct stages in the Stranski–Krastinov growth of the germanium islands, and that the orientation of the silicon substrate has little effect on island orientation. The interaction of molecular oxygen with Ge(111) has been examined in detail with the STM in a series of consecutive exposure images of one sample region.[10] The images in this study show that the nucleation of the oxygen absorption is non-homogeneous in the initial stages. There have also been reports on the atomic-scale modification of the electronic characteristics of the Ge(111) surface on exposure to atomic hydrogen.[11] The Ge(111) surface has even been utilized in *in situ* atomic scale surface modification experiments.[12] Finally, an elevated temperature study of the Ge(111) surface around the c2 × 8 to 1 × 1 phase transition at ~300°C shows that the c2 × 8 "melts" progressively from the domain boundaries as the transition temperature is approached.[13]

The rest of this chapter will review the STM results on the clean (111) and (001) germanium surfaces, the GeSi(111) surface, both Ge(111) and Ge(001) terminated with an arsenic monolayer, DAS phases induced by submonolayer tin on Ge(111), adsorption of molecular oxygen on Ge(111), and the heteroepitaxy of germanium on silicon.

5.2.2 Clean Surfaces of Ge and Alloys

5.2.2.1 Ge(111)–c2 × 8. The clean surface of Ge(111) has a native reconstruction characterized by c2 × 8 symmetry. Early LEED studies had suggested an 8 × 8 symmetry for the Ge(111) surface, but an analysis of the "missing beams" in the diffraction pattern arising from sputtered and annealed samples suggested the existence of multiply twinned 2 × 8 domains[14] rather than a three-fold symmetric structure, such as the 7 × 7 on Si(111). Further investigations pointed out that the LEED patterns were better explained by twinned domains of c(2 × 8) symmetry.[15] Angle resolved ultraviolet photoemission spectroscopy (ARUPS) experiments performed on both the Si(111)–7 × 7 and the Ge(111)–c2 × 8 systems show similar surface bands, with both systems indicating a basic periodicity of two lattice spacings.[16] The evidence from the photoemission studies suggested that, although the symmetry of the diffraction patterns from Si(111) and Ge(111) were different, the basic structural elements were similar. Adding more confusion to the issue was an experiment by McRae *et al.*[17] where the surface of a 2000 Å epitaxial layer of Ge grown on Si(111) displayed the 7 × 7 reconstruction native to silicon, not the c2 × 8 of germanium.

The first study of the Ge(111) surface with the STM was performed in 1985 by Becker *et al.*[2] Figure 1 is a constant current tunneling image of the sputtered and annealed Ge(111)–c2 × 8 surface. This figure has undergone a linear transformation to remove the thermal drift. In this image, rendered as a gray scale where dark regions are lower in apparent height than bright regions, a variety of reconstructions, manifest as protrusions, are visible. In the lower-right is a region of c(2 × 8) symmetry, while the upper-left part of

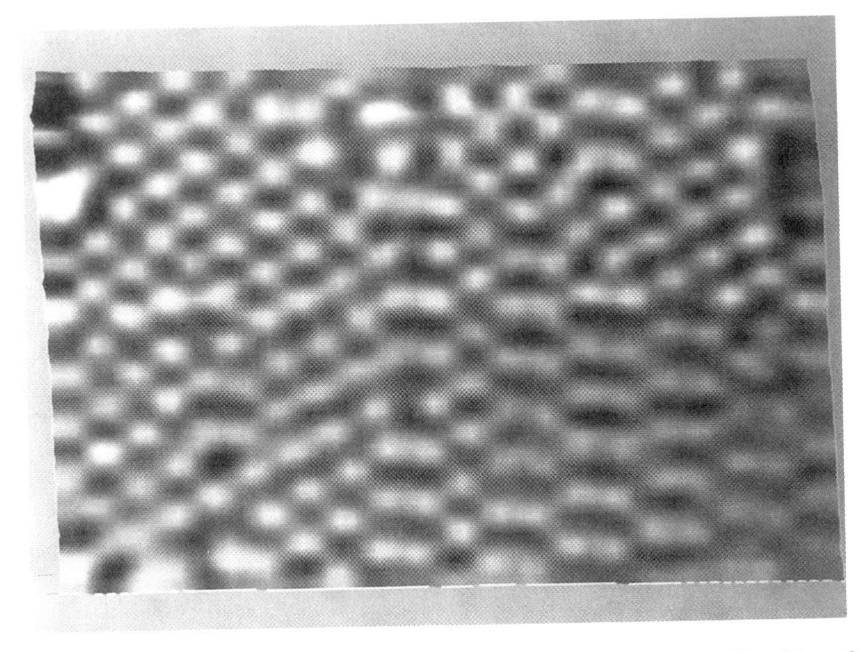

FIG. 1. Tunneling image of Ge(111) c2 × 8, and c4 × 2 reconstructions taken from Figure 2 of Ref. 2. The demanded tunneling current is 0.8 nA at a tip bias of −2.0 V with respect to the sample.

the image shows an array of protrusions in 2 × 2 ordering. A small region in the upper-right is consistent with protrusion ordered similarly to the 2 × 2 phase; these are in local c(4 × 2) symmetry. While this study could not determine the nature of the protrusions (single or multiple atom clusters), fundamental aspects of the surface structure were readily apparent in the STM images. First, the 8 × symmetry of the Ge(111) surface was clearly due to twinned domains of c2 × 8 unit cells. Second, domains of local 2 × 2 and c4 × 2 symmetry coexisted with the c2 × 8. The existence of these secondary surface structures had gone undetected since they shared the same half-order LEED beams with the c2 × 8. Third, the c2 × 8 was constructed of a coherent stacking of 2 × 2 and c4 × 2 cells, alternating row by row, to create one domain of c2 × 8 symmetry. This last feature points to a major advantage of the STM over large area sampling probes in surface studies where two or more phases may coexist.

Following the initial study, a more comprehensive study of the Ge(111) surface structures, and their relation to analogous structures on Si(111) was published in 1989 by Becker *et al.*[5] New techniques employed in this study were the use of dual polarity tunneling images and interrupted feedback-loop current–voltage (I–V) characteristics,[18] both of which are described in

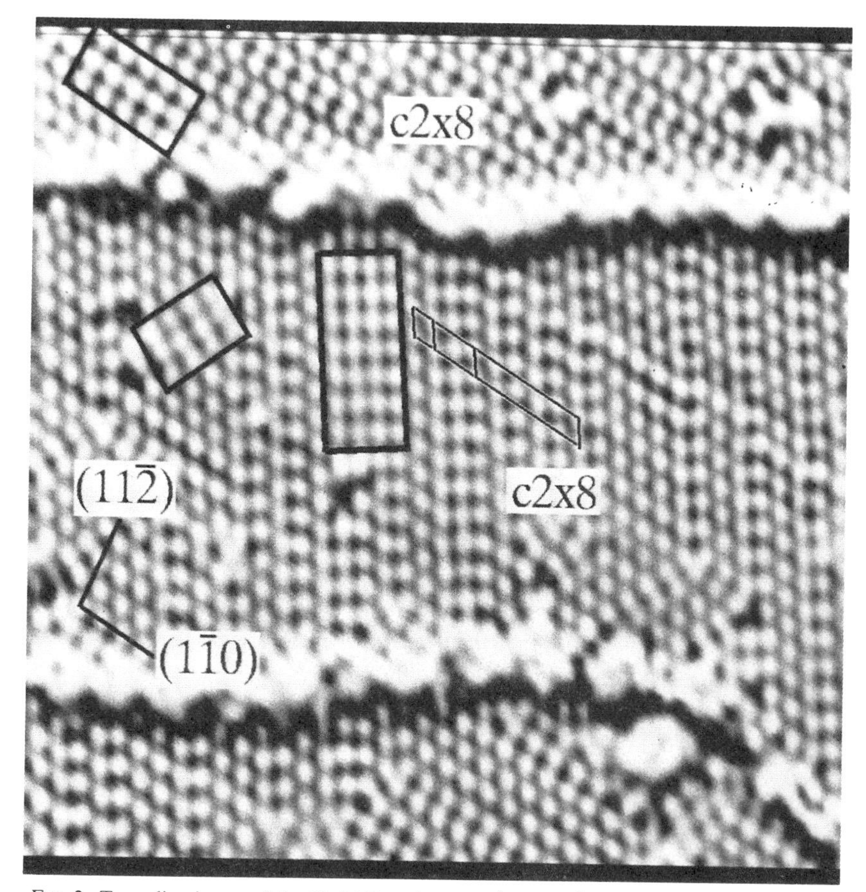

FIG. 2. Tunneling image of the Ge(111) surface, 300 Å × 300 Å, taken from Figure 1 of Ref. 5. The grey scale rendering is keyed on local height, with a dynamic range of ~ 1 Å. The principal crystallographic directions are indicated, as are local regions of 2 × 2, c4 × 2, and c2 × 8 symmetry. Two atomic steps are seen to run across the image, which was acquired tunneling into empty sample states at a bias of 1.0 V at 40 pA demanded tunneling current.

Chapter 4. Most important, however, was the detailed comparison of dual polarity images of the surface structures on Ge(111) to the 5 × 5, 7 × 7, and 9 × 9 DAS structures found on Si(111), along with the simpler 2 × 2 structures found on non-equilibrium, laser-stabilized Si(111).[19] Figure 2 shows a constant current tunneling image of the unoccupied electron states on a Ge(111) surface over a region approximately 300 Å in extent. In this image regions of 2 × 2, c4 × 2, and c2 × 8 symmetry are seen, along with two double-layer atomic steps running laterally across the image. The thin lines designate 2 × 2, c4 × 2, and c2 × 8 unit cells, while the thick lines indicate three twinned domains of c4 × 2. The uppermost terrace is almost entirely

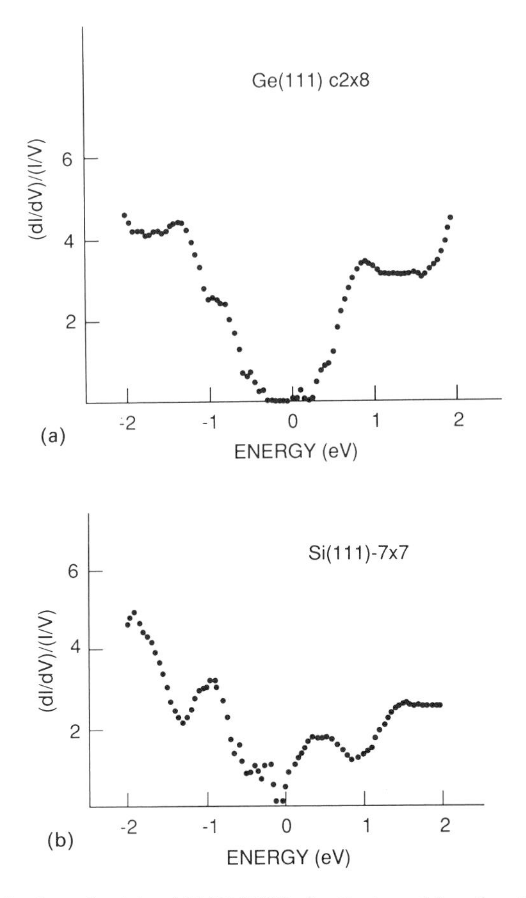

FIG. 3. Normalized conductivity $(dI/dV)(/I/V)$, for the tunnel junction I–V characteristics over (a) Ge(111) c2 × 8 and (b) Si(111)–7 × 7. This data is taken from Ref. 5.

single domain c2 × 8. The most striking feature of this image, paralleling the earlier study,[2] is the tendency for Ge(111) to display multiple phases, in contrast to Si(111)–7 × 7, despite the similarity in both appearance and apparent height (~ 1.0 Å) of the protrusions dominating both surfaces. A further observation is that the atomic steps in Fig. 2 do not follow principal crystallographic directions, nor are they obviously reconstructed as is the case for Si(111)–7 × 7.[20] Together, these features imply that the substrate atomic arrangement below the protrusions are very different for the Ge(111) and the Si(111) surfaces.

Figure 3 shows the junction I–V characteristics for the Ge(111)–c2 × 8 surface, and, for comparison, the Si(111)–7 × 7. The normalized conductivity, $(dI/dV)/(I/V)$, has been suggested as a qualitative measure of the electronic state density[18] (see Chapter 4). Positive values of the bias voltage represent tunneling from the tip into unoccupied sample states, while negative values correspond to tunneling from occupied sample states into unoccupied tip states. The data for the silicon surface is essentially the same as that reported for Si(111)–7 × 7 by Wolkow and Avouris.[21] In that work and earlier work by Hamers *et al.*[22] the peaks have been assigned, going from left to right, as an occupied adatom back bond state, occupied rest atom state, occupied adatom state (near 0V), empty adatom state, and stacking fault state.[23] These state assignments are consistent with UPS studies, and the accepted model for the 7 × 7 reconstruction, the DAS of Takayanagi *et al.*[6] In this model, there are 19 dangling bonds in each 7 × 7 unit cell; 12 adatom, 6 restatom, and 1 corner hole restatom. Theoretical calculations using similar adatom geometries with smaller cells[24] suggest that an electron transfer takes place, completely filling the restatom band, and depressing it to ~0.8 eV below E_F. This leaves a partially occupied adatom band pinning the Fermi level, accounting for the two adatom states detected near OV bias. For the Ge(111) surface, the situation is different. Two occupied states, at −0.8 V and −1.3 V are seen, with a single empty state at +0.6 V. The unoccupied states are close to features reported in ARUPS at −0.85 and −1.4 V.[16] Following the assignments for silicon, the peaks at −1.3 and −0.8 V are assigned to adatom backbond and restatom occupied states, while the single unoccupied state is assigned to an unoccupied adatom dangling bond. Unlike Si(111), no occupied adatom states are detected. Rather, a clear gap is seen, suggesting that the Ge(111) surface is semiconducting.

The situation regarding the surface atomic structure of Ge(111) can be resolved by examining dual polarity tunneling images taken just above and below the surface energy gap. Before this discussion is undertaken, an explanation of the site nomenclature on silicon and germanium (111) surfaces is in order. Figure 4 shows an atomic model illustrating the two adatom sites that are the most energetically feasible. The T_4 (top, four-fold coordinated) geometry is shown on the left, with the similar H_3 (hollow, three-fold coordinated) shown on the right. In the T_4 configuration the adatom is located directly above a backbond atom (the lower of the two in the surface double layer). It is bonded to three top layer restatoms (the "surface" atoms in the surface double-layer) and forms a fourth bond with the backbond atom directly below, causing these two to relax somewhat away from each other. In the H_3 geometry, the adatom is located above a "hollow" in a surface six-membered ring formed by three restatoms and three backbond atoms. Here, the adatom bonds to the three restatoms in the ring. The aforementioned DAS model for the Si(111)–7 × 7 fixes the silicon adatoms in T_4 sites on the surface double layer. Somewhat surprisingly, total energy calculations[24] favor the T_4 site over the H_3 for both $\sqrt{3} \times \sqrt{3}$ and 2 × 2 subunits.

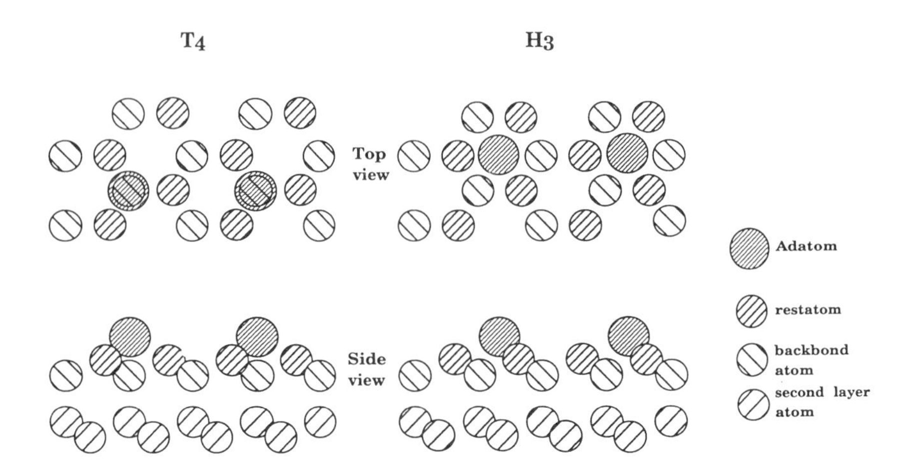

FIG. 4. Diagram of T_4 and H_3 adatom sites on Ge(111) surfaces. Adatoms, restatoms, backbond atoms, and second layer atoms are indicated.

At first glance, the hollow H_3 site seems favorable, with no first double-layer atom directly below the adatom. However, by forming a bond between the adatom and the backbond atom directly below, the T_4 adatom is able to relieve some of the strain inherent in forming the three restatom bonds by pulling them in laterally towards itself, and then relaxing itself and the subsurface backbond atom away from each other.

Figure 5 shows both the c2 × 8 surface, and a region consisting of both 2 × 2 and c4 × 2 symmetry, along with their domain boundaries. The x's, designating high points in the unoccupied state images, fall between apparent high points in the simultaneously acquired occupied state images. Further, it is clear that, unlike the case for Si(111), there is a complete separation between the apparent high points for images taken tunneling into unoccupied states (left panel) and tunneling out of filled states (right panel). Electron states on either side of the surface energy gap do not have the same spatial position, in fact they have the positions expected for adatoms occupying T_4 top sites on a 1 × 1 substrate. This registry may be determined by measuring the bulk orientation using Laue x-ray methods. The difference in symmetry imposed upon the pair of STM images by the choice of T_4 and H_3 adatom sites is demonstrated in Figure 6 where atomic models of the c2 × 8 utilize both "flavors" of adatom site. Turning back to the data in Fig. 5, the high points in the unoccupied state image correspond to the Ge adatoms, while the high points in the occupied state images register the substrate restatoms. This

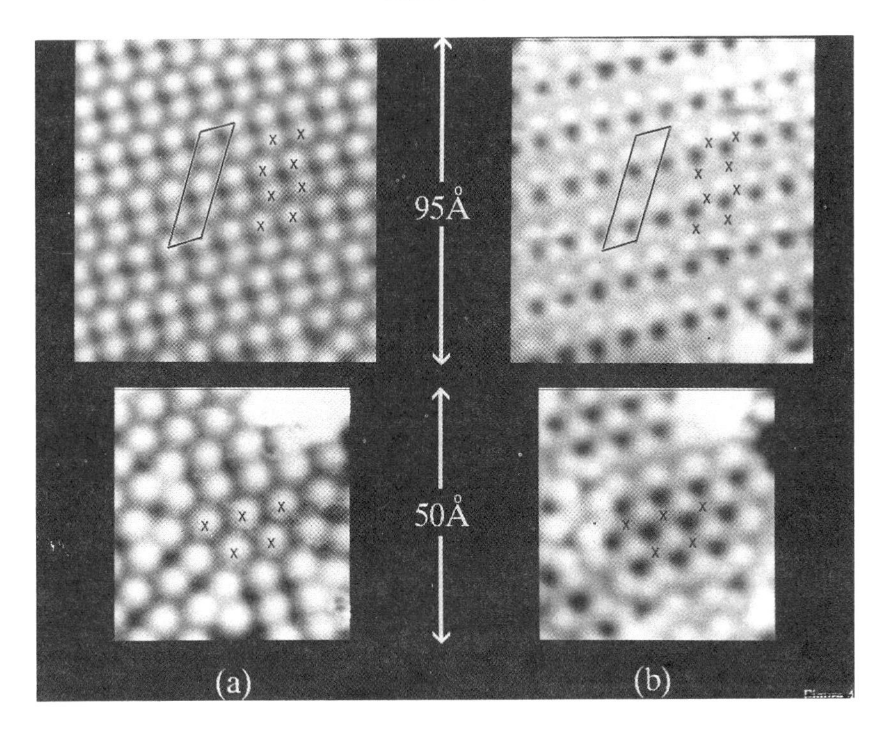

FIG. 5. Dual polarity tunneling images for the Ge(111)–c2 × 8 surface (upper) and the 2 × 2, c4 × 2, and $\sqrt{3} \times \sqrt{3}$ domains (lower). The x's designate equivalent spatial positions in both sets of images. The solid lines denote a single c2 × 8 mesh in the upper set of images. The unoccupied state image (a, left panels) was acquired at +2.0 V bias, while the occupied state images (b, right panels) at −2.0 V bias, both with 1 nA demanded tunneling current. The grey scale is keyed on apparent height, with a range of 0.8 Å in the unoccupied state images, and 0.5 Å in the occupied state images. This data is taken from Ref. 5.

correspondence holds for the c2 × 8 regions, as well as the 2 × 2 and c4 × 2 subunits. Where there are small regions of $\sqrt{3} \times \sqrt{3}$ symmetry (generally at domain boundaries), the high points in the images coincide exactly, as would be expected for T_4 adatoms on a 1 × 1 substrate.[24] The complete separation between tunneling images taken at opposite polarities is due to the fact that all of the phases found on Ge(111) have equal numbers of rest- and adatom dangling bonds in this simple adatom model. This equality allows for a nearly complete transfer of charge from an adatom to the nearby restatoms, emptying the adatom surface band and depressing the restatom surface band as occurs for Si(111)–7 × 7, resulting in a semiconducting surface. The images in Fig. 5 and the *I–V* data in Fig. 3 contradict a proposed dimer–chain model.[25] which employs substrate reconstruction similar to that found at the dimer walls on the DAS 7 × 7. For the dimer–chain geometry there are no

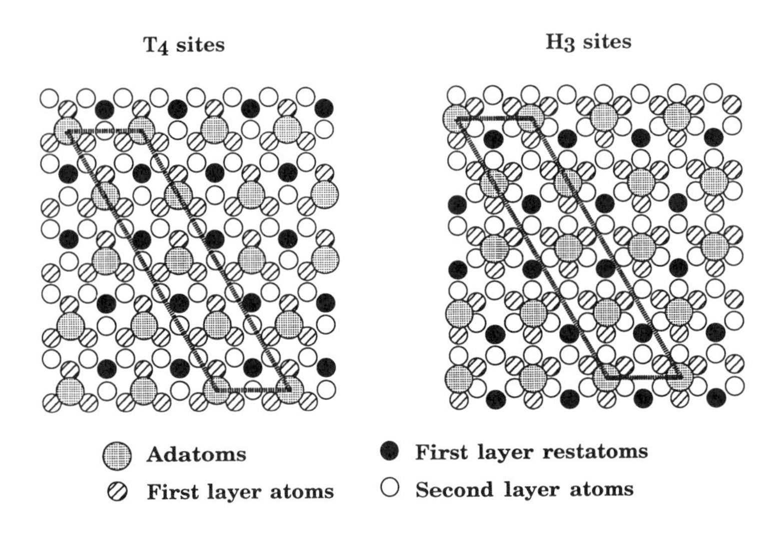

FIG. 6. Schematic of the c2 × 8 adatom reconstruction for Ge(111) using (left) T_4 sites and (right) H_3 sites.

restatoms, only adatoms, implying a partially filled adatom surface band as exists on the DAS 7 × 7. As in the 7 × 7, this requires superposition dangling bond states in dual polarity tunneling images and no surface energy gap, contradicting the experimental evidence.

In a recent series of experiments, Feenstra and Slavin have been using the STM to examine the structure of cleaved and annealed Ge(111).[7] They have shown that single domains of c2 × 8 with extent greater than 500 Å may be obtained by this method. Further, they demonstrated, by imaging the surface post-cleave in the 2 × 1 phase, and after annealing to create the c2 × 8 phase, that double-layer "holes" are created on large terraces far removed from steps. Figure 7 shows a 1.6 micron square region of the Ge(111)–c2 × 8 surface prepared in this manner. The area of the holes (9.8% ± 1.6%) agrees well with the 12.5% greater surface atom density of the c2 × 8, 2 × 2, and c4 × 2 phases over the 2 × 1 π-bonded chain, suggesting these (and nearby steps) as a source for the adatoms in the former phases. This experiment, together with the previous work, confirms the simple adatom phases as accounting for the Ge(111) surface structure.

Feenstra has also studied the c2 × 8 to 1 × 1 transition on Ge(111) by imaging the surface at temperatures a short distance above and below the phase transition temperature of ~ 300°C.[13] He finds that the transition is not first order, but consists of a progressive "melting" of the c2 × 8 at domain

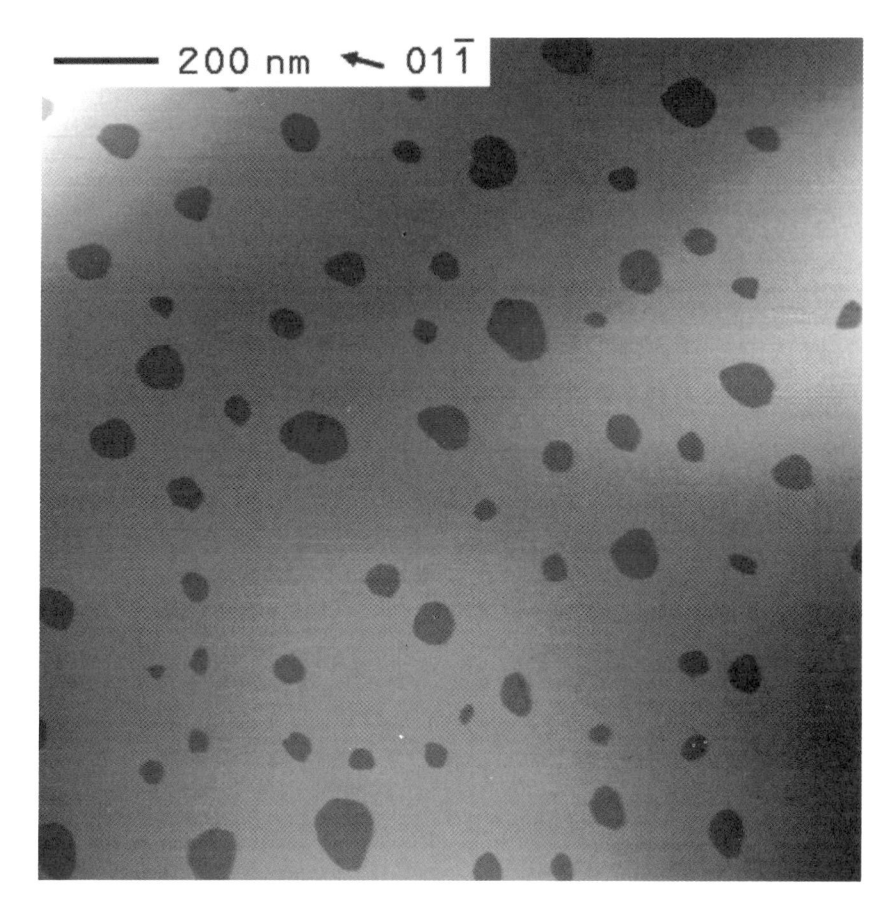

FIG. 7. Long-range tunneling image of the flat Ge(111)–c2 × 8 surface showing the "holes" resulting from transformation from the 2 × 1 reconstruction. The c2 × 8 structure in this image was created by annealing from the cleaved Ge(111)–2 × 1 surface. Taken from Ref. 7.

boundaries, extending further into adjacent domains as the phase transition temperature is approached.

5.2.2.2 Ge(001)–2 × 1. Unlike the Ge(111) surface, whose c2 × 8 reconstruction is markedly different from the 7 × 7 found on Si(111), the clean Ge(001) surface is very similar to the Si(001) in that both display a 2 × 1 LEED pattern at room temperature. Early investigators[26] suggested that this symmetry was a consequence of the dimerization of neighboring surface atoms, reducing the number of unsatisfied dangling bonds by half. This arrangement results in an odd number of electrons per surface atom, generating a partially filled surface band. Theoretically, the surface energy may further be reduced by substituting asymmetric dimers for the symmetric

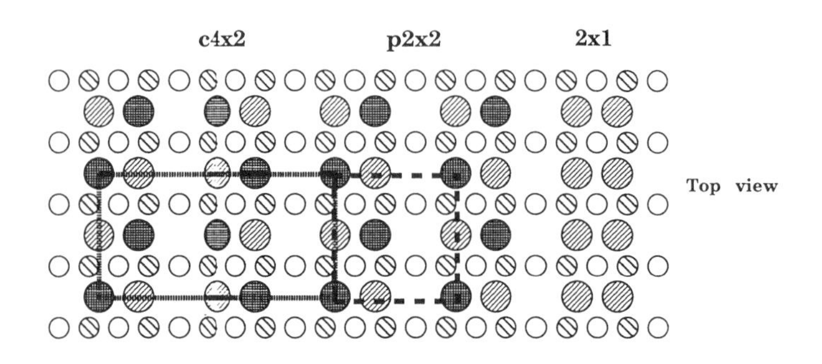

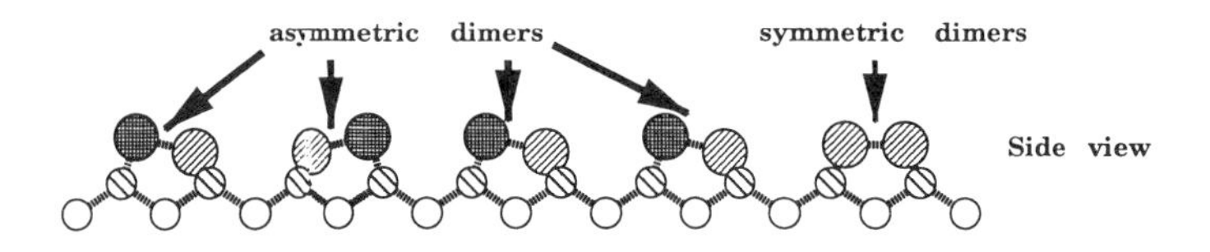

FIG. 8. Schematic of the dimer reconstructions found on the Ge(001) surface. First layer atoms, second layer atoms, symmetric, and asymmetric dimers are indicated.

ones in the 2×1.[27] This distortion, or buckling, where half the surface atoms move towards the substrate and half recede, reduces the energy in the occupied part of the band, further lowering the free energy and opening up a gap between the filled and empty surface bands. Various local arrangements of asymmetric dimers may produce local higher order reconstructions (such as p2×2, c2×2, c4×2, etc.). These configurations have been investigated theoretically for Ge(001).[28] In this study, the p2×2 and c4×2 configurations were calculated to have a significantly lower free energy (60 meV/unit cell) than the symmetric 2×1 and various $N \times 1$ combinations, but are within a few meV/unit cell of each other.

Figure 8 is a schematic atomic model of a dimerized Ge(001) "2×1" surface showing both symmetric and asymmetric dimers, and the larger unit cell, lower symmetry reconstructions (c4×2, p2×2) that result from the allowed pairings of the asymmetric dimers, both inter-row and intra-row. A row of symmetric 2×1 dimers is shown for comparison. Other asymmetric dimer arrangements such as c2×2, 4×1, 8×2, and others are not illustrated since these have not been observed on either Si(001) or Ge(001).

The first investigation of the Ge(001) surface using the STM was by Kubby *et al.*[4] In this work, the Ge(001) samples were cleaned by ion sputtering followed by an anneal at 800 to 850°C for 1 to 20 minutes, and then slowly

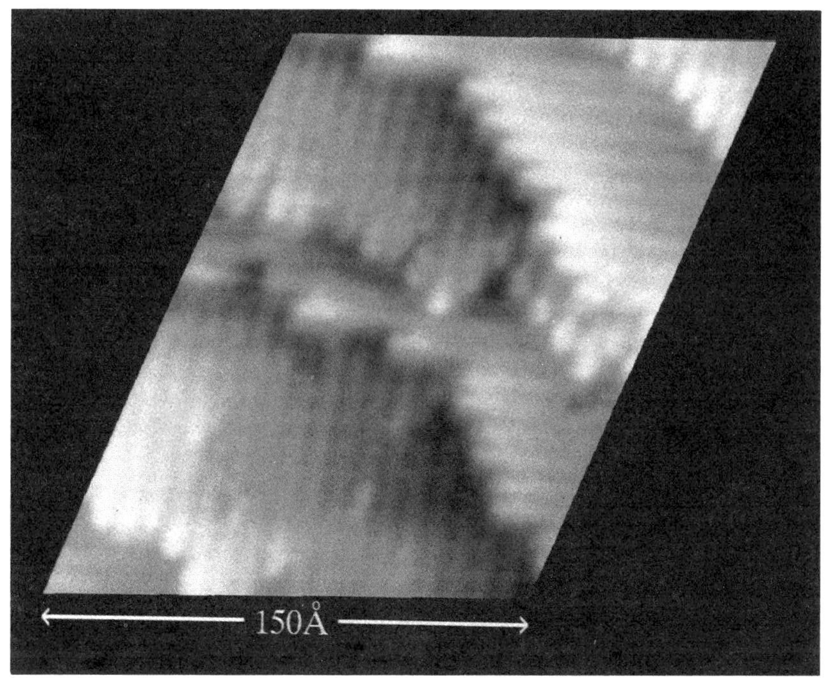

FIG. 9. Stepped terrace region of Ge(001). The contrast has been enhanced by light sourcing to show all terrace levels. Individual terraces are separated by monatomic ⟨110⟩ and ⟨100⟩ steps. The dimer rows run along the ⟨110⟩ direction. Taken from Ref. 4.

cooled to room temperature. Figure 9 shows a 150 Å extent stepped region of the Ge(001)–2×1 surface. In this figure, resolution was not adequate to resolve the individual dimers in each row. Several 1.4 Å high steps are visible, delineating regions of nominal 2×1 symmetry from the adjacent terrace of 1×2 symmetry, analogous to that seen in STM studies of the Si(001) surface.[29] The presence of two domains of 2×1 symmetry on the (001) surface is a consequence of the bulk structure of germanium. Several different step configurations are visible in this image, with step terminations both on and between dimer rows in the lower terrace, and running along both ⟨110⟩ and ⟨100⟩ surface directions. As was found for Si(001),[29] two different atomic configurations for these steps exist, depending on whether the atoms forming the lower step riser participate in dimer bonding or not. Inspection shows that both are visible in this image.

Figure 10 shows the result of interrupted feedback tunneling spectroscopy performed on the Ge(001) surface. The normalized conductivity shows one occupied state at -0.9 eV, and two unoccupied states at $+0.3$ and $+1.0$ eV, with the peak at $+0.3$ not seen in all of the spectra. The peak at -0.9 eV agrees in principle with valence band photoemission results[30] and corres-

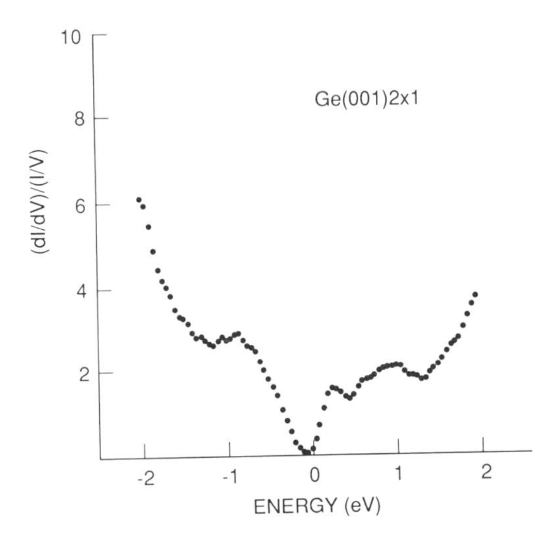

FIG. 10. Normalized conductivity for Ge(001). Taken from Ref. 4.

ponds to the D_{up} band in self-consistent calculations by Pollman *et al.*[31] While the peak at $+1.0$ eV agrees less favorably with the calculated position of the D_{down} unoccupied band, this may be due to the quasiparticle self-energy correction, suggested by Hybertsen and Louie,[32] to the density functional calculations for Ge(001). Both the calculation, and this interpretation of the spectroscopy data are consistent with buckled dimer model, where some charge transfer occurs between the down dimer atom and the raised, or up, dimer atom.

Figures 11(a) and (b) show a set of dual polarity tunneling images depicting a typical region of the Ge(001) surface at a resolution showing individual dimers. The dark lines are used to indicate the registry between the two tunneling images. It can plainly be seen that both asymmetric and symmetric dimers are present, but both this figure, and Fig. 9 show that the missing dimer defects that are present in large numbers on Si(001),[29] and have been suggested as playing a role in stabilizing that surface, are not present in similar numbers on Ge(001). More extensive work on preparing Si(001) surfaces[33] demonstrates that a missing dimer defect density of $\sim 1\%$ may be realized, but it is clear that the Ge(001) surface has a lesser tendency to form this structure. When the unoccupied states tunneling image (11a) is superimposed with the simultaneously acquired occupied state tunneling image (11b), the dimer rows are seen to occupy identical spatial positions. This is in contrast to similar work on Si(001), where a 180° phase shift in the direction of dimerization exists between the occupied and unoccupied state tunneling images. This difference is not currently well understood, but may

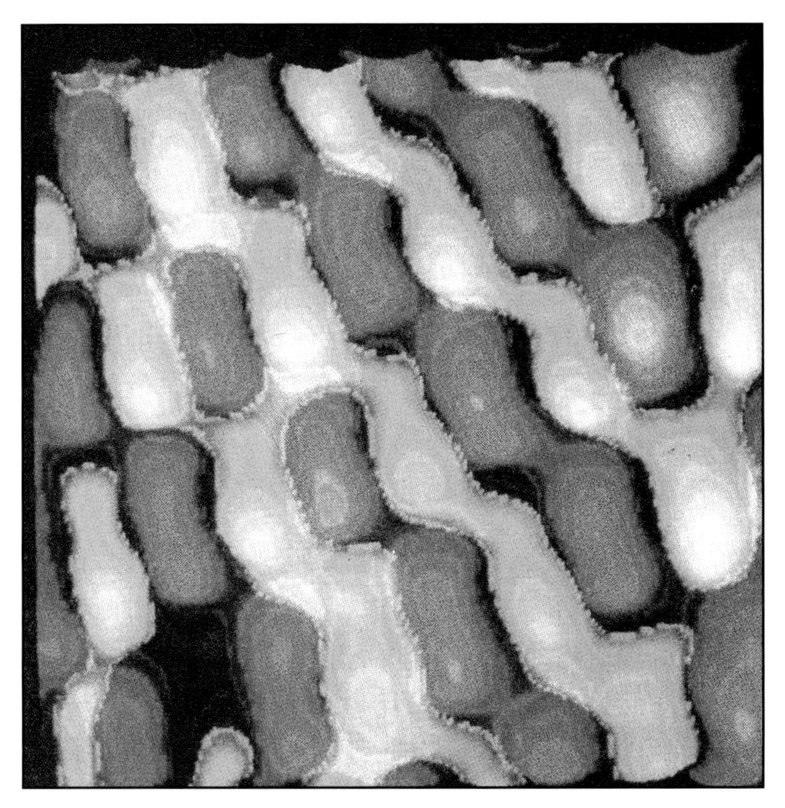

FIG. 5. (Chapter 3) Chemical potential image of a cleaved MoS_2 surface (red or bright spots) superimposed on a simultaneously recorded STM image (green or dark spots). The chemical potential signal is high over the sulphur atoms and low over the molybdenum atoms, in direct contrast to the STM image.

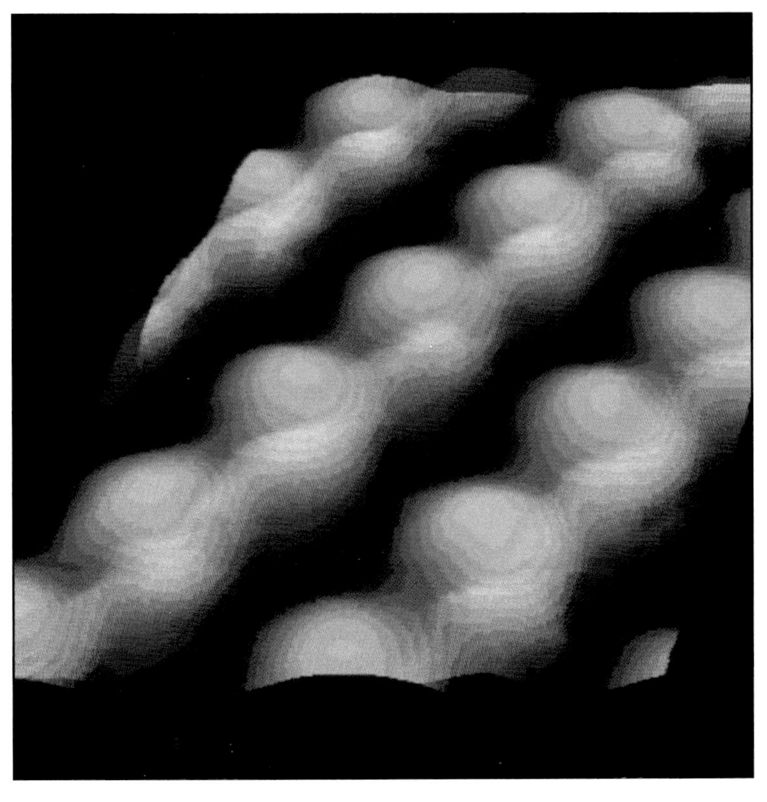

FIG. 8. (Chapter 5, Section 3) Combined color STM images of the GaAs(110) surface, showing the Ga atoms (blue) and As atoms (red). The blue image was acquired by tunneling into empty states that are localized around surface Ga atoms, and the red image was acquired by tunneling out of filled states that are localized around surface As atoms. [From Ref. 4].

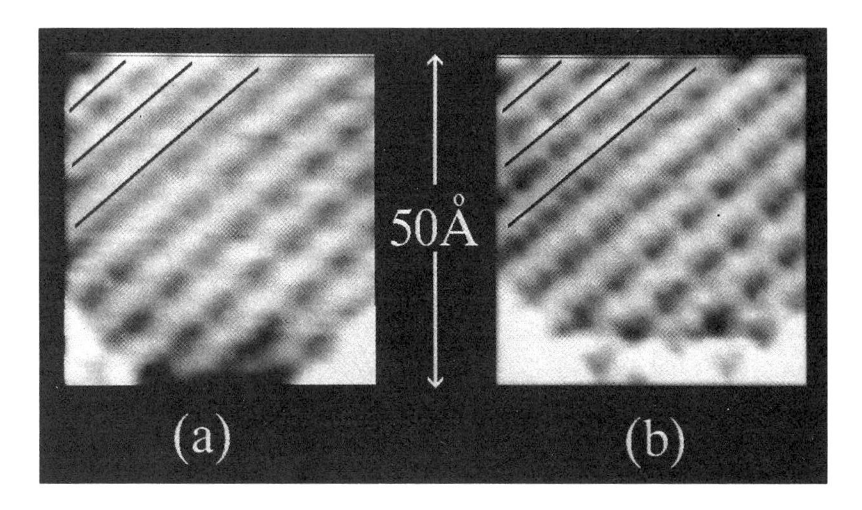

FIG. 11. Dual polarity tunneling images of Ge(001). The left image (a) is of unoccupied sample states at +2.0 V bias, while the right (b) is of occupied sample states at −2.0 V bias, both at 1 nA demanded tunneling current. The dark lines denote equivalent spatial positions in both images.

be related to the differences in core electronic structure between germanium and silicon. Curiously, the image phase shift between up and down dimer atoms is similar for both materials.

Further studies comparing step structures for vicinal Ge(001), comparing the images to those for similar Si(001) surfaces[34], show generally similar features, with a reduced tendency to form double steps. This observation implies that the energy required to form the double step configuration on Ge(001) versus a pair of single steps is relatively greater than that for the Si(001) system.

5.2.2.3 GeSi(111)-5 × 5. Along with the native (111) and (001) faces, an alloy of silicon and germanium has also been imaged at atomic resolution with the tunneling microscope. Early investigations of germanium–silicon alloys showed that the surface of GeSi at roughly a 50–50 stoichiometry exhibited a 5 × 5 reconstruction that was qualitatively similar to that displayed by the 7 × 7 of silicon.[35] Since the c2 × 8 of germanium bore little resemblance to the tunneling images of the 7 × 7, it was not clear what the configuration of the 5 × 5 would display. Further, if the 5 × 5 was a smaller example of the 7 × 7 structure, it would be more amenable to theoretical treatment having half as many atoms per unit cell. Insights gained from modeling the smaller structure might then be applied to the larger with more confidence.

Figure 12 shows an unoccupied state image of the GeSi(111)-5 × 5[36]

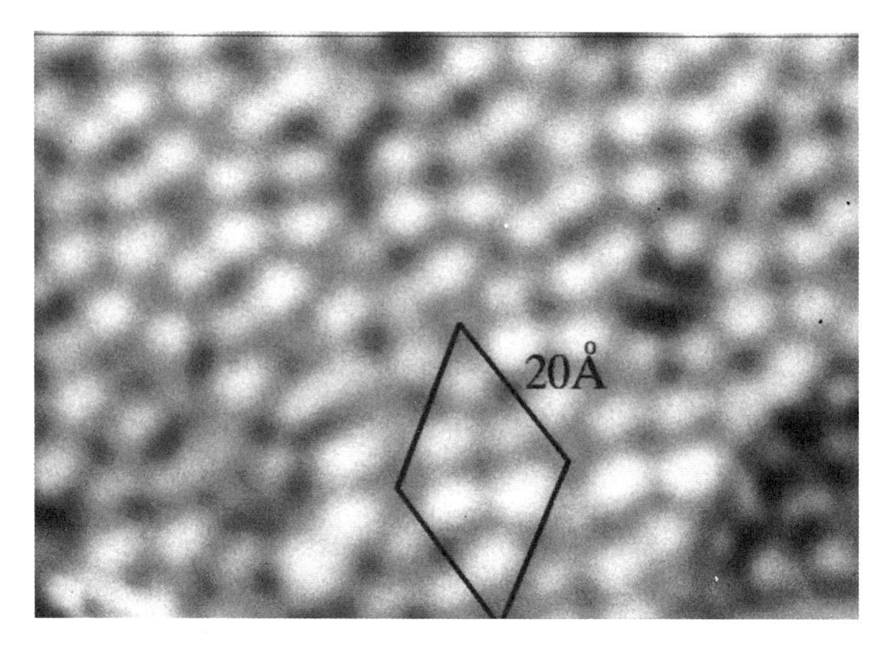

FIG. 12. Tunneling image of GeSi(111)–5 × 5 surface reconstruction. The grey scale is keyed to apparent height, with a dynamic range of 0.5 Å. The black lines denote a unit mesh, with the length scale indicated. Taken from Ref. 36.

surface taken at bias conditions similar to those used by various groups for Si(111)–7 × 7. We see that the 5 × 5, as indicated in the figure, is qualitatively similar in appearance to the 7 × 7. The unit cell, 19.2 Å on edge, consists of a hexagonal array of depressions surrounded by six protrusions. Close examination of the protrusion heights around each depression shows that they alternate in apparent height, with neighboring clusters oriented such that the three protrusions in one half of the rhombus are higher than those in the other half. In terms of relative size and appearance of the protrusions, the images are exactly what would be expected for the DAS structure accounting for the 7 × 7 images of silicon scaled down to 5 × 5. In this model, the six protrusions in the unit cell are the six adatoms predicted by a 5 × 5 version of the DAS model.

The authors concluded that the height variation among the adatoms was likely due to surface segregation of germanium and silicon adatoms. In retrospect, it is unclear whether an ordered surface alloy is in fact needed to account for the STM images. A later series of studies using the STM to compare the features of silicon and germanium adatom surfaces included a study of Si(111)–5 × 5,[5] which may be created using low temperature annealing schedules. In that study, the 5 × 5 structures of silicon demonstrated the characteristic DAS stacking fault state seen in occupied state images of the Si(111)–7 × 7, with considerably less modulation in the corresponding unoccupied state image. It may well be that this modulation of the unoccupied state image due to the partial stacking fault is more pronounced for GeSi, or that both surface segregation and electronic state density are in fact responsible for this feature. Nonetheless, it is quite clear that the DAS structure accounts for the features of the GeSi(111)–5 × 5 phase.

5.2.3 Metal Overlayers on Ge

5.2.3.1 Ge(111) : As–1 × 1. A common problem in semiconductor technology is the fact that the native surfaces of both silicon and germanium are highly reactive under ambient conditions. A large part of the chemistry developed for processing is designed to mitigate unwanted reactivity of the exposed substrates, principally to minimize contamination that leads to poor device characteristics. As mentioned earlier, the fact that the oxide of germanium is unstable contributed to its supplantation in the early 1950s by silicon as the semiconductor of choice. Much of the effort in surface science over the last 30 years has been devoted to searching for methods to reduce the reactivity of free surfaces. One method has been to satisfy dangling bonds by termination with hydrogen or bromine. Another method, demonstrated recently,[37] is to eliminate surface dangling bonds by substituting the semiconductor surface atoms with an appropriate chemical species, leaving a nonreactive surface.

In the diamond lattice of silicon and germanium, where each atom is four-fold coordinated, substitution with arsenic results in an extra valence band electron. In the bulk, this results in n-type doping, whereas at the surface a different result may occur. In principle, arsenic may substitute for the topsite atom in the Ge(111) surface doublelayer, using three of its five valence electrons to bond to the backbond germanium atoms. The other two electrons may be combined into a "lone pair" state, dropping their energy down into resonance with the bulk valence band, and eliminating the dangling bonds which drive the adatom reconstruction of Ge(111). Ultraviolet photoemission studies on both the Si(111) : As–1 × 1 and Ge(111) : As–1 × 1 systems are consistent with this simple model.[37,38] We note that the chemical substitution of arsenic for germanium, which eliminates a dangling bond at each surface site, is more efficient in the destruction of surface dangling bonds than the adatom reconstruction discussed earlier for Ge(111)–c2 × 8, where each adatom saturated three surface bonds, leaving one restatom dangling bond and one adatom dangling bond in place of four dangling bonds.

Figure 13 shows a constant-current tunneling image of the Ge(111) : As–1 × 1 surface, with the inset showing details of the 1 × 1 terraces. The surface

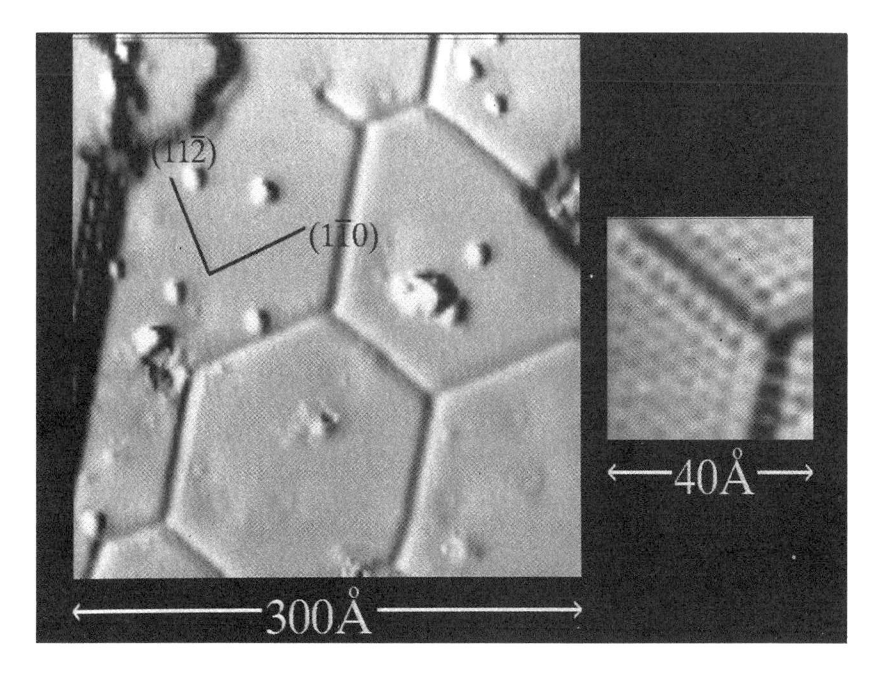

FIG. 13. (Left) Large-scale unoccupied state tunneling image of the Ge(111):As–1 × 1 surface. Domain boundaries are seen as a network of lines oriented along ⟨110⟩ directions as indicated. Two double-layer steps are seen at the upper left. (Right) Small scale tunneling image showing both the 1 × 1 details of the terraces, and a domain boundary intersection. The large-scale image was taken at +2.25 V and 0.1 nA, while the small image was acquired at +2.0 V and 0.1 nA demanded tunneling current. Taken from Ref. 8.

electronic characteristics of the arsenic terminated surface is consistent with quasiparticle calculations,[32] which predict a larger surface energy gap than the native reconstruction. The surface energy gap for Ge(111):As is ~1.0 eV, in good agreement with predictions of 0.8 eV. These calculations also suggest that the unoccupied state image (13b) depicts the location of the backbond Ge atoms, while the occupied state image (13a) registers the topsite As atoms. While the terraces are nominally 1 × 1, the most striking feature of the large scale image is the division of the surface into hexagonal domains, separated by double-layer deep "trenches." The trenches consist of a pair of $\langle\bar{1}\bar{1}2\rangle$ and $\langle 1\bar{1}\bar{2}\rangle$ steps separated by three lattice constants. This appearance of the Ge(111):As surface may be contrasted with that of Si(111):As,[39] where the STM images showed a 1 × 1 surface containing, and possibly stabilized by, a large number of point defects.

In a comprehensive study with the STM, Becker and co-workers[8] showed that the division of the surface into domains ~150 Å in extent was due to the fact that the incorporation of arsenic into the topsites on the Ge(111) surface caused a lateral contraction of 0.025 Å ± 0.013 Å per 1 × 1 unit cell. When the As is substituted for the topsite Ge atoms in the surface double-layer, it relaxes outward and pulls in the three Ge backbond atoms putting the surface into tension. This accommodates both the increased Ge–As bond length and the free configuration of the arsenic bonds, which are closer to 90° rather than the 108° of germanium. This effect has been suggested both by recent cluster calculations for Si(111) : As[40] and surface stress calculations for Ga, Ge, and As substituted into Si(111).[41] Both calculations also showed topsite As atom relaxing outward from the surface, as reported in medium energy ion scattering (MEIS)[42] and x-ray standing wave (XSW)[43] experiments. In the STM study,[8] the details of the domain boundaries were compared to simple calculations of arsenic lone pair superposition for several geometries. No rigorous conclusion was reached, since the occupied state images were somewhat ambiguous in support of any given model geometry. The authors felt that the weight of the evidence supported a rebonded geometry for the $\langle \bar{1}\bar{1}2 \rangle$ step riser, with a simple arsenic termination for the $\langle 11\bar{2} \rangle$ riser.

5.2.3.2 Ge(001) : As. The elimination of dangling bonds by arsenic termination has also been investigated on the (001) surfaces of silicon[44] and germanium[45]. In a similar manner as described for the ⟨111⟩ surfaces, the arsenic forms dimers on the ⟨001⟩ surfaces using two electrons to bond to the substrate silicon or germanium, one electron to bond to the other arsenic in the dimer, and the remaining two are combined into a lone pair. Tunneling images for both Si(001) : As and Ge(001) : As are consistent with this picture, and nearly indistinguishable from each other.[45] Figure 14 shows a set of dual polarity images for Ge(001) : As. Both tunneling into unoccupied states 1.5 eV above E_F and out of occupied electron states 1.5 V below E_F show symmetric dimers, similar to those imaged on Si(001) : As surfaces. The line in the figure register the two images, showing no phase shift along the direction of dimerization, in accordance with the earlier observation for the native Ge(001) surface. This is presumably related to the different core electronic structure, which modulates the charge transfer between the arsenic and the substrate, of the two semiconductors.

5.2.3.3 Ge(111) : Sn. The deposition of metals on germanium, as for other semiconductors, may produce a wide variety of surface phases, depending on the relative amount of material and the deposition conditions (rate, temperature, etc.). Tin, lying just below germanium in the periodic table, is capable of substituting in the diamond lattice, and as such, might be expected to form surface structures beyond simple intermixed layers.

A wide variety of surface phases have been reported for tin deposited on germanium. One phase investigated with the STM[46] has been the DAS 7 × 7 that exists for a range of 0.25–0.4 monolayer of tin deposited on a clean Ge(111) surface at 200°C and subsequently annealed to 600°C[47]. Figure 15

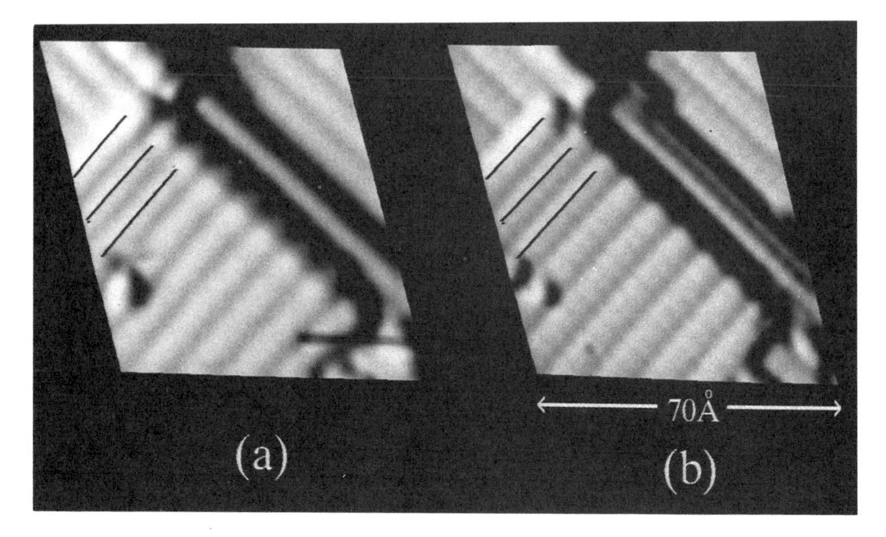

FIG. 14. Tunneling images of the Ge(001):As–2 × 1 surface. unoccupied states images (a) and occupied states images (b) both show symmetric dimers, with individual dimers resolved in the occupied states images.

shows a constant current tunneling image of the snGe(111)–7 × 7 surface. Inspection of the image in the figure shows that there are two populations of adatoms, with the second population ~0.4 Å lower in apparent height. Simply counting the adatoms reveals that 30% of the adatoms on the SnGe(111) surface are low. Rutherford backscattering spectroscopy (RBS) of this sample determined a tin deposition of $2.1 \times 10^{14}/cm^2$ corresponding to 0.29 monolayers. The lack of order in the adatom populations suggest that the tin is incorporated into the germanium surface layer in several sites. Further work with the STM on the unoccupied state electronic structure showed that the SnGe(111)–7 × 7 surface was identical to the Si(111)–7 × 7, suggesting the electronic characteristics of both were derived from a common geometric structure, the DAS model proposed by Takayanagi.[6]

5.2.4 Gas Adsorption on Germanium

5.2.4.1 Oxygen on Ge(111).

Gas adsorption studies on elemental surfaces is a subject that is nearly as old as the study of the intrinsic clean surfaces themselves. Of particular importance is the interaction of molecular oxygen with clean semiconductor surfaces for both fundamental and technological reasons. A considerable body of literature exists on oxygen reactions with silicon surfaces, including several recent STM studies.[48] Since germanium has been technologically less useful, correspondingly less information exists on oxygen reaction with this semiconductor. A recent study

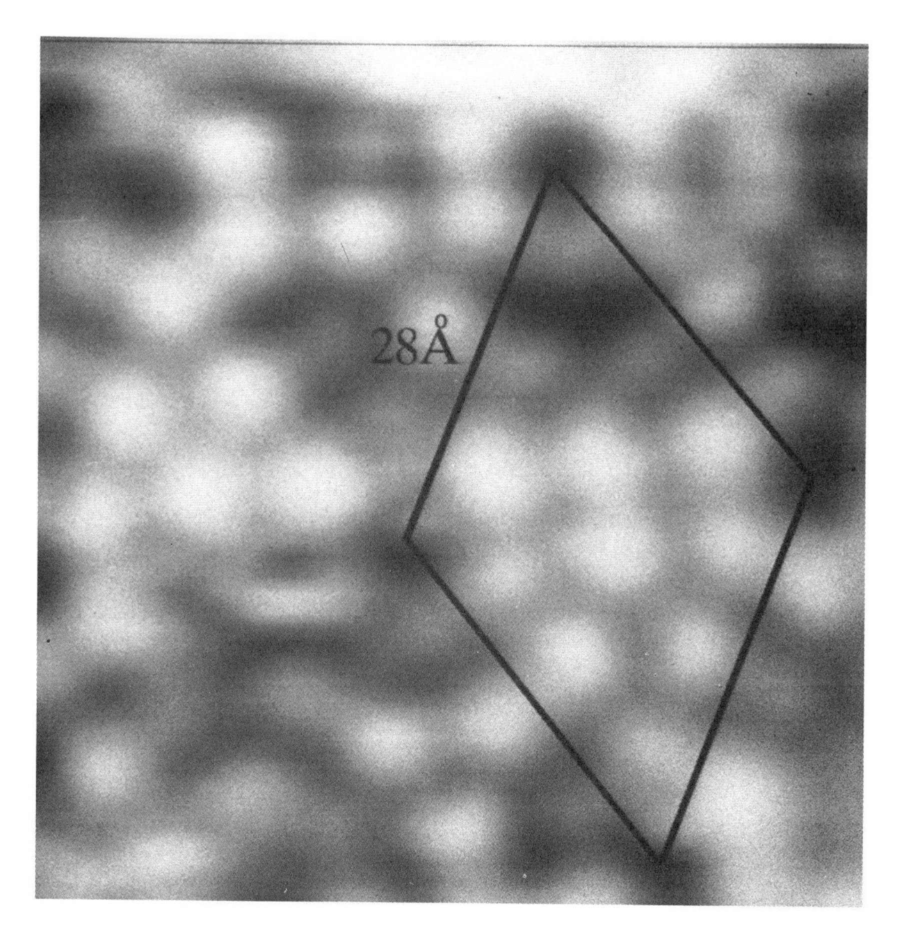

FIG. 15. SnGe(111)–7 × 7 tunneling image. A 7 × 7 mesh is outlined and the scale indicated. Taken from Ref. 46.

with the STM was performed by Klitsner *et al.* where 0_2 was reacted with the Ge(111)–c2 × 8 surface both at room and elevated temperatures.[49] This work showed that oxygen nucleation on Ge(111) was non-homogeneous, occurring primarily at domain boundaries of the c2 × 8, and not at either the step risers or the reconstructed c2 × 8 surface.

STM images indicate that the oxygen adsorption at the domain boundaries resulted in disordering of the germanium adatoms in, and adjacent to, these areas. The domain boundaries on the c2 × 8 surface represent regions where there are a higher number of unsaturated restatom dangling bonds than on the reconstructed surface. By inducing further disordering, the oxygen adsorption should effectively increase the amount of the surface available for

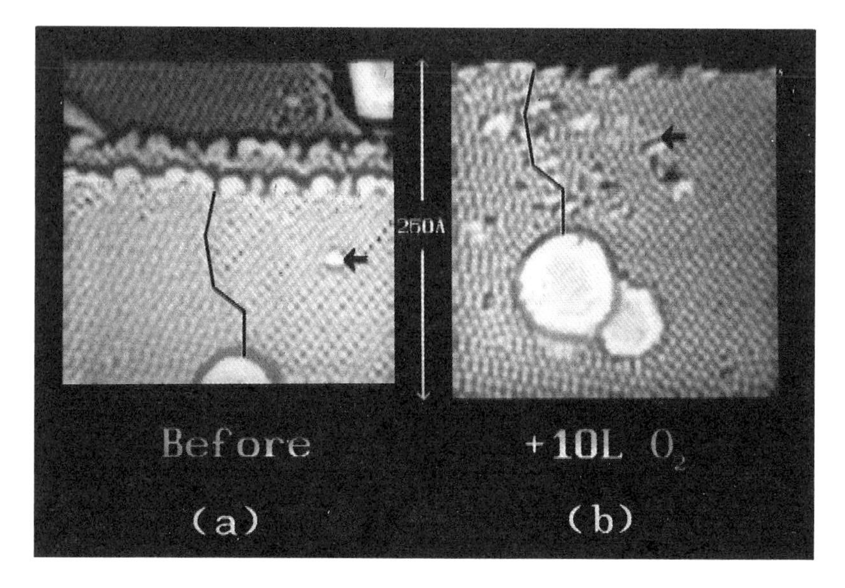

FIG. 16. Ge(111)–c2 × 8 surface before (a) and after (b) exposure to 10 L of molecular oxygen (L = Langmuir = 1×10^{-6} Torr-sec). The black line designates a phase boundary, where much of the initial reaction takes place. Taken from Ref. 49.

reaction. It is unclear, however, if the presence of adsorbed oxygen adjacent to the newly disordered areas affects their further reactivity. In the exposure range studied, no evidence for this was detected. Figure 16 shows tunneling images of the Ge(111) surface before and after exposure to 10 L of 0_2. Here it is clear that the step risers are relatively unaffected, as are the c2 × 8 domains, while the domain boundaries themselves, as indicated by a dark line, are strongly reacted. STM images taken over the same region of the surface after further gas adsorption show that the reacted areas continue to grow at the expense of the neighboring c2 × 8 domains. This accounts for LEED observations[50] that demonstrated that the c2 × 8 beams were unaffected by small amounts of oxygen.

Oxygen adsorption on Ge(111) was also investigated at elevated temperatures (300°C) by these authors. They found that the fundamental difference between the elevated and room temperature studies was the fact that the c2 × 8–1 × 1 phase transition rendered the surface effectively as one large domain boundary. Oxygen nucleation was then homogeneous, and, upon cooling to room temperature, prevented the c2 × 8 from forming, leaving the surface an array of disordered adatoms, accounting for the observed 1 × 1 LEED pattern.

5.2.5 Epitaxy with Germanium

5.2.5.1 Germanium on Si(001). The detailed chemistry and physics of semiconductor epitaxy has been under intense study for a number of years, due principally to the technological importance of this field in device processing. More recently, a large amount of interest has been shown in the heteroepitaxy of germanium on silicon, since superlattices of these materials may be promising as a basis for optoelectronic devices. By using consecutive heteroepitaxial layers of silicon and germanium with appropriate thicknesses, the bandgap and conduction band minimum of these materials may be varied over a range suitable for these devices. In a series of studies with the STM, Mo and co-workers at the University of Wisconsin have investigated the microscopic details of silicon epitaxy on Si(001). They have recently extended these experiments to the epitaxy of germanium on silicon,[9] and found remarkable differences between this and their earlier work on silicon–silicon epitaxy.[51]

In this STM work on the transition from 2D to 3D growth for germanium on Si(001) they find that an intermediate growth phase exists, where the germanium forms small clusters ("huts") with precise facet crystallography and orientation with respect to the silicon substrate. Figure 17 shows a perspective and plan view of a tunneling image of one of these clusters. These hut clusters consist of four reconstructed $\{105\}$ planes, orienting the clusters along two orthogonal $\langle 100 \rangle$ surface directions. Due to the widespread appearance of the hut clusters before the transition to macroscopic clusters, the authors suggest that the hut cluster are an intermediate stage in the heteroepitaxy of germanium on Si(001), providing a more favorable site for accommodation of the incoming germanium atoms than spontaneous nucleation of macroscopic clusters. They go on to speculate that the hut clusters are themselves nucleated at steps forming on the strained Ge(001) : Si(001) surface, since these are a likely spot for strain relief during the epitaxial process. From this kind of work, it is quite clear that the STM will make a significant impact in understanding the details of complex processes during growth.

5.2.6 Conclusion

In the nine years since the introduction of the vacuum tunneling microscope, a wide variety of germanium surfaces have been imaged by a growing number of investigators. While the number of experiments has been less than that for silicon, as in other fields of surface science, the scope has been nearly as great, with the major features of both clean and adsorbate-covered germanium surfaces brought to light. Recent experiments in the heteroepitaxy

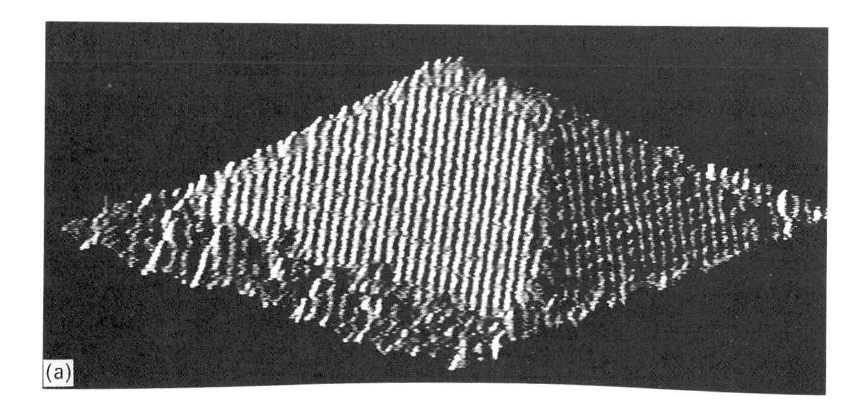

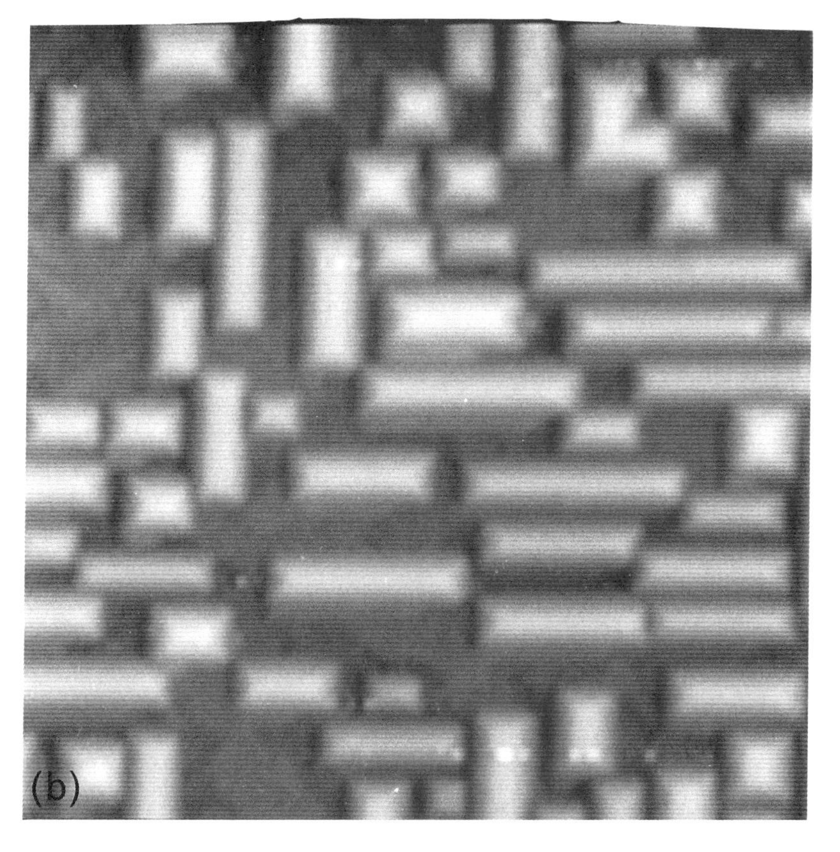

FIG. 17. "Hut" cluster of germanium on Si(001) in perspective (a) (upper) and plan (b) (lower) view. These form during heteroepitaxy of germanium on silicon. The scan area is 400 Å × 400 Å for (a) and 2500 Å × 2500 Å for (b). Taken from Ref. 9.

of germanium on silicon show that interest in this material is increasing after many years of taking a second seat to silicon in basic investigations.

References

1. J. C. Bean, *Science* **230**, 127 (1985); *Silicon-Molecular Beam Epitaxy*, Vols. I and II (E. Kaspar and J. C. Bean, eds), CRC Press, Boca Raton, 1988.
2. R. S. Becker, J. A. Golovchenko, and B. S. Swartzentruber, *Phys. Rev. Lett.* **54**, 2678 (1985).
3. G. Binnig, H. Rohrer, Ch. Gerber, and E. Weibel, *Phys. Rev. Lett.* **50**, 120 (1983).
4. J. A. Kubby, J. E. Griffith, R. S. Becker, and J. S. Vickers, *Phys. Rev. B* **36**, 6079 (1987).
5. R. S. Becker, B. S. Swartzentruber, J. S. Vickers, and T. Klitsner, *Phys. Rev. B* **39**, 1633 (1989).
6. K. Takayanagi, Y. Tanishiro, M. Takahashi, and S. Takahashi, *JVST A* **3**, 1502 (1985).
7. R. M. Feenstra and A. J. Slavin, submitted to *Surf. Sci.* (1991).
8. R. S. Becker, T. Klitsner, and J. S. Vickers, *J. Microscopy* **152**, 157 (1988), R. S. Becker and J. S. Vickers, *J. Vac. Sci. Tech. A* **8**, 226 (1990).
9. Y.-W. Mo, D. E. Savage, B. S. Swartzentruber, and M. G. Lagally, *Phys. Rev. Lett.* **65**, 1020 (1990).
10. T. Klitsner, R. S. Becker, and J. S. Vickers, *Phys. Rev. B* **44**, 1817 (1991).
11. T. Klitsner and J. S. Nelson, *Phys. Rev. Lett.* **67**, 3800 (1992).
12. R. S. Becker, J. A. Golovchenko, and B. S. Swartzentruber, *Nature* **325**, 419 (1987).
13. R. M. Feenstra, A. J. Slavin, G. A. Held, and M. A. Lutz, *Phys. Rev. Lett.* **66**, 3257 (1991).
14. P. W. Palmberg and W. E. Peria, *Surf. Sci.* **6**, 57 (1967).
15. D. J. Chadi and C. Chiang, *Phys. Rev. B* **23**, 1843 (1981); W. S. Yang and F. Jona, *ibid.* **29** (1984); R. J. Phaneuf and M. B. Webb, *Surf. Sci.* **164**, 167 (1985).
16. T. Yokotsuka, S. Kono, S. Suzuki, and T. Sagawa, *J. Phys. Soc. Jap.* **53**, 696 (1984); R. D. Bringans and H. Hochst, *Phys. Rev. B* **25**, 1081 (1982); J. M. Nicholls, G. V. Hansson, and R. I. G. Uhrberg, *Phys. Rev. B* **33**, 5555 (1986); R. D. Bringans, R. I. G. Uhrberg, and R. Z. Bachrach, *Phys. Rev. B* **34**, 2373 (1986); J. Aarts, A. J. Hoeven, and P. K. Larsen, *Phys. Rev. B* **37**, 8190 (1988).
17. E. G. McRae, H. J. Gossman, and L. C. Feldman, *Surf. Sci.* **146**, L540 (1984).
18. J. A. Stroscio, R. M. Feenstra, and A. P. Fein, *Phys. Rev. Lett.* **57**, 2579 (1986).
19. R. S. Becker, J. A. Golovchenko, G. S. Higashi, and B. S. Swartzentruber, *Phys. Rev. Lett.* **57**, 1020 (1986).
20. R. S. Becker, J. A. Golovchenko, E. G. McRae, and B. S. Swartzentruber, *Phys. Rev. Lett.* **55**, 2028 (1985).
21. R. Woldow and Ph. Avouris, *Phys. Rev. Lett.* **60**, 1049 (1988).
22. R. J. Hamers, R. M. Tromp, and J. E. Demuth, *Phys. Rev. Lett.* **56**, 18 (1986), R. J. Hamers, R. M. Tromp, and J. E. Demuth, *Surf. Sci.* **181**, 346 (1987).
23. R. S. Becker, J. A. Golovchenko, D. R. Hamann, and B. S. Swartzentruber, *Phys. Rev. Lett.* **55**, 2032 (1985).
24. J. E. Northrup, *Phys. Rev. Lett.* **57**, 154 (1986).
25. K. Takayanagi and Y. Tanishiro, *Phys. Rev. B* **34**, 1034 (1986).
26. R. E. Schlier and H. E. Farnsworth, *J. Chem. Phys.* **30**, 917 (1959).
27. D. J. Chadi, *Phys. Rev. Lett.* **43**, 43 (1979); D. J. Chadi, *J. Vac. Sci. Tech.* **16**, 1290 (1979).
28. M. Needels, M. C. Payne, and J. D. Joannopoulos, *Phys. Rev. B* **38**, 5543 (1988).
29. R. M. Tromp, R. J. Hamers, and J. E. Demuth, *Phys. Rev. Lett.* **55**, 1303 (1985); R. J. Hamers, R. M. Tromp, and J. E. Demuth, *Phys. Rev. B* **34**, 5343 (1986).

30. J. G. Nelson, W. J. Gignac, R. S. Williams, S. W. Robey, J. G. Tobin, and D. A. Shirley, *Phys. Rev. B* **27**, 3924 (1983); T. C. Hsieh, T. Miller, and T.-C. Chiang, *Phys. Rev. B* **30**, 7005 (1984).
31. J. Pollman, P. Kruger, A. Mazur, and G. Wolfgarten, p. 81, in *Proceedings of the Eighteenth International Conference on the Physics of Semiconductors,* (0. Engstrom ed.), World Scientific, Singapore, 1986.
32. M. S. Hybertsen and S. G. Louie, *Phys. Rev. Lett.* **58**, 1551 (1987).
33. B. S. Swartzentruber, Y. W. Mo, M. B. Webb, and M. G. Lagally, *J. Vac. Sci. Tech. A* **7**, 2901 (1989).
34. J. E. Griffith, J. A. Kubby, P. E. Wierenga, R. S. Becker, and J. S. Vickers, *J. Vac. Sci. Tech. A* **6**, 493 (1988); J. E. Griffith and G. P. Kochanski, *Crit. Rev. Sol. St. and Mat. Sci.,* **16**, 255 (1990).
35. H. J. Gossman, J. C. Bean, L. C. Feldman, and W. M. Gibson, *Surf. Sci.* **138**, L175 (1984).
36. R. S. Becker, J. A. Golovchenko, and B. S. Swartzentruber, *Phys. Rev. B* **32**, 8455 (1985).
37. R. D. Bringans, R. I. G. Uhrberg, R. Z. Bachrach, and J. E. Northrup, *Phys. Rev. Lett.* **55**, 533 (1985).
38. M. A. Olmstead, R. D. Bringans, R. I. G. Uhrberg, and R. Z. Bachrach, *Phys. Rev. B* **34,** 6401 (1986); R. I. G. Uhrberg, R. D. Bringans, M. A. Olmstead, R. Z. Bachrach, and J. E. Northrup, *Phys. Rev. B* **35**, 3945 (1987).
39. R. S. Becker, B. S. Swartzentruber, J. S. Vickers, M. S. Hybertsen, and S. G. Louie, *Phys. Rev. Lett.* **60,** 116 (1988).
40. C. H. Patternson and R. P. Messmer, *Phys. Rev. B* **39**, 1372 (1989).
41. R. D. Meade and D. W. Vanderbilt, *Bull. Am. Phys. Soc.* **34**, 448 (1989).
42. M. Copel, R. M. Tromp, and U. K. Koehler, *Phys. Rev. B* **37**, 10756 (1988).
43. J. R. Patel, J. A. Golovchenko, P. E. Freeland, and H. J. Gossman, *Phys. Rev. B* **36**, 7715 (1987).
44. R. I. G. Uhrberg, R. D. Bringans, R. Z. Bachrach, and J. E. Northrup, *Phys. Rev. Lett.* **56**, 520 (1986).
45. R. S. Becker, to be published.
46. R. S. Becker, B. S. Swartzentruber, and J. S. Vickers, *J. Vac. Sci. Tech. A* **6**, 472 (1988).
47. T. Ichikawa and S. Ino, *Sol. St. Comm.* **27**, 483 (1978); T. Ichikawa and S. Ino, *Surf. Sci.* **105**, 395 (1981).
48. F. M. Leibsle, A. Samsvar, and T.-C. Chiang, *Phys. Rev. B* **39**, 5780 (1988); J. P. Pelz and R. H. Koch, *Phys. Rev. B.* **42,** 3761 (1990); I.-W. Lyo, Ph. Avouris, B. Schubert, and R. Hoffmann, *J. P. Chem.* **94**, 4400 (1990); H. Tokumoto, K. Miki, H. Murakami, H. Bando, M. Ono, and K. Kajimura, *JVST A* **8**, 255 (1990).
49. T. Klitsner, R. S. Becker, and J. S. Vickers, submitted to *Phys. Rev. B* (1991).
50. B. Z. Ol'shanetskii, N. I. Makrushin, and A. I. Volokitin, *Sov. Phys. Sol. St.* **14**, 2713 (1973).
51. Y. W. Mo, B. S. Swartzentruber, R. Kariotis, M. B. Webb, and M. G. Lagally, *Phys. Rev. Lett.* **63**, 2393 (1989).

STM Volume in Methods of Experimental Physics

5.3. Gallium Arsenide

R. M. Feenstra

IBM Research Division, T. J. Watson Research Center, Yorktown Heights, New York

Joseph A. Stroscio

Electron and Optical Physics Division, National Institute of Standards and Technology, Gaithersburg, Maryland

5.3.1 Introduction

Among the III–V semiconductors, GaAs has been extensively studied with the STM, with most of the effort to date focusing on the (110) surface which can be easily prepared by cleavage.[1-22] On other III–V materials, work has now appeared on InP, InSb, and GaP surfaces.[23-25] GaAs is of the zincblende structure, which has fcc translational symmetry with a two atom basis; a Ga atom at (0, 0, 0), and an As atom at (1/4, 1/4, 1/4) of the nonprimitive fcc unit cube, as shown in Fig. 1. There are four nearest neighbor bonds of length 0.245 nm to each atom, with the bonds separated by the tetrahedral angle of 109.47°. GaAs cleaves easily on the (110) family of planes exposing a "non-polar" crystal face; so-called because it contains equal numbers of Ga and As atoms, as shown in Fig. 2. Alternatively, the (100) surface is a "polar" surface, which can be terminated ideally by pure Ga or As atoms (see Fig. 1). In reality, the termination is varied and depends on the precise surface preparation, which can alter the ratio of surface Ga to As atoms. The (111) face of GaAs is also polar. In the bulk crystal, out of the eight (111) planes there are four (111) planes containing only Ga atoms, sometimes known as the (111A) planes as shown in Fig. 3. The other four planes are comprised of As atoms, the (111B) planes. Thus, a $\langle 111 \rangle$ oriented wafer will have a (111A) face on one side, and the other parallel face will be a (111B) face (note this occurs because to generate the alternate face below one layer requires the breaking of three bonds versus one). Sometimes these are also referred to as (111) and $(\overline{111})$. The stoichiometry of the (111) face can also be varied by

METHODS OF EXPERIMENTAL PHYSICS
Vol. 27

ISBN 0-12-475972-6

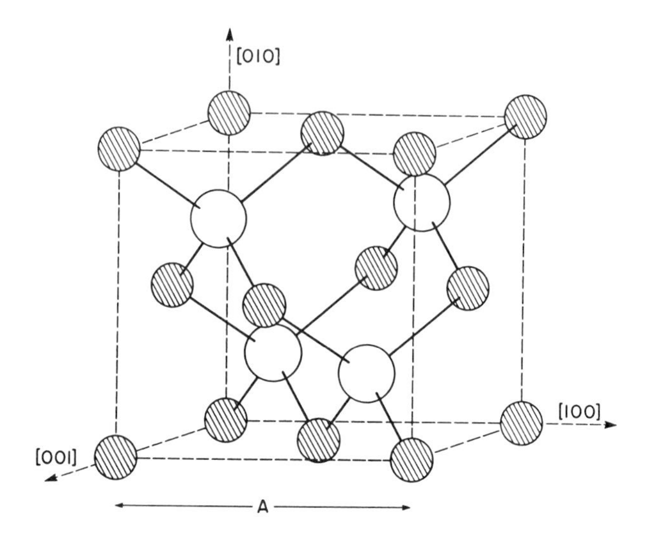

FIG. 1. Conventional unit cube for GaAs, with a volume A^3 that is four times larger than that of a primitive unit cell. [From Ref. 1.]

preparation, producing a variety of surface structures depending on the exact termination.

The bonding in GaAs, and III–V semiconductors in general, can be described as mixed ionic and covalent in character. The 4s and 4p subshells contribute eight electrons per GaAs unit, three from Ga and five from As. The partially ionic character results from the difference in electronegativity

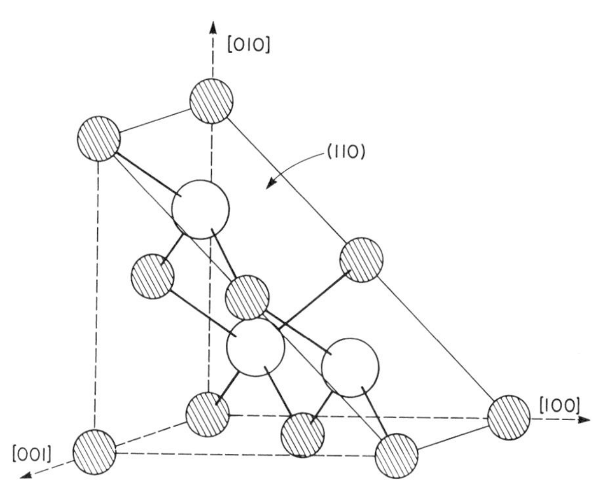

FIG. 2. Bisection of the GaAs unit cube (Fig. 1) by the (110) plane. [From Ref. 1.]

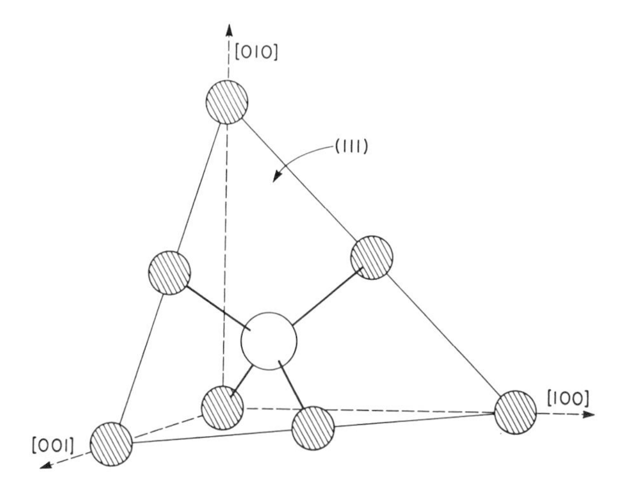

FIG. 3. Truncation of the GaAs unit cube by the (111) plane. Note that a plane of this family contains only one kind of atom. [From Ref. 1.]

of Ga and As and can be seen clearly in the calculated charge density contours shown in Fig. 4. From a number of procedures the charge transfer is estimated to be somewhere between 0.3–0.4 electrons from Ga to As. In the surface properties, the electronegativity difference can be used to formulate an empirical "electron counting" rule, which seeks to minimize the number of dangling bonds by filling missing bonds of the more electronegative species.[3] The partially ionic character of the surface bonds leads to marked voltage dependence in the STM imaging, demonstrated on the (110) surfaces. This selective imaging of the Ga or As species depends on tunneling into empty states, versus tunneling from filled states of the GaAs,[4] as discussed in Chapter 1 of this book.

The transfer of charge from Ga to As atoms on a GaAs surface leads to

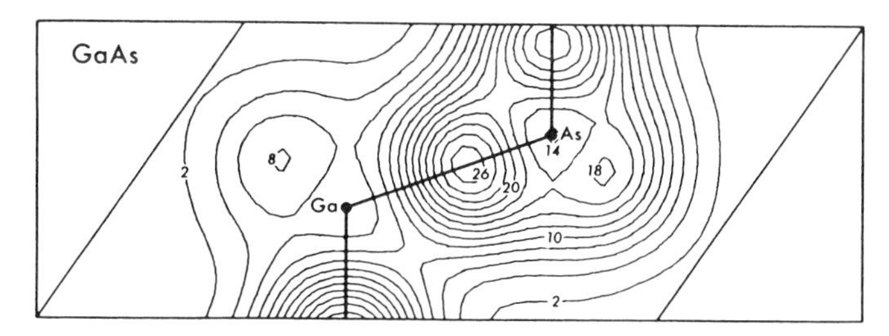

FIG. 4. Valence charge density for GaAs. The contours are in units of $(e/\Omega c)$, where Ωc is the volume of the primitive unit cell. [From Ref. 2.]

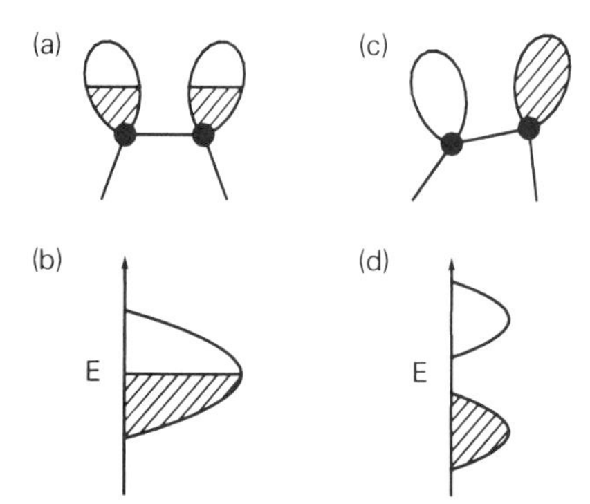

FIG. 5. Schematic view of buckling on a semiconductor surface. Surface atoms with half-filled bonds (a), form a half-filled band of surface states (b). Charge transfer from one bond to the other (c), splits this band of states into a filled band and an empty band (d). [From Ref. 6, reprinted by permission of Kluwer Academic Publishers.]

the formation of relatively large surface state band gaps. To understand this effect, consider two atoms on the surface, which may be two isolated atoms or two members of a chain of atoms on the surface, as illustrated in Fig. 5. We suppose that in the non-relaxed surface, each of these atoms has a dangling bond, and each bond is occupied by a single electron, as shown in Fig. 5(a). A two-dimensional set of such bonds will form half-filled band of states at the surface, as pictured in Fig. 5(b). A more energetically favorable situation is for one of the dangling bonds to transfer charge, creating one filled bond and one empty bond, as shown in Fig. 5(c). This transfer of charge results in a splitting of the band of states into a filled band and an empty band, separated by a band gap, as pictured in Fig. 5(d). Accompanying the charge transfer will be some possible relaxation in the positions of the atoms on the surface. If the two atoms are initially inequivalent (e.g., a Ga and an As atom), then the energy of their dangling bonds is necessarily different, and some buckling of the surface always occurs. If the two atoms are equivalent (e.g., two atoms of the same chemical species forming a symmetric dimer), then the buckling may or may not occur depending on the energetics of the situation. It is important to note that an entire electron may not be transferred from one dangling bond to another. In reality, all the states on the surface are composed of linear combinations of both dangling bonds, and the amount of charge transfer depends on the concentration of spectral weight from the associated bonds.

The buckling of the surface as just described provides a close description of the GaAs(110) surface, which is considered to be a theoretically prototypical surface as discussed in Section 5.3.2.1.[4–9] The other GaAs surfaces are more complicated since extensive reconstruction occurs, as discussed in

Sections 5.3.2.2 and 5.3.2.3.[10-12] The resulting band gap for the (110) surface turns out larger than the bulk band gap (2.4 versus 1.4 eV).[8,9] This unique situation provides one with some enhanced spectroscopic capabilities on this surface. The size of the bulk band gap can be measured, together with its variation due to changes in the bulk semiconductor composition, e.g., by forming the mixed compound AlGaAs. Also, by growing adjoining layers of the semiconductors, one can hope to measure the relative positions of their band edges, thereby obtaining the "band offsets."[13] Measurements of this type are discussed in Section 5.3.3 below. Using the STM spectroscopy, one can perform measurements of the position of the conduction and valence band edges (relative to the surface Fermi level) in the vicinity of surface adsorbates, as was already discussed in Chapter 4. Also, the introduction of states within the band gap due to the presence of individual atoms of small clusters of atoms can be sensitively measured, without any background effects arising from states that exist over the entire surface. This subject is discussed in Section 5.3.4 below. We consider the effects of various metals, including Sb, Au, and Fe, on the GaAs(110) surface.[14-22] A characteristic spectrum of states is found to be introduced into the band gap by these metal atoms and clusters, and this spectrum is interpreted in terms of evanescent states of the semiconductor.

5.3.2. Imaging

5.3.2.1. GaAs(110). While GaAs(110) is not the surface most common in technological applications, it has been the III–V surface in which most theoretical and experimental work has been carried out on. It is the only nonpolar surface and can be easily prepared by cleavage, as described earlier. The creation of an ideal (110) surface leaves two dangling bonds per unit cell directed out of the surface, which results in two electronic states within the fundamental band gap (see Figs. 2 and 5). It was known very early that the ideally terminated surface is energetically unstable.[7] Common to all the III–V (110) surfaces, the Group V atoms move out of the surface plane and the Group III atoms move toward the bulk; this can be characterized by a rotation angle ω of about 30° for all III–V(110) surfaces, as shown in Fig. 6. As described earlier, this relaxation is intimately connected with an electronic rearrangement forming an empty dangling bond at the surface Ga and a doubly occupied dangling bond at the surface As. This results in a surface band gap of 2.4 eV, which is substantially larger than the 1.4 eV bulk band gap.[8,9]

Cleavage of GaAs produces (110) surfaces that are visually almost perfectly flat, and are found with the STM to be devoid of steps over distances on the μm scale. Figure 7 shows an STM image of the GaAs(110) surface. Typical corrugations are on the order of 0.1 Å, but can be observed to be much larger

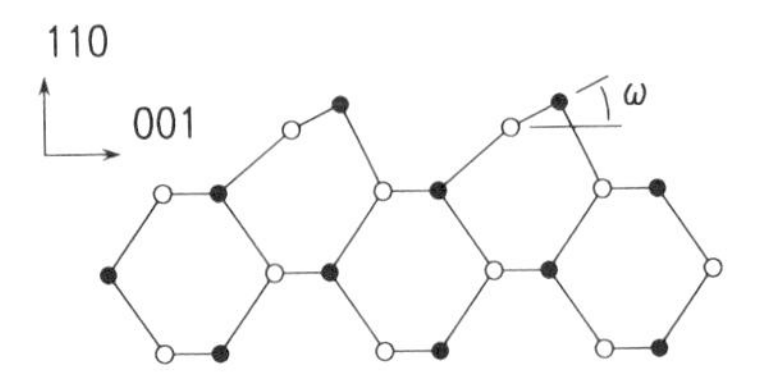

FIG. 6. [1$\bar{1}$0] cross-sectional view of the GaAs(110) surface, illustrating the relaxation of the surface atoms with buckling angle ω. Ga atoms are shown by open circles, and As atoms by closed circles.

at low positive sample bias on p-type material. The most noticeable property of the image in Fig. 7 is the fact that only one maximum appears in the unit cell that contains two atoms, as shown in Figs. 2 and 6. The appearance of only one maxima in the unit cell results from the energy-selective nature of

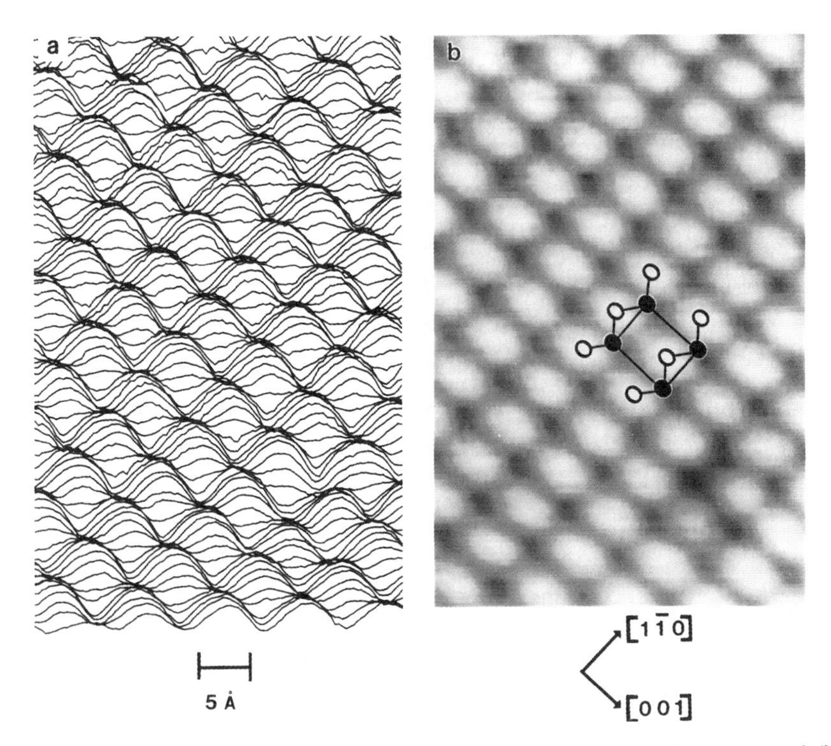

FIG. 7. STM image of the p-GaAs(110) surface showing (a) original line scans, and (b) gray-scale top view. The peak to valley corrugation is ~1.1 Å. The sample voltage was 1.5 V. A schematic top view of the atomic positions is shown in (b), where the open circles represent Ga atoms and the closed circles represent As atoms. The rectangle indicates a surface unit cell. [From Ref. 5.]

FIG. 8. Combined color STM images of the GaAs(110) surface, showing the Ga atoms (blue) and As atoms (red). The blue image was acquired by tunneling into empty states that are localized around surface Ga atoms, and the red image was acquired by tunneling out of filled states that are localized around surface As atoms. [From Ref. 4.] (See color insert.)

tunneling microscopy, which is discussed in Chapters 1 and 4 of this book. For GaAs(110), the surface relaxation previously described results in the filled states mainly localized on As sites and the empty electronic states localized on Ga sites. Thus, tunneling out of filled states yields maxima-selecting As states; Ga state images are obtained by switching the voltage polarity and tunneling into empty sample states.[4,5] The registry of the As and Ga state images can be obtained by recording simultaneous filled and empty state images by interlacing scan lines at alternate biases, as described in Section 4.1. By overlapping the two images, one ends up with a composite image showing the registry between filled and empty states with two atoms now in the surface unit cell, as shown in Fig. 8.

Upon examining the spatial separation between filled and empty states in Fig. 8, one finds a separation perpendicular to the GaAs chains (along the

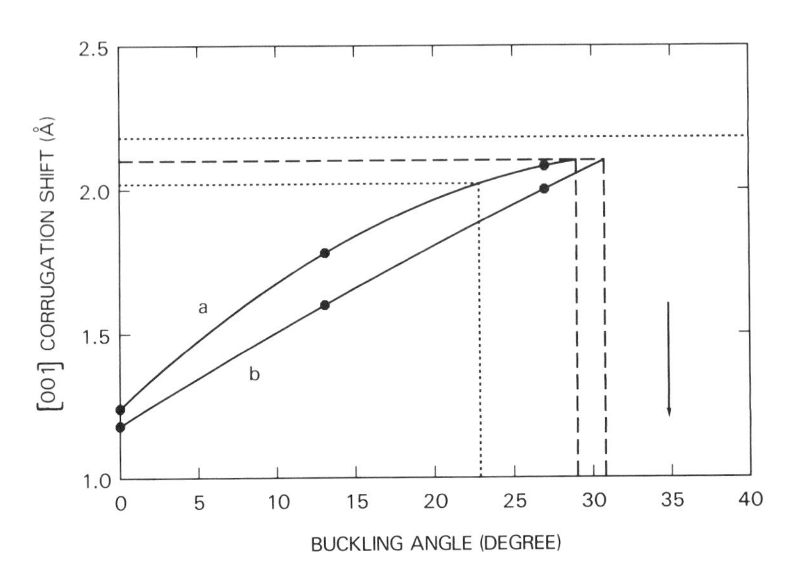

FIG. 9. Shift in lateral position between occupied and unoccupied states (Δ) vs buckling angle (ω). Experimental bounds for Δ are shown by the horizontal dashed line. Theoretical curves are given by the solid lines, which are quadratics fitted to the calculated points shown, for constant state densities of (curve a) 10^{-8} and (curve b) 10^{-10} electron/bohr^3. The vertical dashed lines then give the deduced mean values for the buckling, with an uncertainty bound shown by the vertical dotted line. The arrow indicates the angle (34.8°) where the Ga bonds become planar. [From Ref. 4.]

[001] direction) larger than the physical separation of the atoms (2.1 versus 1.4 Å). This interesting effect arises from the fact that the details of the STM images show spatial extents of surface electronic properties, and in this particular case, the spatial separation is intimately connected with the surface relaxation indicated above. This interplay is described in Fig. 9, which shows the calculated spatial separation between the valence and conduction band states as a function of buckling angle. For small angles, the separation is about 1.2 Å which is close to the physical separation between the atoms. With increasing buckling, the filled and empty surface states move apart in energy, thereby increasing the surface band gap. As shown in Fig. 9, this is also reflected by an increase in the spatial separation between the filled and empty states. In comparison with the experimental separation, we deduce a buckling angle in the range of 29–31°, which is in agreement with other experimental measurements. Similar phase shift measurements have also been applied to InSb(110), which yielded similar values for the buckling angle.[24] The significance of these STM measurements is that we determined what is more or less a vertical displacement, which is not the norm for STM, but was accomplished by an inversion of electronic structure information.

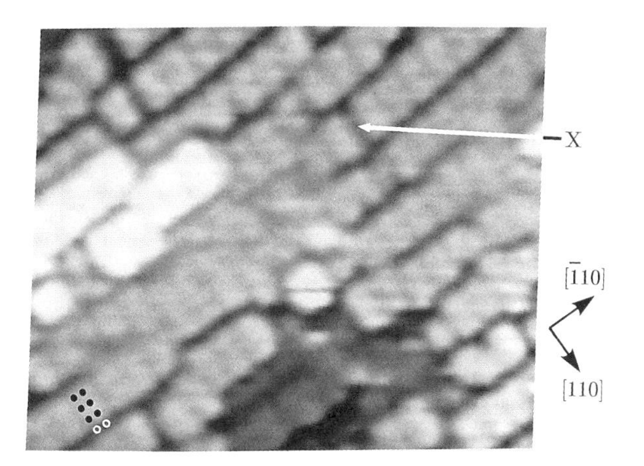

FIG. 10. STM image of the GaAs(100)–2 × 4 surface showing a 165 × 130 Å^2 area including small islands one plane up (light) and one plane down (dark), each having the same structure as the main plane. Disorder in the pairing of arsenic atoms can be seen at X. A superposition of the 2 × 4 structure is shown in the lower-left corner, with the As atom positions marked with filled circles and the missing atom sites marked by open circles. The image was acquired with a sample bias of −2.3 V, and the gray scale cover a vertical height of 7 Å. [From Ref. 10.]

While GaAs(110) shows nearly perfect surfaces, due to the cleavage procedure, the same is not true for the other low index surfaces, which display a variety of surface reconstructions depending on preparation as described next.

5.3.2.2 GaAs(100). Epitaxial growth of GaAs and other III–V compounds is generally performed on the (100) crystal face. Thus, it is of particular importance to understand the geometric structures of this surface. From the viewpoint of the STM field, however, progress on this surface occurred some years after comparable studies of Si and GaAs(110) surfaces.[10,11] The main reason for this delay is due to the difficulty in preparing the (100) face: growth in a molecular beam epitaxy system is necessary to achieve high-quality surfaces. In addition to the preparation difficulties, the surface geometry is complicated by the fact that different structures can form, depending on the relative concentration of Ga to As on the surface. Here, we describe the first STM study of the GaAs(100) surface, performed by Pashley and co-workers.[10] The sample was grown in an MBE chamber, and was capped with a thick layer of arsenic. The sample was then transported through air into the STM chamber, where the excess arsenic was desorbed by heating at 370°C. Surfaces prepared in such a manner tend to have an excess of As remaining on the surface, thus having the 2 × 4 or c(2 × 8) diffraction patterns characteristic of the As–terminated surface.

In Fig. 10 we show an STM image of the GaAs (100) surface, and a

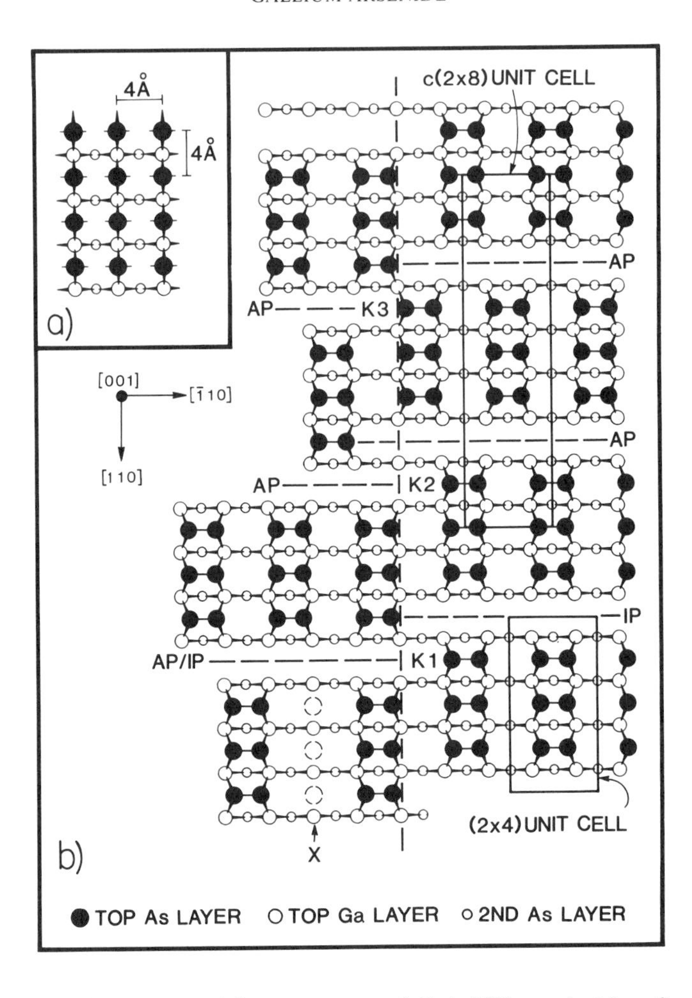

FIG. 11. (a) The structure of the unreconstructed GaAs(100) arsenic-rich surface. (b) The missing-dimer model for the GaAs(100)2×4 surface. The two types of missing-dimer boundary, in-phase (IP) and antiphase (AP), are shown giving rise to 2×4 and c(2×8) structures. Disorder in the arsenic pairing (X) with three missing arsenic atoms (dashed circles) is also shown. [From Ref. 10.]

structural model is shown in Fig. 11. The surface is seen to consist of a series of rectangular shaped corrugation maxima, separated in the [1$\bar{1}$0] direction by dark bands. Each corrugation maxima can actually be resolved in other STM images into three dimers,[10] and the dark bands correspond to missing dimers in the surface layer. In the structural model, the dimers are believed to be composed completely of As. With one missing dimer out of every four, the electron counting at the surface results in a completely filled surface band (composed primarily of the dangling bonds from the As dimers).[3] Depending on the precise arrangement of the missing dimers, the surface structure is either 2×4 or c(2×8), as shown in the structural model.

5.3.2.3 GaAs($\overline{111}$). In addition to the (100) surface, MBE growth can be accomplished on other GaAs surfaces. For the ($\overline{111}$) surfaces a 2×2 diffraction pattern is observed during and after growth in As-rich conditions. Subsequent annealing results in a transition to a $(\sqrt{19} \times \sqrt{19}) - R23.4°$ reconstruction. These structures have been studied using STM by Biegelsen and co-workers.[12] For the 2×2 structure, they propose a model based on As–trimers, with the As atoms in a trimer bonding amongst themselves and to As atoms in the underlying layer. These trimers then form an hexagonal array with 2×2 symmetry. It is found that the electron-counting rule works for this surface, with the As dangling bonds being all completely saturated.

Upon annealing to 500°C for about 10 minutes, the 2×2 surface converts to a $\sqrt{19} \times \sqrt{19}$ structure. STM images of this surface, and a structural model, are shown in Fig. 12. A large area scan is shown in Fig. 12(a), where the dark triangular shaped patches are 3.3 Å (a bilayer step) lower than the surrounding regions. A higher magnification image is shown in Fig. 12(b), and in 12(c) the structural model is overlaid on the STM image. The structural model consists of hexagonal rings of atoms, with six As atoms in the uppermost layer, and twelve As atoms in the next layer. The ring of As atoms gives rise to the circular shapes seen in the STM image. Since the $\sqrt{19} \times \sqrt{19}$ structure necessarily contains an odd number of atoms, the electron-counting rule cannot be completely satisfied for this surface.[12] However, for such a large unit cell (as for the case of Si(111)–7×7), the half-occupied dangling bond which is left over introduces only a relatively small amount of state-density into the band gap, and thus such a structure can still be energetically favorable.

5.3.3 Cross-Sectional Imaging

With the ability to easily prepare the GaAs(110) surface by cleavage in vacuum, one can immediately consider the possibility of *cross-sectional* imaging of structures. Superlattices, consisting of layered arrangements of different III–V compounds (such as GaAs and $Al_xGa_{1-x}As$), are generally grown in the (001) direction, and viewing of a (110) face then permits an examination of the structures in cross-section, exposing the interface between the two materials. From such studies one can gain structural information concerning the superlattice, such as the extent of interdiffusion at the interfaces. Also, since the clean (110) surface is free of surface states in the band gap region, one can also use STM spectroscopy to reveal the bulk band gaps

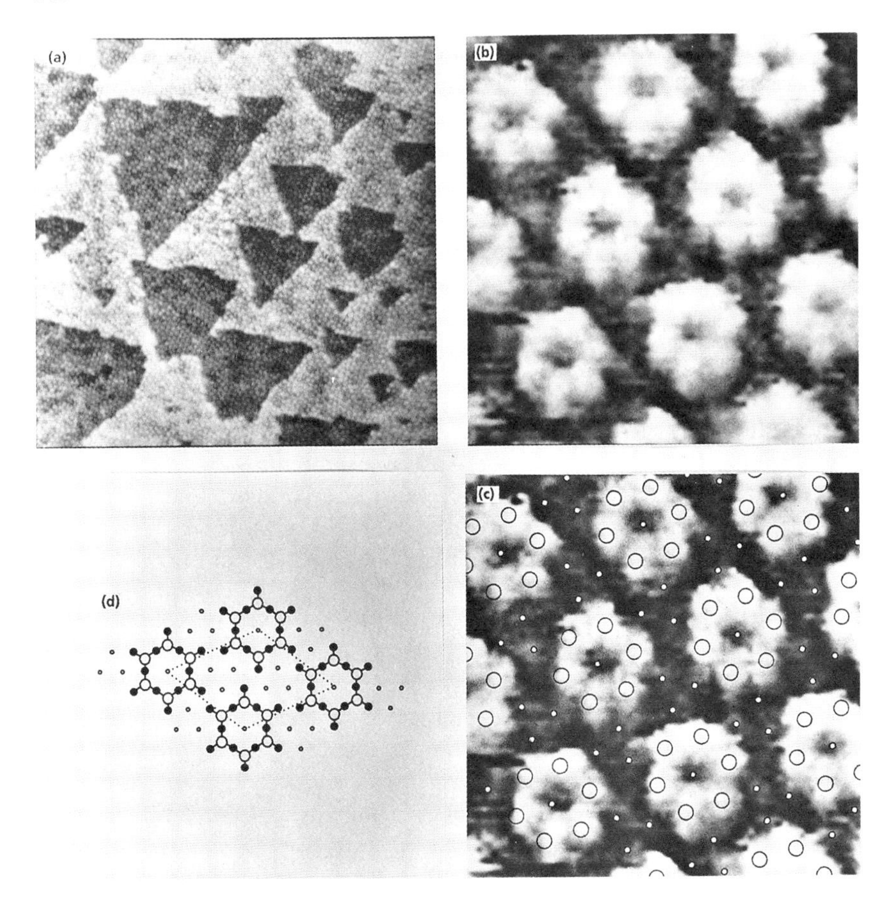

FIG. 12. GaAs$(\overline{1}\overline{1}\overline{1})\sqrt{19} \times \sqrt{19}$ reconstruction. (a) Filled-state image 120-nm-square scan; (b) 4.9-nm-square-scan; (c) same as (b) with overlay of threefold As atoms from unrelaxed model; (d) detailed model; large open circles denote top As atoms; medium closed circles denote second-layer Ga atoms, and small open circles denote third-layer, threefold-coordinated As atoms. The image was acquired with a sample bias of −1.8 V. [From Ref. 12.]

and the relative placement of the band edges in adjacent layers, i.e., the band offsets.

The major problem encountered in performing cross-sectional imaging is to physically locate the STM probe-tip over the region of interest on the cleavage face. This requires having both the ability to conveniently move the sample laterally, and some means of large-range imaging to determine when the tip is correctly placed. These requirements were fulfilled in the work of Salemink and co-workers by using a conventional STM "louse" for sample motion, and a UHV scanning electron microscope for tip positioning.[13]

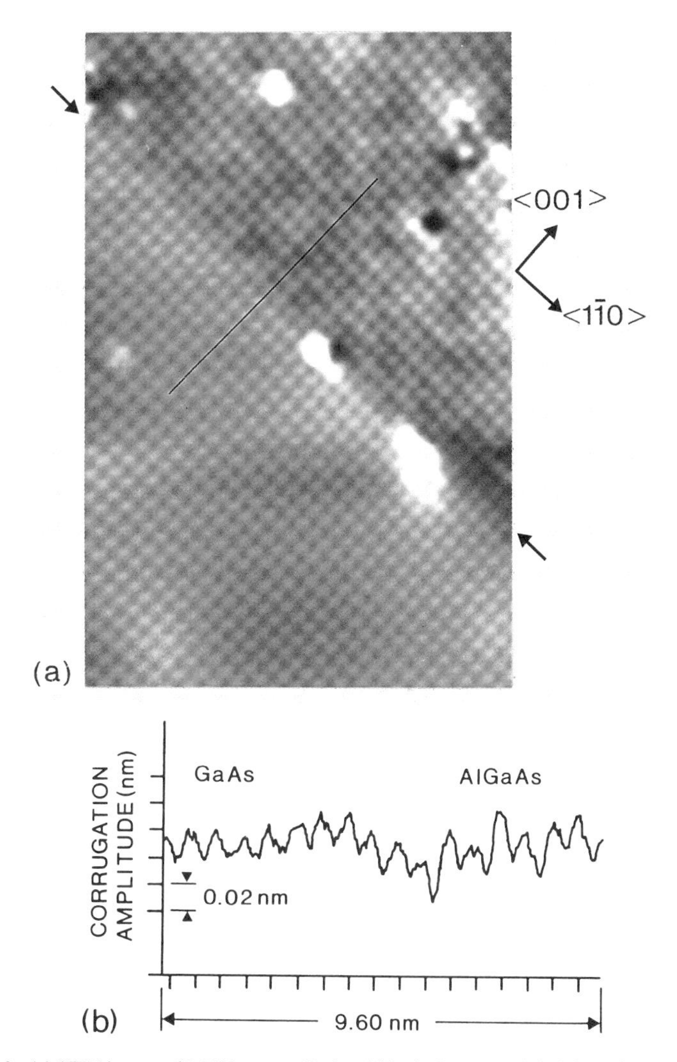

FIG. 13. (a) STM image of MBE-grown GaAs–AlGaAs layers and their interface, imaging the As atomic lattice. Note the very clean GaAs and the presence of oxygen patches on the AlGaAs. The image width is 10.5 nm. The interface is indicated by arrows. (b) Profile of corrugation along section indicated in (a). The interfacial width, as observed in the local density-of-states, is one to two unit cells. [From Ref. 13.]

Figure 13 shows an STM image and cross-sectional scan obtained from a cleaved superlattice with $x = 0.38$ Al concentration. The AlGaAs material is seen near the top of the image, and the GaAs materials is seen at the bottom. The AlGaAs region appears to be topographically slightly lower than the

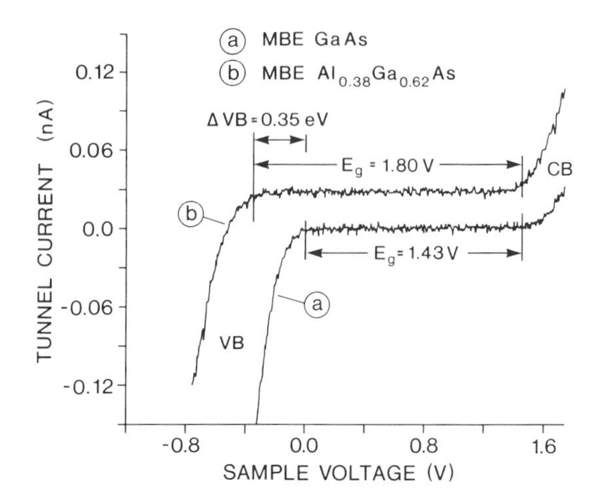

FIG. 14. Current–voltage characteristics in GaAs (curve 1) and in AlGaAs (curve 2), both taken approximately 5 nm from the interface. Curve 2 is displaced vertically by 0.03 nA for clarity. The GaAs band gap is 1.43 eV and its valence band edge is very near the Fermi level (0 V). The measured AlGaAs band gap is 1.80 eV. The measured position of the Fermi level relative to the AlGaAs valence band edge is 0.35 eV. The valence band (VB) and conduction band (CB) are indicated. [From Ref. 13.]

GaAs, probably owing to the larger band gap of the AlGaAs as discussed later. In addition, the AlGaAs regions are found to be more inhomogeneous and "patchy" compared to the GaAs, which is attributed in this experiment to the preferential adsorption of oxygen onto the AlGaAs. The STM image of Fig. 13 was acquired at negative sample bias of -2.3 V, so that As atoms are seen in the image. We see in the image that the registry of the atoms is maintained across the AlGaAs–GaAs interface; the transition from one material to the other occurs over a width of 1–2 unit cells.

Tunneling spectra acquired on both sides of the AlGaAs–GaAs interface are shown in Fig. 14. Each of the *I–V* curves was acquired about 5 nm from the interface. Curve 1, taken on GaAs, is typical of highly-doped p-GaAs, with a gap close to 1.43 eV, and the Fermi located at the bottom of the gap. Curve 2, taken on the AlGaAs, shows a larger gap of about 1.80 eV, close to the 1.90 eV expected for this material. The surface Fermi level (0 V) in the AlGaAs is found to be located 0.35 eV above the valence band edge, as shown in Fig. 14. This value is significantly larger than one expects for an ideal superlattice structure in the absence of any surface effects.[13] This discrepancy is attributed to the presence of the oxygen adsorbates on the AlGaAs, which are known to induce some surface states that push the Fermi level towards mid-gap. Thus, at this stage the direct measurement of the relative positions of the GaAs and AlGaAs band edges has still not been obtained, but additional work may yield that quantity for this AlGaAs–GaAs system and perhaps for other superlattice systems as well.

5.3.4 Spectroscopy of Metals on GaAs(110)

All of the spectroscopic work performed to date with the STM on GaAs have been done on the (110) crystal face.[15–22] As discussed earlier, this face is unique in that the dangling bond states occur at energies outside of the bulk band gap, so that the introduction of states within the band gap due to adsorbates can be conveniently studied. Prior to the development of the STM, such studies were performed primarily by photoemission spectroscopy.[26] Such measurements provide a measure of the spectrum of both valence-band and core level states, and one can also extract the position of the surface Fermi level relative to the valence band maximum. However, for the case of states introduced by sub-monolayer coverage of metal adsorbates, photoemission is unable to directly discern any clear spectrum of states in the band gap region, presumably because the density of such states is too low. Nevertheless, movement of the surface Fermi level in the gap is observed, and it is generally found the Fermi level moves from the band edges to the middle of the band gap as the metal coverage increases up to about a monolayer. This so-called "pinning" of the Fermi level is attributed to the existence of states in the gap, but the nature of such states remains unknown, and a variety of models have been postulated to describe the states.[26–28]

With the advent of the STM, one has the capability to observe very low densitities of states within the band gap. Moreover, by acquiring spatially resolved spectra, these states can be associated with definite geometric structures on the surface. Such studies have been performed on GaAs(110) for a variety of metal adsorbates, and we focus in this section on the results for Sb, Au, and Fe. An introduction to the spectroscopic techniques has already been given in Chapter 4, in which the spectroscopy of clean and oxygen covered GaAs(110) surfaces is discussed. For the metal adsorbate systems, as for oxygen, at a coverage of greater than about 0.01 monolayer ($ML = 8.85 \times 10^{14}$ atoms/cm^2) some amount of band bending exists over the entire surface, so that the "dopant induced" contributions to the tunnel current, seen on the totally clean surface, disappear. Thus, on or near the metal adsorbate we observe their characteristic state-density, and far from the adsorbates we observe zero current in the band gap region of the semiconductor. Charging of the metal adsorbates can occur for low semiconductor doping densities, and this can lead to significant problems in passing current through the metal clusters into the bulk material.[17] High-doping concentrations of 10^{19}–10^{20} cm^{-3} are used to overcome that limitation. Alternatively, one can deposit enough metal so that an interconnected network of clusters

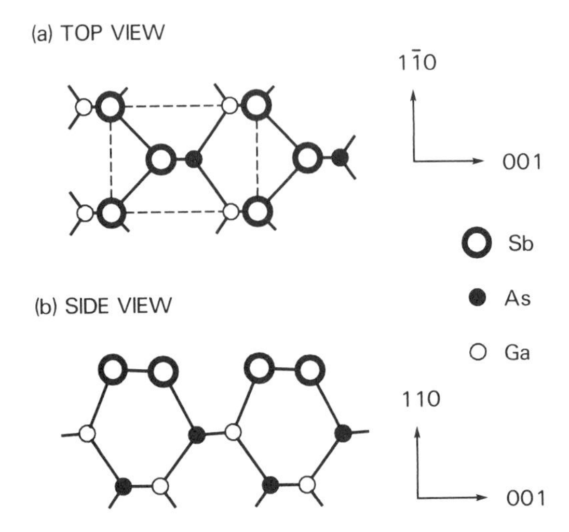

FIG. 15. Schematic diagram of the surface geometry for GaAs(110)1 × 1–Sb as determined by low-energy electron diffraction (Ref. 8). (a) Top view. (b) Side view. [From Ref. 16.]

exists over the surface, and the current can then propagate continuously along the surface.

In Figs. 15 and 16 we show a structural model and STM image for a monolayer of Sb on the GaAs(110) surface.[14–18] In the model, the Sb atoms occupy positions close to those expected by an extension of the bulk lattice. Deposition of Sb at room temperature produces large flat terraces; Fig. 16 shows a portion of such a terrace. A well-ordered part of the Sb layer, displaying 1 × 1 corrugation, appears in the lower part of the image. The black area appearing in Fig. 16 is a hole extending down to the GaAs substrate, and some small Sb clusters residing on top of the ordered Sb layer are also visible in the image. The Sb–GaAs system is one of a class of overlayers known as "passivating" (other members of this class include arsenic on Si(111) and Si(100)). In such systems, formed typically by column-V adsorbates on nonpolar semiconductor faces, the adsorbate layer has one additional electron per atom compared to the bulk terminated semiconductor surface. In that case, the overlayer contains enough electrons to fully saturate the surface dangling bonds. Following these arguments for buckling of a clean surface, we then expect a relatively large surface band gap for such systems. Indeed, it is found for Sb on GaAs(110) that the width of the surface-state band gap is very close to that of the bulk band gap.[16]

Spatially resolved spectroscopy for Sb–GaAs is shown in Fig. 17. The STM image shows an ordered portion of the Sb monolayer, with a hole extending down to the GaAs substrate. Spectra were acquired at the points indicated

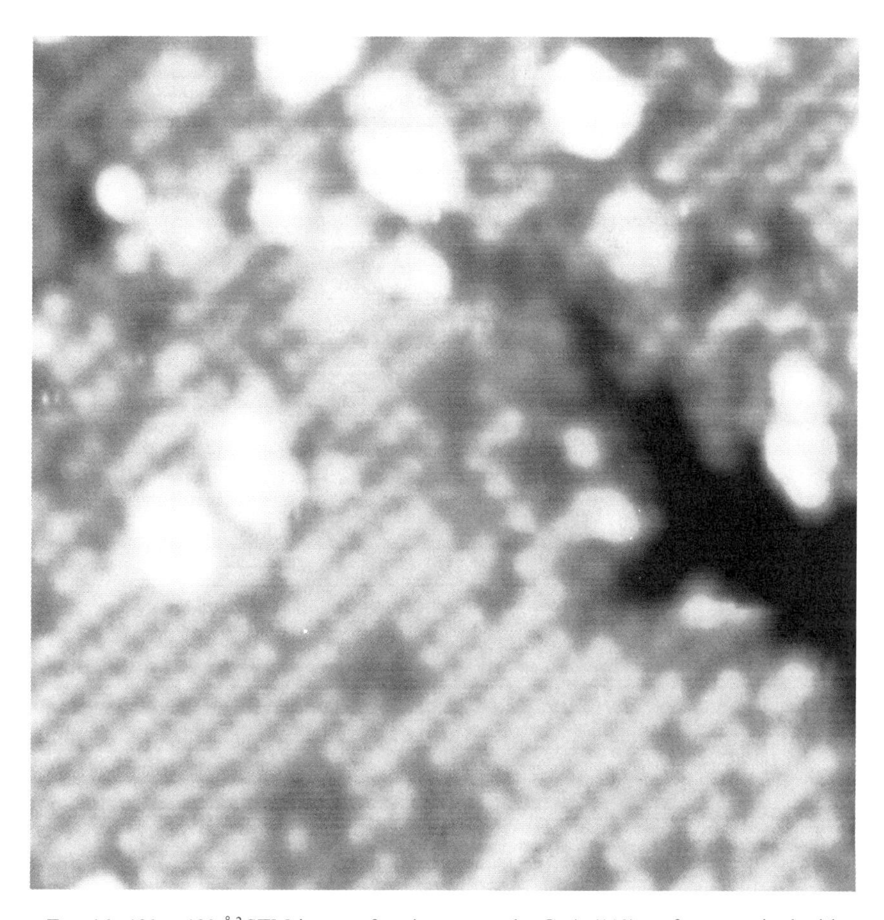

FIG. 16. 120 × 120 Å^2 STM image of antimony on the GaAs(110) surface, acquired with a sample bias of -2 V. The Sb coverage is $\simeq 1.0$ ML. The topographic height is displayed by a gray scale, ranging from 0 Å (black) to 5 Å (white). [From Ref. 15.]

by the checkered markers in the image; points (a) and (b) are located at the edge of the Sb terrace, and points (c) and (d) are located on the well-ordered portion of the Sb layer. In the latter two spectra, a band gap region where the conductivity is close to zero is observed, with edges marked by E_v and E_c in Fig. 17. The width of this region is about 1.4 eV. This band gap arising from the Sb overlayer is found to be positioned almost identically with the bulk band gap.[16] At the terrace edges, spectra (a) and (b), an intense state located at about 0.8 V is observed within the band gap region. The surface Fermi level (0 V) occurs about 0.2 V above the bottom of the band gap. Similar spectra have been acquired from other Sb partial monolayers and isolated clusters, and the general appearance of the spectra is the same as seen

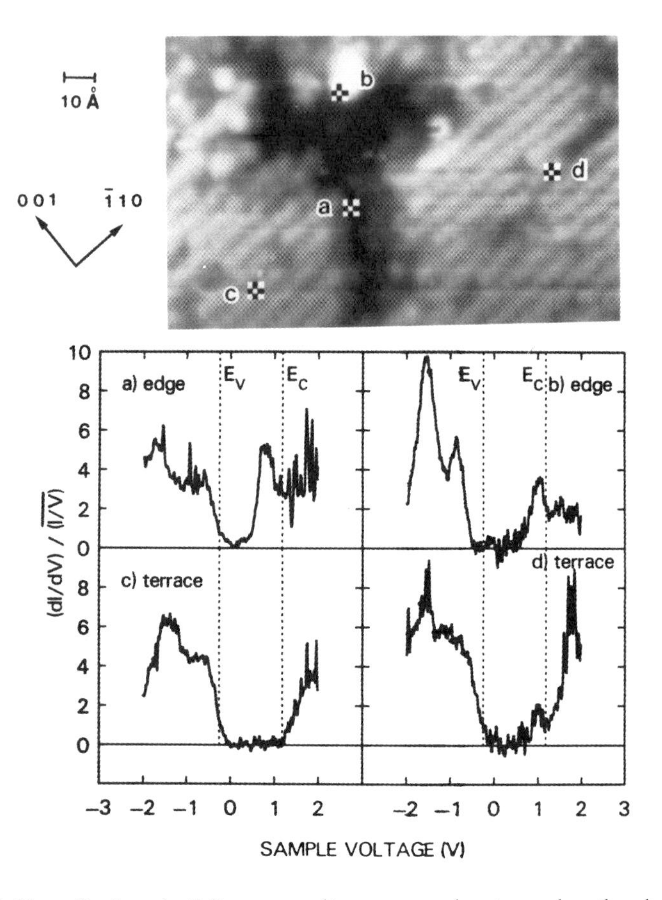

FIG. 17. Normalized conductivity versus voltage, measured on top and on the edge of an Sb terrace. The associated $120 \times 80\,\text{Å}^2$ STM image is also shown, with the locations at which the spectra were acquired indicated by checkered markers. The valence band maximum and conduction band minimum are marked by E_v and E_c, respectively. [From Ref. 17.]

in Fig. 17. A band gap state is observed at about $E_v + 1.0\,\text{eV}$, and, in some cases, a tail of states extending out of the valence band is also seen. The Fermi level is located at a minimum in the conductivity, pinned between these two sets of states.

The spectra displayed in Fig. 16, and similar observations at other Sb coverages, were the first observation of the band gap states which pin the Fermi level for a disordered metal overlayer on GaAs(110). In terms of geometry, it is clear that these states arise from the edges of the Sb terraces, but beyond this nothing more definite can be said concerning the geometry. Thus, it was desirable to find another system in which the formation of the band gap states could be studied, and for that reason the Au on GaAs system

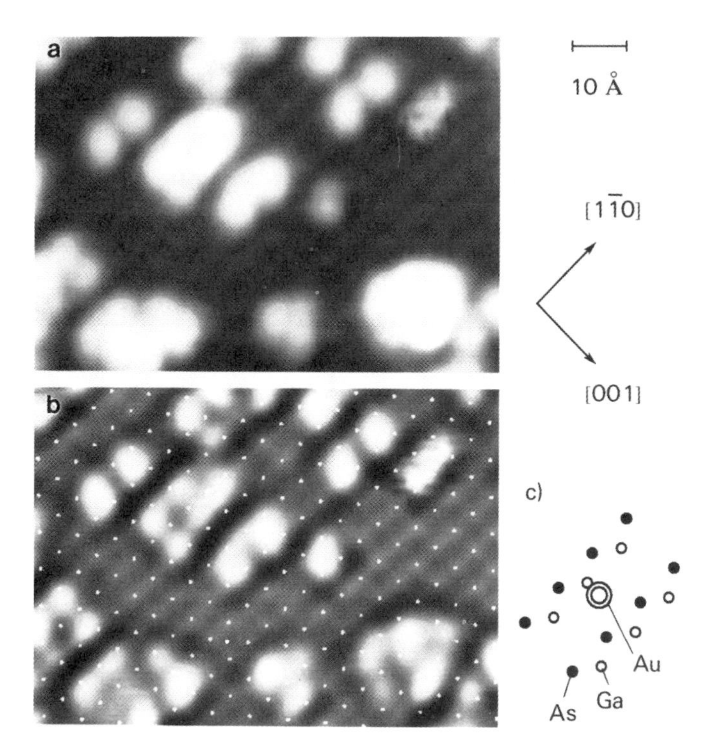

FIG. 18. 90 × 60 Å^2 STM images of 0.25-ML Au on the GaAs(110) surface, acquired with a sample bias of −2.5 V. The gray scale corresponds to (a) surface height, and (b) surface curvature. A top view of the atomic positions is pictured in (c), with a 2.2 × expanded lateral scale. [From Ref. 19.]

was studied.[19,20] There, at a few tenths ML coverage, individual Au atoms and small groups of atoms can be observed in the STM images, as shown in Fig. 18. The image of Fig. 18(a) is shown using a conventional gray-scale corresponding to surface height, and the Au adsorbates have a typical height of about 2 Å. In Fig. 18(b) the contrast of the image is enhanced by plotting the curvature of the surface. The corrugation of the bare surface is then clearly visible, and it reflects the positions of the As atoms for the sample bias of −2.5 V used. A grid of points is overlaid on Fig. 18(b) located at the As atom positions, and the Au atoms are seen to be located equidistant between four As atoms. Thus, it is found that the Au atoms are located about 1.4 Å away from surface Ga atoms, indicating that the Au adsorbate bonds to the Ga atoms.

Spatially resolved spectroscopy on the small Au clusters is shown in Fig. 19. The portion of the surface pictured there is actually the same as that in Fig. 18, although the probe-tip appears to be somewhat blunter in Fig. 19.

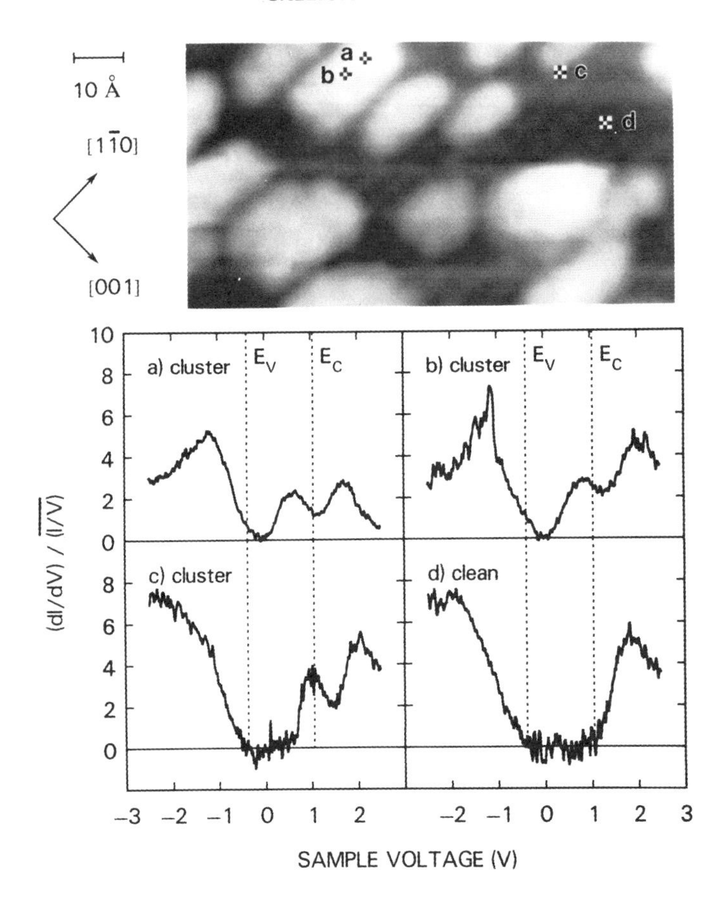

FIG. 19. Normalized conductivity versus voltage, measured on and off the Au clusters. The 100×55 Å^2 STM image of the surface is shown in the top part of the figure, acquired at a sample bias of -2.5 V. Checkered markers give the locations at which the spectra were acquired. [From Ref. 19.]

Four spectra, acquired at the location indicated by the checkered markers, are shown. Spectrum (d), acquired on the clean GaAs surface, shows a band gap region with width of about 1.4 eV. Spectra (a) and (b), acquired on top of a small Au cluster, display an intense peak within the band gap, and also an enhanced state-density in the valence band which extends up into the band gap region. Similarly for spectra (c), although the band gap state is slightly shifted in that case. It was found that the band gap state induced by the Au adsorbates has significant tails that extend spatially away from the Au atom.[19] Subsidiary maxima are observed on neighboring Ga atoms, indicating that "unbuckling" of the GaAs lattice in the vicinity of the adsorbate may be involved in the formation of the band gap states.[18,19] Overall, the appear-

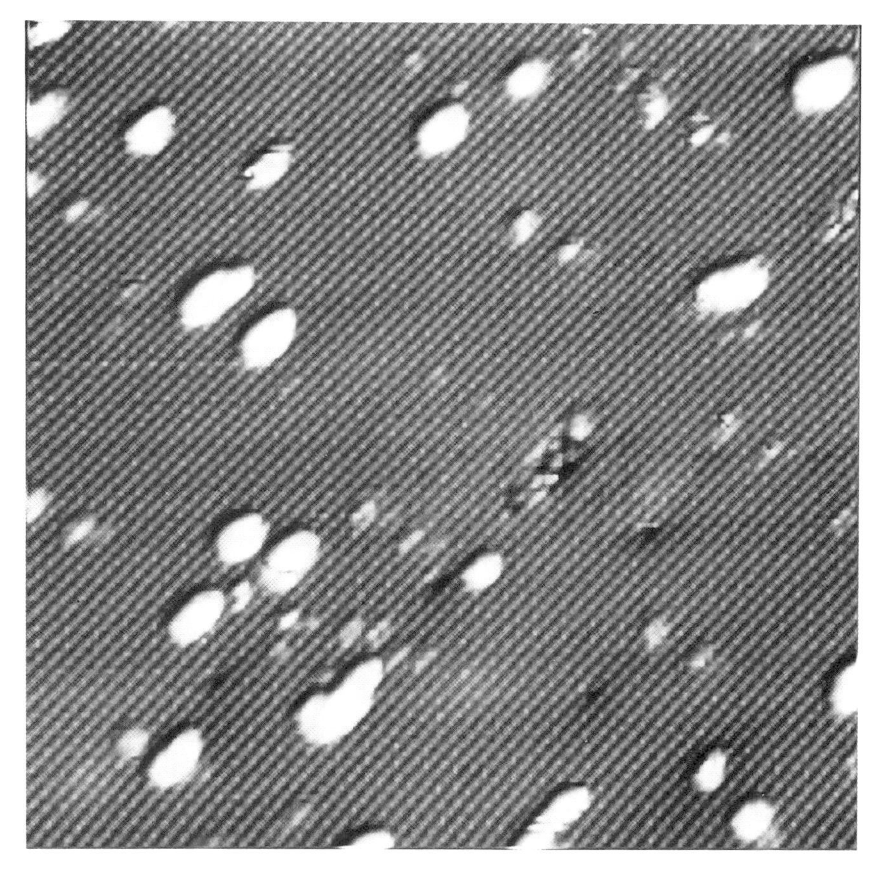

FIG. 20. STM image, 400 × 384 Å^2, of 0.1 Å Fe-p-GaAs(110). The sample bias was −2.5 V. The image is displayed using a gray scale keyed to the gradient of the surface height to increase the dynamic range. [From Ref. 22.]

ance of the spectra on the Au adsorbates is remarkably similar to that seen at the edge of the Sb terraces, even though the chemical identity of the adsorbates and geometry of their bonding is quite different.

A third metal on GaAs(110) which has been studied in detail is Fe.[21,22] When deposited onto the surface, the Fe is relatively mobile and forms three-dimensional clusters, as shown in Fig. 20. Spectroscopic measurements, as a function of cluster size, have been performed. It was found that clusters consisting of greater than about 35 atoms tend to have metallic character. Detailed, spatially resolved spectroscopy have been performed near such clusters. Figure 21 shows an image of an Fe cluster, and I–V characteristics acquired at the points indicated by the circles are shown in Fig. 22. I–V normalization of the spectra has not been performed here, but the spectral

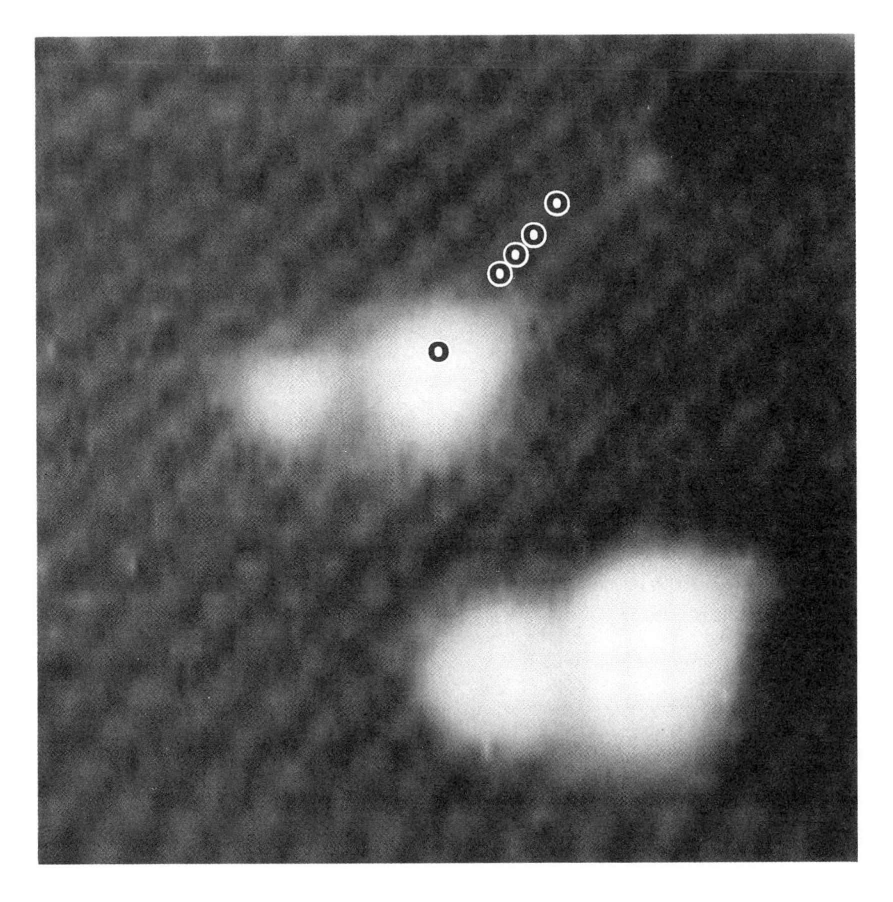

FIG. 21. STM image, 100 × 100 Å^2, of Fe clusters on p-GaAs(110). The sample bias was −1.8 V. *I–V* characteristics were recorded at each pixel in the image. The circles denote the positions of the *I–V* curves shown in Fig. 12. [From Ref. 22.]

features are still evident directly from the raw *I–V* curves. Curve (e), acquired far from the cluster, displays a band gap region with width of about 1.5 eV. Curve (a), acquired on top of the cluster, shows an almost linear characteristic near 0 V, indicating metallic behaviour. Curves (b), (c), and (d), acquired at successively increasing distances away from the cluster, show the decay of the metallic state-density with lateral distance. It is found that the state-density decays exponentially away from the cluster, with decay length that is energy dependent. These observed decay lengths are in qualitative agreement with a one-dimensional model of the evanescent states (with imaginary wave-vector) in the semiconductor band gap. It is also noteworthy that curve (b) in Fig. 22 clearly shows a spectral bump at about 0.9 V, which, in

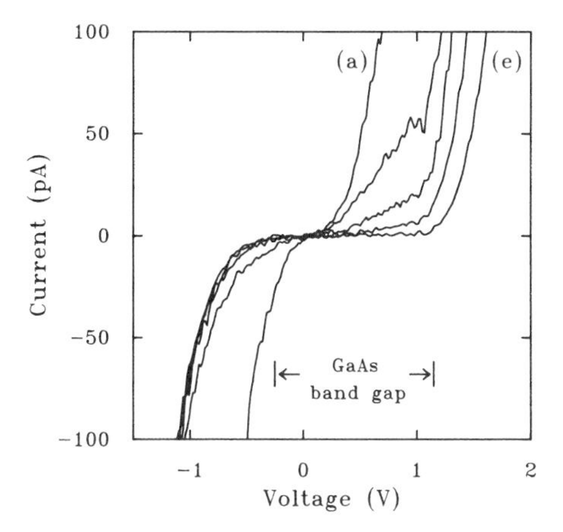

FIG. 22. Tunneling current versus voltage, at various distances from the Fe cluster indicated in Fig. 11. Curve (a) is on the cluster. Curves (b)–(e) correspond to distances from the cluster edge of (b) 3.7 Å, (c) 6.7 Å, (d) 9.6 Å, and (e) 14.3 Å. These tunneling characteristics were obtained with a tip–sample separation decreased by 1.0 Å relative to that used for the topograph. The valence and conduction band edges are at -0.25 and 1.25 V, respectively. [From Ref. 22.]

derivative form, would correspond exactly with the feature seen at 0.8 V in the Sb and Au spectra. Thus, very similar spectra are found for the band gap states induced by all of the Sb, Au, and Fe systems.

Phenomenologically, the observation by STM of the band gap states induced by metal adsorbates on GaAs(110) provides an explanation for the Fermi-level pinning phenomena observed on this surface. At low coverage, the spectrum of states consists of an intense state located near $E_v + 1.0$ eV, together with a tail of states extending out of the valence band. The Fermi level is pinned between these two sets of states. This spectrum appears to be remarkably independent of chemical species or geometry of the metal—the same spectrum is observed at the edge of Sb terraces, on or near small Au clusters, and near larger Fe clusters. For such *disordered* systems, we conclude that the spectral density of band gap states is relatively independent of structure. This result is in complete accordance with the photoemission observation that the Fermi level tends to be pinned near mid-gap, independent of the metal species deposited on the surface.[26] Of course, whenever a two-dimensionally *ordered* arrangement of atoms forms on a surface (such as the ordered Sb terraces), then the spectrum may change, in accordance with whatever bonding arrangement occurs in the ordered layer. Even in those cases, however, it should be noted that a band gap will often form in the surface state spectrum, and the final location of the surface Fermi level is still determined by some disorder-related feature on the surface.

Let us now consider the connection, if any, between the STM results and the various models which previously existed for Fermi-level pinning at surfaces. First, with regard to the "unified defect model" of Spicer and co-workers,[26] we find that the presence of particular surface defects, such as an anti-site defect, is not necessary in order to produce the Fermi-level pinning. Indeed, the presence of such defects is inconsistent with the observed STM structures. For the case of the Sb overlayers, the structural model of an Sb monolayer with no defects or disruption of the underlying GaAs is quite well established from both STM and low-energy electron diffraction studies.[14,16] For Au adsorption, some intermixing between Au and GaAs may occur at higher coverages, but for the individual adsorbates and small clusters discussed here no disruption of the GaAs is seen in the STM images. The spectrum of states observed by STM is generally consistent with that assumed in both the "disorder induced gap states" model[27] and the "metal induced gap states" model.[28] The latter case applies to the case of a fully formed metallic overlayer, such as found for the larger Fe clusters described previously. It is noteworthy that the same spectrum of states observed near the edge of the Fe clusters is also found for the small Au clusters and at the edge of Sb terraces. For the latter two cases, the adsorbate overlayers are *not* metallic, yet they still induce states in the band gap which are characteristic of the evanescent states of the semiconductor.

The STM results described earlier provide us with a good phenomenological understanding of Fermi-level pinning at surfaces. Nevertheless, some questions remain concerning the observed spectra. In particular, it is not clear how such widely different structures (edge of Sb terraces, small Au clusters, and edge of Fe clusters) can give rise to similar spectral densities. Focusing on the case of Au for a moment, it was observed that the Au atoms bind to Ga, thus forming a Au–Ga bond which would be occupied by a single electron. Spectroscopically, such a half-occupied level should form a state located at the Fermi level, yet in the spectra one observes a *minimum* in the state-density at the Fermi level. A possible explanation for this would be a relatively large intra-atomic Coulomb repulsion term, which would split the first and second electron states of this level.[19] Indeed, the location of the Fermi level at a minimum in the state-density is a very general feature of the spectra on all the metal–GaAs systems, thus indicating that such a Coulomb repulsion term might be present in many cases. Shifting our attention to the Fe spectra, the observation of the decay of the state-density, with lateral distance from the cluster, provides a clear connection with the evanescent states of the semiconductor.[21] However, the similarity of the Fe spectra with those of Sb and Au would then lead one to conclude that these evanescent states play a rather important role in all of the spectra. Finally, we comment on the observed feature located near $E_v + 1.0\,\text{eV}$ in all the spectra. It was found for both Au and Fe cases that this state displayed subsidiary maxima on neighboring Ga atoms near the metal clusters, suggesting that unbuckling of the GaAs lattice may be involved in the formation of the band gap states.[18,19] We feel that a more complete understanding of all of these points will be obtained by further experimental and theoretical studies of the metal–GaAs systems. The STM results have prompted a number of theoretical works,[29,30] and good progress has been achieved towards a complete understanding of the band gap states induced by metal adsorbates on the GaAs(110) surface.

References

1. J. S. Blakemore, *J. Appl. Phys.* **53**, R123 (1982).
2. J. R. Chelikowsky and M. L. Cohen, *Phys. Rev. B* **14**, 556 (1976).
3. M. D. Pashley, *Phys. Rev. B* **40**, 10481 (1989).
4. R. M. Feenstra, J. A. Stroscio, J. Tersoff, and A. P. Fein, *Phys. Rev. Lett.* **58**, 1192 (1987).
5. J. A. Stroscio, R. M. Feenstra, D. M. Newns, and A. P. Fein, *J. Vac. Sci. Technol. A* **6**, 499 (1988).
6. R. M. Feenstra, p. 211, in *Scanning Tunneling Microscopy and Related Methods* R. J. Behm, N. Garcia, and H. Rohrer, eds., Kluwer, Dordrecht, 1990.
7. S. Y. Tong, A. R. Lubinsky, B. J. Mrstik, and M. A. Van Hove, *Phys. Rev. B* **17**, 3303 (1978).
8. H. Carstensen, R. Claessen, R. Manzke, and M. Skibowski, *Phys. Rev. B* **41**, 9880 (1990).
9. X. Zhu, S. B. Zhang, S. G. Louie, and M. L. Cohen, *Phys. Rev. Lett.* **63**, 2112 (1989).
10. M. D. Pashley, K. W. Haberern, W. Friday, J. M. Woodall, and P. D. Kirchner, *Phys. Rev. Lett.* **60**, 2176 (1988).
11. D. K. Biegelsen, R. D. Bringans, J. E. Northrup, and L. -E. Swartz, *Phys. Rev. B* **41**, 5701 (1990).
12. D. K. Biegelsen, R. D. Bringans, J. E. Northrup, and L. -E. Swartz, *Phys. Rev. Lett.* **65**, 452 (1990).
13. O. Albrektsen, D. J. Arent, H. P. Meier, and H. W. M. Salemink, *Appl. Phys. Lett.* **57**, 31 (1990).
14. C. B. Duke, A. Paton, W. K. Ford, A. Kahn, and J. Carelli, *Phys. Rev. B* **26**, 803 (1982).
15. R. M. Feenstra and P. Mårtensson, *Phys. Rev. Lett.* **61**, 447 (1988).
16. P. Mårtensson and R. M. Feenstra, *Phys. Rev. B* **39**, 7744 (1989).
17. R. M. Feenstra, P. Mårtensson, and J. A. Stroscio, p. 307 in *Metallization and Metal-Semiconductor Interfaces* I. P. Batra, ed., Plenum, New York, 1989.
18. R. M. Feenstra, P. Mårtensson, and R. Ludeke, *Mat. Res. Soc. Symp. Proc.* **139**, 15 (1989).
19. R. M. Feenstra, *Phys. Rev. Lett.* **63**, 1412 (1989).
20. R. M. Feenstra, *J. Vac. Sci. Technol. B* **7**, 925 (1989).
21. P. N. First, J. A. Stroscio, R. A. Dragoset, D. T. Pierce, and R. J. Celotta, *Phys. Rev. Lett* **63**, 1416 (1989).
22. J. A. Stroscio, P. N. First, R. A. Dragoset, L. J. Whitman, D. T. Pierce, and R. J. Celotta, *J. Vac. Sci. Technol. A* **8**, 284 (1990).
23. J. Nogami, unpublished.
24. L. J. Whitman, J. A. Stroscio, R. A. Dragoset, and R. J. Cellota, *Phys. Rev. B* **42**, 7288 (1990).
25. M. Prietsch, A. Samsavar, and R. Ludeke, *Phys. Rev. B* **43**, 11850 (1991).
26. W. E. Spicer, I. Lindau, P. Skeath, C. Y. Su, and P. Chye, *Phys. Rev. Lett.* **44**, 420 (1980);

W. E. Spicer, P. W. Chye, P. R. Skeath, C. Y. Su, and I. Lindau, *J. Vac. Sci. Technol.* **16**, 1422 (1979).
27. H. Hasegawa and H. Ohno, *J. Vac. Sci. Technol. B* **4**, 1130 (1986).
28. J. Tersoff, *Phys. Rev. Lett.* **52**, 465 (1984).
29. J. E. Klepeis and W. A. Harrison, *Phys. Rev. B* **40**, 5810 (1989).
30. G. Allan and M. Lannoo, *Phys. Rev. Lett.* **66**, 1209 (1991).

6. METAL SURFACES

Young Kuk

Seoul National University, Seoul, Korea

6.1. Introduction

When a surface is created by cleavage or annealing, the total energy of the system increases by the surface energy. While some surfaces undergo reconstruction in order to lower their surface energy, many other surfaces reveal bulk-like terminations at room temperature.[1] For the last two decades, reconstructed metal surfaces with or without adsorbates have been widely studied in surface science. Most of the reported results have been studied by measuring order parameters (short or long-range) by macroscopic surface science tools and fitting the measured results to models. These tools measure surface-sensitive quantities averaged over areas determined by the size of (electron, ion, photon) beam spots. Scanning tunneling microscopy, which was invented by Binnig, Rhorer and co-workers,[2] has become a powerful surface science tool, since it can directly image a surface area of up to several $\mu m \times \mu m$ with atomic resolution. In scanning tunneling microscopy, the three-dimensional variation of charge density at a surface is probed via electron tunneling between a sharp tip and a sample. Therefore not only geometric, but also electronic surface structures can be mapped, since the electron tunneling depends on the electronic density-of-states of the sample and the tip.

While large charge density corrugations (~ 1 Å) are often observed on semiconductor surfaces due to the presence of dangling bonds,[3] those on (1×1) metal surfaces are small (< 0.1 Å), as measured by helium diffraction experiments, unless they are reconstructed or foreign atoms are chemisorbed on them.[4] Metal corrugations are thus 50–100 times smaller than those observed on the Si(111)–(7×7) surface, for example. Atomic imaging of (1×1) metal surfaces, therefore, requires high lateral and vertical resolution as predicted by theory.[5] Soon after the well-publicized Si(111)–(7×7) image,[6] the reconstructed Au(110)–(1×2), a well-understood metal surface, was imaged by the inventors' group,[7] confirming the previously proposed "missing row structure." Structures of reconstructed and chemisorbed surfaces were studied in earlier days, but bulk terminated (1×1) results were frequently reported with development of scanning tunneling microscope

METHODS OF EXPERIMENTAL PHYSICS
Vol. 27

ISBN 0-12-475972-6

(STM) instrumentation. In this chapter, the role of the tunneling tip will be discussed to explain the measured corrugation amplitude on metal surfaces. Scanning tunneling spectroscopy (STS) on metals will be compared with that on semiconductors. Examples of STM studies on metal surfaces including surface structures, chemisorbed surfaces, practical surfaces, and dynamics on metal surfaces will be given. Not all the examples on metal surfaces are included, and examples used in this chapter are not necessarily representative collections.

6.2. Corrugation Amplitudes and the Tunneling Tip

It has been known that a one-atom tip is required to obtain a well-resolved STM image of a bulk-terminated metal surface. However, the role played by the tunneling tip is not well understood. The size, shape, and chemical identity of the tip influences not only the resolution and shape of an STM scan, but also the measured electronic structure. Experimentally, tips have been prepared by mechanical grinding or chemical etching from a variety of materials, most often W. Since a bcc crystal has a lower surface energy on the {110} faces, [100] and [111] oriented tips (single crystal wires) can be sharpened by annealing in the presence of a high electric field,[8–11] producing a pyramid shape of {110} facets. When tip annealing and high electric field are not available, a high tunneling current (100 nA–10 μA) is known to often improve the characteristics of a tunneling tip. These techniques are supposed to produce a sharp, clean, and symmetric tip, but asymmetric or double tips[12] may still be formed, resulting in misleading sample topographics.

One experiment has been reported[9] in which a field ion microscope (FIM) was used to examine the geometry of a tunneling, although other combinations of STM and FIM have since been assembled.[13,14] By taking FIM images of the tunneling tip before and after an STM scan, the character of the STM topograph can be correlated to the tip structure. The dependence of the measured corrugation amplitude on the size of the tunneling tip has been experimentally examined. Figure 1(a) shows an FIM picture of the tunneling tip used in the STM scan of the clean Au(001)–(5 × 20) surface (this surface structure will be discussed later) with corrugation width of 14.4 Å [Fig. 1(b)]. The corrugation amplitude was estimated theoretically[5] as

$$\Delta \propto \exp[-\beta(R + S)], \quad (1)$$

where $\beta \approx (1/4)\kappa^{-1}G^2$. R is the tip radius of curvature (approximated as a hemisphere), S is the gap distance, κ^{-1} is the electron decay length in the vacuum, and G is the smallest surface reciprocal wave vector ($2\pi/a_i$, where a_i is the largest corrugation width). Average measured corrugations of the Au(100)–(5 × 20) and the Au(110)–(1 × 2) reconstructions[15] as a function of

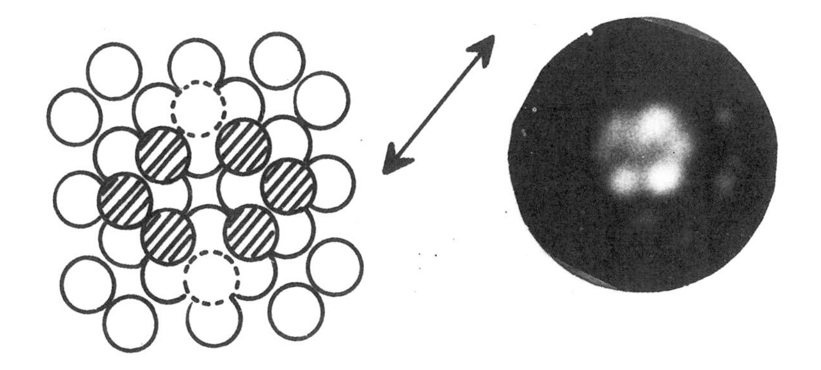

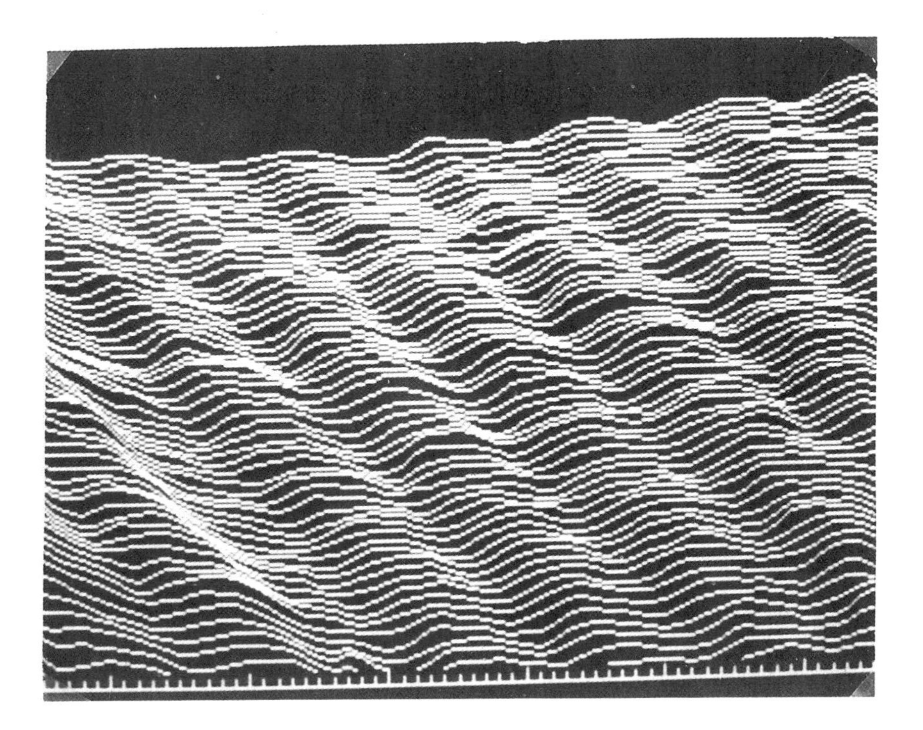

FIG. 1. (a) FIM image of the W(100) tip before and after the STM scan. the arrow indicates the scanning direction with respect to the tip. Filled circles are the first layer atoms. (b) STM topograph of Au(001). Rows along the closed-packed $[1\bar{1}0]$ are separated by ~ 14 Å. Figure from Ref. 9.

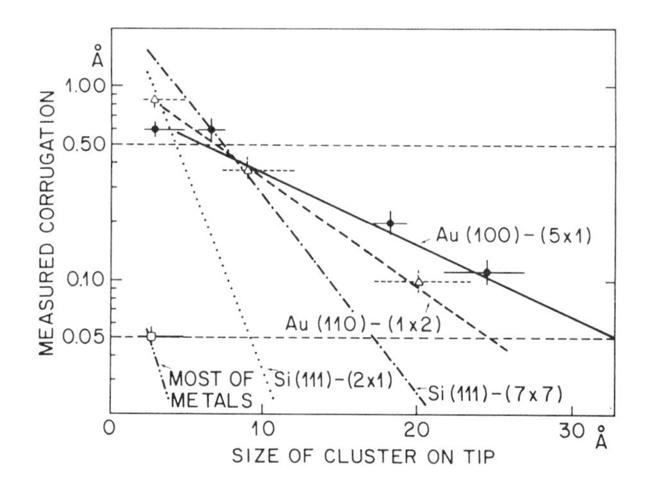

FIG. 2. Dependence of measured corrugation on size of tip for Au(001)–(5 × 20) (solid line and filled circle), Au(110)–(1 × 2) (broken line and triangle), Si(111)–(7 × 7) (– – –), Si(111)–(2 × 1) (dotted line), and most (1 × 1) metals (– – – –). Figure from Ref. 15.

the tip size are summarized in Fig. 2. The measured corrugation for the Au(100) and Au(110) are in good agreement with the values predicted by Eq. 1. Tip size versus measured corrugation for several other surfaces is shown. Broken horizontal lines in Fig. 2 indicate STM noise levels, due to mechanical vibration or electronic noise, for two hypothetical STMs. The significance of the tip size is now apparent from Fig. 2; the tip must be terminated by a single atom in order to detect any reconstruction with the noisier STM. On the other hand, the corrugations of the Au(100), Au(110), and Si(111)–(7 × 7) surfaces could be detected by a 20 Å tip, with an STM noise level of 0.05 Å. In order to image most metals with a small reconstruction unit cell or (1 × 1) surfaces with typically a corrugation level of 0.01–0.05 Å, both a very sharp tip and low STM noise level are required.

6.3. Tunneling Spectroscopy of Metal Surfaces

While STM images of semiconductor surfaces are maps of charge density contours of mainly p-states (valence, conduction band electrons, and dangling bond states), those on metal surfaces can be dominated by s-, p-, d-, and f-state electrons. Because the decay lengths vary with the electronic state, scanning tunneling microscopy and spectroscopy on metals may show substantial variation depending on the governing electron states. Sometimes a surface state can dominate the spectroscopy, other times chemisorption induced states (bonding or antibonding states) can result in a change of the spectroscopy. In this section, tunneling measurements of the local barrier

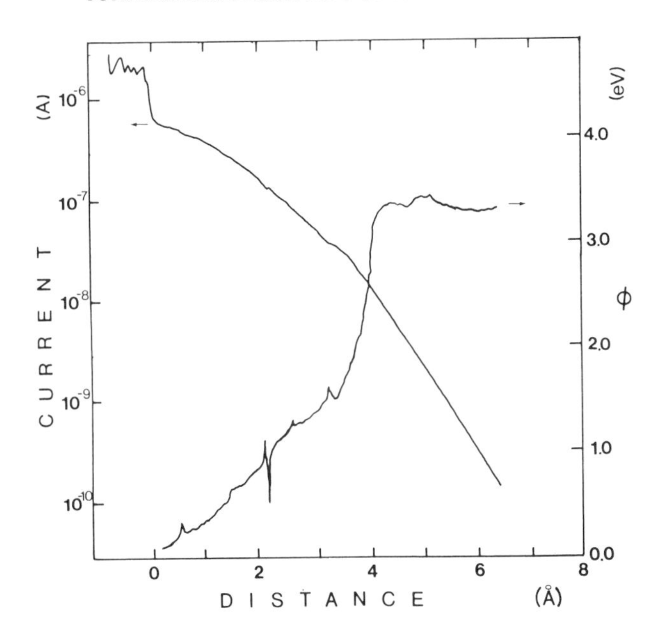

FIG. 3. The tunneling current and local barrier height as a function of gap distance in semilog plot for a Au(100)–(5 × 20) surface. A contact point is defined as 0. The tip voltage is 50 meV. Figure from Ref. 17.

height and electronic density of states will be described to emphasize the difference between metal and semiconductor surfaces (see also Chapter 4 for a discussion of tunneling spectroscopy with the STM).

6.3.1 Current Versus Gap Distance

One direct indication of metal–vacuum–metal tunneling is the exponential dependence of the tunneling current with the tunneling gap.[16] Similar exponential dependencies have been reported on semiconductor surfaces, but more results have been reported on metal surfaces. The continuous exponential variation over four orders of magnitude has been demonstrated by Binnig *et al.*[2] with a good vibration isolation. They reported an average barrier height of ~ 3.2 eV for the junction between a W tip and Pt plate, demonstrating a clean junction.

The I–S relation on a well characterized surface was studied by Kuk and Silverman.[17] Figure 3 shows I–S and ϕ–S spectra on a Au(100)–(5 × 20) surface. As reported earlier by Gimzewski *et al.* on an Ag film[18] and Durig on an Ir film,[19] a near exponential dependence of the tunneling current as a function of the tunneling gap is shown. The contact point ($x = 0$ in Fig. 3) is defined at the abrupt increase in the tunneling current. From the gap distance dependence of this slope, the local barrier height can be obtained from,

$$\bar{\phi} = 0.952(d\ln I/dS)^2, \tag{2}$$

where S is in Å. The result is in good agreement with the theoretical work by Lang[20] (see Chapter 1) for a one-atom Na tip. For $S > 5$ Å, the barrier between the tip and sample is treated as a small perturbation of two independent metallic states, and the local barrier height, the slope in $\ln I$–S, remains constant. But for $S < 5$ Å the perturbation is substantial, and the local barrier height may not remain constant. Mechanisms of unusually high corrugation heights measured on closed packed metal surfaces[21–23] may be closely related to the strong perturbation with low barrier heights. From the contact of the sample and tip, the contact resistance is estimated to be $\sim 24\,\mathrm{k}\Omega$ in good agreement with theoretical prediction.[20,24] The local barrier height and contact resistance are found to be strong functions of the shape and size of the tip, as estimated from the topography after the contact was made.[17] It was reported that the slope variation in $\ln I$–S near contact are different for soft metal[17] and hard metal.[19]

6.3.2 Electronic Structure in Differential Conductivity Measurements

The relation among tunneling current, gap distance, and bias voltage in vacuum tunneling can be simplified by treating both electrodes as free electron-like metals. The vacuum tunneling between the tip and sample can then be divided into two regimes. When the bias voltage is smaller than the work function, the tunneling current is approximated by,

$$I \sim V \exp(-1.05\bar{\phi}). \tag{3}$$

When $V > \bar{\phi}$, the tunneling current dependence on the bias voltage is described by a Fowler–Nordheim relation.[25] In the field emission regime, the tunneling electron experiences the positive kinetic-energy region of the vacuum gap, as shown in Fig. 4. At several bias voltages, the tunneling electrons form standing waves in the positive kinetic energy region, resulting in oscillatory transmission probabilities with bias voltage. Figure 4 shows the oscillations in dI/dV from 5–18 V with a constant tunneling current of 1 nA.[26] The measured result shows good agreement with the calculated oscillatory transmission probability with two planar conducting plates. Recently, field emission resonances have been studied in detail by Kubby *et al.*[27] and Coombs *et al.*[28] to understand the role of topographic features and light emission from a tunneling junction.

While a normal STM image is a convolution of geometric and electronic information of a surface, the electronic structure can be separately

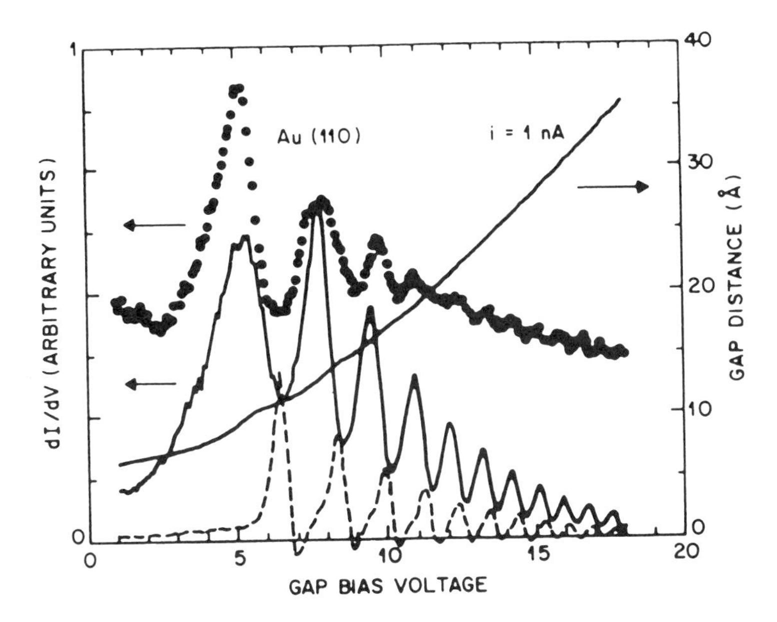

FIG. 4. dI/dV (closed circles: experimental, solid curve: theoretical) and gap distance vs. bias voltage at tunneling current of 1 nA. Figure from Ref. 26.

measured.[29,30] With the tunneling tip poised over a region of interest and the gap fixed by opening the feedback loop momentarily, the bias voltage can be ramped to measure the tunneling current as a function of applied voltage. This measurement can also be performed at each point of topography scan,[31] resulting in spatially resolved I–V relations. Interpretation of these spectra is quite similar to that on metal–insulator–metal tunneling, replacing the insulator with a finite vacuum gap. For small bias voltage ($V < \bar{\phi}$), the tunneling current can be written as,

$$I \sim \int_0^{eV} \rho(E) D(E, V)\, dE, \tag{4}$$

where $\rho(E)$ is the sample local density-of-states and $D(E, V)$ is the transmission coefficient of the barrier at voltage V. D can be estimated by the WKB method for a free electron model; the result is shown in Fig. 5. The transmission coefficient can be approximated by,

$$D(V) = \alpha V + \gamma V^3 + \cdots \tag{5}$$

where the cubic correction term is more apparent at larger voltages in Fig. 5. The local density-of-states can be deduced from dI/dV (differential conductance) in the low voltage limit by,

$$dI/dV \sim \rho(\mathbf{r}, V) D(V), \tag{6}$$

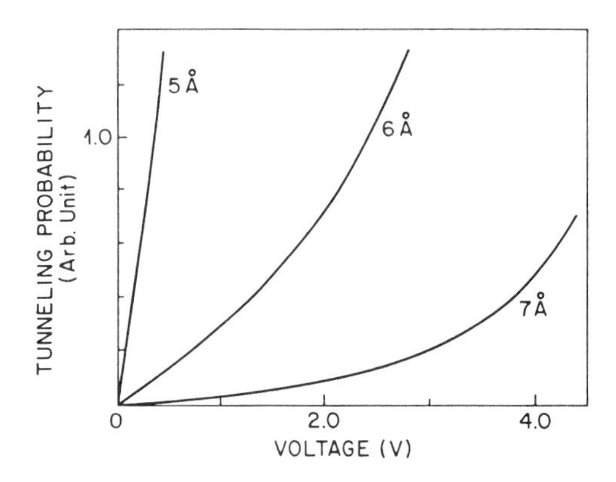

FIG. 5. $D(V)$–V at 3 different gap distances calculated by WKB method. Figure from Ref. 15.

where $\rho(\mathbf{r}, V)$ is the local density-of-states of the sample evaluated at the center of the tip ($\mathbf{r}$). The relation dI/dV versus V is proportional to the electronic density of states in the low voltage limit in metal–insulator–metal tunneling.[32] A plot of $d\ln I/d\ln V$ versus V reduces the parabolic dependence introduced by the cubic term in Eq. 5, so the position of the peak in the curve corresponds approximately to resonances in the tip and sample density-of-states, although the exact strength and position of the peaks will still vary slightly with gap distance.[33,34] When the tunneling gap is small (< 5 Å), the perturbation in the transfer Hamiltonian approach[5] is too large to use Eq. 6, where the bases are two independent solutions of the tip and sample Hamiltonians. As stated earlier, large corrugations observed on several closed-packed metal surfaces may not be explained by this approach.

A series of $d\ln I/d\ln V$–V curves measured on well characterized Au(100)-(5 × 20) at various gap distances is plotted in Fig. 6.[17] The gap distances are estimated from the I–S relation taken after the I–V. The scanning tunneling spectra at different gaps show resemblances, but the peak widths and heights vary slightly with the gap distance. An earlier photoemission measurement for this surface[35] showed a peak near ~ -0.7 eV, near the Brillouin zone boundary. A slight shift of the tunneling peak positions versus the photoemission spectroscopy data has often been observed by several groups, but the reason is not fully understood. The -0.7 eV peak has been ascribed to the bulk sp-band. Although there are other peaks with high intensity from d-bands in photoemission data, there is little evidence of d-band contribution, which would be strongly dependent on gap distance since the electron decay length is much shorter than that in the sp-band. Comparison of scanning tunneling spectroscopy (STS) data to (inverse) photoemission demands special care, since the surface states, overwhelmed by bulk states in

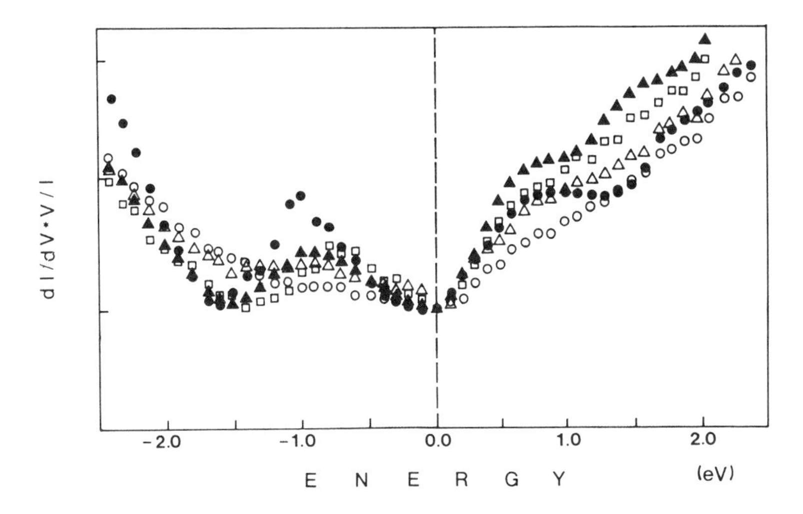

FIG. 6. $V/I*dI/dV$–V spectra obtained on Au(100)–(5 × 20) at various gap distances: 6.0 (filled circle), 6.6 (filled triangle), 7.5 (open square), 8.1 (open triangle), and 8.6 Å (open circle). Figure from Ref. 17.

(inverse) photoemission, can still be probed by tunneling spectroscopy. In addition, STS measures the local density-of-states summed over **k** vectors with varying weighting factors due to different decay lengths.[5] Tunneling spectroscopy of the Au(111) shows a state at ~0.4 eV below the Fermi level. The peak height and width of the surface state vary spatially over the surface. For example, the peak is smaller near the step edge, as shown in Fig. 7. By

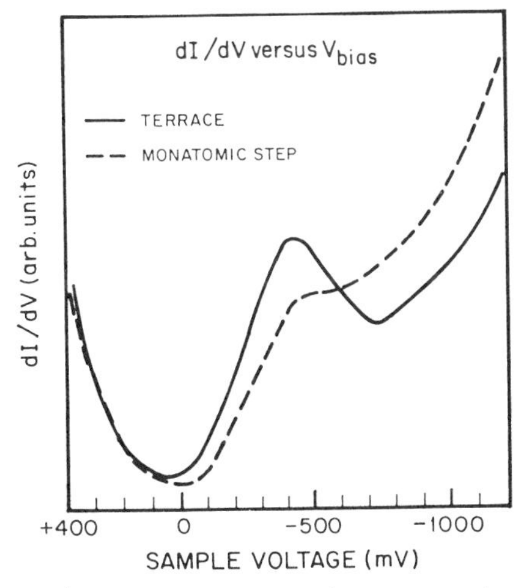

FIG. 7. dI/dV on terraces of the Au(111) surface and near a step edge. Figure from Ref. 35.

scanning at the bias voltage near the surface state, better image of the Au(111) stacking fault reconstruction could be obtained.[36] As these two examples demonstrate, most surface states on metals are not as localized as those on semiconductors, which is mainly due to directional dangling bond states.

6.4. Clean Metal Surfaces

The scanning tunneling microscope was intitially used to simply image surfaces. As tunneling spectroscopy becomes widely used, STS data are employed to refine surface structural models derived from STM images alone. By adding adsorbate sources (gas or solid phase) and temperature control, the study of dynamics on metal surfaces becomes possible.

6.4.1 Surface Structures

One of the most notable achievements of scanning tunneling microscopy is the elucidation of the atomic arrangements of various surfaces. Many disputed surface structural models, which have been studied by various experimental techniques and theoretical calculations, have been resolved by the direct imaging capability of the scanning tunneling microscopy. the first STM study of a metallic surface was performed on the Au(110) surface.[7] The {110} surfaces of Au, Pt, and Ir have been studied by many surface science techniques[37–40] to exhibit a missing-row type reconstruction, resulting in a (1×2) structure. Direct imaging of the surface by scanning tunneling microscopy confirmed the previously proposed missing-row structure.[7,41]

Among several low miller index Au facets, the surface energy on the {111} facet is the lowest. By forming the missing-row structure {111} micro-facets are exposed, such that the surface energy can be lowered. Although the STM has high vertical resolution, it is sometimes not suitable to measure absolute atomic displacements, because the measured corrugation is strong functions of the tunneling tip radius and the electronic structure of the sample. The topograph of Au(110) (Fig. 8) shows the (1×2) reconstruction with alternate $[1\bar{1}0]$ missing rows. It has been reported in LEED studies[42] that as the sample temperature increases, the (1×2) structure undergoes a phase transition to a bulk-like (1×1) structure. Theoretical and LEED studies[42,43] indicated that this is an order-disorder transition of a 2-D Ising universality class. In the disordered (1×1) phase, the top layer atoms are still commensurate with bulk lattice sites, but disordered; the surface is disordered in the lattice-gas sense.[42] By diffraction methods, the transition temperature for the Au(110) was found to be ~ 700 K and is very sensitive to surface impurities. Figure 9 shows a topograph of the quenched Au(110) surface after annealing

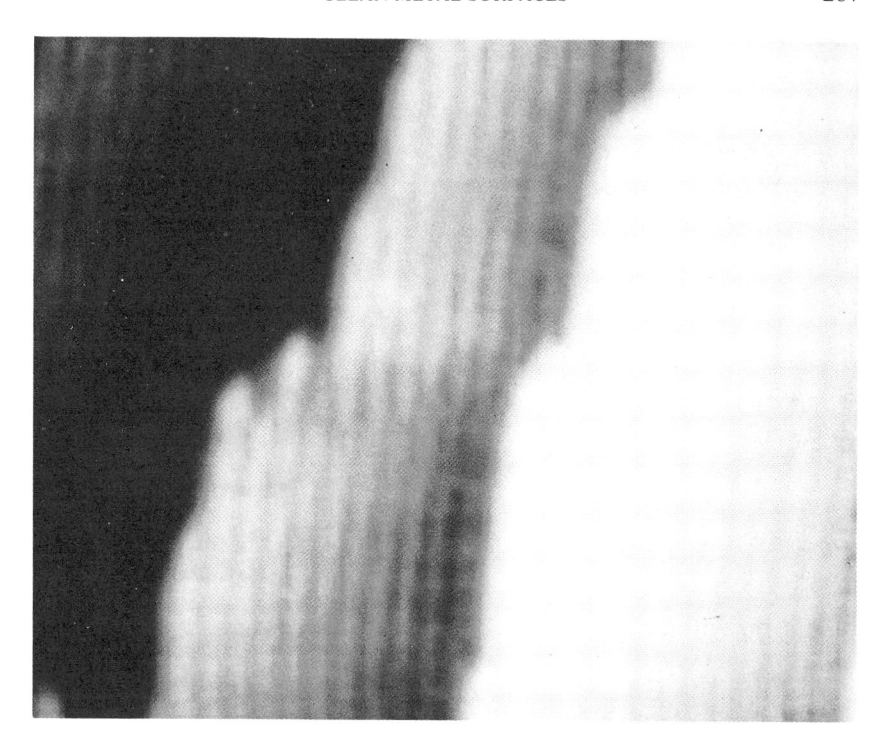

FIG. 8. 240×190-$Å^2$ gray scale topograph of the Au(110)–(1×2) surface after annealing at 600 K. Figure from Ref. 41.

within 10 K of the phase transition temperature in which annealing temperature was determined by the LEED patterns.[44] On this sample, although the STM topograph shows that (1×2) reconstruction with small domain sizes is still present, the half-order spots change to diffused streaks in the corresponding LEED pattern. This is not surprising since the average domain size is 20–40 Å, far less than the usual coherence length of an electron diffraction. The domains are separated not only by steps but also by (1×3)- or (1×4)-type missing-row structures. The corresponding diffraction pattern of the observed STM topograph can be calculated by a Fourier transformation. Figure 10 shows the calculated diffraction intensities on samples annealed just below the phase transition temperature. Since the (1×2) reconstructions in adjacent terraces results in the displacement of the half-order peak from the normal position and a split of the (1, 0) peak. A similar result was observed by x-ray diffraction and explained by the presence of steps.[38] Since this experiment was done on a quenched sample, there is a possibility of re-ordering during quenching. Changes in the LEED pattern have to be

FIG. 9. 225 × 100-Å^2 gray scale topograph of the Au(110)–(1 × 2) surface after annealing at 700 K. Figure from Ref. 44.

carefully monitored before and after quenching. The images quenched from the temperature above T_c shows disorder, confirming LEED results.

The closed-packed {111} plane of fcc metal surfaces such as Au(111)[21,22] and Al(111)[23] have been imaged with atomic resolution (Fig. 11). Unreconstructed (1 × 1) structures of other surfaces have been reported.[45–47] Very high spatial resolution and large corrugations have been observed on these surfaces that cannot be explained by the presently accepted transfer Hamil-

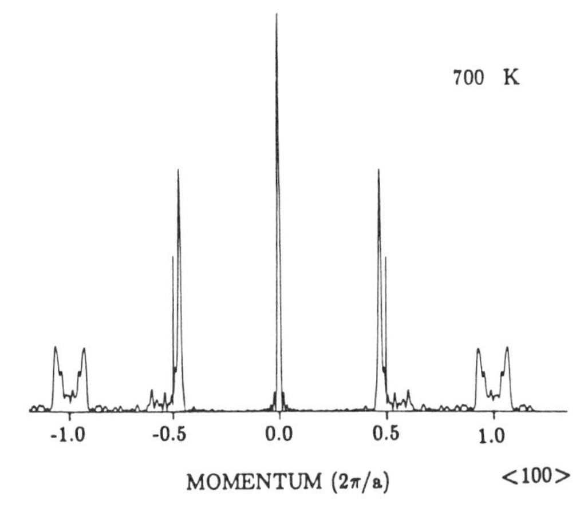

FIG. 10. Diffraction intensity along the [100], calculated from the STM image of Fig. 9. Figure from Ref. 44.

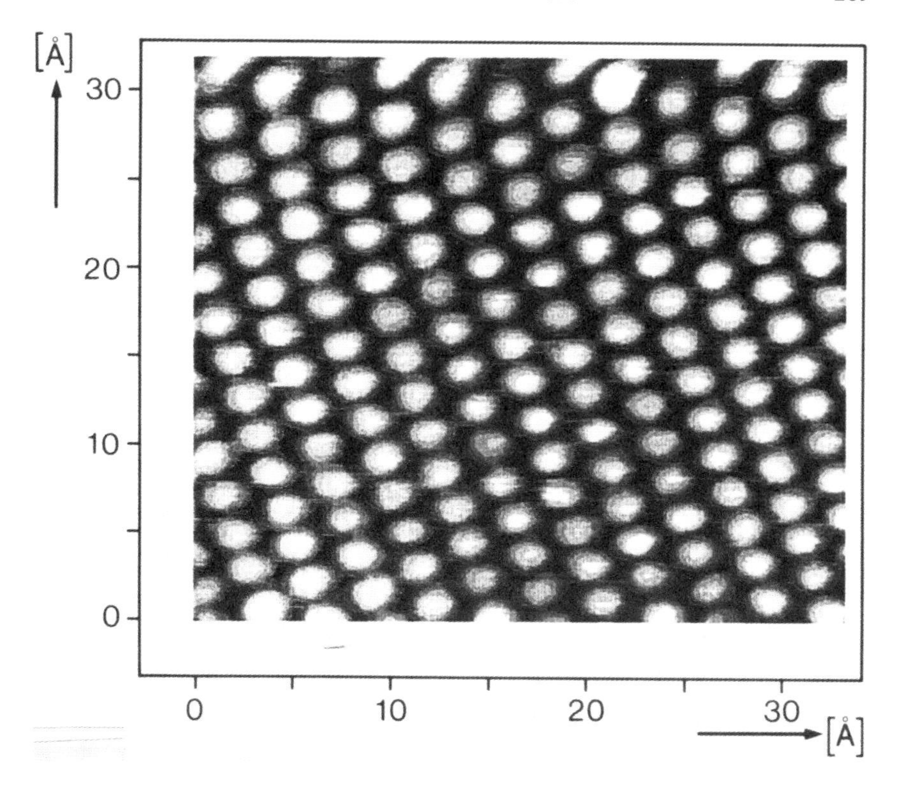

FIG. 11. 34 × 34-Å^2 gray scale topograph of the clean Al(111) surface. Figure from Ref. 23.

tonian approach for tunneling.[5] The corrugation predicted by Eq. 1 is on the order of 0.01 Å at a tip–sample distance of around 5 Å. This can be compared to corrugation values on the order 0.5–1.0 Å, which has been observed at small tunneling resistances approaching tip–sample contact.[48] Several mechanisms have been proposed to explain these anomalous values: 1) localized surface states present near Fermi level, 2) the presence of an atom with an unusual electronic state (for example, an atom with a p_z state) on the apex of the tunneling tip,[49] 3) the influence of atomic forces between the tip and sample. However, these large corrugation values may well be explained by a tunneling theory that includes a strong perturbation, since the large corrugation values are observed at tunneling distances where tunneling resistances are of 10^5 ohms. Experiments at normal resistances of 10^9 ohms, where the Tersoff–Hamann theory is operable, have shown that normal corrugation values are observed at the 0.01 Å level,[50] but of course this level of measurement is a challenge that not all microscopes can achieve.

Figure 12 shows a topograph of the Au(111) surface with atomic resolution. The surface exhibits not only the 1 × 1 unit cell characteristic of the

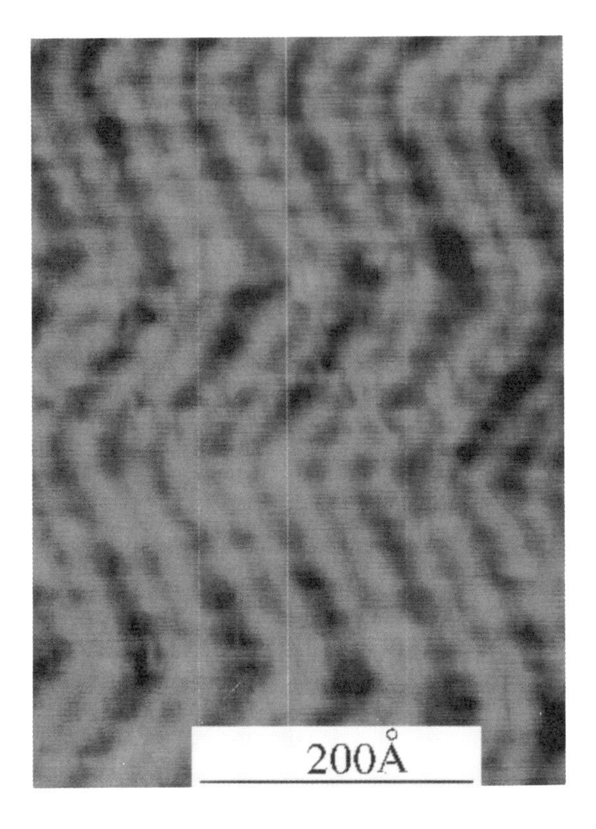

FIG. 12. 3000×2000-Å^2 gray scale topograph of the Au(111)-(22 × $\sqrt{3}$), shown as a herringbone pattern. Figure from Ref. 22.

(111) fcc lattice, but also a long range reconstruction of (22–23 × $\sqrt{3}$), which is caused by stacking faults between fcc and hcp ordering, with the two stackings connected by Shockley dislocations on the surface. These dislocations are found to play an important role in providing nucleation centers as seen in studies of Ni and Fe on Au(111).[50,51] The {100} surfaces of Au and Pt are known to reveal "(5 × 20)" and "(5 × 12)" reconstructions, respectively. These reconstructions are caused by the closed-packed first layer which resembles {111} surfaces. An earlier study by Binnig *et al.*[52] proposed a "(26 × 68)" structure, based on the length calibration of the surface, but recent topographies by the author's group (Fig. 13)[41] show atomic images of the surface, suggesting four domain structures represented by

$$\begin{bmatrix} 5 & 1 \\ 0 & 20 \end{bmatrix}, \begin{bmatrix} 5 & 0 \\ 1 & 20 \end{bmatrix}, \begin{bmatrix} 20 & 0 \\ 1 & 5 \end{bmatrix}, \begin{bmatrix} 20 & 1 \\ 1 & 5 \end{bmatrix}.$$

The domains are separated by regularly arranged misfit dislocations and

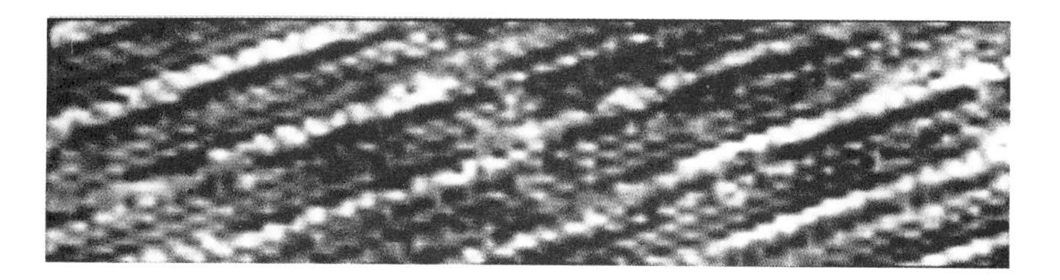

FIG. 13. 160 × 40-Å² gray scale topograph of the Au(100) surface. Figure from Ref. 41.

slight buckling of atomic rows caused by the higher atomic density in the first layer than in the bulk. Higher atomic density in the first layer requires out-diffusion of bulk Au atoms toward the surface layer. It is known that the Au(100)–("5 × 20") reconstruction loses the long-range order above 600°C. The out-diffusion may accelerate above the phase transition temperature and the additional segregated layer results in disorder.[41] STM images above the phase transition temperature shows roughened surface, confirming the model. Pt(100) shows very similar domain structure,[53] since the driving force is the same. In this reconstruction, domains are usually separated by steps instead of forming a phase boundary on a single terrace.

Atomic resolution is not required, nor even desirable, in all scanning tunneling microscopy studies. There are many important and interesting applications for scanning tunneling microscopy with nanometer resolution on practical surfaces. Roughness measurements of metallic layers were studied by some groups.[54,55] For example, silver films condensed at room temperature and 90°K exhibit a difference in roughness, (Fig. 14) that can be explained by diffusion limited aggregation.

6.4.2 Dynamics

Phase transition of metal surfaces can be induced by changing the sample

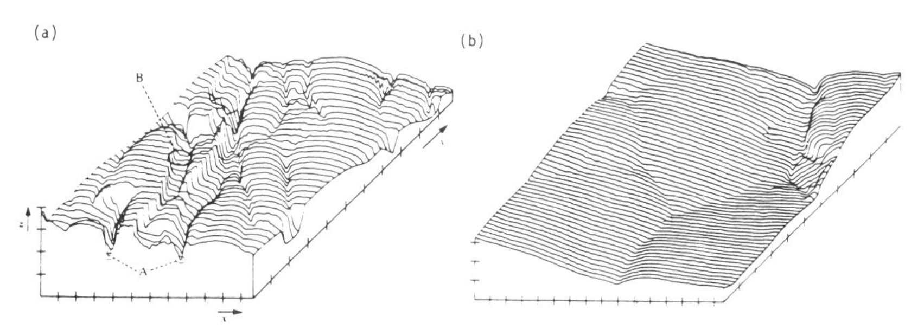

FIG. 14. 600 × 550-Å² topograph of Ag films condensed at (a) 80 K and (b) 300 K. Figure from Ref. 54.

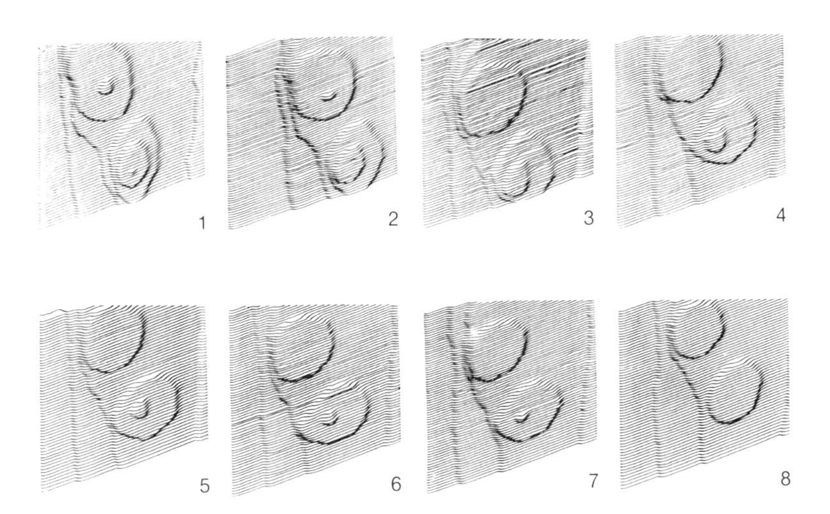

FIG. 15. 400 × 400-Å^2 topograph of Au(111) surface taken every 8 minutes. The indentation was made at frame 1 by the tunneling tip. Figure from Ref. 57.

temperature and introducing adsorbates. Surface phase transitions are receiving increased attention as equilibrium surface structures are better understood. Structural, order-disorder, magnetic, melting, and roughening transitions have been studied by a variety of surface sensitive techniques, which probe long-range orders.[56] Measurement of short-range order has also relied on scattering or diffraction averaged over the beam spot size. In order to study the dynamics by scanning tunneling microscopy, the scanning speed must be shorter than the characteristic time constant of the microscopic atomic motion, and the STM should be operable at the temperature of interest. At present, studies of structural phase transitions, surface diffusion, and absorbate induced transitions have been reported.

Surface diffusion with atomic resolution has been studied by FIM. From these studies activation energies between atomic sites can be deduced on small terraces (usually < 20–30 Å wide) of the FIM tips. Since STM can image wider viewing areas than FIM, not only activation energies but also macroscopic diffusion constants can be obtained. The diffusion of step edges on Au(111) surfaces was observed in UHV and in air.[57] A series of time-lapse STM topographs of indentations or protrusions created by the tunneling tip shows movement of step edges in Fig. 15. With known markers (a circle or a line), the surface diffusion coefficient can be deduced. Since the hopping motion of individual atoms between neighboring sites is faster than the STM scanning speed, the scratch-decay method[57,58] can be used at room temperature to measure diffusion constant on most metals. At low temperature, it is possible to deduce local activation barrier among sites by tracking individual atoms.

Similarly, close observation of step edges by scanning tunneling microscopy at elevated temperatures, permits study of surface roughening transitions.[59] Above a critical temperature, roughening appears as step meandering and step height variation. In many diffraction techniques, determination of the roughening temperature and even proof of its existence is very difficult because of the smooth change of the line shapes. In scanning tunneling microscopy, however, the topography of step edges can be imaged directly. Figure 16 shows gray scale STM images of Ag(115) surfaces at 20, 58, 98, and 145°C. In images at 20 and 58°C, there are large (115) terraces with low step densities. Above 98°C all steps have a large number of thermally generated kinks. The meandering of steps clearly indicates that the Ag(115) surface is already in the roughened state. Although this study only shows snapshots at various temperatures, it demonstrates the feasibility of the study of phase transitions by scanning tunneling microscopy.

6.5. Adsorbate on Metal Surfaces

The study of adsorbate covered surfaces has been one of the important subjects in surface science for many decades.[60] Metallic overlayers on metals have been studied to develop net materials with new magnetic, catalytic, electronic, and geometric characteristics. Information obtained by STM have helped to understand structures and processes of chemisorption.

6.5.1 Metallic Adsorbate

Initial stages of epitaxial growth on metals and semiconductors have been widely studied by STM. These results have yielded not only the structural information of overlayer films but also the dynamics of growth when the overlayer diffusion is sufficiently slow. One example is a commensurate Au overlayer on the Ni(110) surface. While the details of the Au overlayer on Ni(110) could not be determined by other surface science techniques, the (7×4) structure with a $c(2 \times 4)$ subunit structure was clearly resolved by scanning tunneling microscopy, as shown in Fig. 17.[60] A short-range interaction yields the local registry of Au on the Ni(110) results in the $c(2 \times 4)$ unit cell, while the (7×4) long-range structure originates from a lattice mismatch between Au and Ni. Instead of forming a periodic misfit dislocation, [001]

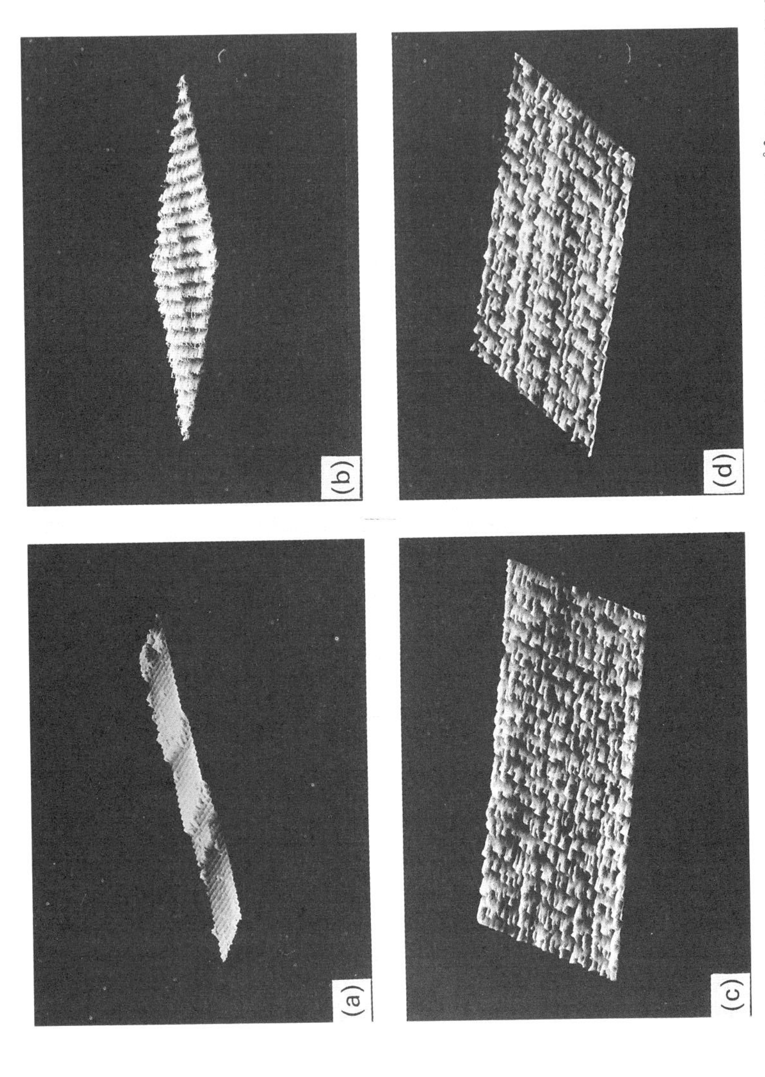

FIG. 16. Gray scale images of the Ag(115) surfaces taken at (a) 20°C (585 × 220-Å^2), (b) 58°C (1180 × 375-Å^2), (c) 98°C (450 × 120-Å^2), and (d) 145°C (465 × 280-Å^2). Figure from Ref. 59.

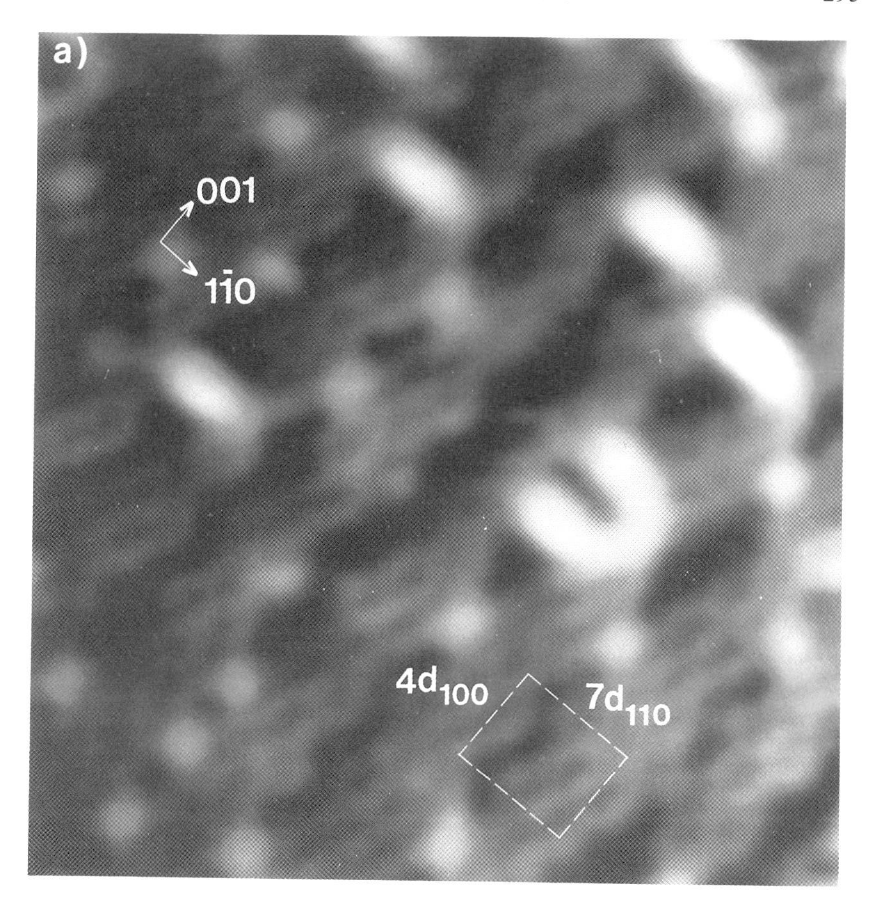

FIG. 17. 120 × 120-Å^2 gray scale topograph of the Au on the Ni(110) surface. A unit cell of the (7 × 4) structure is shown. Figure from Ref. 61.

missing rows appear at every $7d_{\langle 110 \rangle}$ distance to relieve the stress in the Au film.

The initial stages of Ag overlayer growth on the Au(111) have been studied by two groups recently.[61,62] Since the lattice mismatch between Au and Ag is very small (0.24%), the growth is nearly homoepitaxial. Although the Au(111) substrate surface exhibits a 22–23 × $\sqrt{3}$ reconstruction, the Ag overlayer does not show this reconstruction. The Ag layer nucleates around defects, such as voids or step edges. At room temperature, the growth pattern shows a diffusion limited aggregation with local smoothing (Fig. 18). From the study, the surface diffusion coefficient of the Ag overlayer was estimated. In contrast, the early stage of Ni growth on the Au(111) surface shows unique

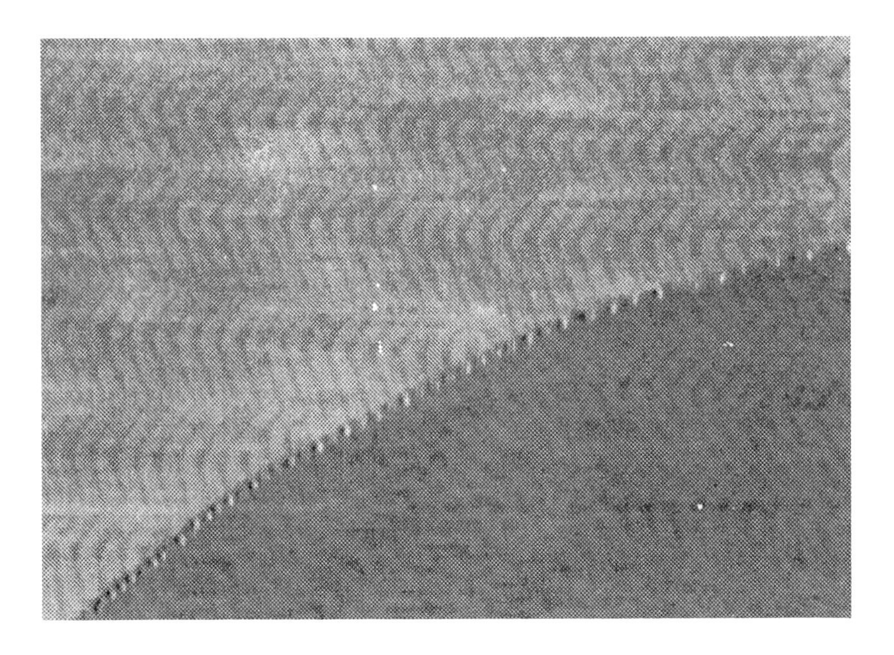

FIG. 18. 1100 × 1100-Å^2 gray scale images of Ag on the Au(111) with V_{tip} = 200 meV, and I_{tunnel} = 0.1 nA. Figure from Ref. 72.

nucleation sites (Fig. 19) determined by the Au substrate reconstruction.[51,63] Since the Au reconstruction is caused by a change of stacking sequence, the Ni overlayer grows preferentially at the areas where two stacking sequences meet to form a surface dislocation. At low coverage, the Ni overlayer forms a highly ordered metal island arrays on an automatically smooth surface, despite of large lattice mismatch between Au and Ni. Similar nucleation characteristics have also been observed with Fe on Au(111).[50]

6.5.2 Adsorbate Induced Structure

Mobility of atomic or molecular adsorbates depends on the interaction between the substrate and the adsorbate or among adsorbates. In general, isolated adsorbates (or small islands) are quite mobile at room temperature. Large islands or a fully covered surface preserves its structure even at room temperature. Some surfaces passivated by adsorbates (i.e., the surface energy is lowered by the adsorption) are stable even at atmospheric pressure.

Sulfur adatoms form a stable p(2 × 1) structure on the Mo(001) surface.[64] The p(1 × 1) surface formed in UHV but imaged in the air was reported recently. On this surface, the passivated Mo(001) substrate is not believed to be reconstructed, the absorbed sulfur adsorbates form a similarly stable

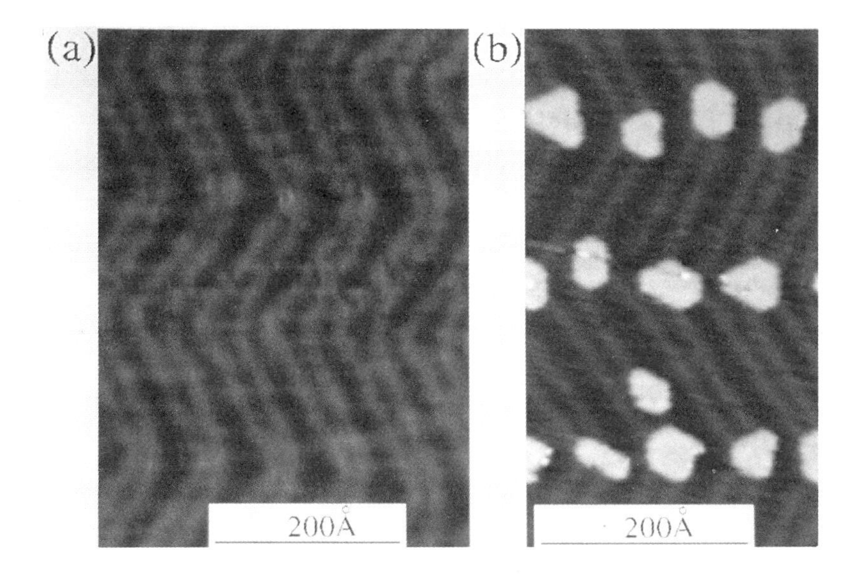

FIG. 19. Nucleation of Ni islands at specific site on the Au(111) surface. The width of the image is 1900 Å. Figure from Ref. 63.

structure on the Re(001) surface.[65] Figure 20 shows a $(2\sqrt{3} \times 2\sqrt{3})R30°$ structure imaged at atmospheric pressure. Hexagonal rings of sulfur atoms is the basic unit of the observed structure. Iodine adsorbates also passivate the Pt(111) surface as shown in Fig. 21. The adsorbate structure reveals $(\sqrt{7} \times \sqrt{7})R19.1°$ structure at the coverage of 3/7 and a (3×3) structure at 4/9 ML coverage.[66] From the difference in the measured corrugation, the registry of the adsorbate iodines was proposed.

On a Cu(11,1,1) surface (vicinal surface of the Cu(100)), sulfur adsorbates form $p(2 \times 2)$ and $c(4 \times 2)$ structures (Fig. 22).[67] These two different structures influence the shape of local step edges. On a single terrace, only one kind of domain can be observed. $[01\bar{3}]$ and $[03\bar{1}]$ directional steps are stable next to $c(4 \times 2)$ terraces, while $[01\bar{1}]$ steps are formed next to $p(2 \times 2)$ terraces. Large molecules can form regular structures on metal surfaces. Benzene and carbon monoxide form a (3×3) structure on Rh(111) surface[68] (Fig. 23). This co-adsorbed overlayer revealed a direct atomic image of a benzene ring for the first time. Three carbon atoms appear with slightly higher electronic charge density than the other three while the bond length between carbons is measured to be 1.5 Å. Close examination of the structure reveals a localized CO $2\pi^*$ orbital. A very localized orbital has not been believed to be possible to image in scanning tunneling microscopy, since the charge density decays quickly from the surface. The structure of copper–phalocyanine was observed by the same group and compared with molecular orbital calculations.[69]

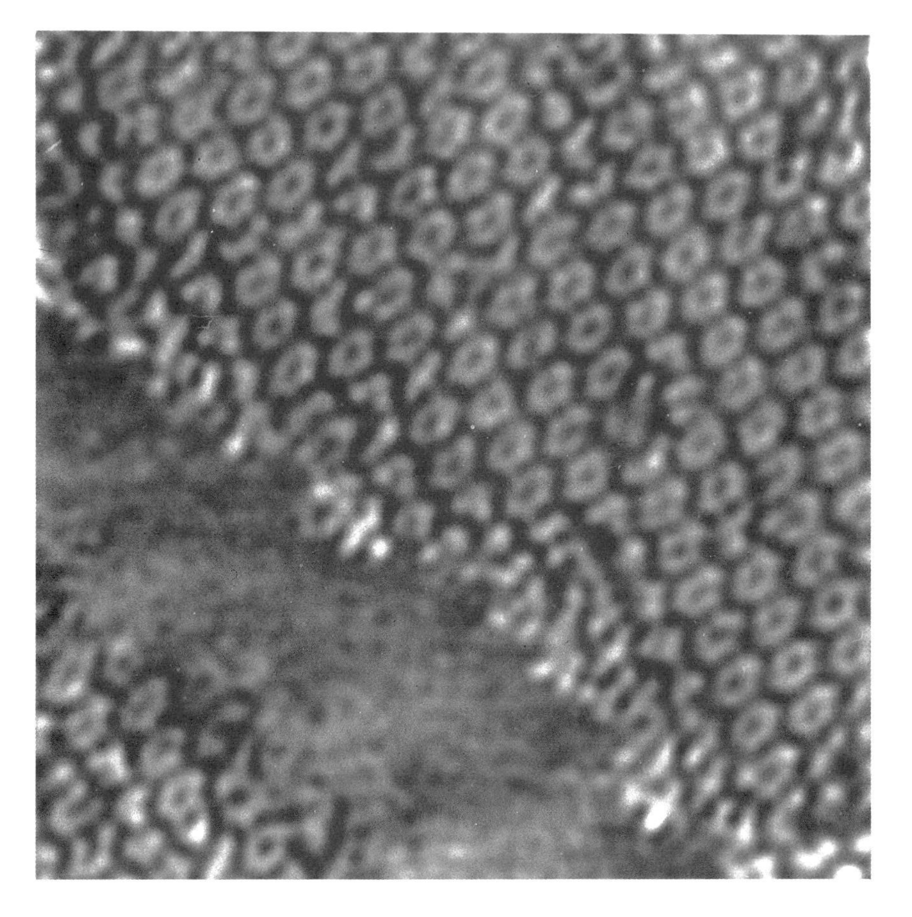

FIG. 20. High resolution image of the $(2\sqrt{3} \times 2\sqrt{3})R30°$ sulfur overlayer on Re(0001), with gray scale. Figure from Ref. 65.

6.5.3 Dynamics

Adsorbates on metal surfaces are in general mobile at room temperature. If the chemisorption rate and surface diffusion are slow enough, dynamic behavior of chemisorption can be studied by scanning tunneling microscopy. If an STM is operated at cryogenic temperature, each hopping motion of an adsorbate can be observed. From such observations, not only the diffusion constant but also the microscopic interaction potential may be deduced.

Oxygen chemisorption on Cu(110) is a model system to observed time-dependent chemisorption processes. Even though the chemisorption process is too fast to be observed by STM (such as precursor states), adsorbed oxygen atomic rows diffuse slow enough at room temperature to allow imaging of

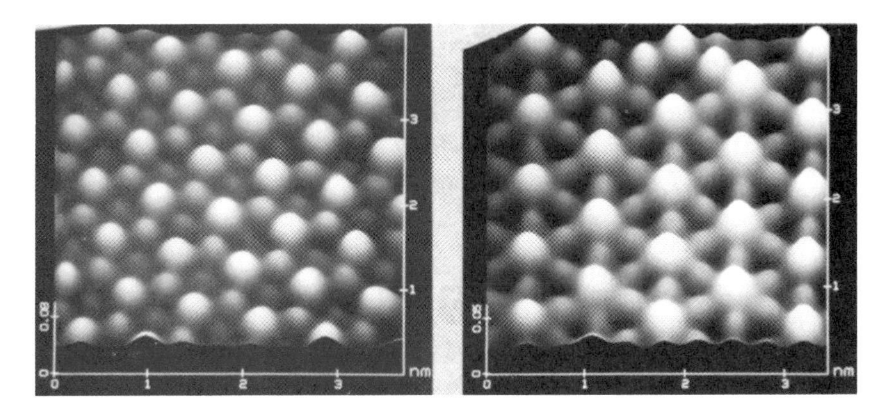

FIG. 21. High resolution images of the $(\sqrt{7} \times \sqrt{7})R19.1°$ and (3×3) structures. Figure from Ref. 66.

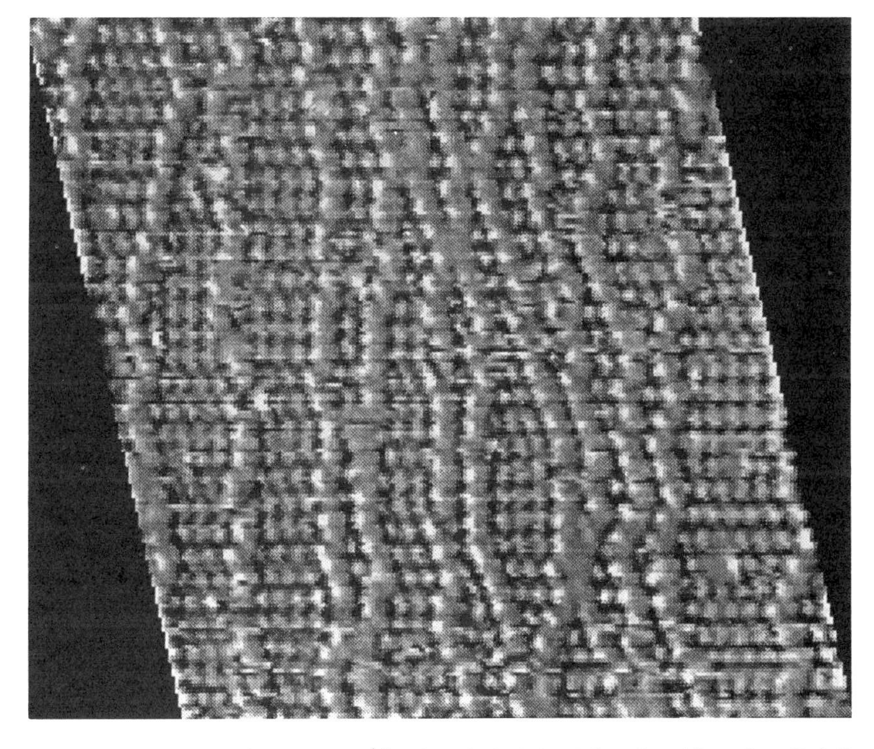

FIG. 22. STM image of a 175×145-$Å^2$ region of a S–Cu(11,1,1) surface. Figure from Ref. 67.

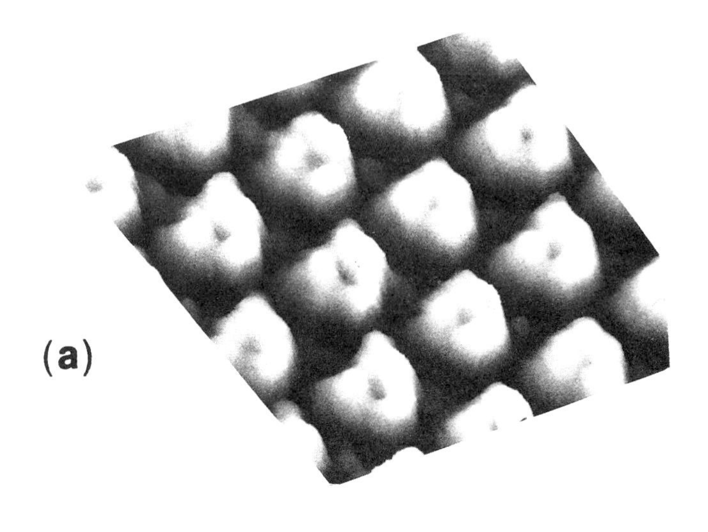

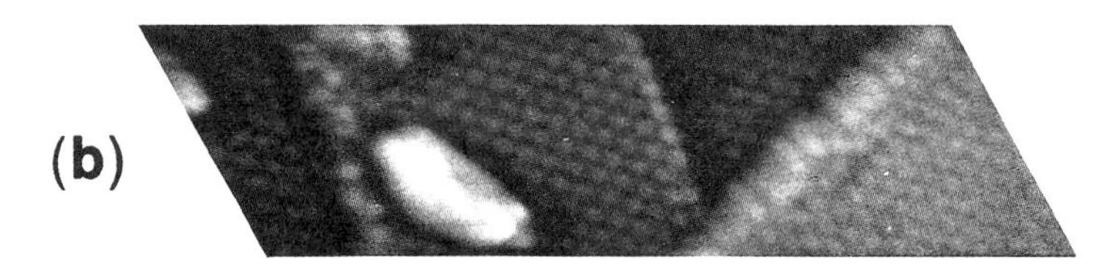

FIG. 23. A three-dimensional view of the benzene molecules co-adsorbed with CO on Rh(111). Figure from Ref. 68.

atomic jumps of the O rows with atomic resolution. Figures 24(a)–(h) show STM images of the same area on a clean Cu(110) surface with increasing O coverage up to 0.4 ML.[46,70] The time interval between consecutive images varied from two to ten minutes, so adjustment of the x-y position was often necessary to compensate for thermal drift. On a clean surface in Fig. 24(a), the step edges appeared to be unusually rough compared to those on other metal surfaces. Step edges can be treated as a fluid-gas interface in a lattice gas model. In this model the concentration of diffusing Cu atoms is determined by the evaporation energy from the step edge onto a terrace. Cu atoms are known to diffuse more rapidly along $[1\bar{1}0]$ than along [001]. Since most observed steps are close to the [001] direction in Fig. 24(a), diffusion of extra Cu atoms from and toward the step region is reduced. In Figure 24(d), the step edges that are terminated by O-induced rows are perfectly straight while

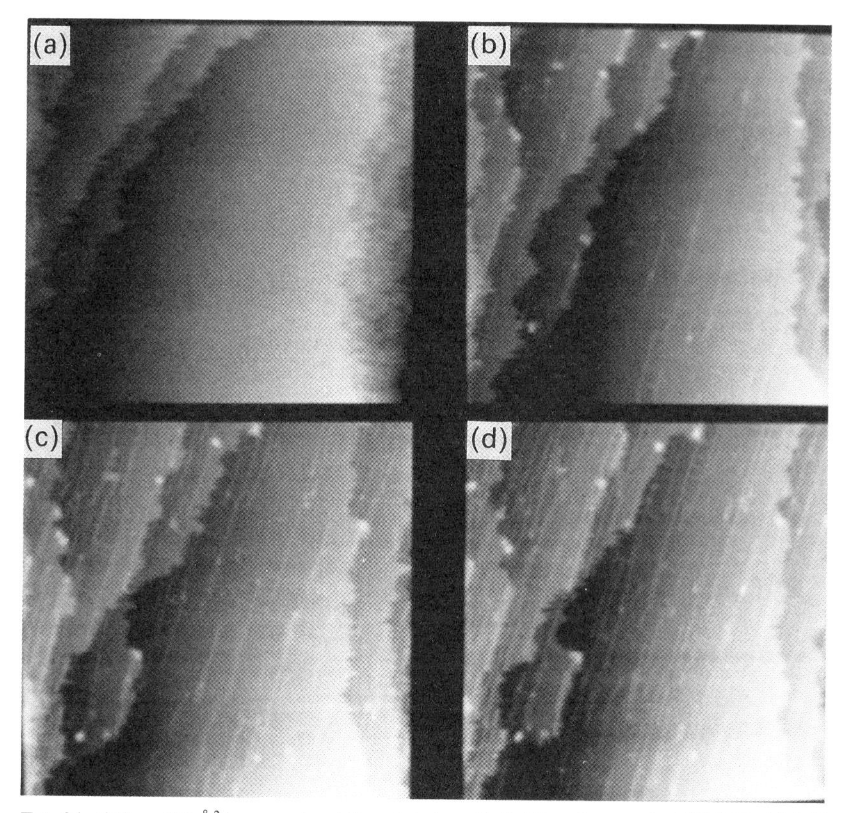

FIG. 24. 600 × 480-Å^2 images at −1 V on (a) clean Cu(110) surface, and at (b) 0.02, (c) 0.06, (d) 0.09, (e) 0.17, (f) 0.20, (g) 0.30, and (h) 0.40 ML coverage of oxygen. Patches are indicated by arrows. Figure from Ref. 70.

those without them remain rough. Fig. 24(b) shows an image taken at 0.025 ML O coverage. Isolated O-induced rows > 25 Å (along the [001]) begin to appear on Cu terraces, but not near step edges. When consecutive STM images were taken of this area at the same coverage, rows shorter than that length seem to coalesce or break apart. That length, therefore, seems to be critical for homogeneous growth. At ~0.3 ML, rectangular patches of missing first layer Cu atoms begin to appear. They increase in size very slowly with additional O exposure. They are also bounded on two sides by the O-induced rows, attesting to the barrier that the reconstruction provides for removal of more Cu atoms. If Cu atoms are not available from step edges on a large terrace, or step edges are terminated by O-induced rows, Cu atoms are

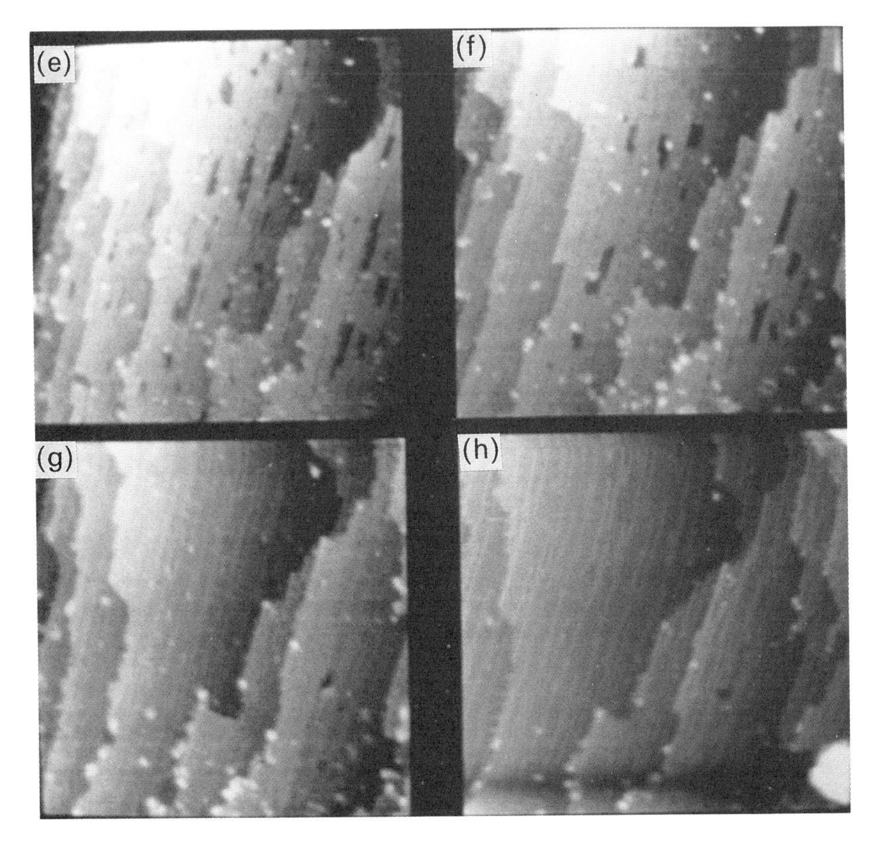

FIG. 24. Continued.

apparently supplied from these patches, though much more slowly than from step edges at low coverage. The measured coverage dependence of the sticking coefficient shows the deviation from Langmuir kinetics, due to the dynamics determined by extra diffusing Cu atoms. This experiment proves that scanning tunneling microscopy can now be used to correlate macroscopic quantities, such as diffusion constant, sticking coefficient, desorption rate, to atomic diffusion.

The formation of $(2\sqrt{2} \times \sqrt{2})R45°$ O islands on Cu(001) is much slower than that on Cu(110).[71] In Fig. 25(a) and (b) the structures of clean Cu(001) and O adsorbed $(2\sqrt{2} \times \sqrt{2})R45°$ are shown. In the report, the diffusion of Cu atoms from the step edges are shown in sequence. Cu atoms now diffuse along the [100] and [010] with the same diffusion rate. They, therefore, form O islands with a 2-fold symmetry. Since the diffusion constant along the $[1\bar{1}0]$ on Cu(110) is faster than that on along the [100] on Cu(001), the formation of O islands is much slower than that on Cu(110). As described in several

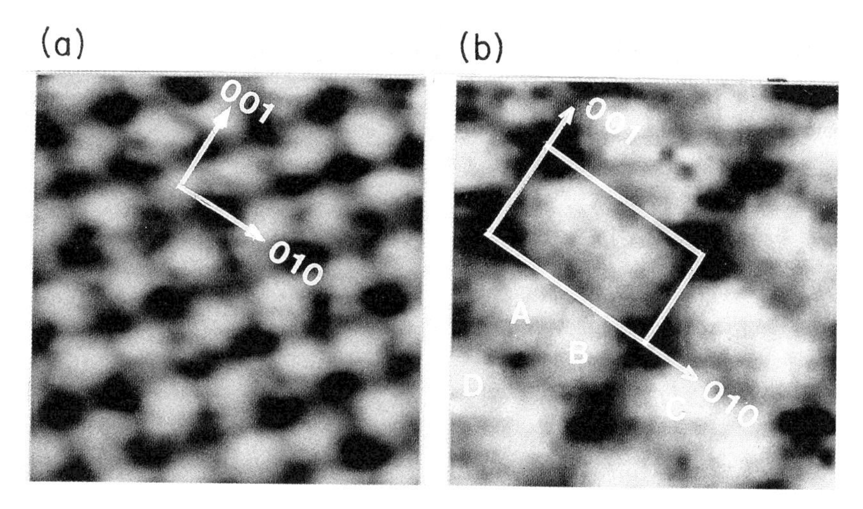

FIG. 25. STM images of 15×15-Å^2 region of (a) (1×1) Cu(001) surface and (b) $(2\sqrt{2} \times \sqrt{2})R45°$–O structure.

examples, scanning tunneling microscopy can now be used not only to identify the adsorbate-induced structure but also to understand the dynamical process of adsorption.

6.6 Conclusion

Metal represents a particular challenge to scanning tunneling microscopy because of the small surface-charge corrugation. Despite this, for the last decade the scanning tunneling microscopy has greatly contributed to the understanding of metal surfaces, including structure, phase transitions, surface diffusion, adsorbate reaction, and epitaxial growth. A UHV STM system in conjunction with other surface science techniques has become a powerful tool for the study of metal surfaces.

Acknowledgment

The author thanks P. J. Silverman, J. A. Stroscio, W. L. Brown, H. Q. Nguyen, H. P. Noh, F. M. Chua, Y. Hasegawa, R. S. Becker, R. J. Behm, D. D. Chambliss, S. Chiang, J. W. M. Frenken, P. K. Hansma, G. K. Gimzewski, R. J. Jacklevic, C. A. Lang, C. F. Quate, M. Salmeron, B. C. Schardt, and S. Rousset.

References

1. G. A. Somorjai, *Chemistry in Two-Dimension Surfaces*, Cornell University Press, Ithaca, 1981.

2. G. Binnig, H. Rohrer, Ch. Gerber, and E. Weibel, *Appl. Phys. Lett.* **40**, 178 (1982); *Phys. Rev. Lett.* **49**, 57 (1982); *Physica* **109/110b**, 2075 (1982).
3. Chapter 5 in this book.
4. T. Engel, "Determination of Surface Struccture Using Atomic Diffraction," in *Chemistry and Physics of Solid Surface V* (R. Vanslow and R. Howe, ed.), Springer-Verlag, Berlin, Heidelberg, 183.
5. J. Tersoff and D. R. Hamann, *Phys. Rev. Lett.* **50**, 25 (1983); *Phys. Rev. B* **31**, 805 (1985).
6. G. Binnig, H. Rohrer, Ch. Gerber, and E. Weibel, *Phys. Rev. Lett.* **50**, 120 (1983).
7. G. Binnig, H. Rhorer, Ch. Gerber, and E. Weibel, *Surf. Sci.* **131**, L379 (1983).
8. Y. Kuk, P. J. Silverman, and H. Q. Nguyen, *J. Vac. Sci. Technol. A* **6**, 524 (1988).
9. Y. Kuk and P. J. Silverman, *Appl. Phys. Lett.* **48**, 1597 (1986).
10. L. W. Swanson and L. C. Crouser, *J. Appl. Phys.* **40**, 4741 (1969).
11. H. W. Fink, *IBM J. Res. Dev.* **30**, 461 (1986).
12. S. I. Park, J. Nogami, and C. F. Quate, *Phys. Rev. B* **36**, 2863 (1987).
13. T. Sakurai, T. Hashizume, I. Kamiya, Y. Hasegawa, A. Sakai, J. Matsui, E. Kono, T. Takahaki, and M. Ogawa, *J. Vac. Sci. Technol. A* **6**, 803 (1988).
14. K. Sugihara, A. Akira, Y. Akama, N. Shoda, and Y. Kato, *Rev. Sci. Instrum.* **61**, 81 (1990).
15. Y. Kuk and P. J. Silverman, *Rev. Sci. Instrum.* **60**, 165 (1989).
16. R. D. Young, J. Ward, and F. Scire, *Phys. Rev. Lett.* **27**, 922 (1971).
17. Y. Kuk and P. J. Silverman, *J. Vac. Sci. Technol. A* **8**, 289 (1990).
18. J. K. Gimzewski and R. Moller, *Phys. Rev. B* **36**, 1284 (1987).
19. U. Durig, J. K. Gimzewski, and D. W. Pohl, *Phys. Rev. Lett.* **57**, 2403 (1986); *Bull. Amer. Phys. Soc.* **35**, 484 (1990).
20. N. D. Lang, *Phys. Rev. B* **36**, 8173 (1987); *Ibid* **37**, 10395 (1988).
21. V. M. Hallmark, S. Chiang, J. F. Rabolt, J. D. Swallen, and R. J. Wilson, *Phys. Rev. Lett.* **59**, 2879 (1987).
22. Ch. Woll, S. Chiang, R. J. Wilson, and P. H. Lippel, *Phys. Rev. B* **99**, 7988 (1988).
23. J. Winterlin, J. Wiechers, H. Brune, T. Gritsch, H. Hofer, and R. J. Behm, *Phys. Rev. Lett.* **62**, 59 (1989).
24. R. Landauer, *Z. Phys. B* **68**, 217 (1987).
25. R. H. Fowler and L. Nordheim, *Proc. Roy. Soc. London A* **11**, 173 (1928).
26. R. S. Becker, J. A. Golovchenko, and B. S. Swatzentruber, *Phys. Rev. Lett.* **55**, 987 (1985).
27. J. A. Kubby and W. J. Greene, *J. Vac. Sci. Technol, B* **9**, 739 (1991).
28. J. H. Coombs and J. K. Gimzewski, *J. Microscopy,* **152**, 841 (1988).
29. J. Hamers, R. M. Tromp, and J. E. Demuth, *Phys. Rev. Lett.* **56**, 1972 (1986).
30. R. S. Becker, J. A. Golovchenko, D. R. Hamann, and B. S. Swartzentruber, *Phys. Rev. Lett.* **55**, 2032 (1985).
31. Chapter 4 in this book.
32. E. L. Wolf, *Principles of Electron Tunneling Spectroscopy*, Oxford Press, Clarendon, Oxford, 1985.
33. C. J. Chen, *J. Vac. Sci. Technol. A* **6**, 319 (1988).
34. N. D. Lang, *Phys. Rev. B* **34**, 1164 (1986).
35. P. Heimann, J. Heimanson, H. Miosga, and H. Neddermeyer, *Phys. Rev. Lett.* **43**, 1957 (1979).
36. M. P. Everson, R. C. Jacklevic, and W. Shen, *J. Vac. Sci. Technol. B* **9**, 891 (1991).
37. Y. Kuk, L. C. Feldman, and I. K. Robinson, *Surf. Sci.* **138**, L168 (1984).
38. I. K. Robinson, *Phys. Rev. Lett.* **51**, 1145 (1983).
39. J. R. Noonan and H. L. Davis, *J. Vac. Sci. Technol.* **16**, 587 (1979).
40. D. Wolf, H. Jagydzinsler, and W. Moritz, *Surf. Sci.* **88**, L29 (1979).
41. Y. Kuk, P. J. Silverman, and F. M. Chua, *J. Microscopy,* **152**, 449 (1988).

42. J. C. Campuzano, M. S. Foster, G. Jennings, R. F. Willis, and W. Unertle, *Phys. Rev. Lett.* **54**, 2684 (1985).
43. M. S. Daw and S. M. Foiles, *Phys. Rev. Lett.* **59**, 2756 (1987).
44. H. Q. Nguyen, Y. Kuk, and P. J. Silverman, *J. de Phys.* **49**, 7988 (1989).
45. Ph. Lippel, R. J. Wilson, M. D. Miller, Ch. Woll, and S. Chiang, *Phys. Rev. Lett.* **62**, 171 (1989).
46. F. M. Chua, Y. Kuk, and P. J. Silverman, *Phys. Rev. Lett.* **63**, 386 (1989).
47. F. Jensen, F. Besenbache, E. Laegsgaard, and I. Stensgaard, *Phys. Rev. B* **41**, 10233 (1990).
48. J. V. Barth, H. Brune, G. Ertl, and R. J. Behm, *Phys. Rev. B* **42**, 9307 (1990).
49. C. J. Chen, *Phys. Rev. Lett.* **65**, 448 (1990).
50. J. A. Stroscio, D. T. Pierce, R. A. Dragoset, and P. N. First, "Proceedings of the 38th National Symposium of the American Vacuum Society," *J. Vac. Sci. Technol.* (to be published).
51. D. D. Chambliss, R. J. Wilson, and S. Chiang, *Phys. Rev. Lett.* **66**, 1721 (1991).
52. G. Binnig, H. Rohrer, Ch. Gerber, and E. Stoll, *Surf. Sci.* **144**, 321 (1984).
53. R. J. Behm, W. Hosler, E. Ritter, and G. Binnig, *Phys. Rev. Lett.* **56**, 228 (1986).
54. J. K. Gimzewski, A. Humbert, J. G. Bednorz, and B. Rehl, *Phys. Rev. Lett.* **55**, 951 (1985).
55. N. Garcia, A. M. Baro, R. Garcia, J. P. Pena, and H. Rohrer, *Appl. Phys. Lett.* **47**, 367 (1985).
56. S. K. Sinha, *Ordering in Two Dimensions*, North-Holland, 1980, New York.
57. R. C. Jacklevic and L. Elie, *Phys. Rev. Lett.* **60**, 120 (1988).
58. R. J. Schneir, R. Sonnenfeld, O. Marti, P. K. Hansma, J. E. Demuth, and R. J. Hamers, *J. Appl. Phys.* **63**, 717 (1988).
59. J. W. M. Frenken, R. J. Hamers, and J. E. Hamers, *J. Vac. Sci. Technol. A* **8**, 293 (1990).
60. E. Bauer, "Chemisorbed Phases," in *Phase Transition in Surface Films*, J. G. Dash and J. Ruvalds, eds, Plenum, New York, 1979.
61. Y. Kuk, P. J. Silverman, and T. M. Buck, *Phys. Rev. B* **36**, 3104 (1987).
62. M. M. Dorek, C. L. Lang, J. Nogami, and C. F. Quate, *Phys. Rev. B* **40**, 11973 (1989).
63. D. D. Chambliss, R. J. Wilson, and S. Chiang, *J. Vac. Sci. Technol. B* **9**, 933 (1991).
64. B. Marchon, P. Berhardt, M. E. Bussell, G. A. Somorjai, M. Salmeron, and W. Siekhaus, *Phys. Rev. Lett.* **60**, 1166 (1988).
65. D. F. Ogletree, C. Ocal, B. Marcon, G. A. Somorjai, M. Salmeron, T. Beebe, and W. Siekhaus, *J. Vac. Sci. Technol. A* **8**, 297 (1990).
66. B. C. Schardt, S-L. Yau, and F. Rinaldi, *Science,* **243**, 1050 (1989).
67. S. Rousset, S. Gauthier, O. Silboulet, W. Sacks, M. Belin, and J. Klein, *Phys. Rev. Lett.* **63**, 1265 (1989).
68. H. Ohtani, R. J. Wilson, S. Chiang, and C. M. Mate, *Phys. Rev. Lett.* **60**, 2398 (1988).
69. P. H. Lippel, R. J. Wilson, M. D. Miller, Ch. Woll, and S. Chiang, *Phys. Rev. Lett.* **62**, 171 (1989).
70. Y. Kuk, F. M. Chua, P. J. Silverman, and J. A. Meyer, *Phys. Rev. B* **41**, 12393 (1990).
71. D. J. Coulman, J. Wintterlin, R. J. Behm, and G. Ertl, *Phys. Rev. Lett.* **64**, 1761 (1990).
72. F. Jensen, F. Besenbacher, E. Laegsgaard, and I. Stensgaard, *Phys. Rev. B* **41**, 10233 (1991).
73. T. Gritsch, D. Coulman, R. J. Behm, and G. Ertl, *Phys. Rev. Lett.* **63**, 1086 (1989).

7. BALLISTIC ELECTRON EMISSION MICROSCOPY

L. D. Bell, W. J. Kaiser, M. H. Hecht

Center for Space Microelectronics Technology, Jet Propulsion Laboratory, California Institute of Technology, Pasadena, California

L. C. Davis

Scientific Research Laboratory, Ford Motor Company, Dearborn, Michigan

7.1 Introduction

The formation of semiconductor interfaces and a complete description of their characteristics is a problem of long standing in solid-state physics.[1,2] Interfaces are primary in determining many aspects of electron transport through semiconductor structures; however, interface formation is still not completely understood. There are several competing theories for interface formation which predict both Schottky barrier heights and semiconductor heterostructure band alignments, but no one of these has been completely successful for all systems.

The investigation of semiconductor interfaces is complicated by the inaccessibility of these interfaces to conventional surface-analytical probes. The final properties of an interface only develop when the interface is fully formed, when processes such as species interdiffusion and bulk band structure development are allowed to occur. Surface-sensitive techniques may be used to study the initial stages of interface formation, but characterization of completed interfaces relies on traditional electrical probes, such as current–voltage or capacitance–voltage measurements, or on optical methods such as internal photoemission.[3,4] In addition, all of these techniques are spatially averaging. In general, semiconductor interfaces are expected to exhibit some degree of heterogeneity, and for some of these techniques the spatial average of interface properties is heavily weighted toward low barrier height regions.

There are several important characteristics that should be addressed by an interface probe. In addition to determining interface barrier height, it is also crucial to obtain a quantitative measure of transport across the interface, which may be affected by band structure, scattering processes, and interface abruptness. The capability for a true energy spectroscopy of carrier transport is therefore necessary. Finally, due to the possibility of interface lateral

METHODS OF EXPERIMENTAL PHYSICS
Vol. 27

ISBN 0-12-475972-6

heterogeneity, it is desirable to characterize interface properties with high spatial resolution.

Scanning tunneling microscopy (STM)[5] has proved to be an extremely valuable and versatile probe of surfaces, which can be used for both spectroscopy and imaging.[6–8] Its capability for atomic-scale characterization of surface topographic and electronic structure has revolutionized surface science. Previous chapters have covered the application of STM to local surface spectroscopy of both semiconductors and metals, and later chapters will emphasize the surface imaging capability. This chapter will discuss a ballistic electron microscopy and spectroscopy which provide similar capabilities for the study of interfaces. The method, ballistic electron emission microscopy (BEEM),[9,10] which utilizes STM in a three-electrode configuration, allows characterization of interface properties with nanometer spatial resolution and enables an energy spectroscopy of carrier transport.

BEEM employs an STM tip as an injector of ballistic electrons into a sample heterostructure. In general, the sample will consist of at least two layers separated by an interface of interest. BEEM operates as a multi-electrode system, as illustrated in Fig. 1, with electrical contact to each layer of the sample surface. Figure 2 shows the corresponding energy diagram of tip, base, and collector for the case of a metal–semiconductor Schottky barrier (SB) system. In this case, the metal base layer serves as a biasing electrode, and the semiconductor functions as a collector of ballistic electron current. All sample structures discussed in this chapter utilize metallic base electrodes.

As a tip-base bias voltage is applied, electrons tunnel across the vacuum gap and enter the sample as non-equilibrium or hot carriers. Since characteristic attenuation lengths in metals and semiconductors may be hundreds of Ångstroms,[11] many of these hot electrons may propagate through the base layer and reach the interface before scattering. If conservation laws restricting total energy and momentum parallel to the interface are satisfied, these electrons may cross the interface and be measured as a current in the collector layer. An *n*-type semiconductor is used for electron collection, since the band bending accelerates the collected carriers away from the interface and prevents their leakage back into the base. By varying the voltage between tip and base, the energy distribution of the hot carriers can be controlled, and a spectroscopy of interface carrier transport may be performed.

Figure 3 shows the simplest spectrum that may be obtained with BEEM at $T = 0$. The theory that describes this spectrum will be discussed later, but the qualitative features may be mentioned here. For tunnel voltages less than the interface barrier height, none of the injected electrons have total energy equal to or greater than the barrier height, and the measured collector current is zero. As the voltage is increased to values in excess of the barrier, i.e.,

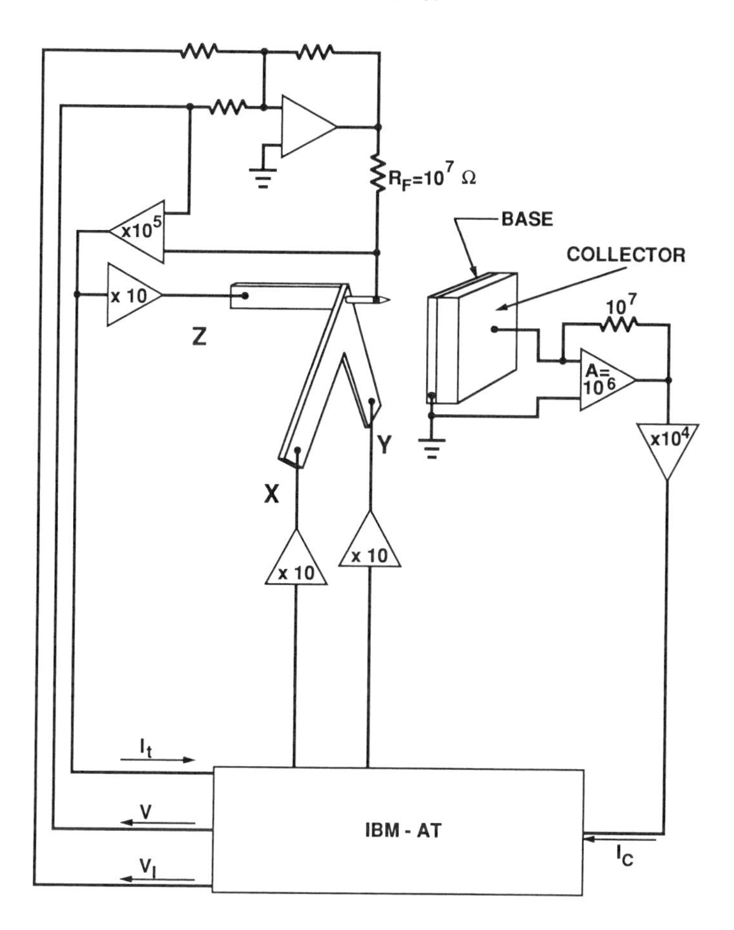

FIG. 1. Block diagram of an STM/BEEM system. The tip is maintained at the tunneling voltage V, and the tunneling current $I_t = V_I/R_F$ is held constant by the STM feedback circuit. The sample base layer is grounded and current into the semiconductor collector is measured by a virtual ground current amplifier.

$eV > E_F + eV_b$, some of the hot electrons cross the interface into the semiconductor conduction band, and a collector current is observed. The location of the threshold in the spectrum defines the interface barrier height. The magnitude of the current above threshold and the threshold spectrum shape also yield important information on interface transport.

In addition to characterizing conduction band structure, it is also important to probe valence band structure at an interface. A complete description of an interface requires separate knowledge of both conduction and valence band Schottky barrier heights and band alignments. A spectroscopy of valence band structure may also be performed with BEEM techniques, but

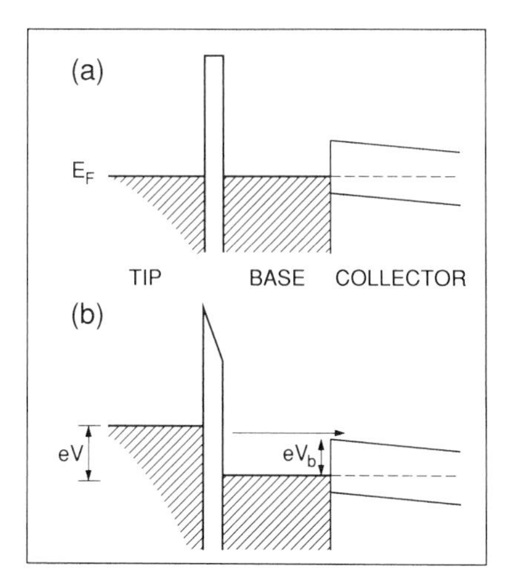

FIG. 2. Energy diagram for BEEM of a metal base–semiconductor collector Schottky barrier system. (a) Applied tunnel voltage of zero. (b) Applied tunnel voltage V, in excess of the interface barrier height V_b. For this case, some of the injected electrons have sufficient energy to enter the semiconductor.

utilizing ballistic *holes* as a probe.[12] Since most p-type barriers are relatively low, ballistic hole spectroscopy has required the development of a low-temperature BEEM apparatus. This apparatus has in turn enabled a study of interface properties with temperature, which provides fundamental insights into interface formation mechanisms.

To date, the main focus of BEEM research has been on the physics of interface transport; however, bulk carrier transport properties are also of

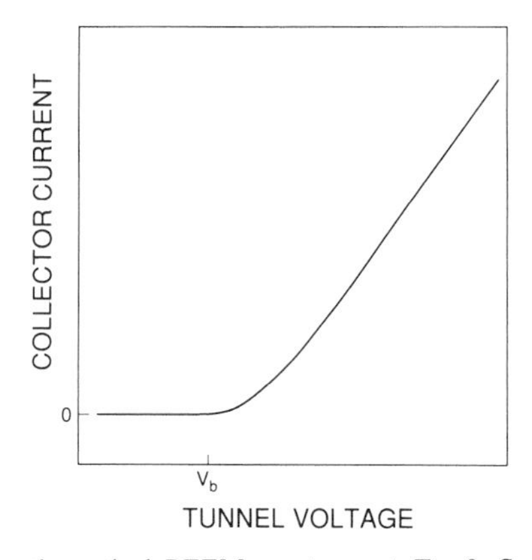

FIG. 3. Example of a theoretical BEEM spectrum at $T = 0$. Such a spectrum displays a threshold at a voltage equal to the interfacial barrier height V_b.

great importance in solid-state physics, and carrier scattering in materials dominates transport properties. BEEM is a valuable technique for the study of ballistic transport and interface structure, but it does not directly provide a means for an analysis of the carriers which scatter in the base electrode and are not collected. However, a related technique, which may be referred to as reverse-bias BEEM, provides a direct spectroscopy of electron and hole scattering. This method is sensitive to carrier–carrier scattering, and allows the first direct probe of the hot secondary carriers created by the scattering process.[13] As a spectroscopy of scattering, it is complementary to BEEM, since it provides a way of analyzing *only* the carriers that scatter, and are not observed by the ballistic spectroscopies. Scattering phenomena in metal–semiconductor systems have been investigated, and a theoretical treatment for the collected current has been developed which yields good agreement with experiment. The observed magnitudes of the currents indicate that the carrier–carrier scattering process is dominant in ballistic carrier transport.

BEEM has been applied to many interface systems, and important examples will be given in this chapter. A theoretical framework for this technique, and for the corresponding ballistic hole process, will be presented. The extension of the theory to the case of the carrier–carrier scattering spectroscopy will also be described. Experimental methods particular to BEEM and related techniques will be discussed. Both spectroscopy and imaging of interfaces is possible with ballistic electron and hole probes, and examples of each will be reviewed. Finally, the initial application of the new scattering spectroscopy will be presented.

7.2 Theory

7.2.1. Ballistic Electron Spectroscopy

Ballistic electron emission microscopy (BEEM) may be understood by using a simple theoretical model, which may be built upon to include more complicated processes. The model treats sequentially the processes of vacuum tunneling, base transport, and interface transport. The effective-mass approximation is used throughout this chapter for each of these processes. The essence of the model is the description of the phase space available for interface transport. Dynamical considerations, such as quantum mechanical reflection at the interface, are not included in this description, although such effects will be mentioned later and may be treated in a more complex theory. The simplest case to consider is that of a smooth interface, which dictates conservation of the component of the electron wave vector parallel to the interface (transverse to the interface normal) k_t. Conservation of total energy across the interface provides a second constraint on tranport. The

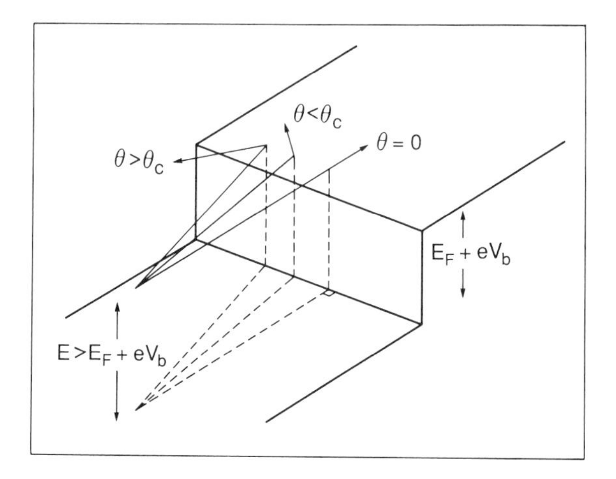

FIG. 4. Schematic diagram representing a particle incident on a potential step of height $E_F + eV_b$, with an initial energy E in excess of this step height. There exists a critical angle θ_c such that a particle incident at an angle in excess of this value is reflected. For an angle of incidence less than θ_c, the particle is transmitted, but is refracted as it crosses the interface.

process is visualized in Fig. 4. This diagram may be taken to represent a simplified metal–semiconductor interface system.

A particle incident on a simple potential step, with total energy in excess of the step height, loses a portion of its kinetic energy as it crosses the step, due to a reduction in k_x, the component of **k** normal to the interface. If the particle is normally incident, the direction of propagation, $\hat{\mathbf{k}}$, is unchanged. For the case of incidence at a non-zero angle to the normal, however, conservation of k_t demands that $\hat{\mathbf{k}}$ changes across the interface; i.e., the particle is "refracted." If the angle of incidence is greater than some "critical angle," the particle is not able to cross the interface and is reflected back. This critical angle may be expressed as[14]

$$\sin^2 \theta_c = \frac{E - E_0}{E}, \tag{1}$$

where the particle total energy is E and the step height is $E_0 = E_F + eV_b$. Another way to specify the criterion for transmission is that E_x must exceed E_0, where E_x is the component of carrier energy associated with k_x, or $E_x = \hbar^2 k_x^2/2m$.

The situation becomes somewhat more complicated for real band structures, if the two regions have dissimilar dispersion relations. In this case, the normal and parallel components of the problem are no longer trivially separable. In particular, if E_x is the energy associated with k_x, $E_x > E_F + eV_b$ is no longer the condition for collection. A change in effective mass, or in the

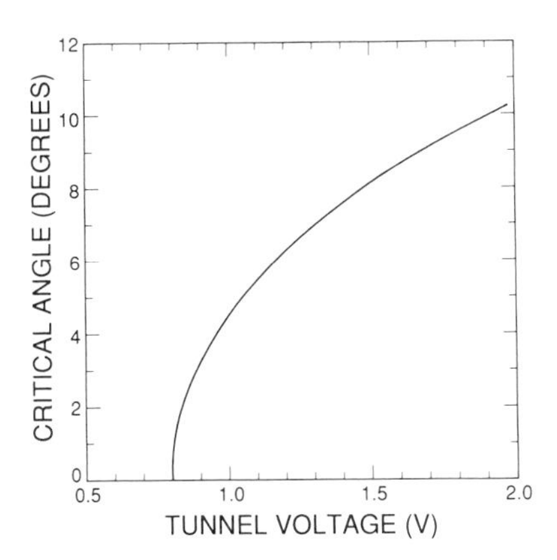

FIG. 5. Plot of critical angle θ_c calculated from Eq. (3) for the case of Au on n–Si(100), with $eV_b = 0.8\,\text{eV}$, $E_F = 5.5\,\text{eV}$, and $m_t/m = 0.2$.

location of the conduction band minimum within the Brillouin zone, alters the critical angle conditions.

For zone-centered conduction band minima in both the metal and semiconductor, we can define this critical angle with respect to the interface normal, in terms of electron energy in the base and the potential step $E_F + eV_b$, as

$$\sin^2\theta_c = \frac{m_{tf}}{m_{xi}} \frac{E - E_F - eV_b}{E + \left(\frac{m_{tf}}{m_{xi}} - \frac{m_{tf}}{m_{ti}}\right)(E - E_F - eV_b)}, \tag{2}$$

where m_{xi} and m_{ti} are the components of effective mass in the metal normal to and parallel to the interface, respectively, and m_{xf} and m_{tf} are the corresponding masses in the semiconductor. For evaporated metal base layers which are polycrystalline or of ill-defined orientation, the approximation of an isotropic free-electron mass for the base is made,* and the expression reduces to[10]

$$\sin^2\theta_c = \frac{m_t}{m}\frac{E - E_F - eV_b}{E} = \frac{m_t}{m}\frac{e(V - V_b)}{E_F + eV}, \tag{3}$$

where the second equality is for an incoming electron with $E = E_F + eV$ and m is the free-electron mass; the second subscript on m_{tf} has now been dropped. This expression, plotted in Fig. 5, predicts that for Si(100) or GaAs, with a small component of effective mass parallel to the interface and for $e(V - V_b) \leqslant 0.3\,\text{eV}$, this critical angle is less than 6 degrees. Critical angle

*Note that in all comparisons of theory and data, both the STM tip and the base layer have been approximated as free-electron metals. For some expressions in this section, however, the more general forms are also given.

reflection has important implications for the spatial resolution of interface characterization, since only electrons incident on the interface at small angles can be collected. Thus single scattering events may decrease collected current, but should not degrade spatial resolution. For a 100 Å base layer, at least 20 Å spatial resolution is expected. Resolution of this order has actually been observed on some metal–semiconductor systems.

The electron injection into the sample by tunneling is treated using a planar tunneling formalism.[15] This description provides simple analytic expressions for the $(E, \mathbf{k})$ distribution of the tunneling electrons and for the total tunnel current. Current across the metal–semiconductor interface is then calculated based upon this initial distribution, by considering the fraction of the total tunnel current which is within the "critical cone." The WKB form of the tunneling probability $D(E_x)$ is written as[16]

$$D(E_x) = \exp\left(-2\int k_x\, dx\right). \tag{4}$$

The vacuum barrier is taken to be square at $V = 0$, with height Φ measured with respect to the tip Fermi level; therefore, Eq. (4) becomes

$$D(E_x) = \exp\left(-\alpha s \frac{2}{3eV}\left[(E_F + \Phi - eV - E_x)^{3/2} - (E_F + \Phi - E_x)^{3/2}\right]\right), \tag{5}$$

where $\alpha = (8m/\hbar^2)^{1/2} = 1.024\,\mathrm{eV}^{-1/2}\,\mathrm{Å}^{-1}$. A common approximation[15] is

$$D(E_x) \approx \exp\left(-\alpha s(E_F + \bar{\Phi} - E_x)^{1/2}\right), \tag{6}$$

with $\bar{\Phi} = \Phi - eV/2$ for a square barrier. Tunnel current is given by the standard expression

$$I_t = 2ea \int\!\!\int\!\!\int \frac{d^3\mathbf{k}}{(2\pi)^3}\, D(E_x) v_x (f(E) - f(E + eV)), \tag{7a}$$

where a is the effective tunneling area, $f(E)$ is the Fermi distribution function, and $v_x = \hbar k_x/m$. This expression may conveniently be written in terms of integrals over E_x and E_t:

$$I_t = C\int_0^\infty dE_x D(E_x) \int_0^\infty dE_t (f(E) - f(E + eV)). \tag{7b}$$

The integration is over all tip states with $E_x > 0$, and $C = 4\pi mae/h^3$. E_x is the energy associated with k_x and E_t is that associated with k_t. (It will be convenient in the remainder of this section to express energies with reference to the bottom of the tip conduction band, unless otherwise noted.) The Fermi function $f(E)$ is defined as

$$f(E) = \left[1 + \exp\left(\frac{E - E_F}{k_B T}\right)\right]^{-1}. \tag{8}$$

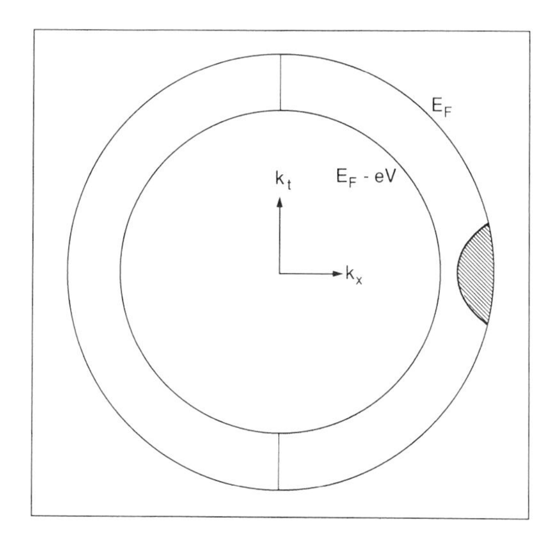

FIG. 6. **k**-space diagram representing the free-electron Fermi sphere of the STM tip, for the case of electron tunneling from tip to sample. For an applied tip–sample voltage V, states within the shell delimited by E_F and $E_F - eV$, and with $k_x > 0$, are eligible to tunnel. These states are shown here in grey. If $eV > eV_b$, a subset of these tunneling electrons, defined by a hyperboloid in **k**-space, satisfy phase space conditions for collection. This volume is also shown by diagonal lines.

For simplicity in modeling the tunneling process, the tip and base are taken to be identical free-electron metals. At $T = 0$, therefore, the appropriate states occupy a half-shell within the tip Fermi sphere[17] between $E = E_F - eV$ and $E = E_F$, as illustrated in Fig. 6.

An expression similar to Eq. (7b) may be written for the collector current, with the allowed phase space within the tip determined by the critical angle conditions. For the case of m_t less than m, these restrictions on tip states are

$$E_t \leqslant \frac{m_t}{m - m_t}(E_x - E_F + e(V - V_b)) \tag{9}$$

and

$$E_x \geqslant E_F - e(V - V_b), \tag{10}$$

where E_F is the Fermi energy of the tip. Taking the equalities as limits $E_t^{\max}$ and $E_x^{\min}$, collector current can be expressed as

$$I_c = RC\int_{E_x^{\min}}^{\infty} dE_x D(E_x)\int_0^{E_t^{\max}} dE_t(f(E) - f(E + eV)). \tag{11}$$

R is a measure of attenuation due to scattering in the base layer, which is taken to be energy-independent for these energies.[18] If $eV_b \gg k_B T$, as with all examples discussed in this chapter, the second Fermi function $f(E + eV)$ in Eq. (11) may be neglected. In both Eqs. (7b) and (11) the integrals over E_t may be performed analytically; for $T = 0$, this is true also of the integrals over E_x.

Equations (9) and (10) define a hyperboloid in tip **k**-space, as shown in Fig. 6. The integration is performed over this hyperboloidal volume, with the cutoff at higher energies provided by the Fermi function $f(E)$ centered at the tip Fermi level. The threshold shape of the BEEM I_c–V spectrum is determined by the behavior of this **k**-space volume with voltage, which in turn is determined by the dispersion relation of the collector conduction band. A parabolic conduction band minimum and the assumption of k_t conservation across the interface therefore result in a parabolic threshold shape to the I_c–V spectrum.

A similar treatment for I_c can be derived for the case of $m_t > m$. In this case, the hyperboloid bounding the appropriate tip states becomes a section of an ellipsoid, centered at $\mathbf{k} = 0$, with the defining relation

$$E_x + \frac{m - m_t}{m_t} E_t = E_F - e(V - V_b). \tag{12}$$

States external to this ellipsoid are allowed for collection, with a cutoff at higher energies again provided by the tip Fermi function.

BEEM spectra are conventionally obtained with the STM operating at constant tunnel current, which normalizes the collected current to the tunnel current and linearizes the BEEM spectrum. In addition, this normalization removes the lowest order effects due to structure in the tunneling density-of-states that may obscure interface structure. Equation (11) for I_c contains factors of s, the tip–sample spacing, which changes with tunnel voltage at constant I_t. Therefore Eq. (11) will not describe accurately an entire I_c–V spectrum. This effect may be included within the theory in two ways. Equation (7b) for $I_t(s, V)$ may be inverted to give $s(I_t, V)$, which is then inserted into Eq. (11). In practice, this inversion must be done numerically; in addition, the prefactor C, which includes effective tunneling area, must be known. The second method is to treat s as a constant s_0, but to normalize $I_c(s_0, V)$ by $I_t(s_0, V)$ for each voltage. This would be an exact normalization if both I_c and I_t were measured at constant s, that is, without feedback control of tunnel current. It is in fact a good approximation even for BEEM spectra measured at constant tunnel current and requires only that the tunnel *distribution* be relatively insensitive to small changes in s. For the barriers considered here, this assumption is valid, producing errors only on the order of a few percent. The expression for I_c then takes the form

$$I_c = RI_{t0} \frac{\int_{E_x^{\min}}^{\infty} dE_x D(E_x) \int_0^{E_t^{\max}} dE_t(f(E) - f(E+eV))}{\int_0^{\infty} dE_x D(E_x) \int_0^{\infty} dE_t(f(E) - f(E+eV))}, \tag{13}$$

where I_{t0} is the (constant) tunnel current at which the BEEM spectrum is measured. This expression is used for the fitting of single-threshold ballistic electron spectra in this chapter. It can be verified that it is a very good approximation to the exact first method discussed earlier.

One consequence of the critical angle effect is the sensitivity of BEEM to higher minima in the collector conduction band structure, rather than just the lowest minimum, which determines the Schottky barrier height. This is due to the opening of additional phase space for electron transport as electron energy exceeds each minimum in turn. This capability is enabled by the control over injected electron energy provided by BEEM.

The conduction band minima of a particular semiconductor are not in general zone-centered; however, the critical angle restrictions are unchanged provided the minimum is "on-axis," i.e., located at $k_t = 0$. The above phase space requirements may therefore be used for a metal on Si(100) or Ge(111), although the wave function coupling into these minima may be different. For GaAs of any orientation, off-axis conduction band minima must always be considered. Similar critical angle requirements exist for these off-axis minima, with the center of the critical cone located at an angle to the interface normal, defined in the base, given by

$$\sin^2\theta_0 = \frac{E_{0t}}{E_F + eV_b}, \qquad E_{0t} = \frac{\hbar^2 k_{0t}^2}{2m}, \tag{14}$$

where k_{0t} is the component parallel to the interface of k_0, the location of the minimum. The criterion for collection may then most conveniently be written in terms of the components of k in the base. If we define $\hat{\mathbf{z}}$ along the interface, and in the direction toward the center of the constant-energy ellipsoid projection onto the interface, then the phase space condition is

$$k_x^2 - \frac{m - m_t}{m_t} k_y^2 - \frac{m}{m_z}(k_z - k_{0z})^2 + k_z^2 > \frac{2m}{\hbar^2}(E_F + eV_b), \tag{15}$$

where $k_{0z} = k_0 \sin\theta_m$, $m_z = m_t \sin^2\theta_m + m_l \cos^2\theta_m$, and m_t and m_l are the masses along the transverse and longitudinal principal axes of the semiconductor constant-energy ellipsoid. θ_m defines the angular position of the conduction-band minimum with respect to the interface. Figure 7 shows the k-space geometry more clearly.

Off-axis minima may be included in the treatment for BEEM, although

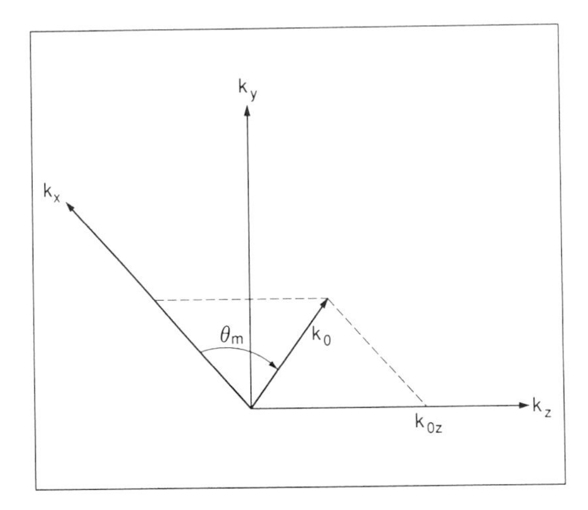

FIG. 7. Diagram representing transmission of electrons across an interface into a conduction band minimum located at k_0 which has a non-zero projection k_{0z} onto the interface plane. The kinematic constraints for collection are determined by this projection of the band structure onto the interface. In the effective mass approximation, this produces constant-energy ellipses, with a parabolic dispersion relation characterized by two masses.

they do not provide an analytic expression for I_c within the theory discussed here. This case is of great importance, however, since collection via these minima provides a powerful probe of fundamental quantities such as the (E, k) tunneling distribution, momentum conservation at the buried interface, and carrier scattering.

The foregoing discussion has assumed that all electrons incident on the interface within the critical angle are collected. This is a result of a purely kinematic model. This classical assumption may not be appropriate for abrupt interfaces, where quantum-mechanical reflection (QMR) must be considered. In this case the integrand of Eq. (11) must be multiplied by the quantum-mechanical transmission factor $T(E, k)$ appropriate to the potential profile of the interface. Using the approximation of a sharp step potential, and for normally incident electrons, this factor may be written as[19]

$$T = \frac{4\,\dfrac{k_{xi}}{m_{xi}}\,\dfrac{k_{xf}}{m_{xf}}}{\left(\dfrac{k_{xi}}{m_{xi}} + \dfrac{k_{xf}}{m_{xf}}\right)^2}, \tag{16}$$

which is the generalization of the more familiar expression in which the masses are equal on both sides of the potential step. k_{xi} and k_{xf} are the magnitudes of the electron wave vector in the base and collector, respectively, and m_{xi} and m_{xf} are the corresponding effective mass components in the normal direction. The subscript x emphasizes that this expression is for normally incident carriers only. In terms of energies referred to the tip conduction band minimum, and for a zone-centered minimum in the semiconductor,

$$T = \frac{4\sqrt{\dfrac{(E_x + eV)(E_x + eV - E_F - eV_b)}{m_{xi}m_{xf}}}}{\left(\sqrt{\dfrac{E_x + eV}{m_{xi}}} + \sqrt{\dfrac{E_x + eV - E_F - eV_b}{m_{xf}}}\right)^2}, \tag{17}$$

which increases as $(V - V_b)^{1/2}$ for V close to the threshold V_b. Within the effective-mass approximation, this expression may be modified to include off-normal propagation to give a generalized expression for T, again in terms of tip-referenced energies:

$$T = \frac{4\sqrt{\dfrac{(E_x + eV)(E_x + eV - E_F - eV_b - \beta E_t)}{m_{xi}m_{xf}}}}{\left(\sqrt{\dfrac{E_x + eV}{m_{xi}}} + \sqrt{\dfrac{E_x + eV - E_F - eV_b - \beta E_t}{m_{xf}}}\right)^2}, \tag{18}$$

where

$$\beta = \frac{m_{ti} - m_{tf}}{m_{tf}}.$$

In the limit of a smooth potential transition from metal to semiconductor, T approaches a step function, and the regime of $(V - V_b)^{1/2}$ behavior becomes small. Alternative expressions for other potentials may also be used. The analyses of data in this chapter do not include quantum-mechanical reflection terms, although later discussions of ballistic hole spectra will address this point further.

7.2.2 Hole Spectroscopy

The previous section dealt with the investigation of ballistic electron transport and its use as a probe of interface conduction band structure. It was mentioned in the Introduction that n-type semiconductors are necessary for electron collection, in order to repel the carriers away from the interface into the collector and prevent leakage back into the base. It is possible to use ballistic *holes*, however, as a probe of valence band structure. The unique aspects of ballistic hole spectroscopy will be discussed in this section.

The implementation of ballistic hole spectroscopy using BEEM techniques is shown in the energy diagram of Fig. 8 for the case of a metal–semiconductor sample structure. Here, a p-type semiconductor serves as a collector of

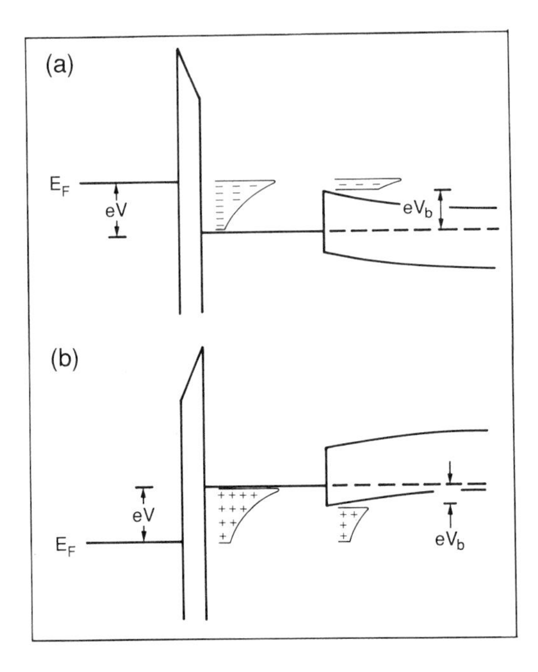

FIG. 8. Energy diagrams of ballistic electron and ballistic hole spectroscopies, from Ref. 12. (a) In ballistic electron spectroscopy of interface conduction band structure, a hot electron distribution is created, of which the most energetic electrons are available for collection. (b) With ballistic hole spectroscopy of interface valence band structure, a hot hole distribution is created in the base by electron vacuum tunneling. The *least* energetic holes are eligible for collection by the semiconductor valence band.

ballistic holes injected by the STM tip. The tip electrode is biased positively; since injection is through a vacuum tunnel barrier, the injection process must be treated in terms of electron tunneling from the base to the tip. This tunneling process deposits ballistic electrons in the tip and creates a ballistic hole distribution in the base, illustrated in Fig. 8. Note that, since the tunneling is strictly by electrons, the energy and angular distributions of the ballistic holes are determined by the vacuum level. The distribution is therefore peaked towards the base Fermi level.

As a positive voltage is applied to the tip, ballistic holes are injected into the sample structure. As in the ballistic electron case, collector current is zero until the voltage exceeds the barrier height; at higher voltages, collector current increases. It is apparent, however, that the peaking of the hole distribution toward the base Fermi level introduces an asymmetry between the ballistic electron and hole spectroscopies. This asymmetry is illustrated in Fig. 8. In BEEM, the portion of the hot electron distribution which is eligible for collection is toward the higher energies, where the distribution is maximum. For the case of holes, the tail of the distribution is collected. This asymmetry introduces a corresponding asymmetry into a ballistic hole spectrum, as will be discussed later.

The threshold behavior of the ballistic hole I_c–V spectrum, however, is the same as for the case of BEEM. For a hole barrier at $E_F - eV_b$, the critical angle condition is

$$\sin^2\theta_c = \frac{m_{tf}}{m_{xi}} \frac{E_F - eV_b - E}{E + \left(\frac{m_{tf}}{m_{xi}} - \frac{m_{tf}}{m_{ti}}\right)(E_F - eV_b - E)}, \tag{19}$$

which may be compared with Eq. (2). For the case of an isotropic free-electron mass for the base, the expression reduces to

$$\sin^2\theta_c = \frac{m_t}{m} \frac{E_F - eV_b - E}{E} = \frac{m_t}{m} \frac{e(V - V_b)}{E_F - eV}, \tag{20}$$

the second equality being for an incoming hole with $E = E_F - eV$. Here, m_t is the *valence band* effective mass parallel to the interface, which Eq. (20) takes as constant and isotropic. The ballistic-hole I_c–V spectrum threshold has a $(V - V_b)^2$ dependence, in agreement with the ballistic electron case, although the appropriate **k**-space volume of integration is most conveniently taken over states in the base, and is quite different from the electron case, as illustrated in Fig. 9. The volume here is represented by the intersection of an ellipsoid, defined by the critical angle condition, and a sphere, determined by the tip Fermi level.

7.2.3 Scattering Spectroscopy

In this section, we present a theory for the generation, in reverse bias, of electron-hole pairs within the base, and their subsequent collection by the collector. To be specific, let us take the collector to be a p-type semiconductor; analogous processes occur with n-type collectors. For simplicity, the theory will treat only the $T = 0$ case. In reverse bias, that is, opposite in sign to that used in ballistic hole spectroscopy, a negative voltage, $-V$, is applied to the tip so that electrons are injected into the base. For $V < V_{bn}$, the position of the semiconductor conduction band edge at the interface, the injected electrons travel ballistically to the interface and are reflected back into the base. Consequently, one might expect that no current will be collected by the semiconductor. However, the injected electrons can scatter off the Fermi sea in the base, exciting electrons below E_F to states above the Fermi energy. The holes that are thereby created may act in the same manner as holes injected by the tip in forward bias, although with a different energy and angular distribution. Those holes that have energy $E < E_F - eV_{bp}$ (V must

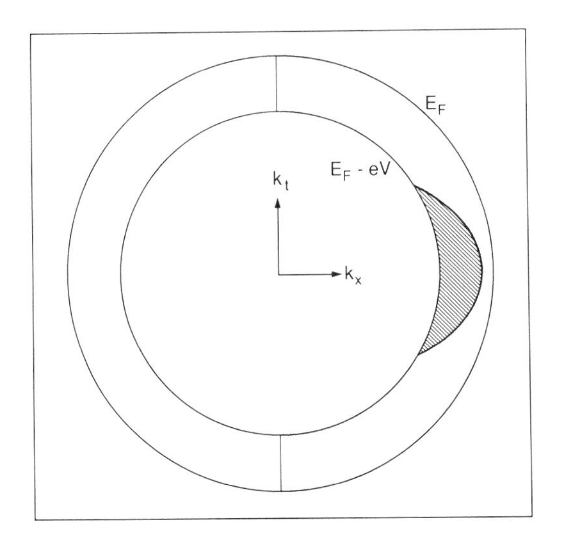

FIG. 9. **k**-space diagram representing the free-electron Fermi sphere of the sample base, for the case of electron tunneling from sample to tip. For an applied sample–tip voltage V, states within the shell (shown in grey) delimited by E_F and $E_F - eV$, and with $k_x > 0$, are eligible to tunnel. If $eV > eV_b$, a subset of these tunneling electrons, defined by the intersection of an ellipsoid and a sphere in **k**-space (diagonal lines), create holes which satisfy phase space conditions for collection.

exceed V_{bp}), and are directed towards the interface within the critical angle θ_c of the normal will be collected, resulting in a collector current I_c. This process is shown in Fig. 10. The corresponding process for hole injection and secondary electron collection with an n-type collector is also shown.

We can construct a theory for these processes along the lines of the theory already developed for forward bias with the addition of a simple but reasonably accurate treatment of the electron–electron (ee) scattering. As in previous sections, the free-electron picture is used, where $E_n = \hbar^2 k_n^2/2m$. An electron injected into the base with energy E_0 can lose energy and scatter to a new state with energy E_1 above E_F. Conservation of energy requires that an electron excited from E to E_2, which must also be above E_F, satisfy the following relationship:

$$E_0 - E_1 = E_2 - E. \tag{21}$$

Likewise, in a free-electron metal, momentum is conserved so that

$$\mathbf{k}_0 - \mathbf{k}_1 = \mathbf{k}_2 - \mathbf{k}. \tag{22}$$

Summing over all states $\mathbf{k}_1$ and $\mathbf{k}_2$ that are above the Fermi energy and which satisfy the conservation laws, we obtain the total rate $R(\mathbf{k})$ at which holes

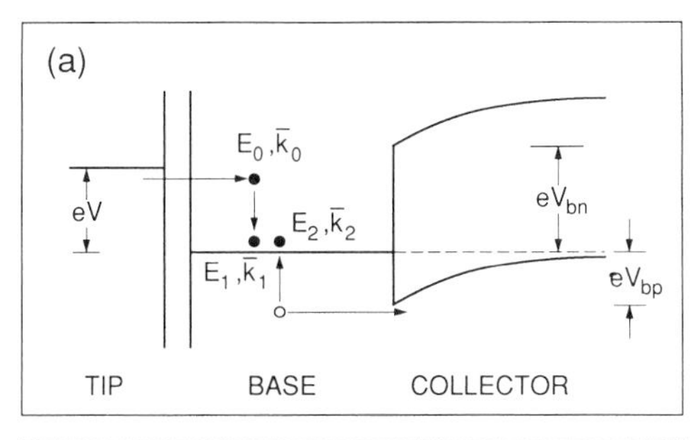

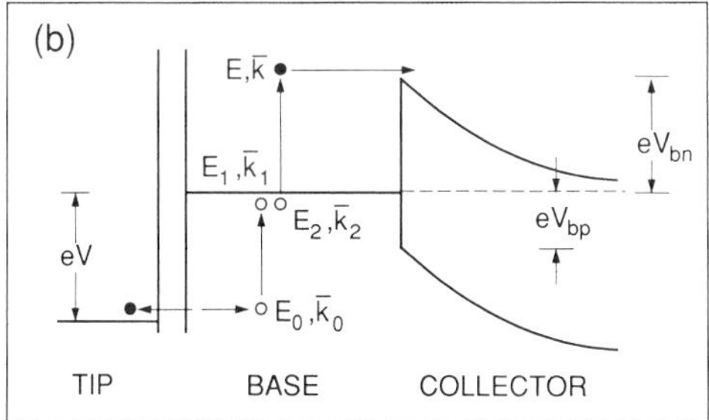

FIG. 10. Energy diagram of the reverse-bias BEEM experiment, which enables a spectroscopy of scattering, adapted from Ref. 13. (a) The process is shown for electron injection and collection, by a p-type collector, of the scattered holes. The hole distribution is produced by scattering of the injected electrons by the Fermi sea in the base electrode, which excites these equilibrium electrons and creates electron-hole pairs. (b) The comparable process for hole injection and secondary electron collection by an n-type collector. In both cases, the incoming carrier with $(E_0, \mathbf{k}_0)$ scatters to final state $(E_1, \mathbf{k}_1)$, exciting the secondary carrier (of opposite type) from $(E_2, \mathbf{k}_2)$ to $(E, \mathbf{k})$.

with momentum $\mathbf{k}$ and energy E are generated:

$$R(\mathbf{k}) = 2\pi/\hbar \sum_{\substack{\mathbf{k}_1,\mathbf{k}_2 \\ E_1,E_2>E_F}} |M|^2 \delta(E_0 + E - E_1 - E_2)\delta_{\mathbf{k}_0+\mathbf{k},\mathbf{k}_1+\mathbf{k}_2}. \tag{23}$$

In addition to energy and momentum conservation, this expression contains a matrix element M, which in general depends upon momentum transfer $\mathbf{q} = \mathbf{k}_0 - \mathbf{k}_1$. For an unscreened Coulomb interaction, $M \propto 1/q$. However, metallic screening removes this singularity, making $M \propto 1/(q^2 + q_{\mathrm{ft}}^2)^{1/2}$. Since q_{ft} is comparable to the Fermi momentum k_{f}, and typically the momentum transfer is small compared with k_{f}, it is reasonable to treat M as a constant. If we make this assumption, Eq. (23) can be evaluated. We find, for $2E_F - E_0 < E < E_F$,

$$R(\mathbf{k}) = \text{constant} \cdot \frac{E_0 + E - 2E_F}{[E_0 + E + 2\sqrt{E_0 E}\cos\theta_{\mathbf{k}\mathbf{k}_0}]^{1/2}}, \tag{24}$$

where $\theta_{\mathbf{k}\mathbf{k}_0}$ is the angle between $\mathbf{k}$ and $\mathbf{k}_0$.

The most significant factor in Eq. (24) is the phase space term $E_0 + E - 2E_F$. Because of it, the lowest energy holes are the least abundant. Since only low-energy holes ($E < E_F - eV_{bp}$) can be collected, I_c is small, and its dependence on $V - V_{bp}$ above threshold is weaker than in forward bias.

The rate $R(\mathbf{k})$ can be related to the lifetime of the injected electron as follows. The inverse lifetime or width Γ consists primarily of two terms. The first, Γ_{ph}, is due to phonon scattering. The second, Γ_{ee}, is due to electron–electron scattering and is expected to be dominant except near the Fermi level.† Γ_{ee} can be found by summing $R(\mathbf{k})$ over all $\mathbf{k}$,

$$\Gamma_{ee} = \hbar \sum_{\mathbf{k}} R(\mathbf{k}), \tag{25}$$

or, transforming to an integral,

$$\Gamma_{ee} = \hbar \int \frac{d^3\mathbf{k}}{(2\pi)^3} R(\mathbf{k}), \tag{26}$$

where the integration is over all $\mathbf{k}$ such that $2E_F - E_0 < \hbar^2 k^2/2m < E_F$. It is straightforward to show that

$$\Gamma_{ee} \propto (E_0 - E_F)^2 \tag{27}$$

for E_0 near E_F, a result obtained from a proper treatment of screening.[20] This gives us confidence that taking the matrix element M to be constant gives reliable results.

The distribution of holes generated can be written in terms of a branching ratio, $\hbar R(\mathbf{k})/\Gamma_{ee}$. This has the advantage that $|M|^2$ cancels out and we do not need to know its magnitude. The physical basis is that every electron injected by the tip decays by the creation of an electron-hole pair. That is, for every electron injected, a hole is created. The probability that the hole has momentum $\mathbf{k}$ is the branching ratio. Hence, if we know the distribution of injected electrons, we can find the resulting distribution of holes.

Note that multiple scattering is neglected. An electron that loses only a small amount of energy may have enough energy remaining to excite an electron (create a hole) from a low-energy state by a subsequent scattering. However, near threshold, such multiple scattering will mainly produce holes too close to E_F to be collected. At higher voltages, this assumption breaks down and multiple scattering should be included.

To proceed with the calculation, let $P(\mathbf{k}_0)$ be the rate at which electrons

†Actually, for the case of electron injection and hole collection in Au/p–Si(100), the electron–phonon and electron–electron scattering rates are comparable for electrons injected just above $E_F + eV_{bp}$, the threshold for hole collection. However, we argue that phonon scattering can be included in the elastic scattering, since a typical energy loss is small compared to electron–hole pair energies, and need not be considered here.

with momentum $\mathbf{k}_0$ are injected into the base. $P(\mathbf{k}_0)$ contains the tunneling probability $\exp(-2\int k_x\,dx)$ and other factors discussed previously in connection with forward bias. All electrons with $\mathbf{k}_0$ such that $E_F + eV_{bp} < E_0 < E_F + eV$ can produce holes that may be collected, provided that the hole energy satisfies $E < E_F - eV_{bp}$ and the angle of incidence of the hole, θ, is within the critical angle θ_c, which is given in Eq. (20). Thus, the collector current is

$$I_c = \sum_{\mathbf{k}_0} P(\mathbf{k}_0) \sum_{\mathbf{k}} \frac{\hbar R(\mathbf{k})}{\Gamma_{ee}}, \tag{28}$$

or, writing the expression as an integration,

$$I_c = 2ea \int \frac{d^3\mathbf{k}_0}{(2\pi)^3}\, v_{0x} D(E_x) \int \frac{d^3\mathbf{k}}{(2\pi)^3} \frac{\hbar R(\mathbf{k})}{\Gamma_{ee}}, \tag{29}$$

with the constraints

$$E_F + eV_{bp} < E_0 < E_F + eV$$

$$\theta < \theta_c$$

$$2E_F - E_0 < E < E_F - eV_{bp}.$$

To simplify the numerical evaluation of Eq. (29), we take $\theta \approx \theta_c$ since $\mathbf{k}$ must be close to normal for collection within the small critical angle. We also include the possibility that the injected electrons may be specularly reflected at the metal–semiconductor interface. Due to the $\cos\theta_{\mathbf{k}\mathbf{k}_0}$ factor in the denominator of Eq. (24), electrons are more effective in producing holes with $\mathbf{k}$ pointing opposite to $\mathbf{k}_0$. Therefore, the reflected electrons make a significant contribution to I_c.

For the results discussed in this chapter, scattering spectra were taken at constant tunnel current, as in the case of the ballistic carrier spectroscopies. This is accounted for by a normalization of I_c by I_t, as previously discussed. In the notation used for Eq. (28), the tip current would be written as

$$I_t = \sum_{\substack{\mathbf{k}_0 \\ E_F < E_0 < E_F + eV}} P(\mathbf{k}_0), \tag{30}$$

which is equivalent to Eq. (7a).

Near threshold, we can easily find the dependence of I_c on $V - V_{bp}$. Including only the important factors, we have

$$I_c \propto \int_{E_F + eV_{bp}}^{E_F + eV} dE_0 \int_{2E_F - E_0}^{E_F - eV_{bp}} dE(E - 2E_F + E_0) \int_0^{\frac{m_t}{m}(E_F - eV_{bp} - E)} dE_t,$$

$$\propto (V - V_{bp})^4. \tag{31}$$

In contrast, the collector current in forward bias varies as $(V - V_{bp})^2$ (for notational convenience, we use $-V$ in reverse bias and $+V$ in forward bias, where V is positive in both cases). The extra power of two in the reverse bias case comes from the hole phase-space factor $E - 2E_F + E_0$ and the additional integration over E.

An expression similar to Eq. (29) may be written for the analogous process of hole injection, secondary hot electron creation, and collection with an n-type semiconductor collector. The same $(V - V_{bn})^4$ threshold dependence is found for this case.

7.3 Experimental Details

The experiments described in this chapter have been performed using a standard STM that has been modified for the particular requirements of BEEM. The most important of these requirements will be discussed in this section.

BEEM is implemented here as a three-terminal experiment; in addition to maintaining tip bias and a single sample bias, individual control of two sample bias voltages is required. This entails controlling base and collector voltages while measuring currents into each of these electrodes. The sample stage on the STM provides contact to both base and collector as shown in Fig. 11. The sample rests on three indium pads, one of which touches and provides contact to the base electrode. Contact to the collector is by spring to a back ohmic contact on the semiconductor. The arrangement requires only minor modifications to the existing STM design. The STM design has been discussed in detail previously.[21] Au tips were used for all measurements.

The I_c–V spectra were obtained at constant tunnel current using standard STM feedback techniques. The value of the tunnel current was normally 1 nA. This method has the advantage of linearization of the acquired spectra, as mentioned in the theoretical discussion of Section 7.2.1. All imaging was also performed at constant tunnel current. This maintenance of gap spacing during imaging avoids artificial variations in collector current, that would result from changes in tunnel current as the tip is scanned across the surface.

It may be mentioned here that modulation techniques are easily utilized with BEEM. There are two straightforward AC techniques for measuring BEEM spectra while under STM feedback control. The tunnel voltage may be coupled with a small AC modulation, and a lock-in amplifier can then be used for harmonic detection. In this way derivative spectra may be acquired directly, although loss of energy resolution results, due to modulation broadening. A second technique involves modulation of the tunnel gap at a constant tunnel voltage, through variation of tunnel current. Collector current I_c is then measured as a function of tunnel voltage with a lock-in

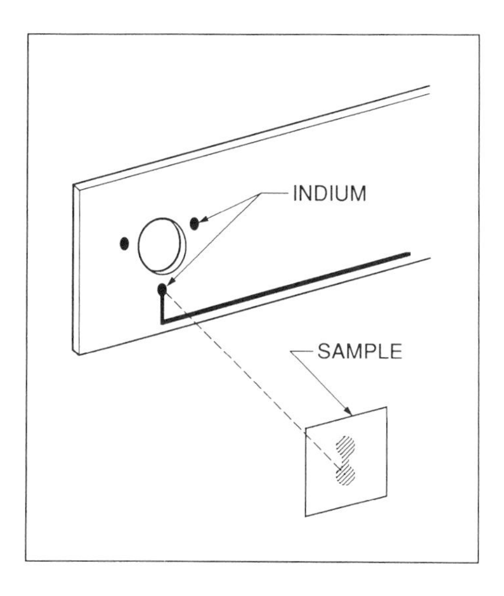

FIG. 11. BEEM sample mounting scheme. The sample rests on three indium contact pads over an aperture on a quartz mounting plate through which the STM tip protrudes. One of these pads contacts one end of the base electrode, which is evaporated in a two-lobed pattern. Tunneling is into the other lobe of the base. The collector current is measured through a back-side ohmic contact.

amplifier. Since I_c is linear with I_t, the first harmonic is proportional to I_c, rather than to its derivative dI_c/dV, so this provides a method of measuring the direct I_c–V spectrum, but with the improved signal-to-noise provided by harmonic detection.

BEEM spectra may also be obtained open-loop; that is, at constant tunnel gap spacing. In this case, also, the collector current spectrum should be normalized by the tunnel current spectrum, in order to remove lowest-order effects due to spectral structure in the tunneling characteristic.

Due to the necessity for measuring small collector currents, a high-gain, low-noise current preamplifier is used, a schematic of which is shown in Fig. 12. The amplifier provides a gain of 10^{11} volts/amp in four stages. The reference input of the amplifier is attached to the base electrode and is maintained at ground potential. Collector current is measured with zero applied bias between base and collector. The effective input impedance of the amplifier is about $10\,\Omega$, which is much smaller than the zero bias resistance R_0 of the diode, which, for reasons discussed later, is always greater than about $100\,k\Omega$. This low input impedance prevents leakage back into the base of electrons that enter the collector. Note that a measurement of collector current by the use of a series resistor would require such a resistor to be at

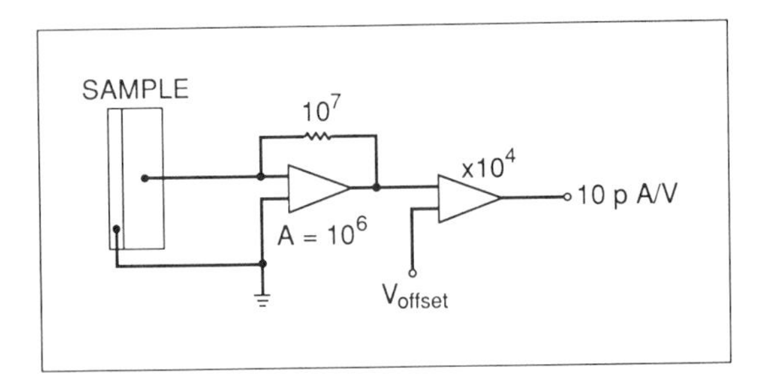

FIG. 12. Schematic diagram of the preamplifier circuit used for measurement of collector current. An electrometer-grade amplifier, the Burr–Brown OPA–128, is used for the initial gain stage. Total gain is 10^{11} volts/amp.

least $10^8\,\Omega$ for adequate sensitivity; this large resistance could cause difficulties due to this same leakage back across the interface.

The current amplifier has an inherent input noise, necessitating a large sample source impedance across its terminals. An amplifier input noise of $100\,\text{nV}/\sqrt{\text{Hz}}$ across a source impedance of $100\,\text{k}\Omega$ produces a noise current of picoamps, which is on the order of the signal to be measured. R_0 must exceed this impedance value for adequate signal-to-noise. R_0 may be increased either by reducing the interface area or by a reduction in temperature. The former method is required if R_0 is low due to ohmic regions at the interface. Low temperature measurement is more effective for samples where R_0 is small due to a low interfacial barrier height and a consequent thermionic current. In the thermionic emission approximation,[22] the differential resistance at zero bias can be written

$$R_0 = \left(\frac{dV}{dI}\right)_{V=0} = \left(\frac{eA^*Ta}{k_B}\right)^{-1} \exp(eV_b/k_B T), \tag{32}$$

where A^* is the Richardson constant and a is the diode junction area. Diode areas are approximately $0.1\,\text{cm}^2$; this requires, at room temperature, a Schottky barrier height of at least 0.75 eV. A reduction in temperature from 293 K to 77 K lowers this value to about 0.2 eV. This dependence of lowest measurable barrier height on temperature is plotted in Fig. 13. To obtain equivalent capability for low barrier height measurements, a reduction in diode area by more than a factor of 10^{10} would be required. In addition to the increase in resistance, low-temperature operation provides increased energy resolution for interface spectroscopy, due to a narrowing of the Fermi edge of the tip. A decrease in Johnson noise of the large tunneling gap resistance also results.

A low-temperature BEEM apparatus, designed for operation at 77 K, was developed for use with low-barrier-height interface systems.[12] This includes

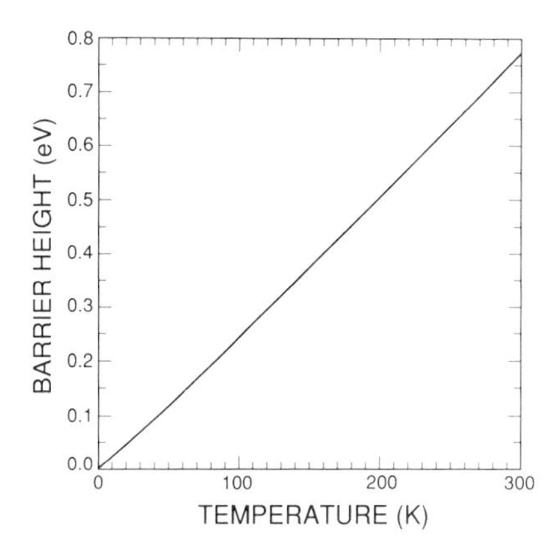

FIG. 13. Minimum measurable barrier height versus temperature calculated from Eq. (32) for n–Si(100). The criterion for measurement by BEEM is $R_0 = 100\,\mathbf{k\Omega}$, as discussed in the text. For this calculation, $A^* = 252\,\text{A/K}^2/\text{cm}^2$ and the diode area is $0.1\,\text{cm}^2$.

most important p-type Schottky barriers; 77 K operation was therefore required for ballistic hole spectroscopy. All p-type Schottky barrier characterization discussed in this chapter was performed at 77 K; for purposes of comparison and for improved energy resolution, selected n-type samples were also characterized at low temperature. Operation at 77 K was accomplished by direct immersion of the STM head in liquid nitrogen, with the entire BEEM apparatus enclosed in a nitrogen-purged glove box.

Sample substrates consisted of both Si ($n = 2 \times 10^{15}\,\text{cm}^{-3}$ or $p = 3 \times 10^{15}\,\text{cm}^{-3}$) and GaAs ($n = 3 \times 10^{16}\,\text{cm}^{-3}$) wafers of (100) orientation. MBE-grown GaAs(100) layers ($n = 5 \times 10^{16}\,\text{cm}^{-3}$ or $p = 3 \times 10^{16}\,\text{cm}^{-3}$), 1 μm thick, were also used. Si substrate cleaning was by growth and strip of a sacrificial thermal oxide followed by a growth of a 100 Å-thick gate oxide. GaAs substrate cleaning consisted of solvent rinsing followed by three chemical oxide growth–strip cycles, terminating with the growth of a protective oxide layer. Final sample preparation was performed in a flowing nitrogen environment, using a non-aqueous spin-etch for removal of oxides prior to metal base layer deposition. Samples were transferred directly into ultra-high vacuum without air exposure. All samples discussed here utilize Au base electrodes. Chamber base pressure was 10^{-11} Torr, and was typically 10^{-9} Torr during Au deposition. Unless otherwise stated, Au thickness was 100 Å as determined by crystal oscillator thickness monitor.

Completed samples were characterized by conventional I–V and transferred by load lock into a glove box that is under constant purge by dry flowing nitrogen, where measurements were performed. The glove box also serves to shield the apparatus from light. Light shielding is important during

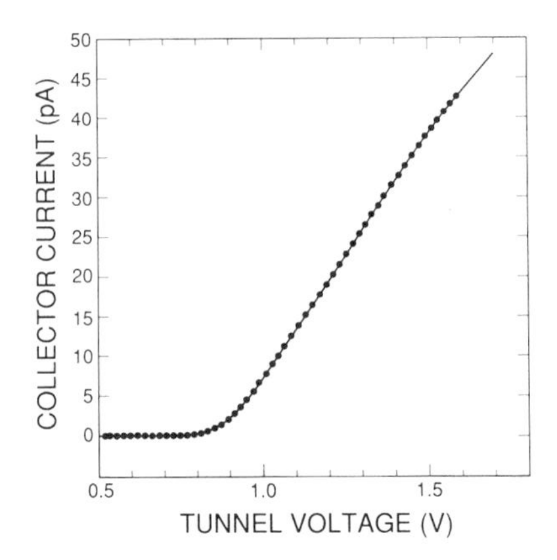

FIG. 14. BEEM I_c–V spectrum (circles) obtained for Au/n–Si(100). Also shown is a fit (line) using Eq. (13). Here, tunnel voltage refers to $V_{sample} - V_{tip}$. Fixed values of $\Phi = 3\,\text{eV}$ and $s = 15\,\text{Å}$ were used for the tunnel barrier. A threshold $V_b = 0.82\,\text{V}$ was derived from the fit to this spectrum.

data acquisition, since photocurrents generated by normal laboratory lighting can be orders of magnitude larger than currents due to collection of tunneling electrons. Checkout of the collector current circuitry and sample contacts is conveniently accomplished by admitting a small amount of light into the box.

7.4 Results

7.4.1 Ballistic Electron Spectroscopy

BEEM and its related spectroscopies have been applied to many different interface systems. Examples given in this section will illustrate important aspects of BEEM capabilities.

The simplest application of BEEM is to the Au/n–Si(100) SB interface system. Au/Si is important from a device standpoint, and it is known to provide high-quality interfaces and reproducible barrier heights. Figure 14 illustrates a representative I_c–V spectrum. A theoretical spectrum that was fit to the data is also shown. The agreement between theory and experiment is excellent; fitting was performed by varying only V_b and R, a scaling factor that includes attenuation in the base. The vacuum gap parameters used for all Au/n-type semiconductor BEEM data are $\Phi = 3\,\text{eV}$, $s = 15\,\text{Å}$ and are not varied during fitting. These values may not be precise, since the theoretical spectrum is not sensitive to small changes in Φ or s; in addition, non-parabolicity of the semiconductor conduction band should add additional curvature to the BEEM spectrum at higher voltages, which may appear as a change in

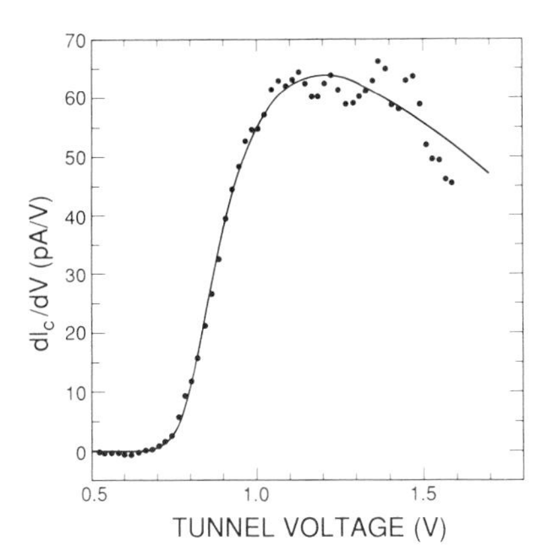

FIG. 15. Derivatives dI_c/dV of the experimental and theoretical spectra of Fig. 14. Comparison of derivatives provides a sensitive test of agreement between theory and experiment.

the effective barrier parameters. These considerations, however, do not change the quality of the fit to the threshold shape or location. Only the on-axis X minimum is considered when calculating theoretical spectra; the angles of the critical cones for the off-axis minima are so large (about 43 degrees) that the tunneling formalism that is used[15] does not provide appreciable current into these large angles.

It can be seen that the threshold shape is especially well fit by theory, and that the predicted quadratic behavior is present. A much more sensitive test of this agreement may be performed by a comparison of the derivatives, dI_c/dV, of the experimental and theoretical spectra. These are shown in Fig. 15 for the spectra in Fig. 14. This agreement provides a powerful test of the assumptions made in the BEEM theory. In particular, k_t conservation dictates a quadratic threshold. One possible interpretation of the removal of this conservation law at the interface is that all carriers of sufficient total energy can be collected. The shape of the spectrum generated by such an assumption is drastically different from that of the conventional theory. In particular, such a spectrum displays a *linear* threshold, in contradiction with experiment. For comparison, the best fit of such a spectrum to the data of Fig. 14 is shown in Fig. 16; the disagreement is pronounced.

Several observations may be made concerning the spatial variation of Au/n-Si(100) spectra. Many individual spectra were obtained for many samples, reflecting a small spread of Schottky barrier heights, from 0.75 eV to 0.82 eV. The variation in R values was also small, with differences between spectra of no more than a factor of three. Some level of variation in the magnitude of I_c (of which R is a measure) is expected, from variations in base

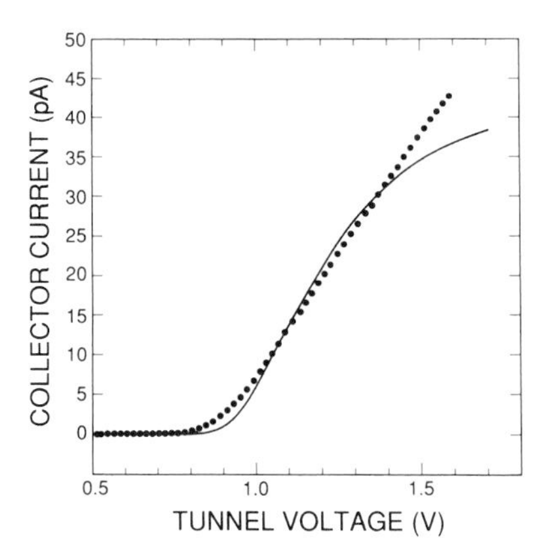

FIG. 16. Best theoretical fit to the data of Fig. 14, under the assumption that k_t is not conserved across the metal–semiconductor interface, and moreover that this implies that all electrons with sufficient energy ($E_{base} > E_F + eV_b$) may be collected. Such an assumption produces a spectrum with a linear threshold, in strong disagreement with the observed spectral threshold.

electrode thickness from point to point. This homogeneity has been probed directly by BEEM interface imaging, and will be discussed in the next section.

One capability which BEEM provides is the ability to determine attenuation lengths of hot carriers in metals and semiconductors. This method provides a straightforward way of determining attenuation lengths that are not weighted toward the thinnest base regions, as would be the case for conventional measurements on large area interfaces. The simplest experiment consists of measuring collector current as a function of base thickness. The effective attenuation length is then given by

$$I_c = I_0 e^{-t/\lambda}, \tag{34}$$

where t is the base thickness and λ is the effective attenuation length. Complete I_c–V spectra of Au/n–Si(100) were obtained and fits performed to obtain R values, which characterize the intensity of the spectra and provide a measure of attenuation in the base. A plot of R versus t is shown in Fig. 17. The expected exponential relationship is apparent, and the derived attenuation length is 128 Å. To obtain the data points that are shown, the R values of many spectra were averaged logarithmically (due to the exponential dependence of Eq. (26)) at each thickness and different samples of each thickness were fabricated, in order to account for spatial variations of base thickness about its average value. Note that this yields a true average that does not weight thin base regions more heavily than others.

While Au/Si provides an interface that is relatively well-characterized, the Au/n–GaAs(100) interface system is considerably more complex. In addition

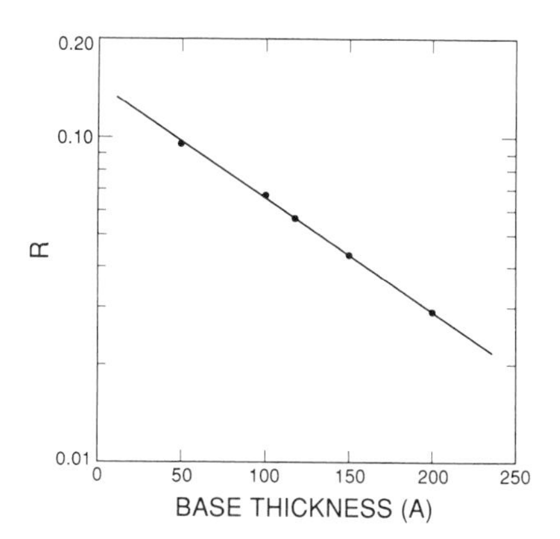

FIG. 17. Plot of collector current, parameterized by R value, for Au/n–Si(100) samples of different Au thicknesses. An exponential R-versus-thickness dependence should appear as a straight line. An effective attenuation length $\lambda = 128$ Å is derived from the slope of this line.

to the direct minimum at Γ, the lowest conduction band has two satellite minima, at the L and X points of the Brillouin zone.[23] A representative BEEM spectrum and fit to theory are shown in Fig. 18. In this case, the data is fit with a three-threshold model. Because the Γ, L, and X points lie at different locations within the Brillouin zone, the current attributable to each depends sensitively on the electron **k** distribution at the interface. Since this is not precisely known, three separate R values are used for the three thresh-

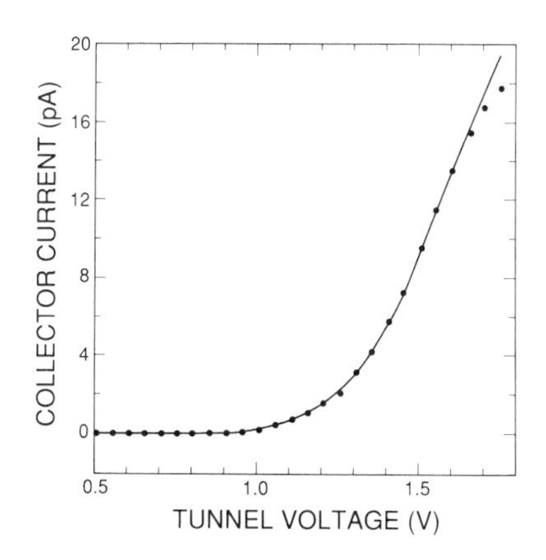

FIG. 18. BEEM I_c–V spectrum (circles) for Au/n–GaAs(100). Plotted tunnel voltage is equivalent to $V_{sample} - V_{tip}$. Also shown is a fit of a three-threshold model to the data. Three different V_b parameters and three separate R values are allowed to vary during the fit. Fixed values of 3 eV and 15 Å were used for Φ and s. The thresholds derived from the fit are at 0.89 V, 1.18 V, and 1.36 V.

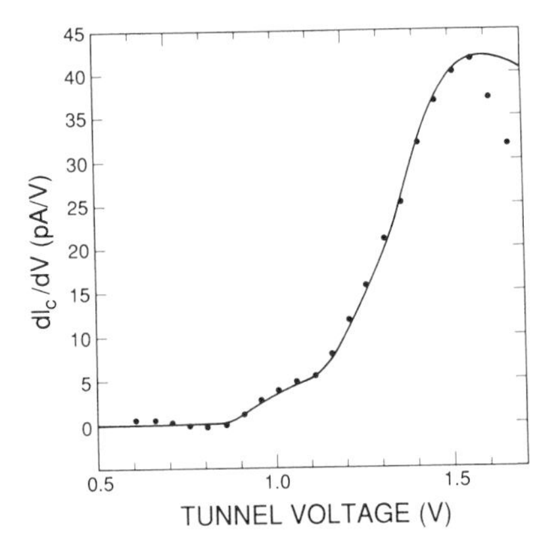

FIG. 19. Derivatives dI_c/dV of the experimental and theoretical spectra of Fig. 18. The multiple-threshold nature of the data can clearly be seen.

olds and are allowed to vary independently. Vacuum barrier parameters $\Phi = 3\,\text{eV}$ and $s = 15\,\text{Å}$ are also used for this case.

The multiple-threshold nature of the Au/n–GaAs spectrum is clear; the derivative of data and theory, given in Fig. 19, makes the three thresholds even more apparent. A free fit of the threshold locations yields $E_\Gamma - E_F = 0.89\,\text{eV}$, $E_L - E_F = 1.18\,\text{eV}$, and $E_X - E_F = 1.36\,\text{eV}$. These values are in agreement with the accepted relative locations $E_\text{L} - E_\Gamma = 0.29\,\text{eV}$ and $E_X - E_\Gamma = 0.48\,\text{eV}$.[23] The relative R values for the three minima indicate a somewhat wider angular distribution than planar tunneling predicts, since the critical cones for the L minima are at about 36 degrees (in the base). In performing the theoretical fit to the spectrum, the L minima were assumed to be on axis; treating these minima in the correct way resulted in much less current than was observed. This observation of a large contribution to the collected current by states with large values of k_t may be due to incorrect modeling of the tunneling process by the use of a planar theory. An alternative explanation may involve elastic scattering of the electrons in the base or at the interface, although the expected R values are difficult to calculate without knowing details of the scattering.

Reproducibility of interface characteristics in the Au/GaAs system is known to be difficult. This difficulty may be reflected in the fact that the spatial variation of BEEM spectra for Au/n–GaAs(100) interface is much greater than for Au/n–Si(100). Measured SB heights for Au/GaAs range from 0.8 eV to 1.0 eV.[9] In addition, the range of R values is as much as two orders of magnitude, much greater than can be explained by a variation in Au thickness. This heterogeneity was also probed by interface imaging, and will be addressed in the following section.

7.4.2 Interface Imaging

Interface imaging is made possible by the capability of the STM to scan a tip across the sample, and by the highly localized electron beam that is injected by the tip. The correspondingly high resolution at the interface has already been discussed theoretically. An interface image is acquired simultaneously with an STM topograph by measuring I_c at each point on the surface during the scan, while at a tip bias voltage in excess of threshold. Examples of the experimental results will be presented here.

It was mentioned in the last section that I_c–V spectra for Au/n–Si(100) shows only a small variation from point to point. This variation was probed directly by BEEM imaging, and a representative surface topograph and interface image for this system are shown in Fig. 20. The average value of I_c over the image is 18 pA, while the RMS variation is only 0.7 pA, which is on the order of the noise level in the acquisition apparatus. This image therefore represents an extremely uniform interface. Spectra were acquired at several locations across the image, and these spectra are plotted in Fig. 21, with their respective locations indicated in the inset. The average value of I_c for these spectra at the imaging voltage of 1.5 V is 18.0 pA, in agreement with the image average. It is notable that the dependence of the collected current on the surface Au topography is quite weak.

In marked contrast to this uniformity is the heterogeneity observed for the Au/n–GaAs(100) interface, for which a typical image pair is shown in Fig. 22. The variation from dark to light in the interface image represents about two orders of magnitude in collected current; the precise ratio is difficult to determine, since there is virtually no detectable current within the darkest areas. This large range of intensity values is typical of this interface system, and is too great to be explained by a simple thickness variation of the base electrode. Moreover, the variation in intensity is due primarily to a variation in R rather than to a change in threshold. Spectra taken within lower current areas, where there is enough signal to determine thresholds, do not indicate a systematic relationship to threshold position.

It has been demonstrated[24] that interfaces formed both on melt-grown and MBE-grown GaAs substrates exhibit this heterogeneity, indicating that bulk defect density does not play a large part in the presence of this interface disorder. The heterogeneity also persists over a wide range of surface preparation conditions. Chemically cleaned GaAs substrates that were exposed to air prior to Au deposition exhibit this behavior; a careful chemical cleaning of the GaAs substrate in flowing nitrogen gas followed by direct transfer to the Au deposition chamber also produces such interfaces, even though photoemission spectroscopy shows them to be oxide-free.[24] This is an indication

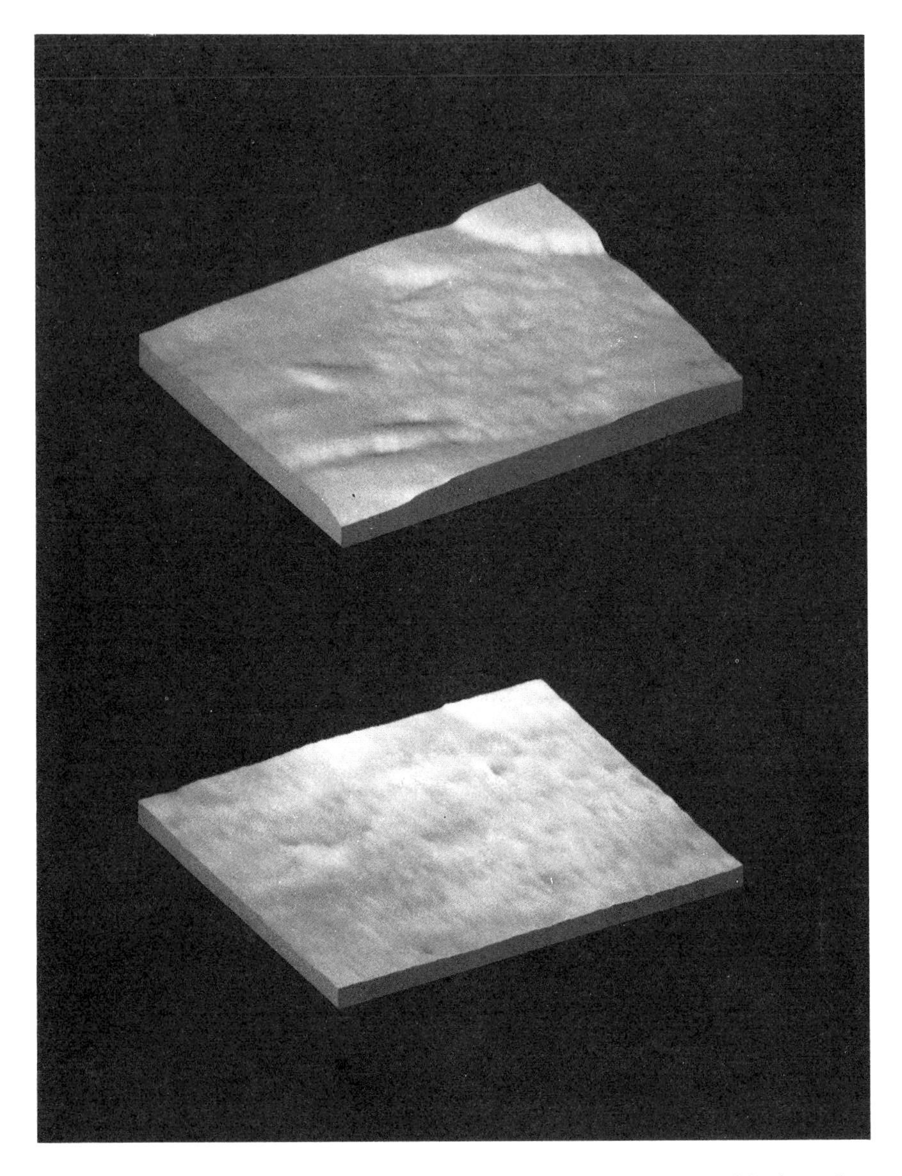

FIG. 20. Images of Au/n–Si(100). (a) Conventional STM topographic image of the Au surface of the sample structure. The image is presented in a light-source rendering. Image area is 510 Å × 310 Å. Surface height range from minimum to maximum is 80 Å. (b) Corresponding BEEM collector current image of the same sample area, obtained at a tunnel bias of 1.5 V and I_t = 1.0 nA. Collector current is shown in grey scale, with largest currents in white. Average current across the image is 18 pA. This image illustrates the uniformity characteristic of the Au/n–Si(100) interface; RMS variation in current is only 0.7 pA, which is on the order of the noise level of the current measurement apparatus.

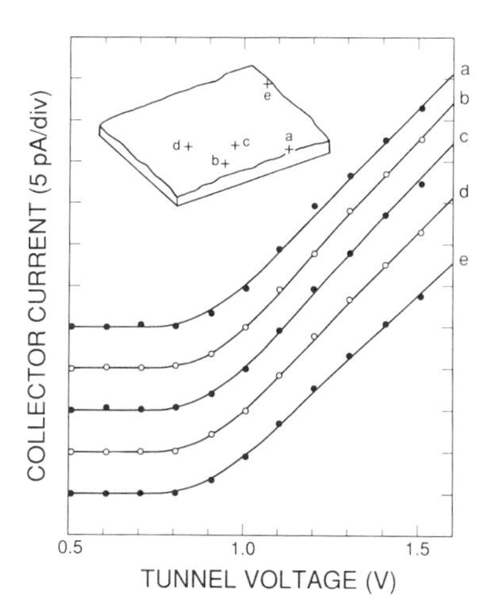

FIG. 21. BEEM spectra I_c versus $V_{sample} - V_{tip}$ obtained at the indicated locations across the image shown in Fig. 20. The spectra are offset for clarity. Average I_c for these spectra, at the tunnel voltage of 1.5 V used to obtain the BEEM image of Fig. 20, is 18 pA, in agreement with the overall BEEM image average. (Copyright 1989 by the AAAs.)

that the interface defect structure is not simply the result of surface contamination.

GaAs is known to dissociate at the interface of a Au/GaAs contact;[25] while the Ga is soluble in Au and tends to migrate to the Au surface, As is insoluble and remains at the interface. The low-current areas actually dominate the Au/GaAs interface, and are interpreted in terms of interfacial islands of As created by GaAs dissociation and Ga migration.[26] The experimental results indicate that this diffusion process dominates the interface formation process in the Au/GaAs(100) system.

This proposed diffusion mechanism was tested by the fabrication of samples with enhanced stability against diffusion.[24] This stability was achieved by incorporating a thin AlAs diffusion barrier at the Au/GaAs interface. AlAs is a good candidate for such a barrier, since it can be deposited epitaxially on a GaAs substrate. In addition, by utilizing precisely calibrated MBE growth techniques, extremely thin, continuous AlAs layers can be grown. For the present work, MBE was used to grow a GaAs buffer layer on a GaAs substrate, which was then annealed in the As flux to promote surface smoothing. This was followed by RHEED-monitored deposition of two monolayers (one unit cell) of epitaxial AlAs(100). The Au base electrode was then deposited to complete the sample. The entire sample was fabricated without exposure to air.

An STM topograph and BEEM image are shown in Fig. 23 for a Au/AlAs/

FIG. 22. Surface topograph–BEEM image pair for the Au/n–GaAs(100) structure, from Ref. 24. The images are presented as in Fig. 20. Image area is 510 Å × 390 Å. (a) Topograph of the Au surface. Range of height across this image is 63 Å. (b) BEEM grey-scale interface image, obtained at $V = 1.5$ V and $I_t = 1.0$ nA. Collector current ranges from less than 0.1 pA (black) to 14 pA (white).

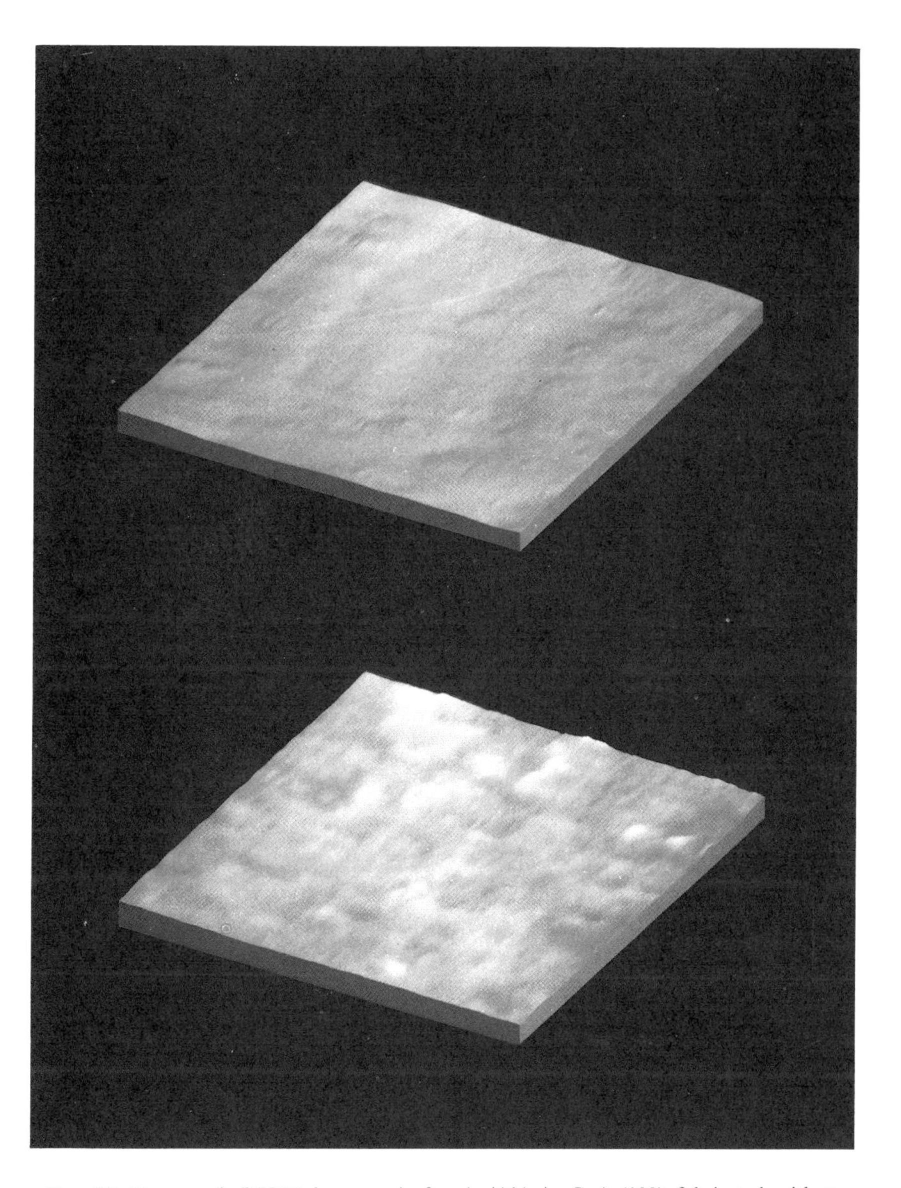

FIG. 23. Topograph–BEEM image pair for Au/AlAs/n–GaAs(100) fabricated with two monolayers of AlAs. Image area is 510 Å × 390 Å. (a) Topograph of the Au surface. Range of height in this image is 24 Å. (b) BEEM interface image, obtained at $V = 1.5$ V and $I_t = 1.0$ nA. Average collector current across the image is 2.0 pA; RMS variation is 0.7 pA, the same as for the Au/Si image of Fig. 20. From Ref. 24.

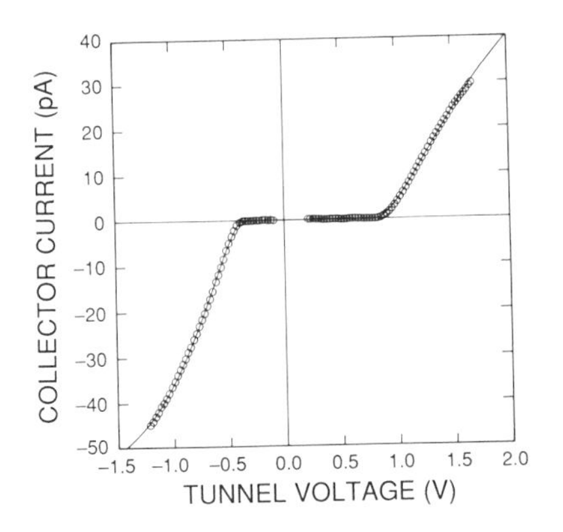

FIG. 24. Ballistic hole spectrum of Au/p–Si(100) (circles), with fit to theory (line), is shown in the lower left quadrant of the figure. Tunnel voltage refers to $V_{sample}-V_{tip}$. This data, and all ballistic hole data, was obtained at 77 K. The fit yields a threshold $V_b = 0.35$ V for the Schottky barrier. As described in the text, the values 3 eV and 8 Å were used for Φ and s. Also shown for comparison in the upper-right quadrant is a ballistic electron spectrum for Au/n–Si(100). It should be emphasized that these two spectra are obtained from two different samples. Adapted from Ref. 12.

GaAs structure. The BEEM image indicates that the heterogeneity present in the Au/GaAs system has been drastically reduced, and the large areas displaying no detectable collector current have been eliminated.

7.4.3 Ballistic Hole Spectroscopy

Ballistic hole spectroscopy was first performed on the SB systems that were previously probed by BEEM, allowing a characterization of the interface valence band structure of these systems. As mentioned in the theoretical discussion, a p-type semiconductor collector is used for ballistic hole spectroscopy, and the typically low barrier heights for p-type semiconductors necessitate low-temperature measurements. A ballistic hole spectrum for Au/p–Si(100) obtained at 77 K is illustrated in Fig. 24. Also shown is a fit of the ballistic hole theory to the data. The sign of the collected current is not arbitrary; a negative sign for the current indicates the collection of holes. The Schottky barrier height eV_b for this system is measured to be 0.35 eV, with only a small spread in this value from point to point. It should be mentioned that the n and p barrier heights measured at 77 K add to 1.19 eV, which agrees well with the Si 77 K bandgap value of 1.16 eV.

Derivatives of both the electron and hole spectra, both obtained at 77 K, are given in Fig. 25. The derivative of the ballistic hole spectrum maximizes and turns over more quickly than that of the electron spectrum. This can be interpreted in terms of the asymmetry between the collected distributions for

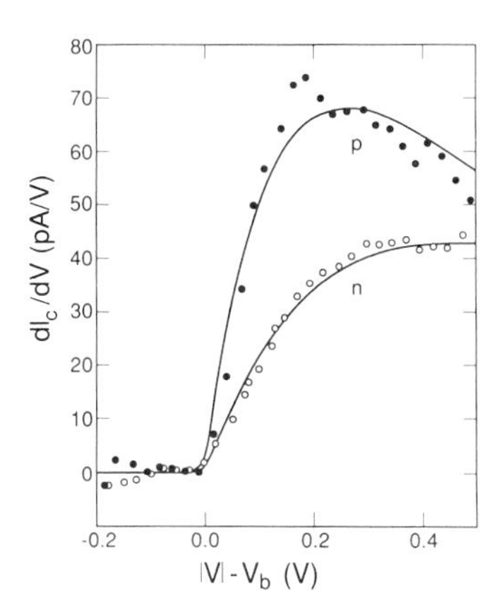

FIG. 25. Derivatives dI_c/dV of the ballistic hole experimental and theoretical spectra of Fig. 24. Also shown for comparison are equivalent derivatives of ballistic electron spectra for Au/n–Si(100), also from Fig. 24. The electron and hole spectra are plotted with a horizontal displacement such that the thresholds coincide. Both experimental spectra were obtained at 77 K. Adapted from Ref. 12.

electron and hole injection. For the hole case, the additional current per unit voltage is decreasing, causing the I_c–V spectrum to inflect more quickly. A second feature to notice is the extremely sharp thresholds in the derivatives, which is provided by the low temperature environment.

For this and other ballistic hole spectra, R values are larger than for the corresponding electron spectra. In addition, the best fits to the data are obtained for the vacuum gap parameters $\Phi = 3\,\text{eV}$ and $s = 8\,\text{Å}$, whereas the values $\Phi = 3\,\text{eV}$, $s = 15\,\text{Å}$ were used for electron spectra. This may be an indication that the tunneling formalism being used does not accurately describe the energy and angular distributions of tunneling electrons. Ballistic hole spectroscopy is more sensitive to the details of the tunneling distribution than is ballistic electron spectroscopy, since the collected carriers come from the tail of the tunneling energy distribution. However, as with BEEM spectroscopy, these considerations do not affect the quality of the fit near threshold or the determination of the interface barrier height.

Although the conduction bands of Si and GaAs are quite different, the valence band structures are similar.[23,27] For both semiconductors, the valence band system is composed of three separate bands, the light-hole, heavy-hole, and split-off bands. The light- and heavy-hole bands are degenerate at the center of the Brillouin zone and define the p-type Schottky barrier; away from the zone center, the mass of the light hole band increases until the two bands are of essentially equal mass, but split by a small energy difference. The third band is non-degenerate at the zone center due to spin-orbit splitting.

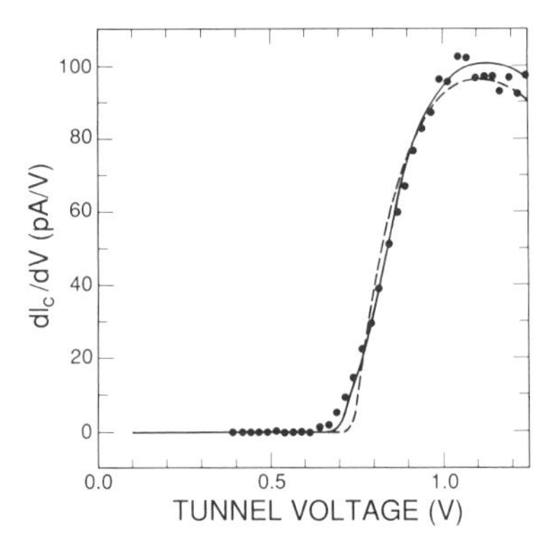

FIG. 26. Derivative dI_c/dV of a ballistic hole spectrum of Au/p–GaAs(100) (circles), with tunnel voltage representing $V_{sample} - V_{tip}$. Also shown are derivatives of two theoretical fits to the data. The dashed line represents the best fit obtainable using a single-threshold model; the disagreement in the threshold region is pronounced. The solid line illustrates a two-threshold fit. As described in the text, the model used for this spectrum incorporates a heavy-hole band, with constant mass 0.51 m, and a light hole band, with mass 0.082 m near the zone center and mass 0.51 m away from zone center. The transition between these masses is taken as abrupt, and the point of transition produces the second threshold. Adapted from Ref. 12.

In the case of Au/p–Si(100), this light-hole/heavy-hole splitting away from $\mathbf{k} = 0$ is only about 30 meV,[27] and is not resolved with the present apparatus; however, this is not the case for Au/p–GaAs(100). A derivative spectrum dI_c/dV for Au/p–GaAs(100) is shown in Fig. 26. Also plotted are two theoretical curves. The first considers only a single threshold at the valence band edge; the second includes the light-hole band, with the approximation that this band changes abruptly from its light-hole mass to the heavy-hole mass away from the zone center. The point of transition is taken to determine the second threshold in the spectrum, and was left as a free parameter. The relative intensities, or R values, were also allowed to vary independently. It is apparent that this second fit agrees well with the threshold shape of the experimental spectrum, while the first fit is poor at threshold. The measured barrier height is 0.70 eV, while the effective splitting of the light- and heavy-hole band away from the zone center is about 100 meV, in good agreement with current values of this splitting obtained by other methods.[23]

It should be noted that the simple phase space model that has been used to analyze electron and hole spectra thus far would not predict the observation of the light hole band in the Au/p–GaAs data. In performing the fit illustrated in Fig. 26, the contributions of two different channels were simply added together. However, the phase space for collection by the light-hole band is completely within that for the heavy-hole band; therefore, there is no additional phase space for collection when the onset of the light-hole band is reached. The assumption of the phase-space model, that all carriers within the critical cone are collected, is therefore inappropriate here; this would not allow additional current due to the light-hole band. The additional current is expected, however, if there is quantum-mechanical reflection at the interface that allows only a portion of the carriers within the critical cone to be collected. A second channel for collection would then produce an increase in current. The particular form of the QMR depends on the potential profile and mass change across the interface. In fact, an abrupt mass change alone will produce a constant QMR term, even in the absence of a change in potential. The observation of both light and heavy hole bands is most simply explained by noting that one channel (heavy hole) involves only holes with $|m_j| = 3/2$ while the other (light hole) corresponds to $|m_j| = 1/2$. The observation of both bands in the ballistic hole spectrum, then, is a direct indication of QMR for this interface system.

A footnote to this discussion is that the split-off bands for both Si and GaAs should, in principle, be observable. However, the masses of these bands are smaller than those of the heavy-hole bands, and the increase in current due to them is difficult to detect. Their contribution to the I_c–V spectrum is not observed, but AC detection methods may resolve these bands.

7.4.4 Scattering Spectroscopy

The first application of the carrier scattering spectroscopy discussed previously was to the interface systems that have been investigated by BEEM: Au/Si(100) and Au/GaAs(100).[13] The importance of these interfaces has already been emphasized; a characterization of transport within the electrodes themselves is also of great importance to a complete understanding of the entire transport process. In addition, analysis of the results of this scattering spectroscopy is aided by the previous characterization of interface transport and the collection process, provided by the ballistic spectroscopies that have already been discussed.

Figure 27 displays two spectra obtained for Au/p–Si(100). The spectrum at positive tip voltage is a ballistic hole spectrum similar to that in Fig. 24, with a threshold that yields a Schottky barrier height of 0.35 eV. Shown at negative tip bias is a spectrum of holes created by carrier scattering. In this case, the negative tip voltage injects ballistic electrons into the sample, some of which scatter with equilibrium electrons in the base and result in the creation of secondary electron-hole pairs. The secondary electrons eventually decay to the Fermi level, but some of the secondary holes may be collected by the Si valence band before they thermalize. The spectrum shown is due to collection of these holes. Also shown is an expanded version of this scattering spectrum. It is clear that the spectral shapes and the magnitudes of the

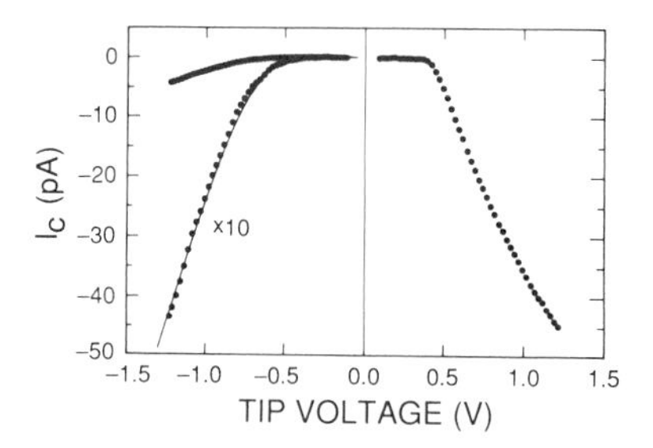

FIG. 27. Carrier–carrier scattering spectrum obtained for the Au/p–Si(100) system, adapted from Ref. 13. Also shown at positive tip bias is a ballistic hole spectrum for the same sample. For clarity, the scattering spectrum is also plotted on an expanded scale, with a fit (line) to the theory described in Section 7.2.3. The value of $V_b = 0.35$ V, obtained from the ballistic hole spectrum, was used as a fixed parameter for the fit; Φ and s were fixed at 3 eV and 15 Å. The only adjustable parameter was overall collector current magnitude.

currents are quite different for the ballistic and scattering spectra. A theoretical curve is also shown, which was fit to the data by adjustment of only an overall scaling factor reflecting the magnitude of the current. Agreement with experiment is excellent. The scattering spectrum also exhibits a threshold at approximately 0.35 eV, the value of the Schottky barrier height for this system. This observation of collected current well below the Si conduction band edge ($eV_b = 0.82$ eV) rules out any processes involving transport in the Si conduction band.

A similar experiment may also be performed for Au on p–GaAs(100). Since the barrier height for this interface is roughly twice that of Au/p–Si, it provides a good test of theoretical description developed for the process. As shown in Fig. 28, the spectra are qualitatively similar to those for Si. This is

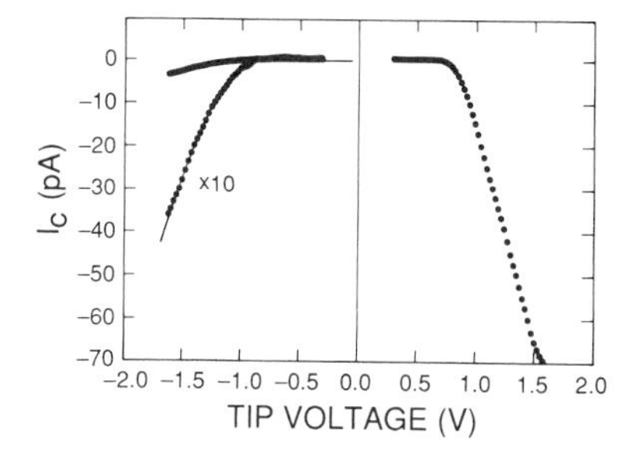

FIG. 28. Scattering spectrum obtained for the Au/p–GaAs(100) system. Also shown at positive tip bias is a ballistic hole spectrum for the same sample. For clarity, the scattering spectrum is also plotted on an expanded scale, with a fit (line) to the spectrum. V_b was fixed at 0.70 V, the value obtained from the ballistic hole spectrum. Φ and s were fixed at 3 eV and 15 Å. From Ref. 13.

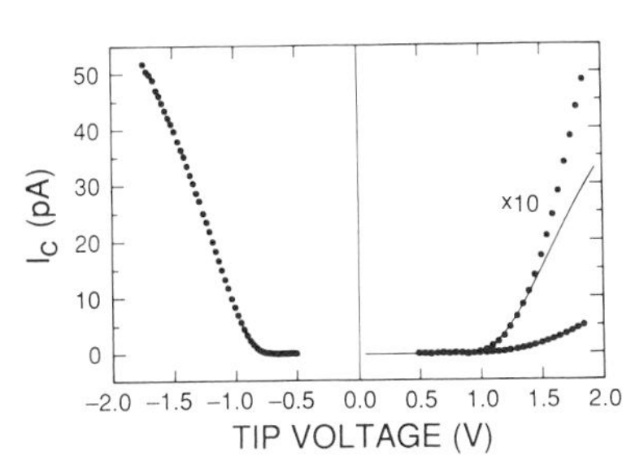

FIG. 29. Scattering spectrum obtained for a Au/n–Si(100) sample, from Ref. 13. Also shown at positive tip bias is a ballistic electron spectrum for the same sample. For clarity, the scattering spectrum is also plotted on an expanded scale, with a fit (line) to the spectrum. V_b was fixed at 0.82 V, the value obtained from the ballistic electron spectrum. In this case, only data at tip voltages less than 1.4 V were used for the fit. Φ and s were fixed at 3 eV and 8 Å.

to be expected, since this spectroscopy is primarily a probe of processes within the base layer, independent of the collector. However, the measured spectra do reflect aspects of the particular collector electrode used. Most noticeably, the threshold is determined by the Au/p–GaAs Schottky barrier height, 0.70 eV. In addition, the current of scattered carriers for this system is smaller than that for Au/p–Si, for equivalent voltages in excess of threshold. This is due to the smaller fraction of phase space available for scattering and collection, and the ratio of the currents for the two systems agrees well with that predicted by theory. This agreement indicates that there is no large interface-specific contribution to the scattering process itself.

This scattering spectroscopy has also been performed for n-type collectors. Here, injected ballistic holes may scatter with equilibrium electrons in the base and create electron-hole pairs, and the secondary electrons are collected in the conduction band of the n-type semiconductor. Figure 29 shows ballistic and scattering spectra for Au/n–Si(100). The threshold reflects the n-type Schottky barrier height of 0.82 eV. In this case, there is good agreement of the theory with the data near threshold; however, the measured collector current increases more rapidly at higher voltages than the theory predicts. The overall magnitude of the current is also high, about two orders of magnitude higher than expected from the theory when only the on-axis conduction band minimum is included. Preliminary Monte Carlo calculations indicate that the inclusion of the off-axis conduction band minima raises the predicted current by an order of magnitude, reducing the discrepancy somewhat. This is due to the fact that the secondary carrier distribution produced by the inelastic scattering process is much less strongly peaked than the incoming distribution, and provides significant current at larger angles. Monte Carlo studies also indicate that the inclusion of elastic scattering, including phonon scattering, reduces the discrepancy still further. The remaining disagreement may indicate that the initial electron distribution is not correctly described by planar tunneling, a possibility that was also raised earlier in connection with ballistic hole spectroscopy.

7.5 Conclusions

This chapter has covered the basic concepts of ballistic carrier spectroscopies using the STM, as well as an extension to the study of non-ballistic processes. BEEM and its related techniques are extremely powerful probes of subsurface interfaces, and of carrier transport through materials and across these interfaces. BEEM provides the most precise method available for the determination of Schottky barrier heights, and in addition is sensitive to other characteristics of the interface band structure, such as effective mass and satellite conduction band minima. The precision of barrier height measurements allows the determination of small changes in barrier height due to such factors as strain or temperature change. Such measurements are one important means for evaluating theories of interface formation.

The implementation of BEEM using STM is a vital aspect of the method, since the highly localized electron injection enables a corresponding spatial resolution of interface properties. This implementation therefore provides an interface imaging capability, which has been used to reveal hitherto unknown heterogeneity at semiconductor interfaces. The theoretical description of BEEM spectroscopy has been discussed, as well as the implications of this theory for interface imaging resolution.

BEEM also provides information on fundamental questions regarding carrier transport. For instance, spectroscopy of interfaces formed on semiconductors with off-axis minima should yield insight concerning the tunneling distribution, in E and $\mathbf{k}$, of electrons in a tip-plane geometry, as well as the much-argued notion of conservation of k_t across a real interface. Two of the methods for probing carrier transport through materials have been presented in this chapter. The first, using conventional BEEM, involves the measurement of collector current as a function of base thickness and a determination of carrier attenuation lengths. This measurement may also be performed for the case of off-axis collector band minima, which should elucidate the contributions of elastic and inelastic scattering to the measured attenuation length.

The second method for the investigation of carrier transport involves the use of carrier–carrier scattering spectroscopy, or reverse–bias BEEM. This spectroscopy allows the first direct probe of carrier–carrier scatttering in materials, and can also be performed as a function of base thickness, temperature and with off-axis collectors to provide a wealth of data on carrier scattering mechanisms. The theoretical treatment for this process yields good agreement with experiment and, as in the case of BEEM theory, may be built upon to incorporate more complicated processes.

BEEM is currently being pursued by a growing number of researchers. Although too recent to be discussed in this chapter, current areas of interest may be cited here. Research continues on interface formation involving metals on the III–V and II–VI semiconductors.[28,29] The role of scattering in the transport process has also spurred both experimental and theoretical studies.[30–34] In addition, interfaces which are thought to be near-perfect, such as the silicide/silicon systems, are being investigated.[32,33,35,36] Such systems are promising for a variety of device structures, and should also elucidate the roles of bulk and interface scattering, as well as metal band structure effects, in the interface transport process. First principles theoretical work, especially emphasizing the problem of transport through epitaxial interfaces, has also been performed.[37,38]

Acknowledgment

The research described in this paper was performed by the Center for Space Microelectronics Technology, Jet Propulsion Laboratory, California Institute of Technology, and was jointly sponsored by the Office of Naval Research and the Strategic Defense Initiative Organization/ Innovative Science and Technology Office, through an agreement with the National Aeronautics and Space Administration.

References

1. L. J. Brillson, *Surf. Sci. Rep.* **2**, 123 (1982), and references therein.
2. G. Le Lay, J. Derrien, and N. Boccara, *Semiconductor Interfaces: Formation and Properties*, Springer-Verlag, Berlin, 1987, and references therein.
3. W. G. Spitzer, C. R. Crowell, and M. M. Atalla, *Phys. Rev. Lett.* **8**, 57 (1962); C. R. Crowell, W. G. Spitzer, and H. G. White, *Appl. Phys. Lett.* **1**, 3 (1962).
4. M. P. Seah and W. A. Dench, *Surf. Interface Anal.* **1**, 2 (1979).
5. G. Binnig, H. Rohrer, Ch. Gerber, and E. Weibel, *Phys. Rev. Lett.* **49**, 57 (1982).
6. R. S. Becker, J. A. Golovchenko, D. R. Hamann, and B. S. Swartzentruber, *Phys. Rev. Lett.* **55**, 2032 (1985).
7. R. M. Feenstra, W. A. Thompson, and A. P. Fein, *Phys. Rev. Lett.* **56**, 608 (1986); J. A. Stroscio, R. M. Feenstra, and A. P. Fein, *Phys. Rev. Lett.* **57**, 2579 (1986).
8. R. J. Hamers, R. M. Tromp, and J. E. Demuth, *Phys. Rev. Lett.* **56**, 1972 (1986).
9. W. J. Kaiser and L. D. Bell, *Phys. Rev. Lett.* **60**, 1406 (1988).
10. L. D. Bell and W. J. Kaiser, *Phys. Rev. Lett.* **61**, 2368 (1988).
11. C. R. Crowell, W. G. Spitzer, L. E. Howarth, and E. E. LaBate, *Phys. Rev.* **127**, 2006 (1962).
12. M. H. Hecht, L. D. Bell, W. J. Kaiser, and L. C. Davis, *Phys. Rev. B* **42**, 7663 (1990).
13. L. D. Bell, M. H. Hecht, W. J. Kaiser, and L. C. Davis, *Phys. Rev. Lett.* **64**, 2679 (1990).
14. C. R. Crowell and S. M. Sze, *Solid State Electron.* **8**, 673 (1965).
15. J. G. Simmons, *J. Appl. Phys.* **34**, 1793 (1963).
16. C. B. Duke, *Tunneling in Solids*, p. 34, Academic Press, New York, 1969.

17. E. L. Wolf, *Principles of Electron Tunneling Spectroscopy*, pp. 35–36, Oxford Univ. Press, New York, 1985.
18. S. M. Sze, C. R. Crowell, G. P. Carey, and E. E. LaBate, *J. Appl. Phys.* **37**, 2690 (1966).
19. S. Gasiorowicz, *Quantum Physics*, p. 77, John Wiley, New York, 1974.
20. J. J. Quinn and R. A. Ferrell, *Phys. Rev.* **112**, 812 (1958).
21. W. J. Kaiser and R. C. Jaklevic, *Rev. Sci. Instrum.* **59**, 537 (1988).
22. S. M. Sze, *Physics of Semiconductor Devices*, 2nd ed., p. 256, John Wiley, New York, 1981.
23. J. S. Blakemore, *J. Appl. Phys.* **53**, R123 (1982).
24. M. H. Hecht, L. D. Bell, W. J. Kaiser, and F. J. Grunthaner, *Appl. Phys. Lett.* **55**, 780 (1989).
25. P. W. Chye, I. Lindau, P. Pianetta, C. M. Garner, C. Y. Su, and W. E. Spicer, *Phys. Rev. B* **18**, 5545 (1978).
26. J. L. Freeouf and J. M. Woodall, *Appl. Phys. Lett.* **39**, 727 (1981).
27. J. S. Blakemore, *Semiconductor Statistics*, rev. ed., p. 63, Dover, New York, 1987.
28. A. E. Fowell, R. H. Williams, B. E. Richardson, A. A. Cafolla, D. I. Westwood, and D. A. Woolf, *J. Vac. Sci. Technol. B* **9**, 581 (1991).
29. M. Prietsch and R. Ludeke, *Phys. Rev. Lett.* **66**, 2511 (1991).
30. L. J. Schowalter and E. Y. Lee, *Phys. Rev. B* **43**, 9308 (1991).
31. H. D. Hallen, A. Fernandez, T. Huang, R. A. Buhrman, and J. Silcox, *J. Vac. Sci. Technol. B* **9**, 585 (1991).
32. A. Fernandez, H. D. Hallen, T. Huang, R. A. Buhrman, and J. Silcox, *Phys. Rev. B* **44**, 3428 (1991).
33. A. Fernandez, H. D. Hallen, T. Huang, R. A. Buhrman, and J. Silcox, *J. Vac. Sci. Technol. B* **9**, 590 (1991).
34. E. Y. Lee and L. J. Schowalter, *Phys. Rev. B* **45**, 6325 (1992).
35. W. J. Kaiser, M. H. Hecht, R. W. Fathauer, L. D. Bell, E. Y. Lee, and L. C. Davis, *Phys. Rev. B* **44**, 6546 (1991).
36. Y. Hasegawa, Y. Kuk, R. T. Tung, P. J. Silverman, and T. Sakurai, *J. Vac. Sci. Technol. B* **9**, 578 (1991).
37. M. D. Stiles and D. R. Hamann, *Phys. Rev. Lett.* **66**, 3179 (1991).
38. M. D. Stiles and D. R. Hamann, *J. Vac. Sci. Technol. B* **9**, 2394 (1991).

8. CHARGE-DENSITY WAVES

R. V. Coleman, Zhenxi Dai, W. W. McNairy, C. G. Slough, and Chen Wang

Department of Physics, University of Virginia, Charlottesville, Virginia

8.1. Transition Metal Chalcogenides

The transition metal chalcogenides form a large group of compounds of the type MX_2 and MX_3, where M is a transition metal such as Ta, Nb, V, and Ti, and X is a chalcogen such as Se or S. Most of these compounds can be grown as highly perfect single crystals and they do not react with oxygen. They form quasi-two-dimensional or quasi-one-dimensional crystal structures, which can be easily cleaved along given crystal planes. These characteristics make them ideal for studies with the scanning tunneling microscope (STM) or the atomic force microscope (AFM). The layer structure crystals of the transition metal chalcogenides often grow in a number of different phases depending on the growth temperature and the stability of a given phase in a specific temperature range. For example, TaS_2 and $TaSe_2$ can be grown fairly easily in the 1T, 2H, and 4Hb phases. The 1T phase exhibits octahedral coordination between the metal and chalcogen atoms and has one sandwich layer per unit cell. The 2H phase exhibits trigonal prismatic coordination between the metal and chalcogen atoms and has two sandwich layers per unit cell. The 4Hb phase has alternating sandwich layers of octahedral and trigonal prismatic coordination with a total of four sandwich layers per unit cell. In the case of the Ta compounds the 1T phases are stable in the temperature range $> 900°C$, the 4Hb phases are stable in the range of $\sim 700°C$ and the 2H phases are stable below $\sim 600°C$. The different phases can be stabilized at low temperature by quenching the crystals from the growth temperature to room temperature. Further experimental details are given in Section 8.3 and details of the crystal structures and phase diagrams for a wide range of layer structure dichalcogenides can be found in the review by Wilson and Yoffe.[1]

The STM and AFM results reported in this review cover data obtained on the 1T, 2H, and 4Hb phases of representative layer dichalcogenides. There are several other phases (polytypes) that exist for specific compounds and these are also described in Ref. 1, but have not yet been studied by STM or AFM.

METHODS OF EXPERIMENTAL PHYSICS
Vol. 27

ISBN 0-12-475972-6

The linear chain transition metal chalcogenides grow as long fibrous crystals that are characterized by chains of metal atoms surrounded by trigonal prismatic cages of chalcogen atoms. The different compounds grow with crystal structures that can be classified according to the number of chains per unit cell. The structures for the compounds studied by STM are given in detail in Section 8.6. Crystal structures of the quasi-one-dimensional crystals have also been reviewed by Meerschaut and Rouxel.[2]

In addition to the strong anisotropy exhibited by these crystal structures, many of the compounds exhibit transitions to a charge-density wave (CDW) phase. The modulation of the local density-of-states (LDOS) at the Fermi level produced by the formation of these CDWs can easily be detected by both the STM and AFM. The detailed CDW formation and structure in a wide range of the layer structure compounds have been reviewed by Wilson *et al.*[3] using electron diffraction data and transport measurements. The CDW structure observed in the quasi-one-dimensional compounds has been reviewed in a number of recent books listed as Refs. 4 and 5.

The STM and AFM data obtained on a representative group of both the dichalcogenides and trichalcogenides are presented in this chapter. The results have confirmed the CDW structures deduced from diffraction studies, and have also been able to give detailed information on the local charge distributions. In a number of cases new information has been obtained on the local structure of the CDW.

8.2. Charge-Density Wave Formation

The formation of CDWs results from the electron–phonon interaction which produces an interaction between electrons and holes at $2\mathbf{k}_F$. The CDW state is most favorable when there are a large number of states connected by the same $2\mathbf{k}_F$ wavevector. This instability in the Fermi sea is most likely to occur in quasi-two-dimensional or quasi-one-dimensional metals where the Fermi surface (FS) is cylindrical or planar since larger areas of the FS can be separated by the same wavevector. The phenomenon is generally known as FS "nesting" and it produces a long-range coherent quantum state of the electron gas in real space, which can be described as a transfer of electrons into a standing wave. The resulting charge modulation can have a wavelength that is two or more lattice vectors and can be either commensurate or incommensurate with the underlying atomic lattice.

The one-dimensional electron gas coupled to the phonon system was originally discussed by Peierls and Fröhlich. They showed that such a system becomes unstable at low temperatures and undergoes a phase transition to an insulating state. In a quasi-one or quasi-two dimensional crystal the phase transition still occurs, but need not produce a completely insulating state. In either case the transition produces a condensate of electrons such that the ground state is characterized by a charge density modulation for small lattice distortions given by

$$\Delta\rho = \rho_1 \cos(2k_F x + \Phi), \tag{1}$$

where ρ_1 is the amplitude and Φ is the phase of the electron condensate.

The ground state is, in terms of a mean field theory, similar to the superconducting ground state as described by the BCS theory. Electron-hole pairs are formed with a total momentum of $q = 2k_F$ and the energy spectrum has a gap for charge excitations from the condensate. CDW transitions can occur at relatively high temperatures compared to superconductivity. The CDW phase is generally competitive with the superconducting phase, which also involves the electron–phonon interaction. In the latter case the electron–phonon interaction pairs electrons at the Fermi level of opposite spin and momentum, and forms a long-range quantum state in momentum space. Although the presence of a CDW reduces the onset temperature for superconductivity, these two phases of the electron gas can coexist at sufficiently low temperatures as observed in $NbSe_2$ below 7 K. Both the CDW and superconducting flux structure have been studied by STM and will be reviewed in Chapter 9.

The onset of a CDW transition and the associated charge transfer is accompanied by a lattice distortion since the ions move to screen the longer range electronic charge modulation induced by the CDW. This periodic lattice distorition (PLD) can be detected by electron, neutron, or x-ray diffraction techniques.[2,3] Electron transport and susceptibility are also strongly influenced by the CDW formation due to their direct dependence on FS topology. Wilson *et al.*[3] have reviewed electron diffraction and resistivity measurements for a wide range of transition metal dichalcogenides. These techniques average over macroscopic volumes of the crystal, while the STM and AFM can look at local areas on an atomic scale. The latter techniques can therefore detect subtle electronic effects occurring over distances of one to several hundred atoms and can also confirm the CDW structure deduced from the more macroscopic measurements. The first STM observations of CDWs were reported by Coleman *et al.*[6] and a comprehensive review of STM results on a wide range of transition metal chalcogenides has been published by Coleman *et al.*[7] This chapter contains results of more complete up-to-date STM studies with substantially more detail on the amplitude of the STM response to CDWs as well as new results on spectroscopy and the determination of CDW energy gaps. New results on the first detection of CDWs with the AFM are also included.

8.3. Charge-Density Waves in Transition Metal Chalcogenides

The transition metal dichalcogenides exhibit CDWs with a variety of onset temperatures and structures depending on the particular compound and crystal phase. The strongest CDWs form in the 1T phases of $TaSe_2$ and TaS_2 which have CDW onset temperatures of ~600 K. The 2H phases of $TaSe_2$ and TaS_2 show substantially weaker CDW formation with much lower CDW onset temperatures. The mixed phase 4Hb crystals support independent CDWs in alternate sandwiches and these CDWs maintain a structure closely related to that found in the related pure phases.

The linear chain transition metal trichalcogenides also exhibit CDW transitions and several of these have been successfully studied by STM. In contrast to the layer structure crystals, the CDWs in the linear chain compounds can be set in motion by applied electric fields above some threshold value. This leads to extensive nonlinear transport effects that have been studied in great detail.[8,9] So far, only the static CDW structure in $NbSe_3$ and TaS_3 has been studied[10,11,12] by STM. Preliminary results on detecting motion of the CDW by STM have not yielded much detail as yet. In the case of $NbSe_3$ the STM results[12] have yielded substantial new information on the specific charge distributions that exist on the individual chains of the unit cell. These results have suggested that the CDWs form on all three chain pairs of $NbSe_3$ rather than on just two of the pairs as previously thought.

The CDW structure in orthohombic TaS_3 has also been well-resolved by STM studies.[11] The charge distribution on the individual chain pairs has suggested a strongly coupled charge modulation forming on more than one pair of chains. This observation has confirmed some aspects of the single CDW forming below 214 K in orthorhombic TaS_3. The STM results on linear chain transition metal trichalcogenides will be reviewed in Section 8.9.

8.4. Experimental STM and AFM Response to CDW Structures

The STM detects the LDOS at the Fermi level as measured at the position of the tip. The formation of the CDW gaps (annihilates) a portion of the FS and can substantially modify the DOS at the Fermi level. The extent to which this affects the LDOS, as measured by the tunneling current as a function of position, has not been analyzed with specific models related to the band structures in the CDW state of the various materials. However, Tersoff[13] has developed a qualitative model for the effects of FS collapse on the LDOS in semiconductors or semimetals of low dimensionality. Whenever the FS collapses to a point at the corner of the surface Brillouin zone, the STM image will, in effect, correspond to an individual state with a nodal structure corresponding to the periodicity of the unit cell involved in the FS collapse.

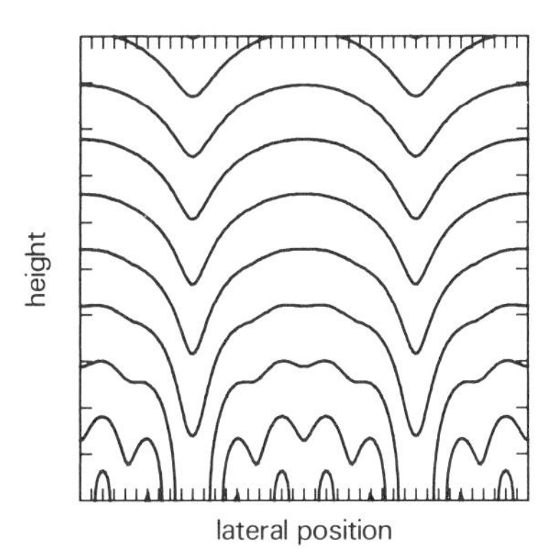

FIG. 1. Contours of constant LDOS for partial FS elimination and strong mixing of wavefunctions at $\mathbf{k}_{\parallel}$ and $(\mathbf{k}_{\parallel} + \mathbf{G})$ showing the relative amplitudes of superimposed atomic and CDW modulations. (From Ref. 7.)

In the general case, the strength of this nodal structure will depend on the degree of FS obliteration and on the strength of mixing between $\mathbf{k}_{\parallel}$ and $\mathbf{k}_{\parallel} + \mathbf{G}$ where $\mathbf{G}$ is a reciprocal lattice vector. A calculated curve for partial FS obliteration and strong mixing is shown in Fig. 1. The experimental STM data to be discussed in this chapter show a systematic variation of CDW amplitude in different materials proportional to the strength of the CDW transitions and to the degree of FS obliteration as deduced from other experiments as well as from band structure models. The Tersoff[13] model can easily be adjusted to fit this systematic trend and further examples of calculated nodal structure can be found in Refs. 7 and 14.

Other mechanisms can also play a role in the STM amplitude response, and although the experiments show a systematic CDW amplitude variation proportional to the expected CDW strength, a number of contributions can be folded into the overall STM response. The CDW is accompanied by a PLD proportional to the CDW strength. In the strongest CDW transitions this can be on the order of ~ 0.2 Å and the surface S or Se atoms will be modulated in height. This will, in turn, modulate the conduction electron wavefunction at the position of the tip, and the tunnel current will respond to the resulting variations in electron density above the surface. The STM response may also depend on the response of the tunneling matrix elements to the specific components of the conduction electron wavefunctions[15,16] above the surface. From an experimental point of view, this may also introduce a complex dependence of the observed STM amplitude on distance between the tunneling tip and the surface. This will complicate the quantitative comparison of CDW amplitude measured for a wide range of materials. By measuring many STM scans on each material, the systematic trends of CDW structure and amplitude can be established, but absolute amplitudes with a direct relation to the intrinsic CDW structure such as charge transfer, FS obliteration, and PLD have not been established.

The CDW amplitudes are measured in the constant current mode where the z-deflection of the piezoelectric element measures the height variation required to maintain a constant tunneling current as the surface is scanned. By measuring dI/dz, an effective barrier height can be calculated for a given tip–sample combination, and the low voltage approximation is given by

$$\Phi = 0.95 \left[\frac{d \ln I}{dz} \right]^2 . \qquad (2)$$

Experimental measurements[14] of Φ as a function of relative tip to sample distance show that this effective barrier is a rapid function of distance for the layer structures used in these experiments. Absolute amplitudes measured for a given material can only be systematically compared if the set points of the different tip–sample combinations can be normalized in an appropriate fashion. The effective barrier heights derived from Eq. (2) also vary by an order of magnitude for the different CDW materials, the strongest CDW materials show the lowest effective barrier heights. In addition to this intrinsic trend, anomalous changes in effective barrier heights, due to impurities or tip geometry, can influence amplitude response through the formation of zero bias anomalies even though the grey scale scans show highly resolved structure. Spectroscopy measurements taken simultaneously with the grey scale scans will detect such anomalies and allow such effects to be eliminated from any quantitative measurements. More discussion of these experimental problems can be found in Ref. 14.

The spectroscopy mode can also be used to measure the energy gap associated with the CDW and the accompanying partial or total obliteration of the FS. In most cases studied here, the materials remain as semimetals below the CDW transitions so that the FS obliteration is only partial. The gap structure observed in the I versus V or dI/dV versus V curves will, therefore, be weaker or stronger depending on the degree of FS obliteration and the resulting change in DOS at the gap energy. The spectroscopy measurements have been successful in detecting a well-defined gap structure in most of the CDW materials studied and these measurements will be systematically reviewed in the various sections. The gaps measured for most of the CDW materials indicate that the values of $2\Delta/k_B T_C$ are large, relative to the weak coupling BCS value of 3.5. The electron–phonon interaction producing the CDW transitions would therefore be characterized by a coupling constant in the strong coupling regime, rather than the weak coupling BCS model.

The AFM responds entirely differently to the CDW structure with respect to the detection of amplitude. The CDW amplitude is an order of magnitude smaller than observed in the STM scans on the same material. This is related to the fact that the AFM responds to the total charge density at the surface, which is dominated by the surface Se or S atoms. Although the STM responds to the modulation of the conduction electron wave function by the surface Se or S atoms, it also measures any change in the LDOS at the Fermi level, which is dominated by the CDW, forming in the conduction electrons originating from the metal atom layer below the surface. The observation of the order of magnitude decrease in CDW amplitude, relative to atomic amplitude for the AFM versus STM, suggests that the STM response to the CDW structure is dominated by changes in the LDOS at the Fermi level. The enhanced CDW amplitudes observed in the STM scans would therefore appear to be directly related to electronic structure modifications induced by the CDW. Use of the AFM to study CDW structure has only recently been successful and extensive data is not yet available. The preliminary results are reviewed in this chapter.

8.5. Experimental Techniques

The AFMs and STMs used in these investigations range from home-built to commercial models. Some of the home-built STMs are designed to operate in a cryogenic liquid environment. Operating with the microscope submersed in a cryogenic liquid has produced a high degree of thermal stability, and has greatly aided in the detection of the many subtle electronic effects observed. The cryogenic STMs are all based on a similar design described later. A tube microscope design,[17] also described here, is the basis for the AFMs and some of the room temperature STMs.

For the cryogenic STMs two separate piezoelectric elements are used to produce the necessary X, Y, and Z motions. An X-Y translator, shown in Fig. 2(a), is used to produce the X-Y motion of the tip. The translator is formed by cutting two triangular sections from a single block of piezoelectric material, so as to define the X and Y arms. A 0.5 mm $Pt_{0.8}Ir_{0.2}$ tip is mounted in a brass collet at the intersection of the X and Y arms. The X arms are poled in opposite directions, so that by application of a voltage, one arm contracts and the other expands, thereby moving the tip. There is only one Y arm, which moves similarly by expansion or contraction.

A piezoelectric bimorph is used to produce the Z motion. The bimorphs are constructed from epoxying together similarly poled faces of two circular pieces of piezoelectric material. The resultant sandwich is then cut into two semicircular bimorphs. Application of a voltage perpendicular to the bimorph face results in the expansion of one semicircular disk, and contraction of the other semicircular disk parallel to the disk's radius. The resultant

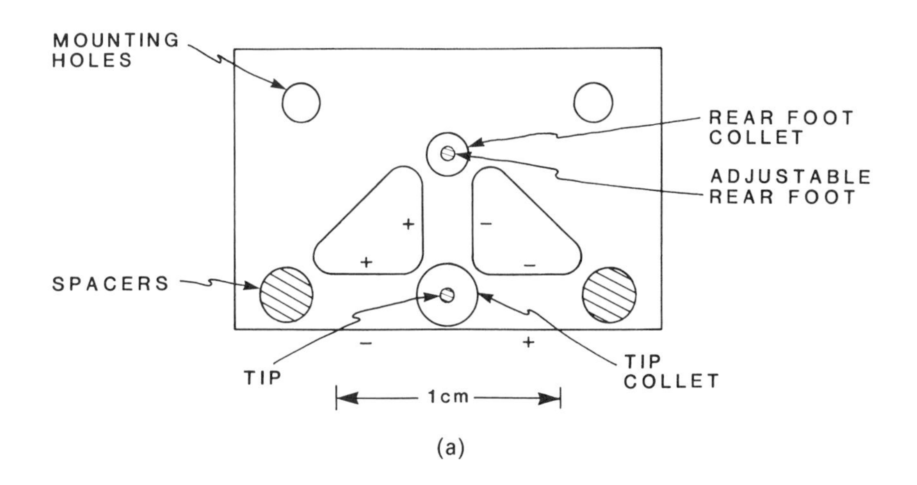
X-Y TRANSLATOR
MOUNTING HOLES
REAR FOOT COLLET
ADJUSTABLE REAR FOOT
+
+
−
−
SPACERS
−
+
TIP
TIP COLLET
1 cm

(a)

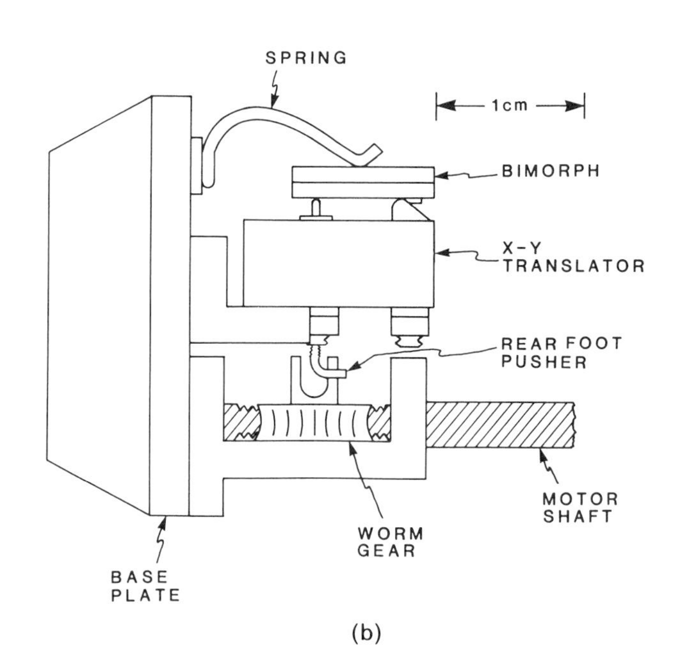
LHE MICROSCOPE
SPRING
1 cm
BIMORPH
X-Y TRANSLATOR
REAR FOOT PUSHER
MOTOR SHAFT
WORM GEAR
BASE PLATE

(b)

motion is parallel to the direction of the applied voltage. Conducting silver paint is used to mount the sample to an insulating pad, which is then glued to the edge of the bimorph. The bimorph is clamped to the X-Y translator by means of a Be–Cu spring clamp. It rests on three spacers positioned at the far end of the X and Y arms on the X-Y translator. The resonant frequency of the assembled microscope, shown in Fig. 2(b) is on the order of 10,000 Hz.

In other versions of the STM, a tip is mounted on a single piezoelectric tube which is then used for X, Y, and Z motions. Two coarse advance screws and one fine advance screw are used for bringing the tip within tunneling distance of the sample. The coarse adjust screws are turned by hand, whereas the fine adjust screw is turned by a stepping motor.

The same setup can be modified to operate as an AFM.[18] In the AFM the sample is mounted on a single piezoelectric tube. A diamond or Si_3N_4 tip mounted on the end of a Si_3N_4 cantilever is used as the AFM tip. A laser beam is reflected off a mirror mounted to the back of the cantilever and is detected by a spot detector that is able to sense cantilever deflections of less than 0.1 nm.

All images produced by the cryogenic STMs were taken in the constant current mode. The tunneling currents were typically in the range of 1–10 nA and the tunneling bias voltages were in the range of 10–30 mV. The tunneling current between the tip and the sample is detected across a series resistor. A PAR 113 preamplifier then amplifies the signal for processing by a home-built logarithmic integrating error amplifier. The error signal is sent to an IBM PC/AT computer that is used to assemble the image. A 12 bit analog-to-digital converter (ADC) with a range of ± 10 V is used to digitize the error signal. This provides a nominal resolution of 0.02 Å at 4.2 K for the LHe microscope and 0.09 Å at 77 K for the LN_2 microscope. The computer also provides the voltages that drive the X-Y translator. Signal averaging can be accomplished through the computer acquisition program by varying the number of samples taken at each data point. Typically 50 to 100 samples are taken.

Analog drives and a storage oscilliscope are used to search for a good scan area before hooking up the more sensitive computer drive. The microscopes are vibrationally isolated from building vibrations by either rubber tires or elastic cords. The resonant frequency of the system is ~ 0.5 Hz.

The transition metal dichalcogenides are grown by the method of iodine

FIG. 2. (a) The X-Y translator used in the cryogenic STM. The plus and minus signs indicate the direction of poling on the arms of the translator. The bimorph rests on the two front spacers and on the adjustable rear foot. (From Ref. 7.) (b) Side view of the cryogenic STM. The tip–sample gap is controlled by advancing or retracting the rear foot pusher using the worm gear assembly, thereby raising or lowering the rear foot of the bimorph. The bimorph then pivots on the front feet, thus changing the tip–sample distance. (From Ref. 7.)

vapour transport. Stoichiometric quantities of powder were sealed in quartz tubes and pre-reacted by sintering at 900°C. The sintered powders were then placed in another quartz tube with iodine gas, and the tube was then sealed. The exact temperatures and temperature gradients differ for each compound. The Ta-based octahedral and 4Hb crystals must be quenched from high temperatures (~900°C for the 1T phase, ~700°C for the 4Hb phase), whereas the Ta-based trigonal prismatic, and Ti and V crystals are slowly cooled from temperatures around 700°C. High quality crystals resulted from these growth procedures, as confirmed by residual ratios ($\rho_{300}/\rho_{4.2}$) of 60 to 300, and the observation of large magneto-quantum oscillations in the 2H and 4Hb compounds. Reference 1 contains a more detailed analysis of many properties of the transition metal dichalcogenides.

The linear chain trichalcogenides are also grown in evacuated quartz tubes using powders initially sintered at 800–1000°C. However, iodine is generally not used as a transport vapor. The crystals are grown for one to two weeks at a temperature of ~750°C in a temperature gradient of approximately 10°C/cm.

8.6. 1T Phase Transition Metal Dichalcogenides

The 1T phase transition metal dichalcogenides have one sandwich per unit cell with octahedral coordination between the metal atoms in the center layer and the Se or S atoms in the top and bottom layers. The crystal structure of the 1T phase exhibits trigonal symmetry and can be stabilized in a number of different compounds. Those that exhibit CDW transitions are in general metals or semimetals, and a wide range of CDW transitions can be found in the different compounds.

The Group V materials have received intense experimental and theoretical study and include the compounds TaS_2, $TaSe_2$, VSe_2, and $NbSe_2$. 1T–$TaSe_2$ and 1T–TaS_2 can be stabilized at low temperature by quenching the crystals from 900°C to room temperature. 1T–VSe_2 is the equilibrium phase at room temparature and crystals can be cooled slowly to room temperature. $NbSe_2$ does not exhibit a 1T phase.

Band structure calculations indicate that these compounds are all good d band metals in the undistorted phase. Below the CDW transition, the FS topology is modified and the conductivity can show varying degrees of change, depending on the amount of FS gapping and the strength of the CDW.

In the case of 1T–$TaSe_2$ and 1T–TaS_2 the CDW is extremely strong, and large changes in the FS topology and the DOS at the Fermi level are expected to occur. This is confirmed by the observation of substantial decreases in the electrical conductivity at the CDW transitions. The STM results to be dis-

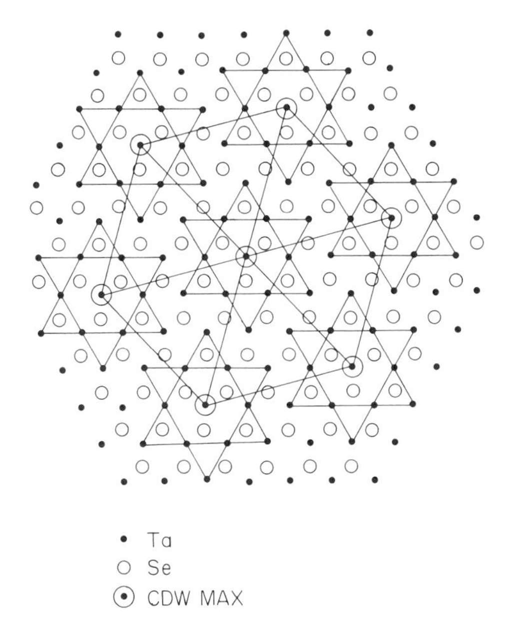

FIG. 3. CDW supercell and star-of-David thirteen Ta atom clusters in $1T–TaSe_2$ and $1T–TaS_2$ layer compounds. The supercell connects the charge hills at the center of each thirteen-atom cluster and is rotated from the underlying atomic lattice by 13.9°. The commensurate superlattice vector is $\sqrt{13}\,a_0$ in length.

cussed here also confirm this conclusion by the observation of a very large amplitude modulation at the CDW wavelength.

The CDWs forming in $1T–TaSe_2$ and $1T–TaS_2$ are similar in the low-temperature commensurate phase and can be described by a six-pointed star-of-David cluster of thirteen Ta atoms as shown in Fig. 3. Charge is transferred toward the center atom in each of these thirteen atom clusters, resulting in a superlattice as outlined by the large hexagon in Fig. 3. The commensurate superlattice vector is $\sqrt{13}\ a_0$ in length. $1T–TaSe_2$ remains commensurate up to 473 K while $1T–TaS_2$ makes transitions to a number of different, nearly commensurate, phases before reaching a final incommensurate phase above 350 K. This difference between $1T–TaSe_2$ and $1T–TaS_2$ is clearly evident in both STM and AFM studies at room temperature and will be summarized later.

In contrast to the 1T phase compounds of Ta, $1T–VSe_2$ exhibits a relatively weak CDW. Conductivity changes are small and the observed STM modulations at the CDW wavelength are small. The formation of the CDW is accompanied by a PLD on the order of 0.1 to 0.2 Å in the Ta layer and smaller distortions of the surface Se or S atoms normal to the surface. This

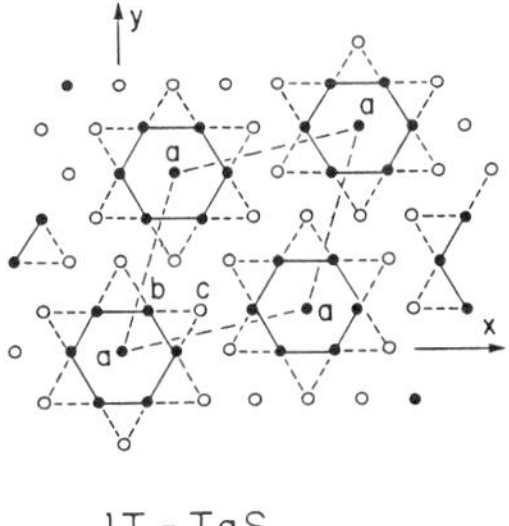

FIG. 4. Ta atom pattern contributing to the CDW charge transfer in 1T–TaS_2. The thirteen Ta atoms centered on the CDW maxima for a six-pointed star-shaped cluster. The electronic charge maximum occurs on the a-type atoms. Charge is transferred from the c-type atoms toward the b- and a-type atoms. (From Ref. 19.)

will produce a change in the LDOS at the position of the tip, but the total z-deflection due to the CDW is more than an order of magnitude larger than the PLD. This suggests that additional mechanisms are involved in the STM response to the CDW structure.

8.6.1 STM of 1T–TaS_2

8.6.1.1 STM of 1T–TaS_2 at 77 and 4.2 K. In the commensurate CDW phase of 1T–TaS_2 at low temperatures, the $\sqrt{13}\,\mathbf{a}_0 \times \sqrt{13}\,\mathbf{a}_0$ unit cell contains three types of atom labeled a, b, and c in Fig. 4. Smith *et al.*[19] have developed a band structure model to describe the effects of the CDW formation and to interpret experimental results from angle-resolved photoemission spectroscopy. Their analysis of the CDW phase shows a collapse of the Ta d band into three subband manifolds separated by gaps, and gives the total number of occupied electrons per atom on each type of atom. The calculated numbers are $n_a = 1.455$, $n_b = 1.311$, and $n_c = 0.611$. The calculation indicates that the CDWs have a fairly uniform amplitude within the seven-atom cluster centered on the a-type atom, while the amplitude decreases strongly in the interstitial regions near the c-type Ta atoms.

An STM scan of 1T–TaS_2 at 77 K is shown in Fig. 5(a) and a profile recorded along the track indicated by the black and white line is shown in Fig. 5(b). The total z-deflection is ~ 3 Å and is dominated by the CDW modulation at a wavelength of $\sqrt{13}\,a_0$. The atomic modulation contributes ~ 0.5 Å to the z-deflection. These observations confirm that the charge transfer associated with the CDW must be very large, although the total z-deflection indicates that some enhancement mechanism may be amplifying the STM response. The charge transfer calculated by Smith *et al.*[19] is extremely large, however, the modulation of the conduction electron wavefunction above the surface would not be expected to extend to 3 Å. Nevertheless, the

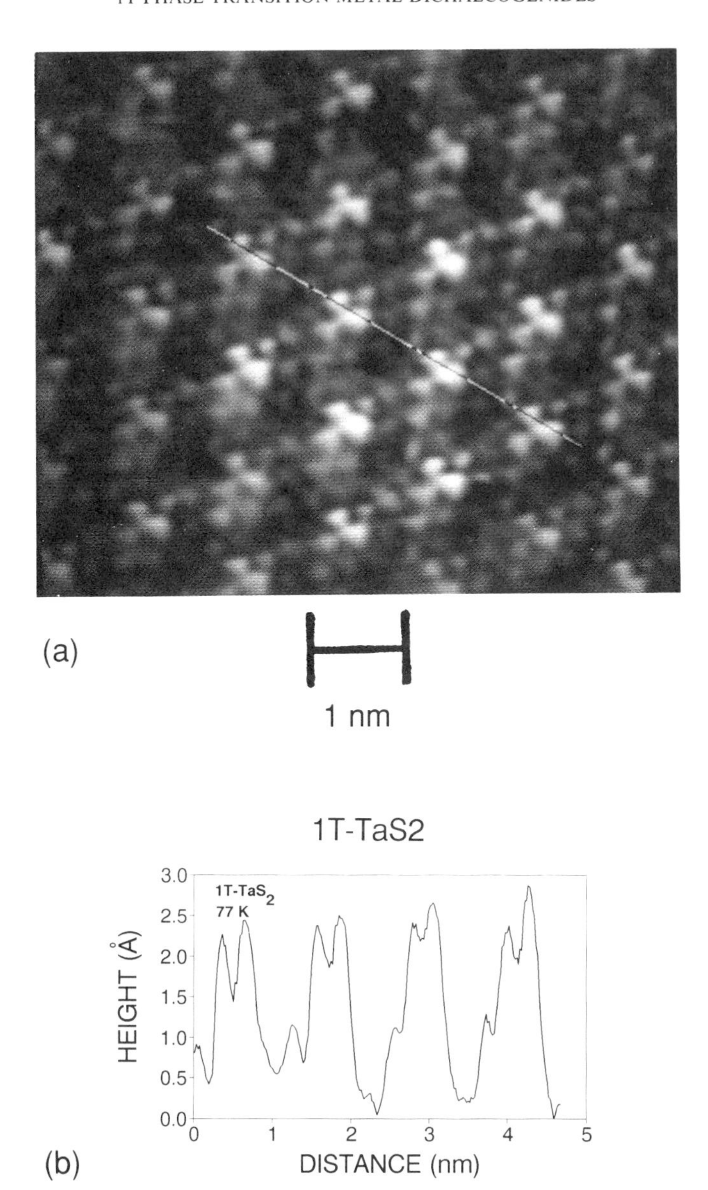

FIG. 5. (a) Grey scale STM image of atoms and CDWs on the surface of a 1T–TaS_2 crystal at 77 K. ($I = 2.2$ nA, $V = 2$ mV). This shows the $\sqrt{13}\,a_0 \times \sqrt{13}\,a_0$ CDW superlattice which is rotated 13.9° from the atomic lattice. (b) Profile of z-deflection along the track indicated by the black and white line in (a). The total z-deflection is ~ 3 Å and is dominated by the CDW modulation at a wavelength of $\sqrt{13}\,a_0$. The atomic modulation contributes ~ 0.5 Å to total the z-deflection of ~ 3 Å. (From Ref. 7.)

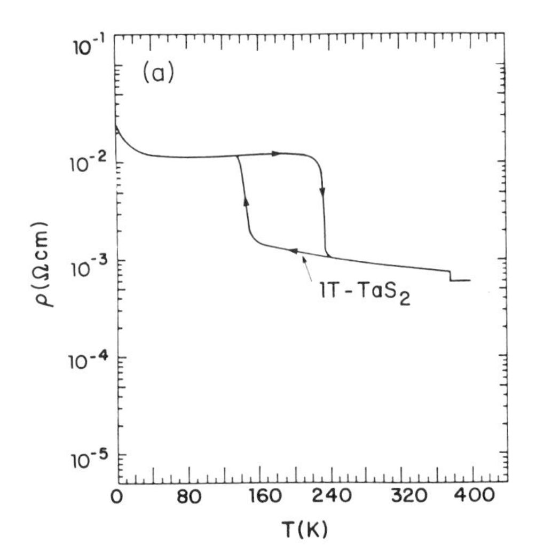

FIG. 6. Resistivity versus temperature for 1T–TaS_2 measured parallel to the layers. 1T–TaS_2 shows a small discontinuity at 350 K where rotation of the CDW superlattice begins. This is followed by a large hysteric discontinuity centered around 200 K which corresponds to a lock-in of the CDW into a commensurate phase. (From Ref. 57.)

STM results indicate that the charge transfer and the accompanying FS modification are extremely large in 1T–TaS_2, and a similar result is found for 1T–$TaSe_2$ as described in Section 8.6.2.

The resistance versus temperature curve measured for 1T–TaS_2 is shown in Fig. 6 and exhibits a complicated structure associated with the various CDW subphases that form in specific ranges of temperature. Two distinct incommensurate phases have been identified in 1T–TaS_2 which can be characterized as nearly commensurate (NC). The NC $1T_2$ phase occurs on cooling between 350 and ~ 190 K, but on warming exists only between ~ 280 and 350 K. The T phase exists upon warming between ~ 220 and 280 K, and indicates a breaking of three-fold symmetry of the wave vector positions observed in the $1T_2$ phase. Above 350 K the final incommensurate phase is formed with the CDW superlattice aligned along the atomic lattice. The NC phase is characterized by higher harmonics of the fundamental CDW wave vectors in electron and x-ray diffraction experiments, thereby suggesting that the CDW superlattice is forming some type of two-dimensional domain structure. This structure of the NC phase existing at room temperature has been studied extensively by STM and will be reviewed later.

At 4.2 K and below, the commensurate CDW phase of 1T–TaS_2 exhibits a drive towards a metal-insulator transition. This may be associated with the CDW modification of the band structure in which a very narrow band forms

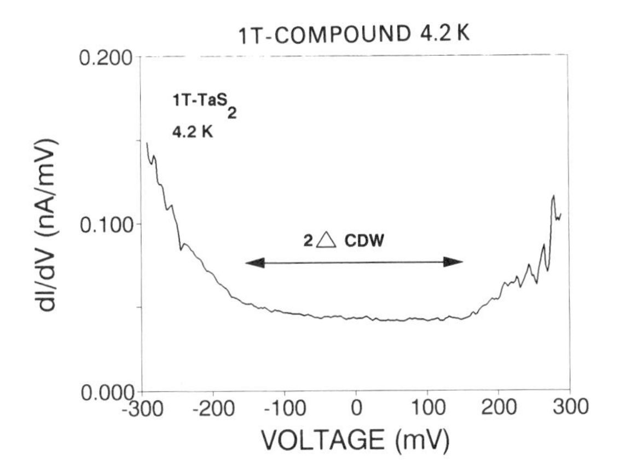

FIG. 7. Conductance (dI/dV) versus bias voltage (V) curve measured on $1T\text{–}TaS_2$ at 4.2 K. The sharp decrease in conductance below ~150 mV is consistent with a large reduction of the FS area due to CDW gap formation. (From Ref. 39.)

leading to a possible Mott–Anderson transition. STM images at 4.2 K are very hard to obtain and give very poor definition. This behavior may be associated with the random onset of the metal-insulator transition and a low density-of-states at the Fermi level.

Although the regular CDW superlattice gives rather poor definition at 4.2 K, the spectroscopy mode which measures dI/dV versus V shows a well-defined CDW gap structure at ~150 mV as shown in Fig. 7. The strong change of conductance above ~150 mV is consistent with a large gapping of the FS. The value of $2\Delta_{CDW}/k_B T_C = 5.8$, calculated using the onset temperature of ~600 K, is in the strong coupling regime, but is substantially smaller than that observed in the 2H phase materials to be discussed in Section 8.7.

8.6.1.2 STM of $1T\text{–}TaS_2$ at 300 K. STM scans of $1T\text{–}TaS_2$ at room temperature show a two-dimensional structure due to a long-range modulation of the CDW amplitude. The period of this long range modulation is ~6 CDW wavelengths and corresponds to an amplitude variation of ~46% as a percentage of the maximum CDW amplitude. Wu and Lieber[20] were the first to report the observation of a two-dimensional domain-like structure which can be readily observed by adjusting the look-up-table (LUT) so that the amplitude variation is exaggerated by the black and white contrast as shown in Fig. 8. A profile through the centers of several domains in the original gray scale image along the track shown in Fig. 8 is shown in Fig. 9. The profile demonstrates the continuous modulation of the CDW amplitude, which follows an approximately sinusoidal variation. A nonsinusoidal profile due to higher harmonics can be expected, and a limited nonsinusoidal

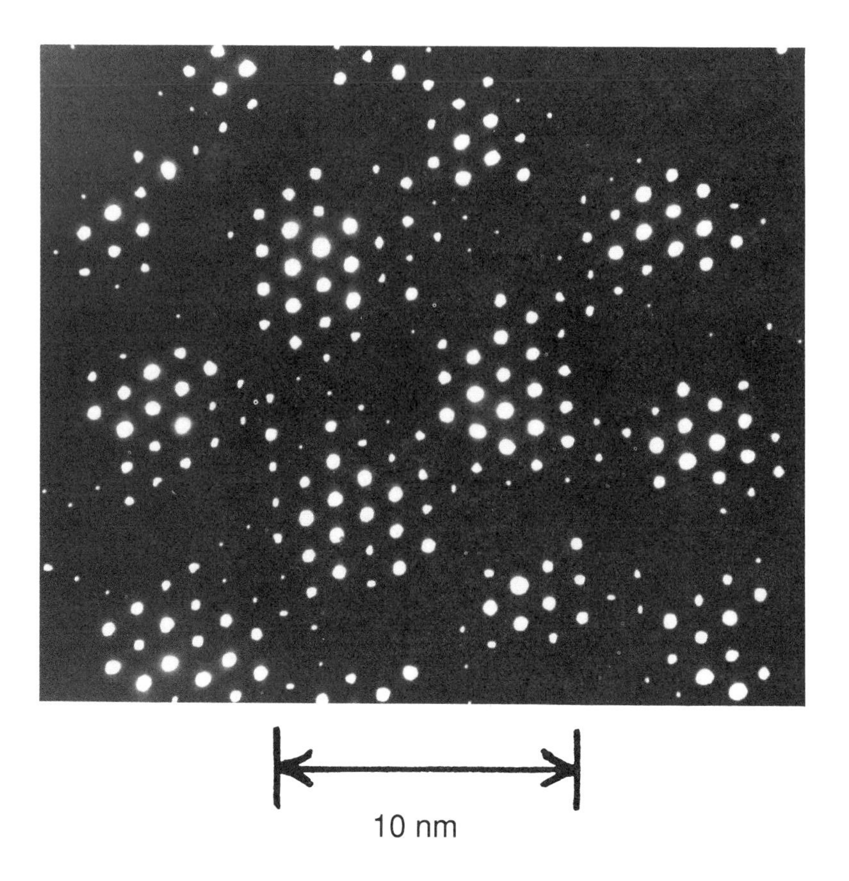

FIG. 8. Grey scale STM image of 1T–TaS_2 taken at room temperature with the LUT adjusted to emphasize the two-dimensional domain-like structure resulting from a variation of the CDW amplitude ($I = 2.2\,\text{nA}$, $V = 25\,\text{mV}$). (From Ref. 29.)

behavior consistent with a wide discommensuration is indicated in the averaged profiles in Fig. 10. However, present experimental error does not allow a precise calculation of the higher harmonic component. The STM scan shown in Fig. 8 is completely dominated by the CDW amplitude with the atom modulation an order of magnitude smaller than the CDW amplitude. Any beat structure or interference modulation by the atomic structure is negligible. When the atomic modulation is appreciable, the beat structure makes it difficult to separate the true CDW amplitude modulation, although this interference effect does not account for the major part of the observed long-range amplitude modulation observed in such scans. The interference or beat structure occurs between the CDW modulation and the modulation produced by the surface S atoms. These two modulations are out of phase so

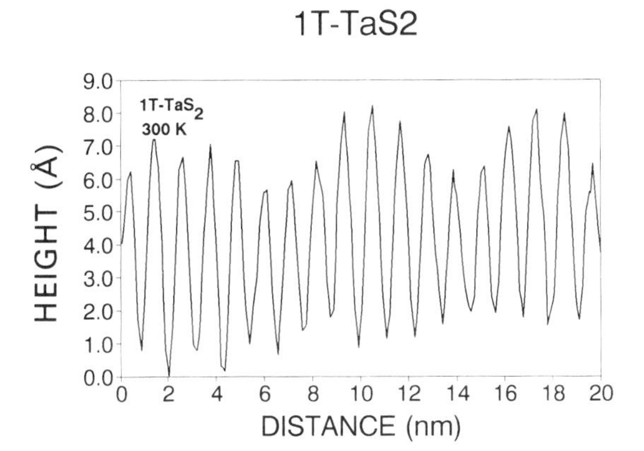

FIG. 9. Profile of z-deflection measured along the black and white track aligned along a row of CDW maxima as shown in Fig. 8. The CDW amplitude undergoes a continuous ~ sinusoidal modulation with a period of ~ 6 CDW wavelengths.

that any amplitude modulation of the CDW due to interference will be small. In fact, such interference modulation, although repeating at ~ 6 CDW wavelengths, reaches a minimum at the center of the CDW domain where the CDW maximum is in phase with the Ta atom in the metal layer below the surface. Therefore, any two-dimensional domain-like structure created by the interference would not be in phase with the true CDW amplitude modulation structure, which dominates the data shown here. These effects were considered by Coleman *et al.*[21] in discussion of measurements of amplitude modulation in 1T–TaS_2 at 300 K. The presence of beat structure in STM scans with appreciable atomic modulation superimposed on the CDW modulation makes it difficult to determine reliable phase information from graphical construction of the real space data. A small superimposed atomic modulation will generate apparent localized phase slips due to interference, which do not reflect a true phase slip between the CDW and atomic lattices. However, the amplitude modulation introduced by beating will remain extremely small in scans where the CDW amplitude is large and dominant. The observed amplitude of the CDW modulation also scales with the enhancement of the CDW amplitude and not with the atomic amplitude. Asymmetries in the CDW domain amplitude due to the out-of-phase interference component are also negligible.

Nakanishi and Shiba[22] originally predicted the existence of a two-dimensional domain-like structure in 1T–TaS_2 based on a free energy model calculation. This model was used to calculate domain structures, and predicted specific phase and amplitude variations of the CDWs, based on the existence of higher harmonics indicated by diffraction and photoemission data. The

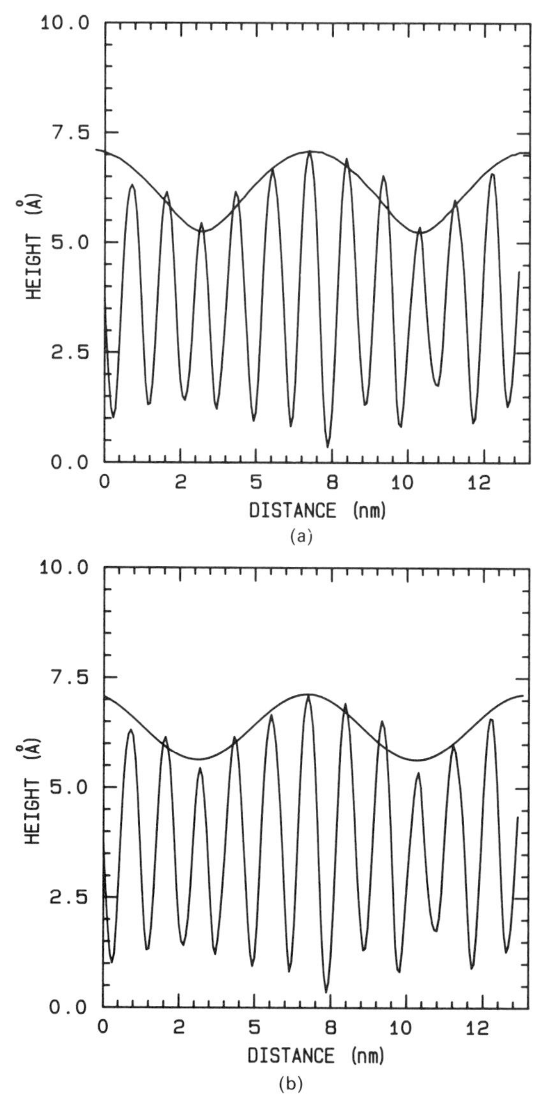

FIG. 10. (a) Profile of CDW maxima obtained by averaging profiles across 4 separate domains in the original grey scale image from which Fig. 8 was generated. The solid line represents the amplitude variation of the maxima of one CDW component of the Nakanishi and Shiba[22] domain model as calculated by Burk *et al.*[23] The calculated solid curve has been scaled to match the experimentally observed maximum CDW amplitude in the center of the domain. The experimentally observed relative amplitude variation is 0.54 and the calculated variation is 0.63. (b) Experimentally observed profile of the CDW amplitude compared to a sine wave (solid line). At the center of the domain boundary the amplitude is consistently less than that required to match a sine wave.

domains would consist of more nearly commensurate regions in the center of the domain, with phase shifts and amplitude variations across the domain. The model is consistent with wide discommensurations, in which the CDW amplitude modulation would follow a continuous variation with the maximum amplitude occuring at the center of the domain. The amplitude modulation would deviate from sinusoidal due to the presence of higher harmonics and a non-uniform phase variation.

Figure 10 shows a profile of CDW maxima obtained by averaging the profiles across four separate domains in a given STM scan of 1T–TaS_2. The solid line in Fig. 10(a) represents the variation of the maxima of one CDW component in the Nakanishi and Shiba[22] model as derived from a calculation by Burk *et al.*[23] The absolute CDW amplitude in the calculated curve has been scaled to match the experimentally observed maximum CDW amplitude, measured from maximum to minimum at the center of the domain. This maximum CDW amplitude is measured from the zero of Fig. 10 and not from the minimum observed in the profile. A profile through a row of CDW maxima is offset from the corresponding absolute CDW minima by $\sim a_0$. The solid line represents the predicted variation of the CDW maxima along one of the CDW component directions. The experimentally observed variation is extremely close to the calculated profile from the Nakanishi and Shiba model, which contains wide discommensurations (domain boundaries). Figure 10(b) shows a sine wave fit to the profile of the CDW maxima as indicated by the solid line. The profile calculated from the Nakanishi and Shiba model is a slightly better fit than the sine wave. The experimental profile indicates a depressed CDW amplitude at the minimum of the domain boundary consistent with the presence of higher harmonics. However, higher experimental accuracy would be required to deduce quantitative data on the higher harmonic content. The observed amplitude variation is certainly consistent with the Nakanishi and Shiba[22] model. Commensurate domains with narrow discommensurations at the boundaries would require a much stronger non-sinusoidal amplitude modulation than is observed. However, the amplitude modulation data alone do not confirm the complete domain model, since no local phase variation information can be accurately determined from the real space data. The local phase variation would change over a number of CDW wavelengths and measurements of CDW maxima, relative to atomic positions using graphical methods, cannot be used to accurately determine the local variations in phase, angle, or wavelength of the CDW. However, substantial regions at the commensurate angle of 13.9° do not appear to be present. The domains and CDW amplitude modulation result from local variations in charge transfer, CDW wavevector, and phase across the domain. The observation of a significant modulation of the CDW amplitude requires that the local CDW charge transfer, wave vector, and PLD change non-uniformly across the domain. The detailed phase variation has to be measured by other techniques. Burk *et al.*[23] have recently obtained wide-area STM scans in the constant height mode which enable them to resolve satellites in the Fourier transforms. By using the location and intensity of these satellites, a fit to the Nakanishi and Shiba[22] model can be made that determines both CDW amplitude and phase as calculated parameters. The results are in approximate agreement with the original Nakanishi and Shiba[22] model fit.

The STM and AFM are clearly the only experimental techniques that can verify this structure in local detail, but the complete analysis of the STM and AFM results is still under development. The continuous variation of amplitude as shown in Figs. 9 and 10 suggest a wide discommensuration with a phase variation extending over a number of CDW wavelengths. A localized discommensuration is not observed, although the STM does not give the precise phase or angle of the CDW with respect to the atomic lattice. The alignment of domain centers is observed by STM to make an angle of $\sim 5.6°$ with respect to the CDW superlattice. The domains in the STM scans show a degree of irregularity when comparing the centers of adjacent domains, and possible variations of the domain superstructure have been discussed by Wilson.[24] The present STM observations indicate an average angle of $\sim 12°$ between the CDW superlattice and the atomic lattice. If commensurate regions exist near the center of the domains, they cannot extend beyond one or two CDW wavelgnths, and the phase shift will occur over fairly wide discommensurations comprising at least 50% of the domain area. Local variations in domain size, alignment, and amplitude are also observed. Some of this may be connected with defects, although point defects produce very localized modifications of the CDW amplitude.[25]

The model of Nakanishi and Shiba[22] with the phase shifts across three or four CDW wavelengths can certainly be used to fit the STM amplitude data within experimental error. This detailed structure of 1T–TaS_2 at room temperature has also been examined by AFM and further discussion will be included under Section 8.6.1.3. A measurement of the precise phase variation can be made from high resolution Fourier transforms obtained from large area STM or AFM scans covering many domains.[23]

Initial studies of 1T–TaS_2 at 225 K have been carried out by Thomson *et al.*[26] after warming the sample from 143 K. The data was obtained with the STM operating in the constant height mode, so that profiles and amplitudes could not be measured. However, apparent domain boundaries were detected and the geometry suggested long narrow domains with a width of 65–70 Å. This striped domain structure was oriented at $\sim 26°$ to the CDW superlattice, and within each domain the pattern of atoms and CDWs suggested commensurate regions. More detailed study of the T phase by both STM and AFM will be required before a complete analysis can be made.

8.6.1.3 AFM of 1T–TaS_2 at 300 K. The atomic force microscope has also been used to study the CDW modulated domain-like structure existing in 1T–TaS_2 at room temperature.[27] The AFM detects the total charge density rather than the DOS at the Fermi level and this results in a substantially different amplitude response to the CDW superlattice than is observed with the STM. In the case of the STM, the CDW modulation was observed to dominate the z-deflection with the atomic modulation contributing only a small fraction of the total modulation. In the case of the AFM scans, the CDW and atomic modulations make equal contributions, and the total z-deflection observed with the AFM is substantially less than that observed with the STM. The equal amplitudes of the CDW and atomic modulations make the AFM scans more suited to carrying out Fourier transform analysis of the relative orientations of the CDW and atomic lattice.

A typical AFM scan of 1T–TaS_2 at 300 K is shown in Fig. 11. The surface S atoms are very well-resolved and the CDW modulation produces enhanced seven-atom clusters with variable enhancement amplitudes that repeat over distances of ~ 6 CDW wavelengths. The seven-atom clusters near the CDW maxima show a continuous variation in the amplitude and asymmetry of the individual atom enhancements. This indicates an incommensurate CDW structure that appears to show a nearly continuous amplitude modulation similar to that observed in the STM scans. A profile across one of the domains is shown in Fig. 12. The profile is taken along a line of CDW maxima, and a profile of the corresponding CDW minima show similar amplitude variations. Both the maxima and minima of the CDW show a continuous amplitude variation, thereby giving rise to the conclusion that the overall CDW amplitude reaches a maximum at the center of the domain and a minimum at the boundary. Abrupt amplitude changes indicating the existence of localized discommensurations are not observed. However, the precise magnitude of the CDW modulation cannot be easily corrected for the beat structure effects introduced when the atomic and CDW amplitudes are of equal magnitude. The total amplitude modulation due to beat effects and true CDW amplitude modulation will then be comparable. However, the interference beat structure will be out-of-phase with the CDW amplitude modulation and will produce interference minima at the center of the CDW domain. This is a relatively weak effect since the surface S atoms and the CDW maxima are never in phase. The AFM profiles appear to reflect the CDW amplitude modulation, but the absolute CDW amplitude cannot be determined accurately. The AFM profiles indicate the same continuous modulation of the CDW amplitude as observed in the more accurate STM profiles, but would have to be corrected for the interference contribution. The AFM

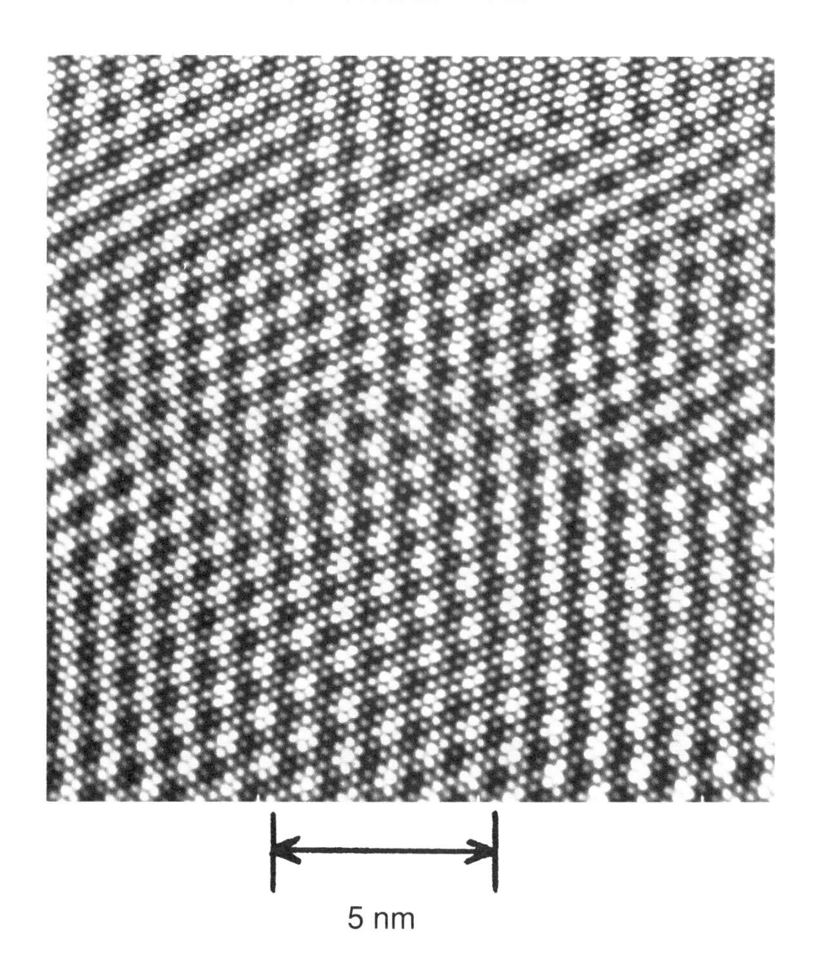

FIG. 11. Grey scale AFM image of 1T–TaS_2 taken at room temperature. The incommensurate CDW produces a clearly visible modulated structure. The atoms and CDWs contribute approximately equal amplitudes to the AFM image (constant force mode with force adjusted to $\sim 10^{-8}$ newtons). (From Ref. 29).

profiles are consistent with the more accurate STM determinations of the CDW amplitude modulation, but the AFM results alone cannot be used for an accurate determination of the absolute CDW amplitude. The discommensurations extend over a number of CDW periods and the precise phase shifts cannot be determined from either the STM or AFM scans. The data is consistent with the wide discommensuration model of Nakanishi and Shiba[22] provided that the phase shifts are spread over a number of CDW periods.

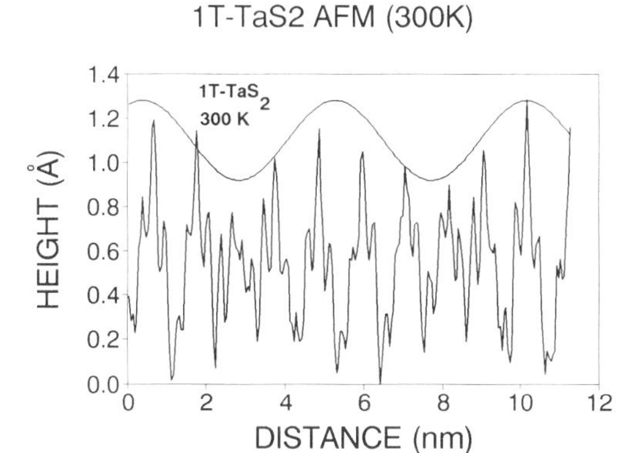

FIG. 12. Profile of the z-deflection along a row of CDW maxima in Fig. 11. It shows a continuous amplitude variation which approximately follows the sine wave shown by the solid curve.

Regions of completely commensurate CDWs would be small, and exist only at the center of the domain, if they exist at all.

The Fourier transform (FT) of the full 27 nm × 27 nm scan of $1T\text{-}TaS_2$, from which a smaller area was shown in Fig. 11, is shown in Fig. 13. It gives

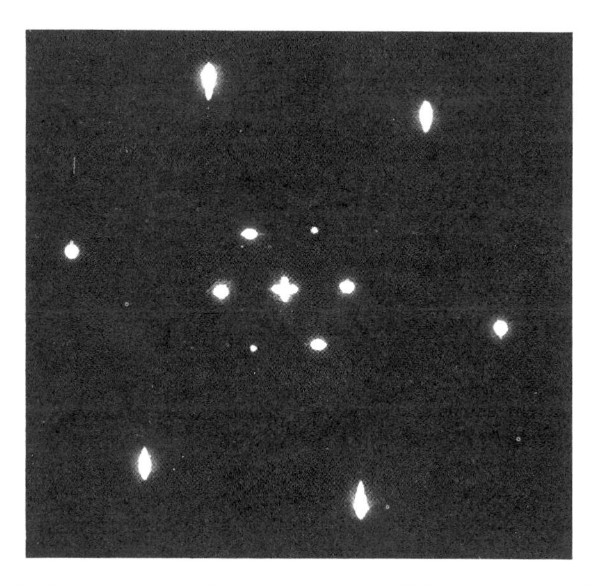

FIG. 13. Fourier transform of the original data set from which a subset was used for the AFM image shown in Fig. 11. The two hexagonal sets of spots represent the CDW superlattice and the atomic lattice. The relative angle of rotation measured using pairs of spots is 11.8° ± 0.4°, close to the average incommensurate value of 11.9°. (From Ref. 29.)

sufficient resolution to show that the angle between the CDW superlattice and the atomic lattice is $11.8° \pm 0.4°$, and that the first order FT spot is therefore at the incommensurate position. No second-order satellite spots can be resolved sufficiently in this FT for any measurement of the position or strength of higher-order Fourier coefficients. Even with higher resolution there are probably not enough domains in this scan to make a clear analysis of second-order structure. A much larger area scan with good definition will probably be required to bring out satellites due to the CDW modulation structure. Calculations of exact phase variations within the CDW domain could then be attempted.

8.6.2 STM of 1T–$TaSe_2$

An STM scan of 1T–$TaSe_2$ at 77 K recorded by computer and presented as contours of constant z-deflection is shown in Fig. 14. The CDW charge maxima appear as white plateaus centered on a $\sqrt{13}\,\mathbf{a}_0 \times \sqrt{13}\,\mathbf{a}_0$ hexagonal lattice with an experimental spacing of 12.4 ± 0.2 Å, again in agreement with the expected CDW superlattice spacing. The contours of constant LDOS at the Fermi level show no evidence of modulation by the surface layer of Se atoms and the interstitial positions between the charge maxima show three deep minima and three saddle points symmetrically located relative to the maxima of the CDW superlattice. The entire pattern is therefore dominated by the CDW, although the deep minima and saddle points must reflect the inequivalent surface Se atom arrangements as well as the CDW modulations of the LDOS at these positions. As shown in Fig. 15 the saddle points are centered over a Se atom, while the deep minima occur at positions between surface Se atoms that also have no Ta atom located in the layer below. The three pairs of interstitial positions should not be quite equivalent due to the 13.9° rotation of the CDW superlattice, provided the surface Se atoms contribute to the modulation of the conduction electron density. In general, the charge density contours show a small inequivalence at these interstitial positions, but it is close to the limit of resolution for the present STM operation. The calibrated z-deflection from the deep minima to the maxima in the image of Fig. 14 is very large with a value of 2.4 ± 0.2 Å.

1T-$TaSe_2$ in the CDW phase at room temperature is a semi-metal with a resistivity of $2 \times 10^{-3}\,\Omega\,\mathrm{cm}$ parallel to the layers. The resistivity shows a monotonic decrease below room temperature, reaching a value of $10^{-4}\,\Omega\,\mathrm{cm}$ at 4.2 K. This increase in conductivity does not appear to have any major effect on the STM images obtained at 4.2 K. The STM scans at 4.2 K show patterns that are essentially the same as those observed at 77 K. The average maximum to minimum z-deflection observed at 4.2 K is 3.5 ± 1.4 Å and although constant during a given run, it varies in the range 2.5 to 5 Å for

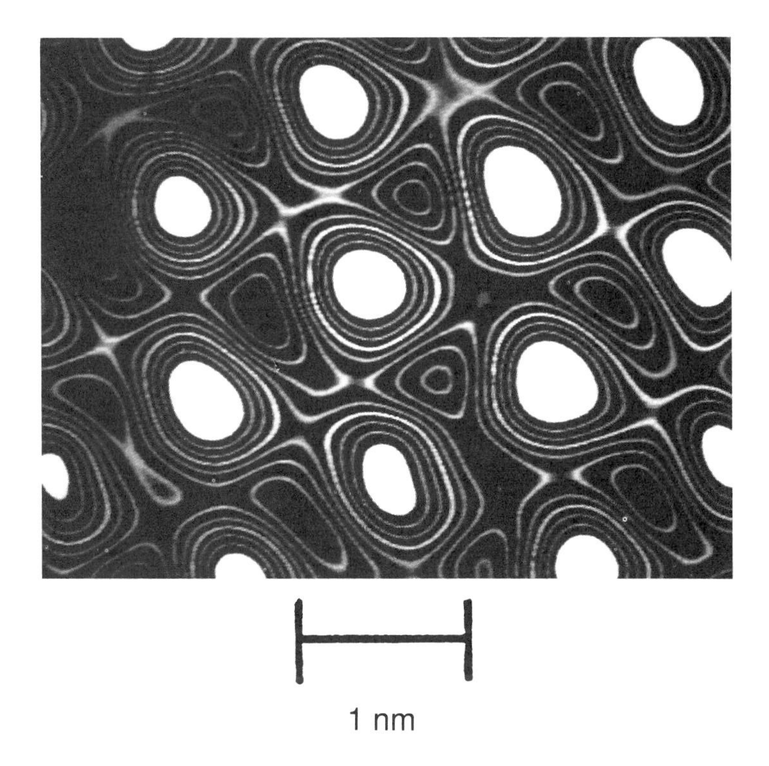

FIG. 14. An STM image of 1T–$TaSe_2$ taken at 77 K using a contour LUT which shows contours of constant z-deflection corresponding to contours of constant LDOS. The white plateaus indicate the CDW maxima centered on a $\sqrt{13}\,\mathbf{a}_0 \times \sqrt{13}\,\mathbf{a}_0$ hexagonal superlattice with an experimental spacing of 12.4 ± 0.2 Å, in good agreement with the value of 12.54 Å measured by electron diffraction. (From Ref. 7.)

different crystals or cleaved surfaces. The deflection observed at 4.2 K is on average larger than that observed at 77 K, but at both temperatures the deflection is anomalously large relative to the expected spatial modulation of the LDOS from an extended FS.

The relative strength of the atomic modulation and the degree of atomic resolution are extremely sensitive to the STM response. Even when the atomic modulation is clearly resolved it represents only a small fraction of the total z-deflection, ~0.2 Å out of a total of ~2 Å. Figure 16 shows a grey scale scan and a profile recorded at 4.2 K. The atoms are well-resolved in Fig. 16(a) but the CDW amplitude completely dominates the profile, as shown in Fig. 16(b).

The FS modifications of the 1T–$TaSe_2$ band structure have not been studied in detail and at present no estimates of the charge transfer have been

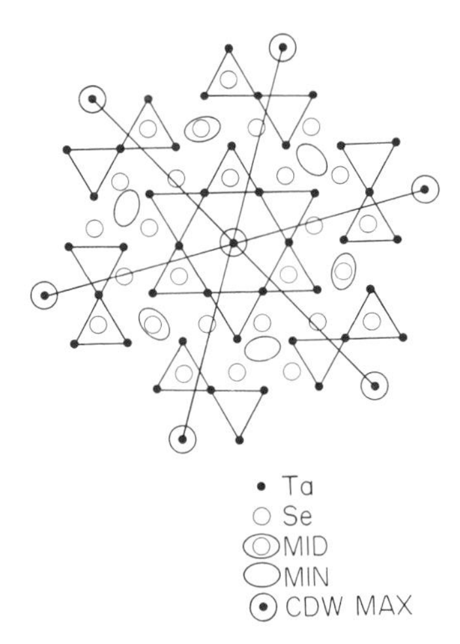

FIG. 15. Diagram showing positions of deep minima and saddle points of the CDW relative to the 1T–$TaSe_2$ atomic lattice. Deep minima occur between groups of three surface Se atoms which have no Ta atom in the layer below and saddle points over Se atoms. The ellipses indicate the location of minima and saddle points. (From Ref. 7.)

developed. The STM response to the 1T–$TaSe_2$ CDW is similar in magnitude to that observed for 1T–TaS_2, except that the atomic modulation is often absent in the 1T–$TaSe_2$ scans at both 77 and 4.2 K. The CDW in the 1T phase of TaS_2 and $TaSe_2$ both form commensurate $\sqrt{13}\,\mathbf{a}_0 \times \sqrt{13}\,\mathbf{a}_0$ superlattices at low temperature. The presence of a weak atomic modulation in the STM scans of 1T–TaS_2 and the substantially weaker atomic modulation in many STM scans of 1T–$TaSe_2$ must reflect a subtle difference in the FS and electronic structure. The low temperature resistivity certainly reflects this difference, although no information is presently available on the detailed differences in the two Fermi surfaces that result from the band folding.

The high-temperature Fermi surfaces are different based on the band calculations of Woolley and Wexler.[28] The 1T selenide has a pancake-shaped region around Γ while the sulphide does not. The calculated k dependence for the selenide is greater than that in the sulphide and this may enhance the importance of a nesting vector component parallel to the **c** axis. This FS difference is not due to an enhanced intersandwich interaction but to the accidental fact that the Fermi level of 1T–$TaSe_2$ falls in the middle of the approximately t_{2g} (the g subscript refers to geräde) triplet band whose splitting due to the trigonal distortion is k dependent.

These differences in the high temperature FS can certainly propagate into a difference of FS topology in the CDW superlattice state and consequently produce a difference in the CDW modulation of the LDOS. The STM

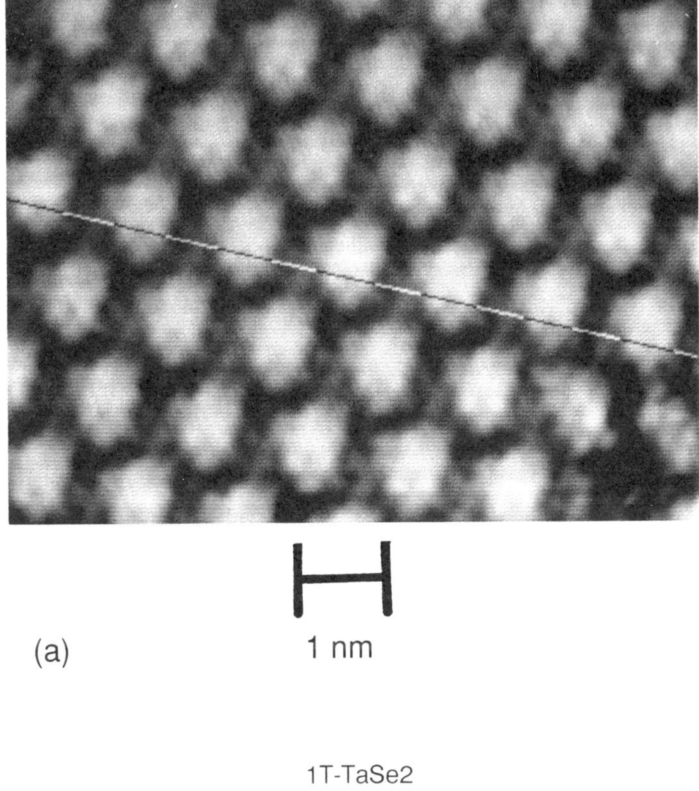

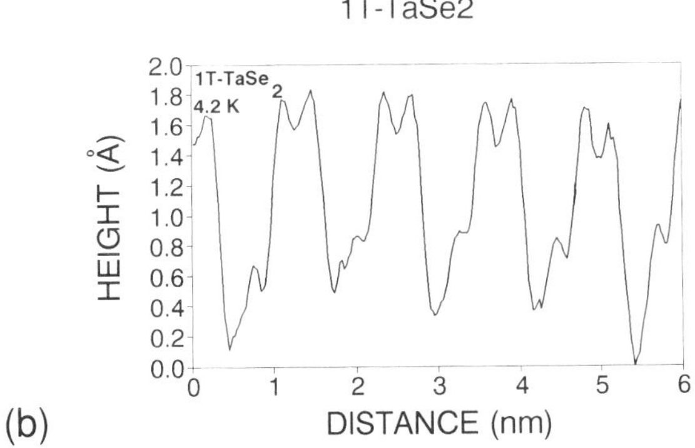

FIG. 16. (a) STM grey scale image of 1T–$TaSe_2$ taken at 4.2 K. The surface Se atoms are resolved and appear superimposed on the CDW modulation of much larger amplitude. White areas with a weak superimposed atomic modulation are centered on the CDW maxima. (b) Profile of z-deflection taken along the black and white track shown in (a). The total z-deflecction of ~2.0 Å is dominated by CDW modulation with the atoms contributing ~0.3 Å. (From Ref. 14.)

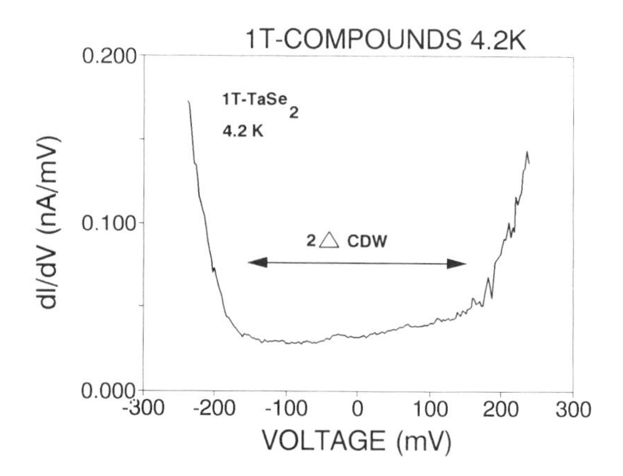

FIG. 17. Conductance versus bias voltage curve for $1T\text{–}TaSe_2$ measured at 4.2 K. A sharp increase in conductance at ± 150 mV can be identified with the CDW gap and the resulting change in DOS above the gap edge. (From Ref. 39.)

response suggests such a difference, but no detailed model has been worked out as yet. Differences in the CDW induced FS topology can be reflected in the spatial modulation of the LDOS through a mechanism proposed by Tersoff.[13]

The spectroscopy mode gives a measure of CDW gap structure that is in reasonable agreement with the gap measured for $1T\text{–}TaS_2$. A dI/dV versus V curve taken at 4.2 K is shown in Fig. 17 and indicates a CDW gap at ~ 150 meV. Using a CDW onset temperature of ~ 600 K this gives $2\Delta_{CDW}/k_B T_C = 5.8$.

8.6.2.1 AFM of $1T\text{–}TaSe_2$ at 300 K. The AFM scans of $1T\text{–}TaSe_2$ at 300 K show a uniform pattern of atoms and CDWs. An example is shown in Fig. 18. The uniformity of this pattern indicates a constant phase relation between the atoms and the CDW maxima. Enhanced seven-atom clusters look identical and are spaced on a superlattice of wavelength $\sqrt{13}\,a_0$ as expected. The total z-deflection is on the order of 1 Å as observed in the AFM scans of $1T\text{–}TaS_2$. The atomic and CDW amplitudes make equal contributions of ~ 0.5 Å each. The profiles of z-deflection show a uniform amplitude over many CDW wavelengths, in agreement with the profiles expected for a uniformly commensurate superlattice.

The AFM scans of $1T\text{–}TaSe_2$ form an excellent data set for a Fourier transform analysis of the relation between the atomic lattice and the CDW superlattice. A Fourier transform of a 13 nm × 13 nm AFM scan of $1T\text{–}TaSe_2$ at 300 K is shown in Fig. 19. Using the three pairs of spots, the measured angle between the atomic lattice and the CDW superlattice is

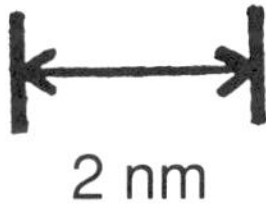

FIG. 18. Grey scale AFM image of 1T–$TaSe_2$ taken at room temperature. A uniform pattern of atoms and CDWs that maintain a constant relative phase indicates a commensurate structure as observed in the STM images (constant force mode with force equal to $\sim 10^{-8}$ newton). (From Ref. 29.)

$14.2° \pm 0.5°$, thereby indicating good agreement with the expected commensurate angle of 13.9°. More details can be found in Ref. 29.

8.6.3 STM of 1T–VSe_2

1T–VSe_2 is a good d band metal, and band structure calculations carried out by Myron[30] and Woolley and Wexler[28] predict high-temperature FS cross sections that are close to those calculated for 1T–$TaSe_2$. A significant feature of the similarity is the occupied electron pocket in the center of the zone at the Γ point, unlike the FS found for 1T–TaS_2, where these states are unoccupied. The FS in 1T–VSe_2 also shows a greater k_z dependence than that in the 1T sulphides.

In spite of this similarity between the high-temperatures FSs of 1T–$TaSe_2$ and 1T–VSe_2, the CDW formation is entirely different. The $\mathbf{q}$ vectors in the metal-metal plane of 1T–VSe_2 are $\mathbf{a}_0^*/4$ rather than $\mathbf{a}_0^*/\sqrt{13}$ as found in 1T–$TaSe_2$. The STM scans of 1T–VSe_2 at both 77 and 4.2 K show a $4\mathbf{a}_0 \times 4\mathbf{a}_0$

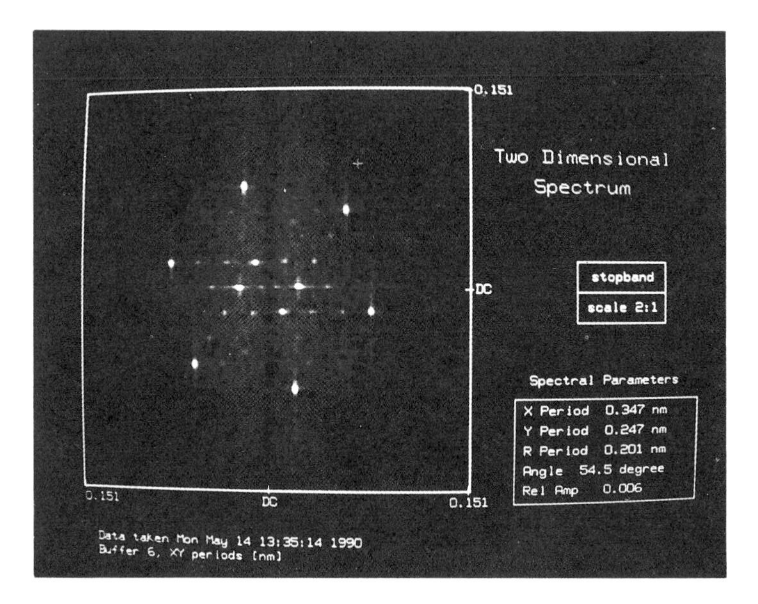

FIG. 19. Fourier transform of a 13 nm × 13 nm AFM scan of 1T–$TaSe_2$ taken at 300 K. The two hexagonal sets of spots represent the CDW superlattice and the atomic lattice. The relative angle of rotation between the CDW superlattice and atomic lattice is $14.2° \pm 0.5°$ as measured using the three pairs of spots. This value is in good agreement with the expected commensurate angle of 13.9°. (From Ref. 27.)

superlattice within the layer plane. A typical grey scale image recorded at 4.2 K is shown in Fig. 20(a) and exhibits a profile in which the atomic modulation dominates as shown in Fig. 20(b). In this respect, the STM response to the CDW is completely different than observed for 1T–$TaSe_2$ or 1T–TaS_2. The total z-deflection can vary in the range 0.8 to 4 Å, but in all cases the CDW modulation is less than, or at most equal to, the atomic modulation. This major difference in CDW formation and STM response is probably related to the presence of a strong p–d hybridization in 1T–VSe_2, as compared to 1T–$TaSe_2$. In the latter the d-based band is separated by a large gap (~ 0.1 Ry) from the chalcogen p-based valence band. In 1T–VSe_2 a significant overlap of the d and p bands occurs and can lead to a more complicated pattern of FS instabilities. In addition, the STM scans can show widely different amplitudes for different components of the triple **q** CDW. This appears to reflect a competition between a double and triple **q** structure.

Diffraction studies by a number of authors including Williams,[31] Tsutsumi *et al.*,[32] Van Landuyt *et al.*,[33] and Fung *et al.*[34] have suggested a variety of CDW superlattice structures, that form upon cooling the crystal below the initial incommensurate CDW phase forming at 140 K. Recent electron and x-ray diffraction studies by Tsutsumi,[35] Yoshida and Motizuki,[36] and

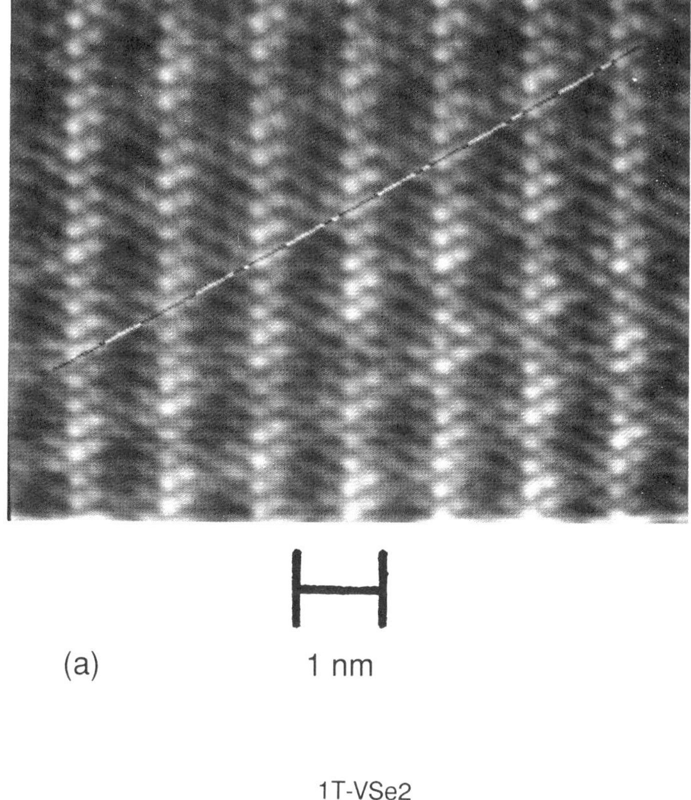

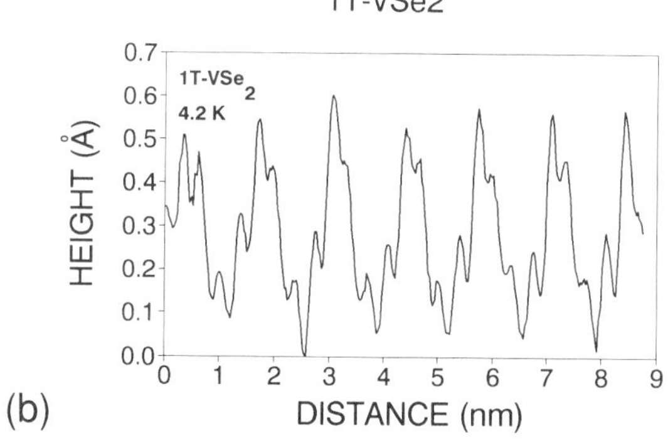

FIG. 20. (a) Grey scale image of a STM scan on 1T–VSe_2 recorded at 4.2 K. The weak $4\mathbf{a}_0 \times 4\mathbf{a}_0$ CDW superlattice is superimposed upon the pattern of surface Se atoms ($I = 2.2$ nA, $V = 25$ mV). (b) Profile of z-deflection along the line shown in (a). In the STM profiles of 1T–VSe_2 the atomic modulation and the CDW modulation are comparable.

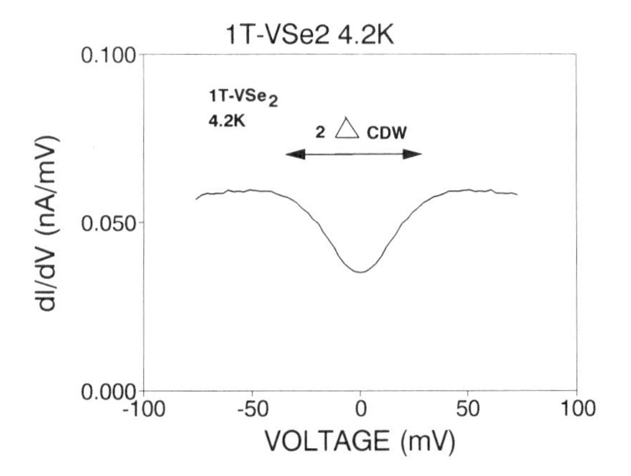

FIG. 21. Conductance versus bias voltage curve measured on $1T–VSe_2$ at 4.2 K. A slope change is observed at ± 40 mV which can be identified with the CDW gap edge. (From Ref. 39.)

Eaglesham *et al.*[37] generally agree on a low-temperature CDW superlattice with a commensurate component of $4\mathbf{a}_0$ parallel to the layer and an incommensurate component of $\sim 3\mathbf{c}_0$ perpendicular to the layer. Eaglesham *et al.*[37] observed a $3q$ to $2q$ transition below 80 K, and the STM scans at 77 and 4.2 K tend to confirm this possibility. Above 80 K a number of different possible CDW structures have been reported involving either domain structures of the $3q$ or $2q$ variety or different $\mathbf{q}$ vectors. These observations may be a function of the specific sample or cooling history and have not yet been completely resolved.

Conductance data from $1T–VSe_2$ shows a much smaller energy gap consistent with the lower CDW onset temperature of 110 K. At 4.2 K this CDW is commensurate with a wavelength of $4a_0$. A conductance versus bias voltage curve for $1T–VSe_2$ is shown in Fig. 21 with a gap structure identified at ~ 40 mV. This gives a value of $2\Delta_{CDW}/k_B T_C \simeq 8.4$, a value midway between the 1T phase Ta compounds and the 2H phase Ta compounds. The CDW in $1T–VSe_2$ behaves more like the CDWs observed in the 2H phase compounds in that it has a relatively low amplitude and remains aligned along the atomic lattice. In this respect, all of the CDW gaps measured by the STM show a systematic trend[38,39] consistent with the general qualitative features observed for each CDW. CDW gaps measured in the 2H phase compounds will be described in later sections and more details can be found in Refs. 38 and 39.

8.6.4 STM of $1T–TiSe_2$

The materials in the Group IV transition metal compounds are more covalently bonded than the Group V compounds and tend to be semiconductors rather than metals. Many experiments and calculations suggest that 1T–$TiSe_2$ is an exception, with a small indirect overlap between a Se-based s/p band near Γ and a Ti-based d band near L. Small pockets of holes and electrons result from this indirect overlap and a second-order phase transition is observed at 202 K in electron diffraction experiments.[40,41] This is identified with the formation of a $2\mathbf{a}_0 \times 2\mathbf{a}_0 \times 2\mathbf{c}_0$ CDW superlattice. The CDW forms in the commensurate state, and it has been suggested that the CDW involves an electron-hole interaction in which a zone-boundary phonon is driven soft. Angular resolved photoemission data, electron and x-ray diffraction data, magnetic susceptibility data, resistivity data, and infrared reflectivity data by Woo *et al.*[40] and Brown[42] generally support models based on band overlap. However, Stoffel *et al.*[43] have used photoelectron spectroscopy data in an analysis which concludes that $TiSe_2$ is a very narrow band semiconductor in which the Se 4p and Ti 3d bands have a strong interaction leading to CDW formation. The STM results show a CDW with medium intensity, but it involves substantially less charge transfer than occurs in the CDWs of the Group V 1T phases of $TaSe_2$ and TaS_2.

An STM scan of 1T–$TiSe_2$ recorded at 77 K is shown in Fig. 22(a) and shows a well-resolved pattern of surface Se atoms. The CDW superlattice modulation is also detected as an enhancement of z-deflection on every second row of atoms. A profile along one of the atomic rows shown in Fig. 22(b) indicates that the atomic and CDW modulations generate approximately equal z-deflections of ~ 1 Å each. In the magnified contour plots, the CDW maxima show a phase shift relative to the seven-atom clusters of surface Se atoms surrounding the maxima. This would place the CDW maxima close to the Ti atom positions, but the exact charge distribution will depend on the relative roles of the Ti d band and Se s/p valence band in the CDW formation and charge transfer.

A band Jahn–Teller mechanism has been developed by Motizuki *et al.*[44] and extended by Suzuki *et al.*[45] This model includes a wave vector and mode-dependent electron-lattice interaction as well as a FS nesting, both of which contribute to CDW formation. This model explains the stability of the triple **q** CDW state as well as many of the experimental results. Nevertheless, considerable controversy on the exact electronic structure exists, with some models and interpretations[40,42] confirming the indirect band overlap, while other results[43] suggest a narrow band semiconductor.

Spectroscopy of 1T–$TiSe_2$ with the STM shows structure that could be identified with a narrow gap. A dI/dV versus V curve recorded at 4.2 K is shown in Fig. 23, and for positive bias a flat region in the DOS is observed between 0 and 0.2 V. Defects due to non-stoichiometry can produce an n-type semiconductor, which would account for the shift in the possible gap structure to the positive bias side of the curve. This structure in the conductance

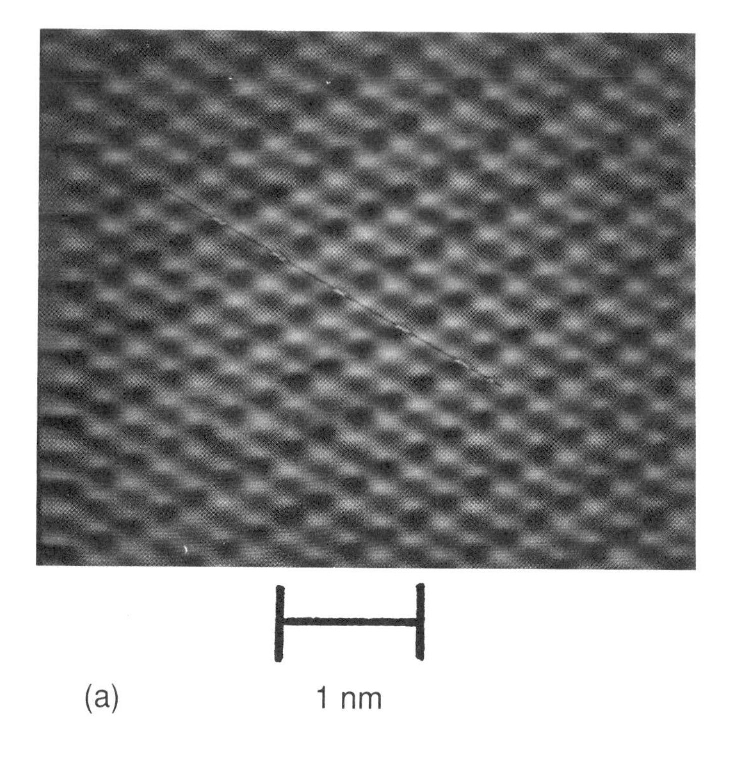

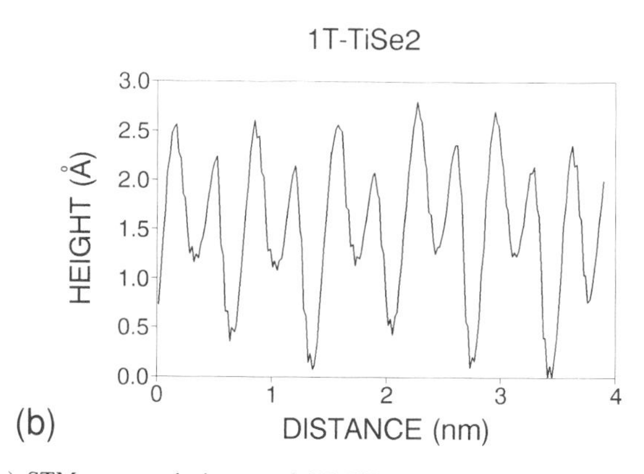

FIG. 22. (a) STM grey scale image of 1T–$TiSe_2$ at 77 K showing the CDW and atomic modulations ($I = 8.6$ nA, $V = 50$ mV). (b) Profile of z-deflection along the line shown in (a). The $2a_0$ modulation due to the CDW is clearly resolved with the atoms and CDWs contributing equal amplitudes. (From Ref. 7.)

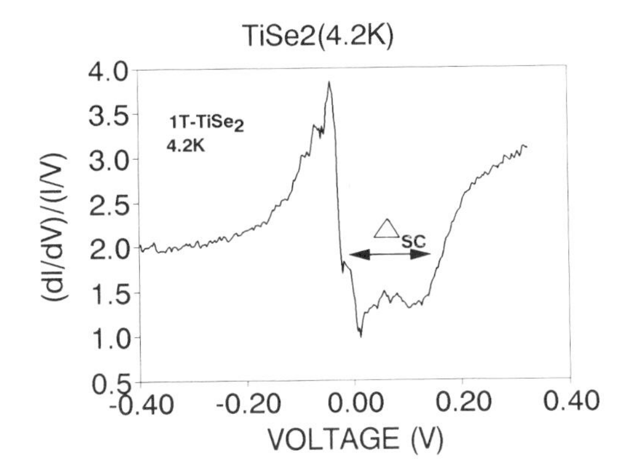

FIG. 23. Normalized conductance curve obtained on 1T–$TiSe_2$ at 4.2 K. A flat region between 0 and 0.2 V indicates the presence of a weak semiconducting gap.

versus voltage curve is very similar to that observed in the STM spectroscopy of 1T–TiS_2, which is definitely a narrow band semiconductor and is discussed in the next section. This tunneling spectroscopy data provides some confirmation of the narrow band semiconductor analysis presented by Stoffel *et al.*[43]

8.6.5 STM of 1T–TiS_2

There is general agreement that 1T–TiS_2 is a small band gap semiconductor with a gap on the order of 0.2 to 0.3 eV. A STM scan of 1T–TiS_2 at 4.2 K is shown in Fig. 24(a) and shows well-defined atomic structure. Many of the STM scans show a long-range modulation in z-deflection, but the individual atomic modulation amplitudes remain fairly constant at ~2 Å. This modulation can be fairly regular with a period of 7–8 atomic spacings, but it can also show random variations, suggesting surface undulation rather than intrinsic modulation of the electronic structure. Fig. 24(b) shows a profile along a row of atoms and the long-range undulation is clearly detected as a height variation of ~1 Å.

No evidence of CDW formation is observed, but the STM response to the atomic structure is very similar to that observed in 1T–$TiSe_2$. The atomic modulation in both cases is 1 to 2 Å and suggests some enhancement of z-deflection, independent of the presence or absence of a CDW. The 2H phase layer compounds, which are good metals with weak CDWs, show much smaller z-deflections in the STM scans.

Spectroscopic measurements have also been carried out on 1T–TiS_2 at 4.2 K and a typical dI/dV versus V curve is shown in Fig. 25. It shows a structure characteristic of a semiconductor with a gap of ~0.3 to 0.4 eV. The

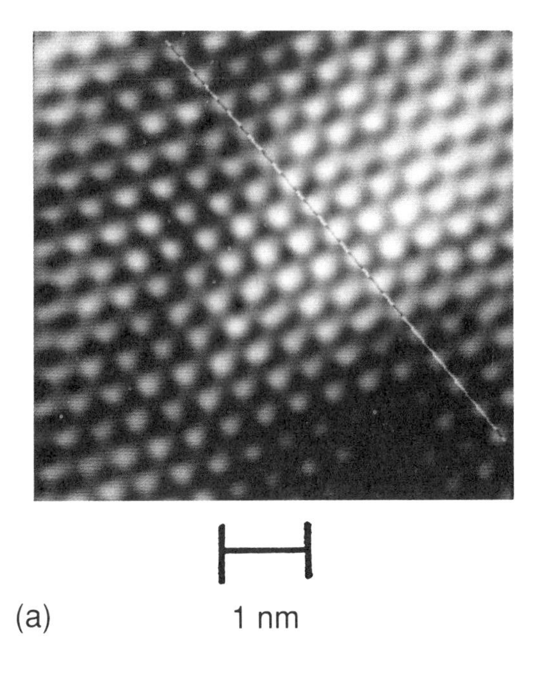

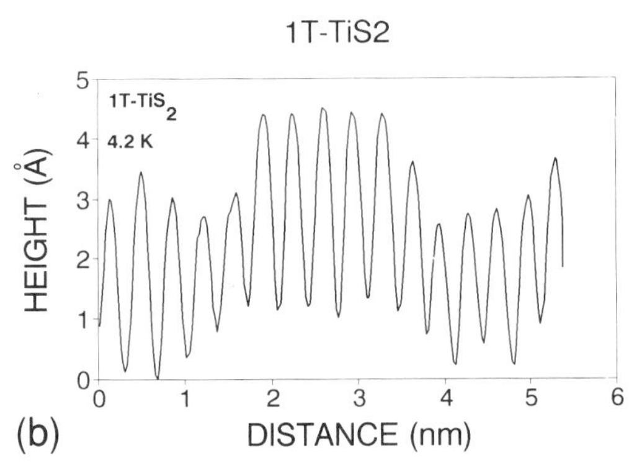

FIG. 24. (a) STM grey scale image of 1T–TiS_2 at 4.2 K showing only an atomic modulation ($I = 2.2$ nA, $V = 25$ mV). (b) Profile of z-deflection along the line shown in (a). Note the long-range modulation of the surface atomic structure which suggests a possible two-dimensional supercell structure of the surface atoms. (From Ref. 7.)

asymmetry suggests that the 1T–TiS_2 sample is an n-doped semiconductor with E_f near or in the Ti d conduction band due to excess Ti or defects. The flat regions in the conductance curve for positive tip bias therefore occur as the Fermi level of the tip sweeps across the gap in the TiS_2 substrate. This

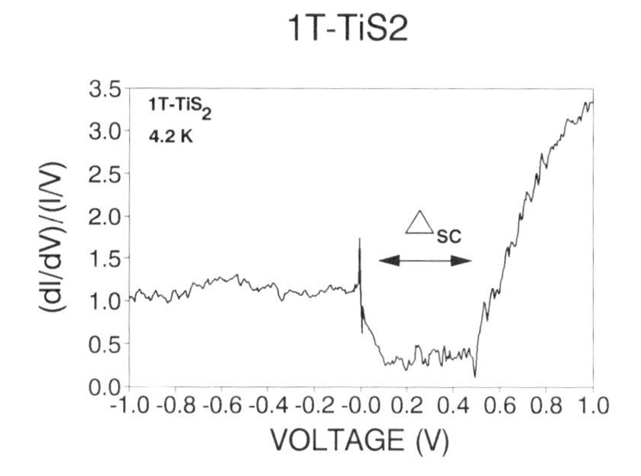

FIG. 25. Normalized conductance curve recorded on $1T–TiS_2$ at 4.2 K. The flat region of low conductivity can be associated with a semiconducting gap of ~0.3 to 0.4 eV. (From Ref. 39.)

spectroscopy result gives reasonable confirmation of the predicted semiconductor band structure of $1T–TiS_2$.

8.7. 2H Phase Transition Metal Dichalcogenides

The 2H phase materials have two three-layer sandwiches per unit cell, and the transition metal atoms within each sandwich exhibit trigonal prismatic coordination with the chalcogen atoms. When compared to the 1T phase materials, the CDW transitions are weaker and occur at much lower temperatures. The electrical conductivities of the 2H phase crystals show only small changes at the CDW onset, thereby indicating that the FS changes do not involve large charge transfers although substantial changes in FS topology can occur.

The STM scans are consistent with the presence of weak CDWs since, in all cases studied so far, the atomic modulation amplitude has been equal to or stronger than the superimposed CDW modulation amplitude. Extensive STM data has been collected on $2H–NbSe_2$, $2H–TaSe_2$, and $2H–TaS_2$.

The CDWs form an incommensurate $3\mathbf{a}_0 \times 3\mathbf{a}_0$ superlattice at the CDW onset temperature. The CDW in $2H–TaSe_2$ becomes commensurate at low temperatures, while the CDWs in $2H–NbSe_2$ and $2H–TaS_2$ remain incommensurate down to the lowest temperatures measured. The FS topology resulting from the refolding of the band structure into the new Brillouin zone (BZ), forming below the CDW onset, generates a large number of FS sections with reduced cross sections. However, these are fairly extended within the new BZ and some show a substantial dispersion in the z direction. In general this topology, although substantially modified by the CDW, will mean a weaker $\mathbf{k}_{\parallel}$ and $(\mathbf{k}_{\parallel} + \mathbf{G})$ mixing and a rearrangement, rather than annihilation, of FS area. The LDOS will therefore be expected to show a weaker nodal structure than can occur in the 1T phase compounds.

8.7.1 STM of 2H–$NbSe_2$

The CDW in 2H–$NbSe_2$ has an onset temperature of ~ 33 K and can most easily be detected by measuring the change in sign of the Hall effect. The resistivity shows only a weak anomaly at the CDW onset and 2H–$NbSe_2$ remains a good metal below the CDW onset. A superconducting transition occurs at 7.2 K and below this temperature CDWs and superconductivity coexist. The CDWs form a triple $\mathbf{q}$ structure with an approximate $3\mathbf{a}_0 \times 3\mathbf{a}_0$ superlattice. The wave vectors are incommensurate and are given by $\mathbf{q} = (1 - \sigma)\mathbf{a}_0/3$ with $\sigma = 0.025$ at 33 K and $\sigma = 0.011$ at 5 K as determined by neutron diffraction.[46]

STM scans[47] of 2H–$NbSe_2$ at 4.2° K show a pattern dominated by the atomic modulation with an amplitude of ~ 0.4 Å. At high magnification, as shown in Fig. 26(a), the superimposed CDW enhancement appears on every third row of atoms and represents an additional z-deflection of ~ 0.2 Å as shown in the profile of Fig. 26(b). The CDW modulation is centered on the Nb atoms, which are displaced from the atomic modulation of the surface Se atoms by $a_0/\sqrt{3}$. The superposition of this CDW modulation causes the atomic modulation pattern of Se atoms near the CDW maxima to be distorted, as shown in the LDOS contour scan of Fig. 27. The appearance of enhanced three-atom clusters in the grey scale plots, such as Fig. 26(a), is due to this superposition of atomic and CDW modulations. The total z-deflection of the STM is approximately 0.6 Å, as shown in Fig. 26(b), with equal amplitudes contributed by the atoms and the CDWs. The CDW and atomic modulations are both enhanced, relative to the intrinsic spatial oscillation of the electron density expected from the periodic lattice disorder, or from the atomic modulation of the electron wavefunction. The possible sources of this enhancement will be discussed in the summary and conclusions.

The CDW forming below 33 K in 2H–$NbSe_2$ gaps only a small part of the Fermi surface so that the change in the density-of-states of the conduction electrons is relatively small. Nevertheless, the CDW gap can be detected with the STM by measuring dI/dV versus V and detecting the change in the slope at the gap edge. A dI/dV versus V curve for 2H–$NbSe_2$ at 4.2 K is shown in Fig. 28 (middle curve) with the gap edge clearly identified at 34 mV. This gives a very large value of $2\Delta_{\mathrm{CDW}}/k_{\mathrm{B}}T_{\mathrm{C}} = 23.9$, indicating that a short coherence length model as suggested by McMillan[48] is required.

Below 7.2 K the superconducting phase of 2H–$NbSe_2$ can also be studied by STM. 2H–$NbSe_2$ is a very anisotropic type II superconductor and the

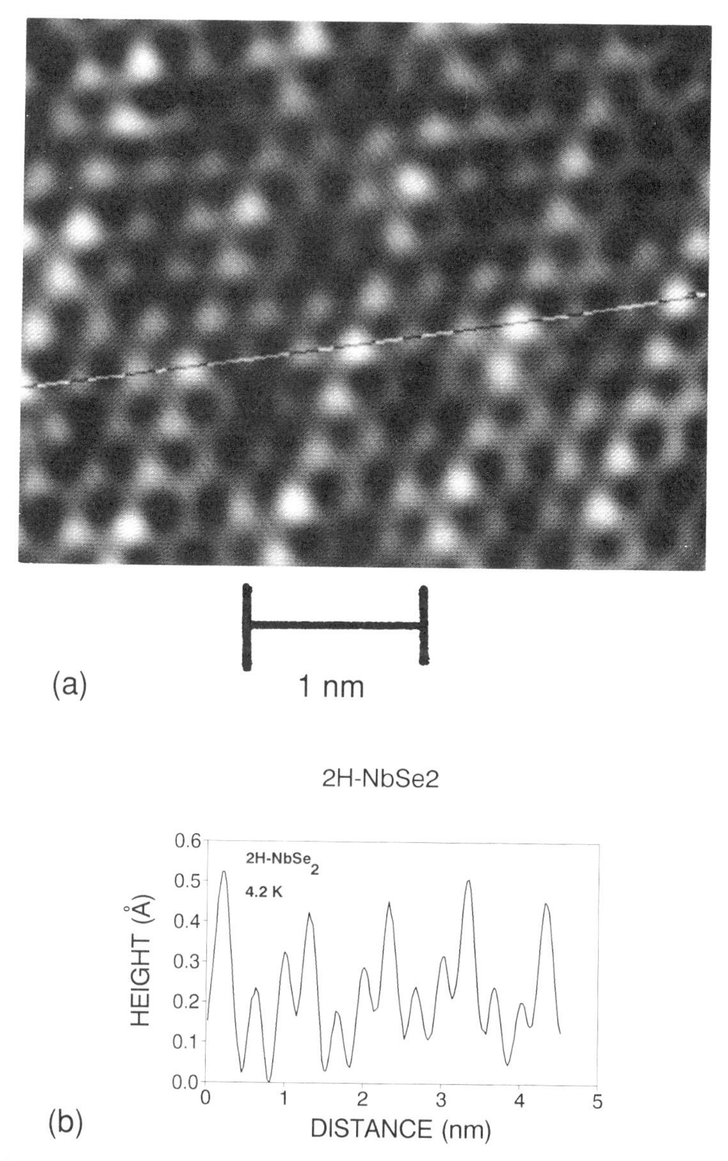

FIG. 26. (a) STM scale image of 2H–NbSe$_2$ at 4.2 K ($I = 2.0$ nA, $V = 25$ mV). The $3\mathbf{a}_0 \times 3\mathbf{a}_0$ CDW modulation is clearly resolved at this magnification. The CDW modulation is centered on the Nb atoms which are displaced by $a_0/\sqrt{3}$ from the atomic modulation due to the surface Se atoms. This gives rise to the asymmetric three atom clusters near the CDW maxima. (From Ref. 7.) (b) Profile of z-deflection along the line shown in (a). The atomic and CDW modulation amplitudes are comparable.

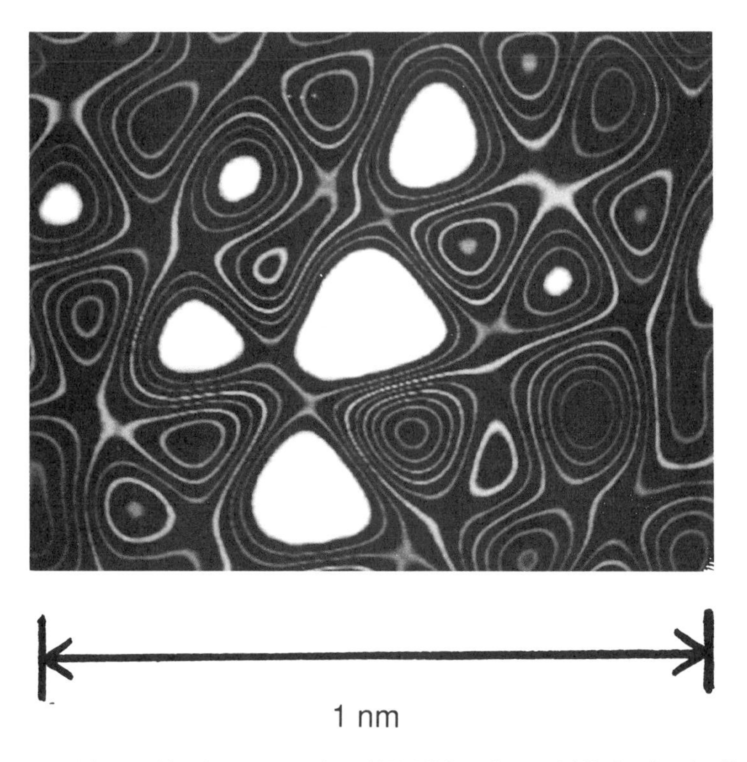

FIG. 27. High-magnification contour plot of 2H–$NbSe_2$ taken at 4.2 K showing detail of the three-atom cluster near a CDW maximum ($I = 2.0$ nA, $V = 25$ mV). (From Ref. 7.)

magnetic field penetrates the superconducting phase to form an Abrikosov vortex lattice. This has been studied in detail by Hess *et al.*[49] using the STM. Images can be formed by measuring dI/dV versus position rather than using the constant current mode as in the scans showing the CDWs. The quantity dI/dV is proportional to $|\Psi_s(eV, r)|^2 n_s(eV)$ and will detect variations in the superconducting DOS. Fig. 6 in Chapter 9 shows a dI/dV STM scan of 2H–$NbSe_2$ at 0.3 K in a magnetic field of 10 kG, applied perpendicular to the layers recorded at an applied bias voltage of 1.3 mV, which is just above the superconducting gap edge of 1.1 mV. The black areas are the vortex cores containing normal electrons, while the white regions represent the strongly superconducting areas with an enhanced DOS just above the gap edge.

8.7.2 STM of 2H–$TaSe_2$

2H–$TaSe_2$ is very similar to 2H–$NbSe_2$, but forms a stronger CDW with an onset at 122.3 K and goes into a superconducting phase only at temperatures below ~ 0.2 K. At 122.3 K a triple **q** CDW forms in an incommen-

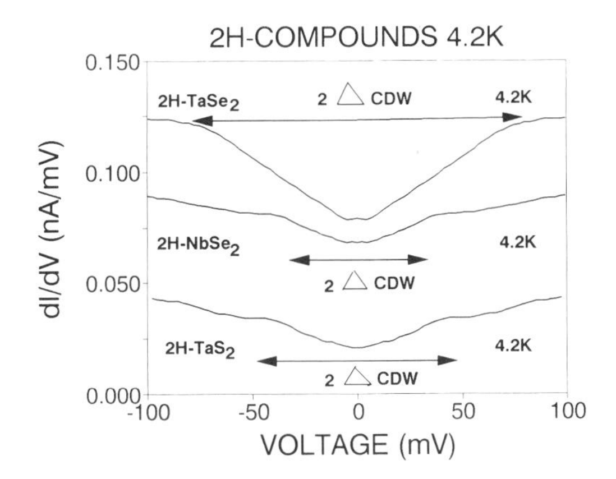

FIG. 28. dI/dV versus V plot recorded on 2H–$TaSe_2$, 2H–$NbSe_2$, and 2H–TaS_2 at 4.2 K. Slope changes can be identified with the CDW gaps and associated density-of-states (DOS) changes near the gap edges. The average gaps are estimated to be ± 80 mV, ± 34 mV, ± 50 mV respectively as indicated by arrows which indicate the region of width 2Δ centered on zero bias. (From Ref. 39.)

surate state characterized by a second-order (or nearly second-order) transition followed by a first-order transition ~ 90 K, below which the $3\mathbf{a}_0 \times 3\mathbf{a}_0$ CDW superlattice becomes commensurate. The STM scans at temperatures of 77 and 4.2 K show the presence of a weak CDW modulation forming a $3\mathbf{a}_0 \times 3\mathbf{a}_0$ CDW superlattice. Figure 29(a) shows an STM scan taken at 4.2 K with a weak enhancement of the z-deflection present on every third row of atoms. A profile along the track drawn in Fig. 29(a) is plotted in Fig. 29(b). The total z-deflection is only 0.2 ± 0.05 Å with a CDW enhancement of ~ 0.1 Å on every third atom. This enhanced z-deflection due to the CDW is larger than the periodic lattice distortion of the surface Se atoms, which is estimated to be about 0.02 Å perpendicular to the surface. The FS and band structure undergo substantial rearrangement, but many extended sections of FS remain, as detected by magnetoquantum oscillations of the magnetoresistance[50] and susceptibility.[51] Strong $\mathbf{k}_\parallel$ and $(\mathbf{k}_\parallel + \mathbf{G})$ mixing may occur at selected points on the FS, but FS obliteration and singularities in the DOS should be much weaker than in the 1T phase of $TaSe_2$. The observed z-deflection is an order of magnitude smaller than observed for 1T–$TaSe_2$, and in this respect follows the expected trend.

Doran and Woolley[52] have carried out band structure calculations for the $3\mathbf{a}_0 \times 3\mathbf{a}_0$ CDW state of 2H–$TaSe_2$. They start with results obtained for the high-temperature band structure calculated by Wexler and Woolley,[53] and show that stabilization of the CDW comes from destruction of the peak in

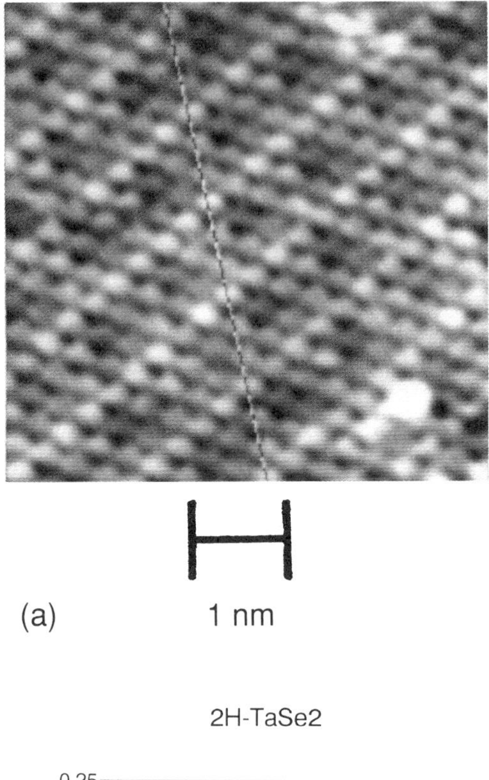

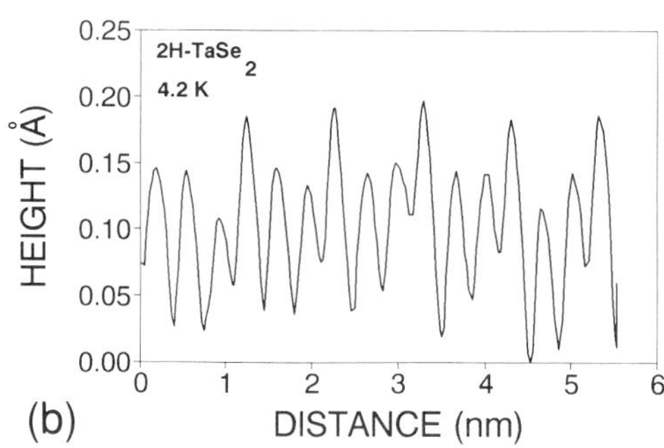

FIG. 29. (a) STM grey scale image of 2H–$TaSe_2$ at 4.2 K ($I = 2.2$ nA, $V = 25$ mV). (b) Profile of z-deflection along the line shown in (a) revealing a weak CDW superlattice enhancement on every third atom. (From Ref. 7.)

the DOS lying just below the Fermi level. Further details can be found in the original papers.

Smith *et al.*[19] have also developed a model band structure calculation in order to analyze photoelectron data. The model allows them to estimate the number of Ta d electrons residing on each inequivalent Ta atom in the distorted CDW state. These atoms were labeled "a," "b," and "c" in Fig. 4 and charge transfer occurs toward the center atom "a" located at the center of a seven-atom cluster, although in this case the "c" atoms are not part of a separate 13-atom cluster as occurred in the 1T phase. The numbers calculated for the Ta d electrons on these inequivalent sites are $n_a = 1.014$, $n_b = 1.005$, and $n_c = 0.977$. This charge transfer is relatively weak compared to the 1T phase and the STM results certainly reflect this, although an exact quantitative comparison cannot be made.

The spectroscopy of 2H–$TaSe_2$ is very similar to that observed for the other 2H phase compounds. A dI/dV versus V curve is shown in Fig. 28 (upper curve). A relatively abrupt change of slope is observed at ± 80 mV and we have identified this with the onset of tunnel conductance above the CDW gap edge. Using a value of $\Delta_{CDW} = 80$ meV and $T_C = 122.3$ K gives a value of $2\Delta_{CDW}/k_B T_C = 15.2$. This is again very large and requires the short coherence length model of McMillan,[48] a conclusion reached for the CDW in all 2H phase compounds so far measured.

8.7.3 STM of 2H–TaS_2

The 2H phase of TaS_2 forms a $\sim 3\mathbf{a}_0 \times 3\mathbf{a}_0$ incommensurate CDW superlattice at low temperatures. The resistivity shows a very small anomaly at the CDW onset, suggesting only a small change in the conduction electron DOS. 2H–TaS_2 has a peak in susceptibility at 80 K, a break in resistivity at 75 K, and a change in sign of the Hall resistivity between 65 and 55 K. Careful resistivity measurements by Tidman *et al.*[54] show an abrupt change in $\rho_{\parallel}$ at 75.3 K, leading to the conclusion that the Fermi surface modification occurs at or below this temperature.

Electron diffraction experiments[55] show the CDW to be incommensurate down to about 14 K. The wave vector is given by $\mathbf{q}_\delta = (1 + \delta)\mathbf{a}_0^*/3$ where $\delta = 0.017 \pm 0.009$ at 70 K. A possible slight decrease in δ as temperature decreases is suggested by the data. The departure from commensurability for 2H–TaS_2 is roughly of the same magnitude, although opposite in sign, as that of the incommensurate phases of 2H–$NbSe_2$ and 2H–$TaSe_2$.

Nuclear magnetic resonance (NMR) studies[56] of 2H–TaS_2 powder at 4.2 K indicate a locally commensurate CDW superlattice. The authors conclude that at low temperatures there is a discommensurate CDW state for 2H–TaS_2, where regions of locally commensurate CDWs are separated by domain walls or discommensurations.

At 4.2 K the STM scans show a well developed hexagonal $3\mathbf{a}_0 \times 3\mathbf{a}_0$ superlattice as shown in the grey scale image of Fig. 30(a). The charge maxima are exceptionally well defined and show a high atom at each vertex

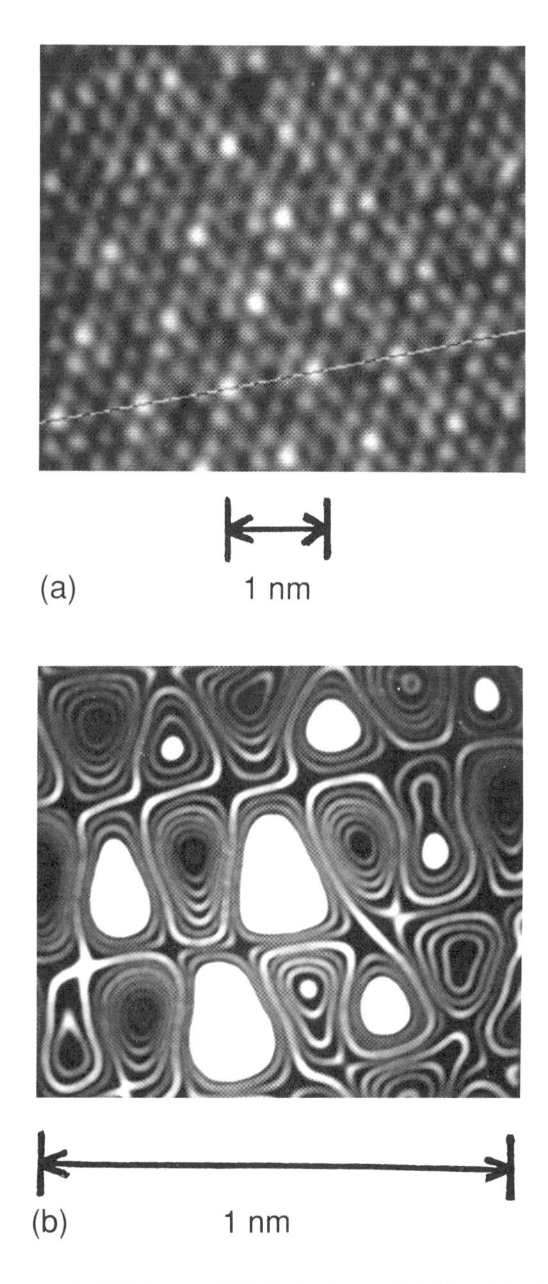

FIG. 30. (a) Grey scale STM image of 2H–TaS_2 taken at 4.2 K ($I = 2.2$ nA, $V = 25$ mV). (b) Contour plot showing the asymmetric seven-atom cluster of surface S atoms located above the CDW maximum occuring on the Ta atom cluster. (From Ref. 7.)

of the triple **q** CDW pattern. The seven-atom cluster of S atoms near each charge maximum still shows the asymetry connected with the phase difference due to the $a_0/\sqrt{3}$ displacement between the CDW and atomic modulations. This is again clearly demonstrated in the high-magnification contour plot of Fig. 30(b). This charge asymmetry, due to the superposition, is characteristic of all three 2H phases studied. In 2H–TaS_2 the total z-deflection is 0.6 ± 0.1 Å compared to the value 0.2 ± 0.05 Å observed for 2H–$TaSe_2$.

Shubnikov–de Haas measurements of the FS cross sections have been carried out by Hillenius and Coleman[57] and show a range of cross sections that are more than an order of magnitude smaller than those observed for 2H–$TaSe_2$. Eleven frequencies in the range 0.04 to 6.4 MG were detected. The two highest of frequencies in 2H–TaS_2 match the two lowest frequencies in 2H–$TaSe_2$. The larger extended sections observed in 2H–$TaSe_2$ are completely absent in 2H–TaS_2. The rest are systematically smaller, thereby leading to a total FS area in 2H–TaS_2, which is more than an order of magnitude smaller than observed in 2H–$TaSe_2$.

The high-temeprature FSs constructed from the band structure calculations of Wexler and Woolley[53] show that the high-temperature FS sections in the 2H phases of TaS_2 and $TaSe_2$ have nearly identical areas. Therefore, we conclude that the CDW in 2H–TaS_2 obliterates a much larger portion of the FS than occurs in the CDW phase of 2H–$TaSe_2$. Within the Tersoff[13] model this would be expected to enhance the nodal structure at the CDW wavelength. It is also likely that a number of the extremely small FS cross sections may be located at the BZ corners and edges, leading to greater mixing of $\mathbf{k}_{\parallel}$ and $(\mathbf{k}_{\parallel} + \mathbf{G})$. This latter point cannot at present be confirmed, since no detailed band calculations or models of the $3\mathbf{a}_0 \times 3\mathbf{a}_0$ CDW phase of 2H–TaS_2 have been worked out. However, the overall evidence from the FS experiments and comparison of the STM profiles shows that these subtle electronic differences are detected in the STM scans.

The spectroscopy results on 2H–TaS_2 follow the systematic behavior observed for the three 2H phase compounds measured and presented in Fig. 28. The dI/dV versus V curve for 2H–TaS_2 (lower curve) shows a weak slope change indicating that the CDW gap edge is located at a bias of $\leqslant 50$ mV. With a $T_C = 75.3$ K this gives a value of $2\Delta_{CDW}/k_B T_C = 15.4$, approximately the same magnitude as observed for 2H–$TaSe_2$.

8.8. 4Hb Phase Transition Metal Dichalcogenides

The 4Hb phase polytypes of $TaSe_2$ and TaS_2 have four three-atom-layer sandwiches per unit cell arranged with alternating sandwiches of trigonal prismatic coordination and octahedral coordination. The CDW transitions observed in these phases occur independently in the sandwiches of different coordination, and maintain the characteristics associated with the pure octahedral (1T) phase and the pure trigonal prismatic (2H) phase. The CDW transitions are modified to some extent, but onset temperatures in the trigonal prismatic layers are still ~ 600 K. The octahedral sandwiches show a very strong $\sqrt{13}\mathbf{a}_0 \times \sqrt{13}\,\mathbf{a}_0$ CDW superlattice, while the trigonal prismatic sandwiches show much weaker $3\mathbf{a}_0 \times 3\mathbf{a}_0$ CDW superlattices. The STM scans indicate that the strong CDW amplitudes observed for the 1T sandwiches have been reduced relative to the pure 1T phase, but they still provide the dominant z-deflection in the STM scans of the 1T surface. This reduction of CDW amplitude does, however, allow the STM response to provide a better resolution of the surface atomic modulation. Details of the STM results for 4Hb–$TaSe_2$ and 4Hb–TaS_2 will be discussed next.

8.8.1 STM of 4Hb–$TaSe_2$

The CDWs forming in 4Hb–$TaSe_2$ are similar in structure to those forming in the pure octahedral (1T) and pure trigonal prismatic (2H) phases. They maintain the same CDW superlattice orientation as observed in the pure phases, but only the CDW in the octahedrally coordinated layer becomes commensurate with lockin to a $\sqrt{13}\,\mathbf{a}_0 \times \sqrt{13}\,\mathbf{a}_0$ superlattice at 410 K after an onset at ~ 600 K. The trigonal prismatic layer develops an approximate $3\mathbf{a}_0 \times 3\mathbf{a}_0$ superlattice at ~ 75 K, which remains incommensurate down to 10 K, as determined by neutron diffraction data.

The CDWs in the trigonal prismatic sandwiches show no order in the **c** axis direction, and therefore exhibit two-dimensional correlation in the plane of the trigonal prismatic sandwich. On the other hand, the much stronger CDW in the octahedral sandwiches shows long range order along the **c** axis, indicating that it interacts across the intervening trigonal prismatic layer to form a three-dimensional CDW superlattice.

The STM scans can be carried out on a crystal with either an octahedral surface sandwich or a trigonal prismatic surface sandwich obtained by cleaving a single crystal and scanning both sides of the cleave. The STM z-deflection shows that the strength of each independent CDW has been reduced in the mixed 4Hb phase compared to the corresponding pure phases, with the result that the atomic modulation is more easily resolved on the octahedral sandwich. However, the $\sqrt{13}\,\mathbf{a}_0 \times \sqrt{13}\,\mathbf{a}_0$ CDW superlattice still makes a dominant contribution to the STM deflection.

An STM scan on the octahedral sandwich of 4Hb–$TaSe_2$ is shown in Fig. 31(a) and shows the dominant $\sqrt{13}\,\mathbf{a}_0 \times \sqrt{13}\,\mathbf{a}_0$ superlattice. The profile of z-deflection is shown in Fig. 31(b) and indicates an anomalously large CDW amplitude of ~ 5 Å similar to that observed in the pure 1T phase. The strength of the CDW in the octahedral sandwich can also be seen in STM

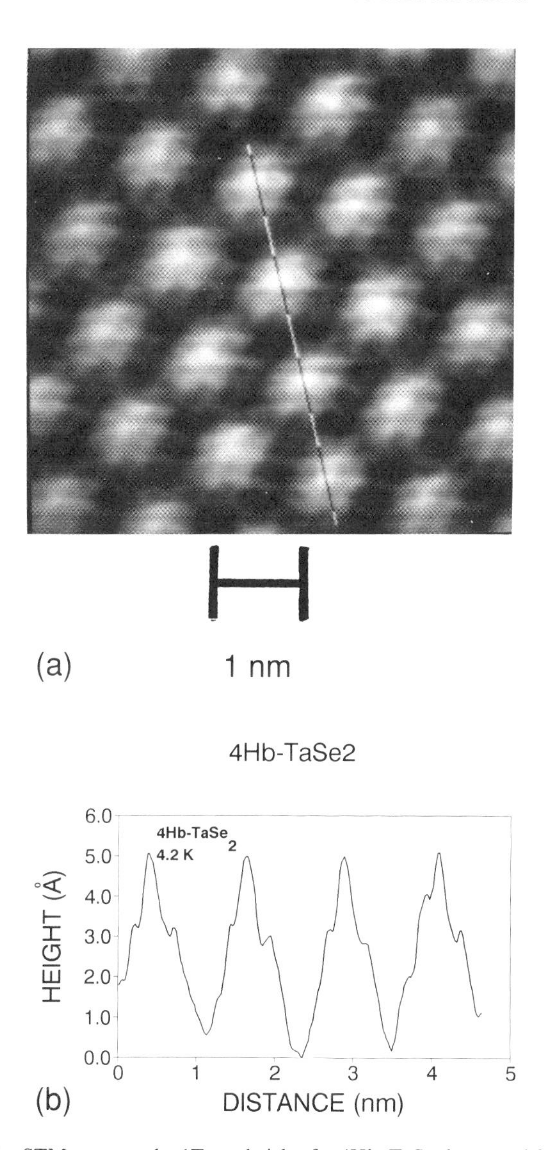

FIG. 31. (a) An STM scan on the 1T sandwich of a 4Hb–$TaSe_2$ layer at 4.2 K showing both the CDW and atomic modulations ($I = 2.2$ nA, $V = 25$ mV). (b) Profile of z-deflection along the line shown in (a). The z-deflection of ~ 5 Å is dominated by the CDW in the 1T layer, but is reduced in amplitude from that observed in the pure 1T phase.

scans on the trigonal prismatic sandwich of 4Hb–$TaSe_2$. As shown in the composite of STM scans in Fig. 32, the CDW can show a large contribution by enhancing sunflowers of surface Se atoms, even though it resides in the octahedral sandwich that lies ~ 6 Å below the surface sandwich. Adjustment of the STM scan can bring out the $\sim 3\mathbf{a}_0 \times 3\mathbf{a}_0$ CDW superlattice modulation in the surface sandwich, but this remains weak relative to the atomic modulation of the surface Se atoms. (See the series of scans in Fig. 32.)

By careful adjustment of the STM, it is possible to simultaneously detect both CDW modulations when scanning the trigonal prismatic sandwich. An example is shown in Fig. 33(a) where the two CDW superlattices generate coincident maxima on a super-superlattice of size $3\sqrt{13}\,\mathbf{a}_0 \times 3\sqrt{13}\,\mathbf{a}_0$. This super-superlattice is outlined by the large rhombus drawn in Fig. 33(a) and gives rise to an enhanced intensity on every third major peak in the profile of Fig. 33(b).

8.8.2 STM of 4Hb–TaS_2

Crystals of 4Hb–TaS_2 show a CDW structure in which two related $\sqrt{13}\,\mathbf{a}_0 \times \sqrt{13}\,\mathbf{a}_0$ commensurate superlattices appear in the octahedrally-coordinated layers after an initial CDW onset temperature of ~ 600 K. Friend *et al.*[58] compared **q** vectors of the CDWs found in 4Hb–TaS_2 to those observed in the pure 1T and 2H phases and concluded that the transfer of ~ 0.12 electrons per tantalum atom from the octahedral to the trigonal prismatic layers could account for the required FS changes. Below the CDW resistive anomaly at 315 K, the 1T sandwiches are believed to become semiconducting while the 1H sandwiches remain metallic below the CDW transition at 22 K. Data published by Wattamaniuk *et al.*[59] on the temperature dependence of resistivity perpendicular and parallel to the layers support this conclusion. The two CDW transitions give rise to discontinuities in $\rho_{\parallel}$, but the behavior remains metallic down to 4.2 K. $\rho_{\perp}$ is dominated by the semiconducting 1T layers and rises to $\sim 1200\,\mu\Omega$ cm at 4.2 K.

At 4.2 K the STM scans on the octahedral sandwiches show a strong CDW superlattice superimposed on the surface S atom pattern as shown in Fig. 34(a). As shown in the profile of Fig. 34(b) the total z-deflection is ~ 2 Å with a dominant CDW modulation amplitude of ~ 1.5 Å. The CDW forms a $\sqrt{13}\,\mathbf{a}_0 \times \sqrt{13}\,\mathbf{a}_0$ hexagonal superlattice of charge maxima and deep minima owing to the strong charge transfer and LDOS modification caused by the CDW. The superlattice pattern is oriented at the commensurate angle of 13.9°, as found in the commensurate phase of the pure 1T crystals. However, in the STM scans on the 1T sandwiches of the 4Hb phase the CDW maxima are systematically lower, and the atomic modulation consistently larger than observed in the pure 1T phase where the CDW is completely dominant.

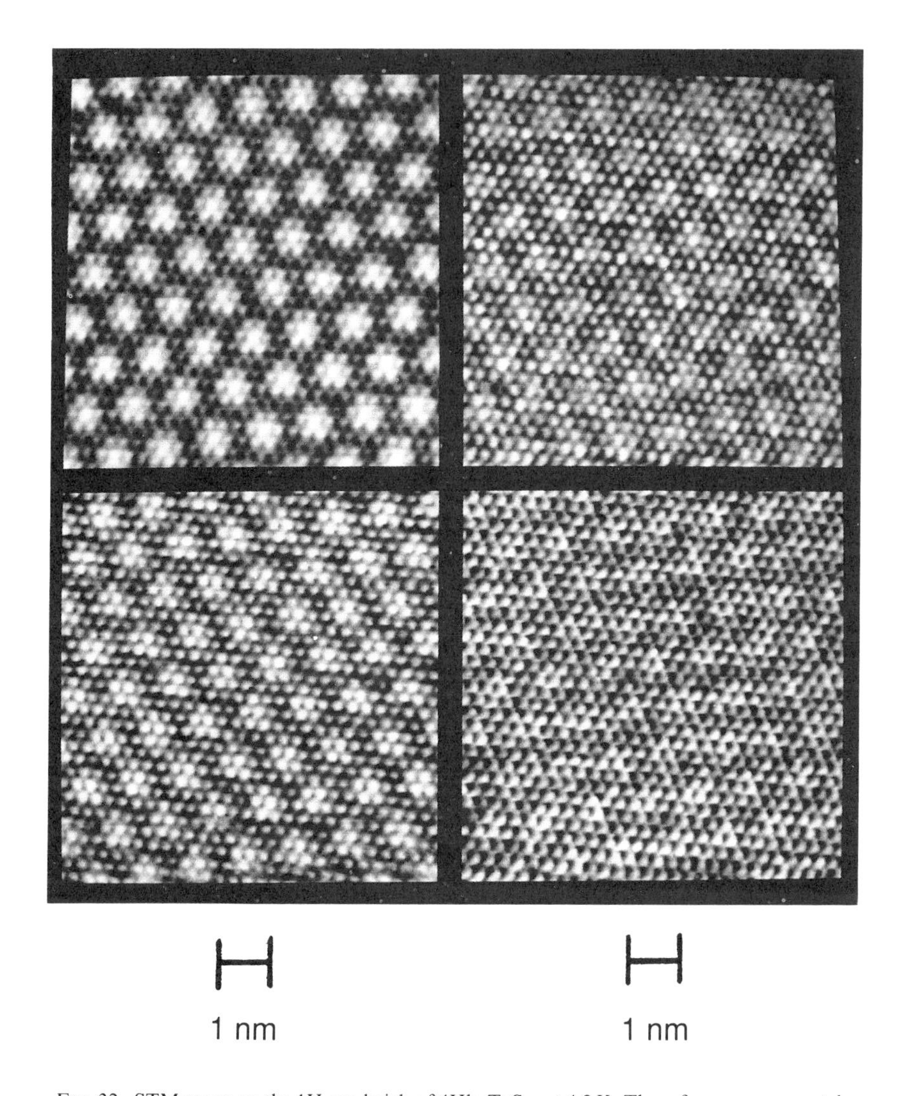

FIG. 32. STM scans on the 1H sandwich of 4Hb–$TaSe_2$ at 4.2 K. These four scans were taken during a period of ~ 2 hours and scan the crystal at identical current and bias voltage settings ($I = 2.2$ nA, $V = 25$ mV). These images demonstrate the variance in the CDW pattern seen on the 1H sandwich of 4Hb–$TaSe_2$ due to the complex superposition of the atomic modulation, the $3\mathbf{a}_0 \times 3\mathbf{a}_0$ CDW modulation from the surface 1H layer, and the $\sqrt{13}\,\mathbf{a}_0 \times \sqrt{13}\,\mathbf{a}_0$ CDW modulation from the underlying 1T layer. Each image shows good atomic resolution, along with a variable contribution from the CDW modulations. The CDW contribution to each image (described in the order in which they were taken) is as follows: lower-right, a $\sqrt{13}\,a_0$ modulation in one direction (slightly counterclockwise from the horizontal) and a weak $3a_0$ modulation along the atom rows closest to the vertical; lower-left, an intermediate strength $\sqrt{13}\,a_0$ triple CDW; upper-right, both a triple $\sqrt{13}\,a_0$ and a triple $3a_0$ CDW; upper-left, a strong $\sqrt{13}\,a_0$ triple CDW. (From Ref. 7.)

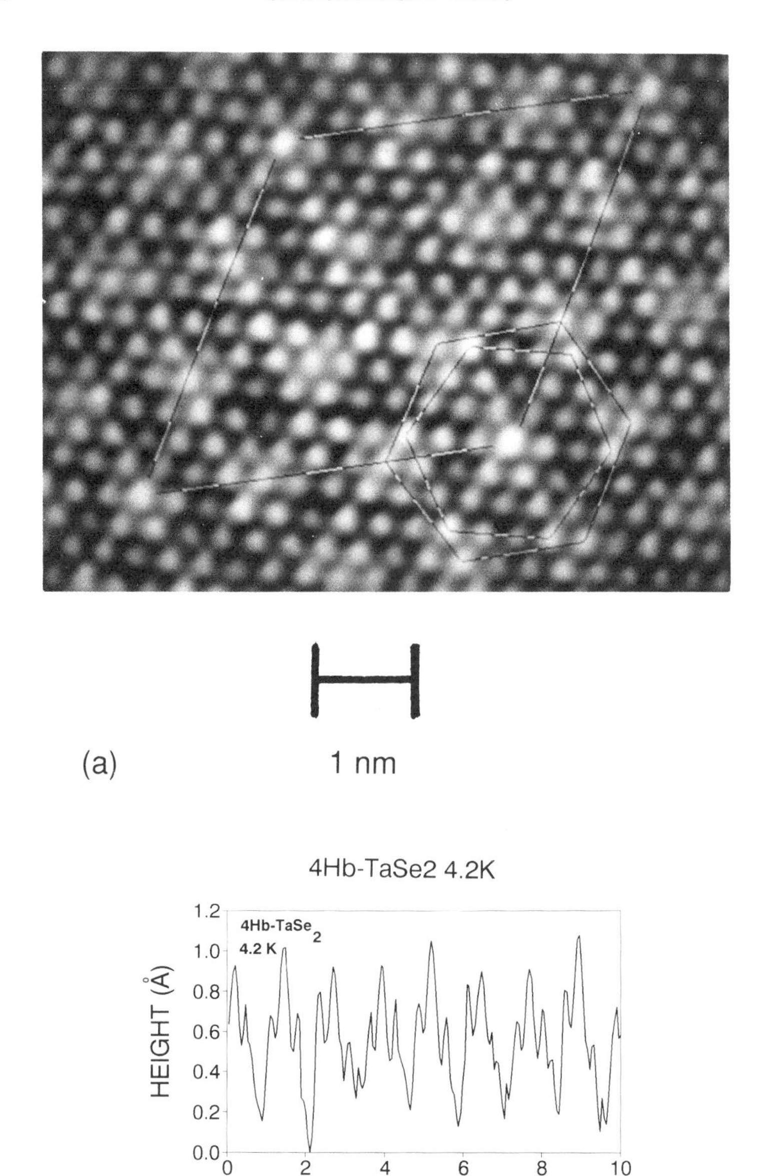

FIG. 33. (a) Grey scale image from a STM scan on the 1H sandwich of a 4Hb–$TaSe_2$ crystal at 4.2 K. Two hexagons outline the $3\mathbf{a}_0 \times 3\mathbf{a}_0$ and $\sqrt{13}\,\mathbf{a}_0 \times \sqrt{13}\,\mathbf{a}_0$ superlattices which are rotated 13.9° with respect to each other. The rhombus outlines the unit cell of the $3\sqrt{13}\,\mathbf{a}_0 \times 3\sqrt{13}\,\mathbf{a}_0$ super-superlattice generated by the superposition of the $3a_0$ and the $\sqrt{13}\,a_0$ modulations ($I = 2.2$ nA, $V = 25$ mV). (b) Profile of z-deflection along the edges of two adjacent $3\sqrt{13}\,a_0$ super-supercells. The three highest peaks each occur at an atom located at the corner of a $3\sqrt{13}\,a_0$ rhombus. These maxima are approximately 0.1 Å higher than the other $\sqrt{13}\,a_0$ maxima and indicate the points of phase match between the two CDW superlattices. (From Ref. 7.)

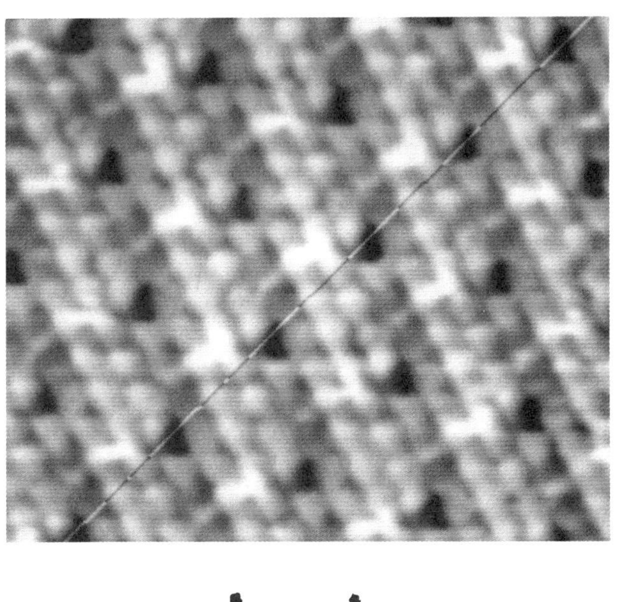

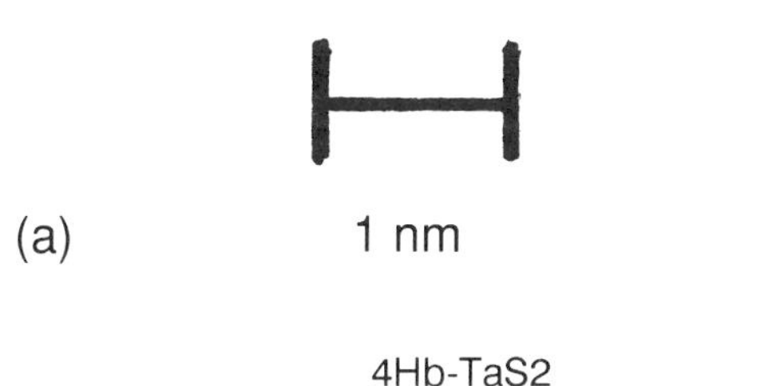

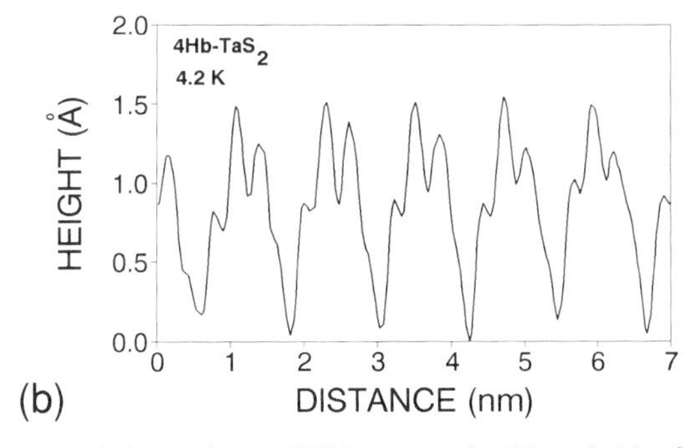

FIG. 34. (a) Grey scale image from a STM scan on the 1T sandwich of a 4Hb–TaS_2 crystal at 4.2 K. Both the CDW and atomic modulations are detected and the $\sqrt{13}\,\mathbf{a}_0 \times \sqrt{13}\,\mathbf{a}_0$ CDW superlattice generates relatively deep minima ($I = 2.2$ nA, $V = 25$ mV). (b) Profile of z-deflection along the line shown in (a). The majority of the z-deflection is due to the deep minima. (From Ref. 7.)

The presence of the much higher conductivity of the trigonal prismatic sandwich with a more extended FS can moderate both the effect of the strong CDW in the 1T phase layers and the possible nodal structure in the LDOS associated with CDW formation. The presence of a complex FS is confirmed by Shubnikov-de Haas oscillations observed by Fleming and Coleman[50] that detect the presence of at least 15 frequencies in the range 0.044–8.2 MG. Wexler *et al.*[60] and Doran *et al.*[61] have used the layer method to calculate band structure in 4Hb–TaS_2. Although the results are close to a superposition of the bands characteristic of the pure 2H and 1T phases, hybridization and electron transfer is found to occur between the two types of sandwich and could easily account for the subtle changes in STM response.

The CDW transition in the trigonal prismatic layers exhibits a reduced onset temperature of 22 K compared to the $\sim$75 K observed in the pure 2H phase. STM scans on the 1H layer of 4Hb–TaS_2 detect clear evidence of a very weak $3\mathbf{a}_0 \times 3\mathbf{a}_0$ CDW superlattice. However, in some cases the stronger $\sqrt{13}\,\mathbf{a}_0 \times \sqrt{13}\,\mathbf{a}_0$ CDW superlattice from the 1T sandwich 6 Å below is detected as enhanced "sunflowers," showing a random variation of intensity in the scan. The overall behavior is similar to the STM response observed for 4Hb–$TaSe_2$, but the CDWs appear to give a weaker modulation of the tunneling current. These results are consistent with the general trends expected from the electronic differences between 4Hb–TaS_2 and 4Hb–$TaSe_2$, but represent only qualitative observations.

8.8.3 Determination of Energy Gaps in 4Hb–$TaSe_2$ and 4Hb–TaS_2

Structure in the dI/dV versus V curves associated with the two separate CDWs has been observed at 4.2 K in both 4Hb–$TaSe_2$ and 4Hb–TaS_2. However, the relative structure introduced by the CDW gaps is considerably different in the 4Hb selenide versus the sulfide. A representative dI/dV versus V curve for 4Hb–$TaSe_2$ at 4.2 K is shown in Fig. 35(a). Two regions, characterized by a rapid onset of conductance increase followed by a peak, are clearly observed. These are characteristic of a gap edge followed by a peak in the DOS just above the gap edge. Both the STM amplitude and the associated gap structure in the dI/dV versus V curve remain well-defined indicating that the CDW strength has not been changed significantly from that observed in the pure 1T and 2H phase. Both the 1T and 1H layers remain metallic at the lowest temperatures. Using the peak positions in the curves of the type shown in Fig. 35(a), we have determined the two CDW gaps in 4Hb–$TaSe_2$ to be $\Delta_{1T} = 140$ meV and $\Delta_{1H} = 60$ meV. Both are reduced from those observed in the pure 1T and 2H phases where the measured values were $\Delta_{1T} = 150$ meV and $\Delta_{2H} = 80$ meV. The calculated values of $2\Delta_{CDW}/k_B T_C$ are 5.4 and 18.5 respectively, comparable to the same values observed in the pure

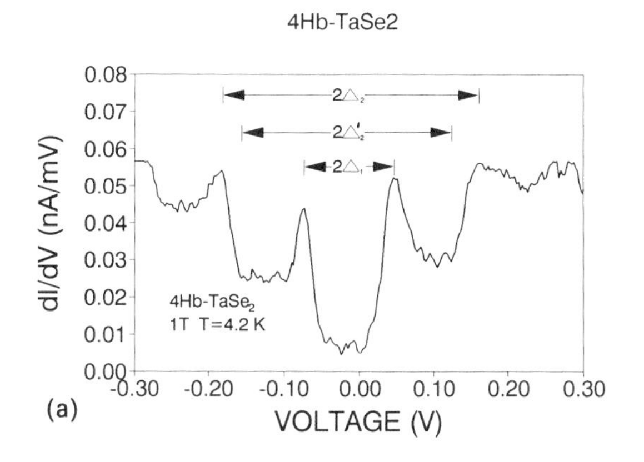

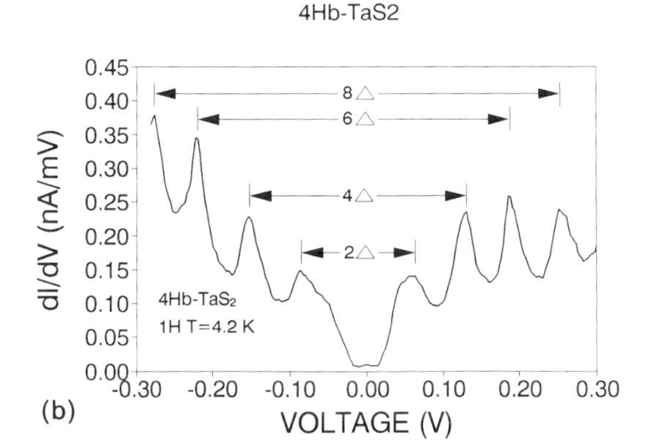

FIG. 35. (a) dI/dV curve for 4Hb–$TaSe_2$ measured at 4.2 K. Sharp peaks in the conductance associated with the CDW gaps in the crystal are apparent at ~60 mV and ~180 mV. In order to compare the high temperature CDW gap value with data on pure 1T–$TaSe_2$, the gap structure can be measured as existing at the point where the rapid rise in conductance occurs. This conduction gives a value of ~140 mV. (b) A dI/dV versus V curve measured at 4.2 K for 4Hb–TaS_2. The conductance peaks are identified with the high-temperature CDW gap that generates peaks at the fundamental of $\Delta = 72$ mV plus harmonics of the fundamental. The low-temperature CDW gap occurs within the minimum near zero bias and is not well-resolved.

phases. Extremely strong coupling is observed in the 1H sandwich which requires a short coherence length model, while the 1T sandwich shows a substantially smaller value of $2\Delta_{CDW}/k_B T_C$.

These results are consistent with the overall behavior observed in the STM response and confirm the electronic modifications expected relative to the separate pure phase measurements. The sharp gap structure is consistent with the reasonably good resolution obtained for the two essentially independent

CDW superlattices in 4Hb–$TaSe_2$, as well as their simultaneous resolution in the CDW super-superlattice detected in Fig. 33(a).

In contrast to the 4Hb selenide, the 4Hb sulfide shows a much stronger modification of the CDW structure from that observed in the separate 1T and 2H phase compounds. The $3\mathbf{a}_0 \times 3\mathbf{a}_0$ CDW superlattice onset temperature is reduced to 22 K and, as discussed in the previous section, the STM amplitude is extremely weak. The dI/dV versus V curves in which good CDW gap structure is observed, are completely dominated by the larger CDW gap associated with the 1T layer. An example is shown in Fig. 35(b). The high-temperature CDW gap structure gives a fundamental peak at ~ 72 mV and this is followed by a harmonic series showing very strong amplitudes. The low-temperature gap structure is not well-resolved and lies in the region of minimum conductance near zero bias. The value is estimated from the weak structure in a number of runs to be $\Delta_{1H} \leqq 30$ meV. The low-temperature CDW gap appears to be reduced substantially from the value of 50 meV observed in the pure 2H phase, and this is consistent with the major reduction of CDW onset temperature in the 1H sandwich from 75 to 22 K. The values of $2\Delta_{CDW}/k_B T_C$ maintain the relative difference observed for the CDWs in the octahedral versus the trigonal prismatic coordination sandwiches of all layer structure dichalcogenides, but the measured values of Δ_{1H} and Δ_{1T} are not sufficiently accurate for quantitative comparison.

The behavior of the spectroscopic response in 4Hb–TaS_2 compared to 4Hb–$TaSe_2$ may arise from the semiconducting nature of the octahedral layer in 4Hb–TaS_2. If tunneling is required between the octahedral layers as proposed by Wattamaniuk *et al.*[59] in their analysis of $\rho_\perp$, then the tunneling electrons may couple very strongly to the CDW gap in the octahedral layers. The existence of multiple tunnel junctions in series can easily lead to strong harmonic amplitudes.

8.9 Linear Chain Transition Metal Trichalcogenides

Three linear chain compounds in the transition metal chalcogenide group have been studied at 77 K by Slough *et al.*[10,11,62] using STM. Two of these, $NbSe_3$ and TaS_3, exhibit CDW transitions while the third compound, $TaSe_3$, does not. Gammie *et al.*[63] have published preliminary STM data on $NbSe_3$ and TaS_3. Dai *et al.*[12] have reported the most complete STM data on $NbSe_3$ at 4.2 K in which the CDW modulations of the two separate CDWs are resolved in great detail. These quasi-one-dimensional metals are built up from a chalcogen trigonal prismatic cage surrounding a metallic chain. The MX_6 trigonal prisms are stacked on top of each other to form MX_3 chains. The chains usually run parallel to the **b** axis of a monoclinic unit cell and neighboring chains are translated by $b_0/2$. The prisms are irregular due to the

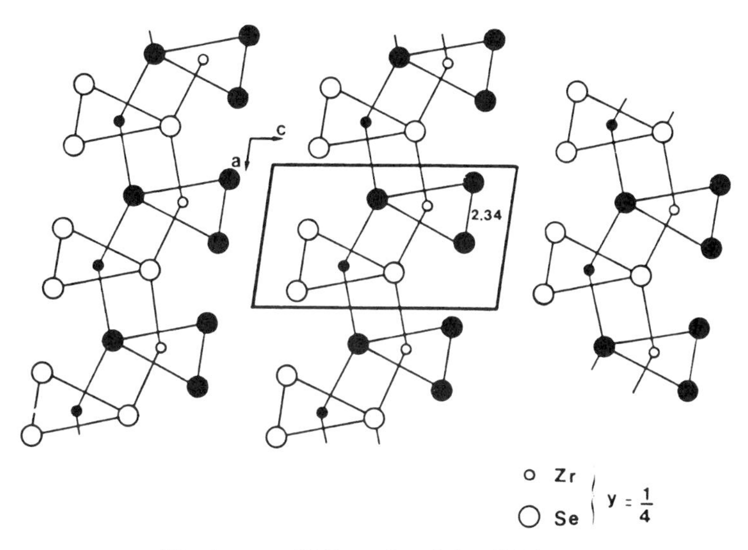

The structure of $ZrSe_3$ projected along [010].

(a)

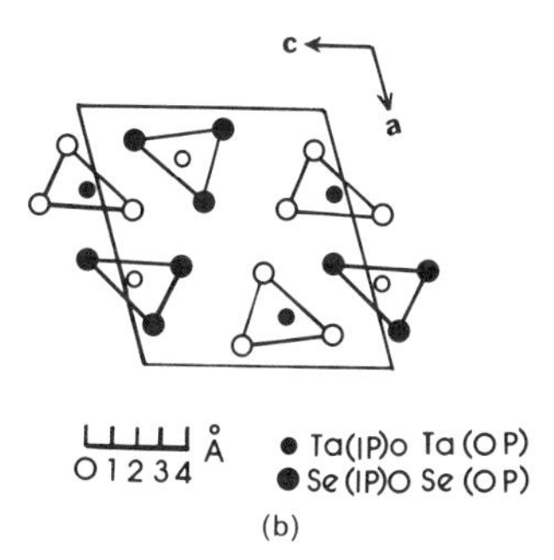

(b)

FIG. 36. Cross sections perpendicular to the chain axes of the unit cells in (a) $ZrSe_3$, (b) $TaSe_3$, and (c) $NbSe_3$. These unit cells are characteristic of the three types of chain structures found in Group IV A trichalcogenides. (From Ref. 2.)

presence of chalcogen-chalcogen pair, and bonding is allowed between one metal atom and chalcogen atoms belonging to neighboring chains. The chalcogen pair can act as an electron reservoir that adjusts the density of electrons on the adjacent metallic chain. Cross sections of the unit cells in $ZrSe_3$, $TaSe_3$, and $NbSe_3$ are represented in Fig. 36, and are characteristic of the three types of chain structure found in the Group IV A transition metal trichalcogenides.

The structure of the trichalcogenides can be described according to the number of chain types found per unit cell. The compounds with one type of

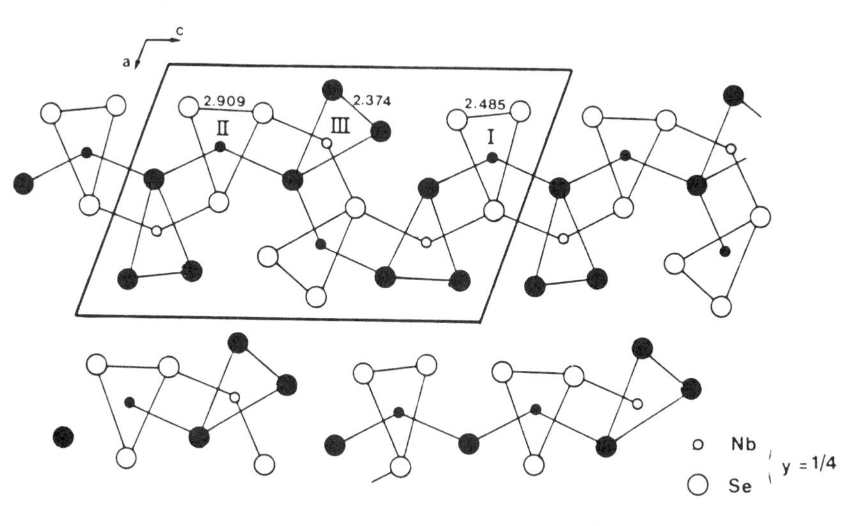

The structure of $NbSe_3$ projected along [010].

(c)

FIG. 36. Continued.

chain per unit cell are represented by $ZrSe_3$ which has two variants, A and B, that differ in the details of bond lengths. In either type the chain is repeated twice in the unit cell. These materials are usually semiconductors and have not been studied by STM or AFM. This type also includes NbS_3, which exists in several forms. One form may exhibit CDW transitions, although further experiments will be required in order to confirm the complete details.

$TaSe_3$ has two types of chains per unit cell with a total of four chains in a monoclinic unit cell. The resistivity data show no evidence of CDW transitions down to 2 K at which point the crystal goes superconducting. A resistivity anomaly has been observed just above the transition to superconductivity, but this is generally thought to be related to the onset of superconductivity. Tajima and Yamaya[64] have investigated this anomalous resistance rise that occurs within 1 K of the superconducting transition. STM studies have been carried out by Slough *et al.*[62] at 77 K and Dai *et al.*[12] at 4.2 K and will be discussed later. More details of the crystal structure and electronic properties can be found in Ref. 2.

The most important compound having three types of chain per unit cell is $NbSe_3$. The three different chains are repeated twice in the monoclinic unit cell and are distinguished by different Se–Se bond lengths of 2.37, 2.48, and 2.91 Å in their triangular basis. This implies considerable electronic exchange between the metallic chains, which can only be treated in detail by band structure calculations such as carried out by Shima and Kamimura.[65] Blocks of four chains, similar to those of $TaSe_3$, are separated by groups of two chains of the type found in $ZrSe_3$. The four-chain block in $NbSe_3$ is not repeated through an inversion center as found in $TaSe_3$, but is interrupted by the two-chain unit that has often been conjectured to have little or no conduction electron density, and therefore not involved in the CDW transitions.[66,67] However, detailed band structure calculations by Shima and Kamimura[65] suggest that this two-chain unit carries appreciable conduction electron density and contributes to the low-temperature of CDW formation. STM results discussed later confirm this result. There are two independent CDW transitions observed, one with an onset at 144 K and one with an onset at 59 K. The CDWs in $NbSe_3$ have been studied extensively by STM, and the CDW structure on each individual chain pair has been resolved.[12] The STM results clearly demonstrate that all three types of chain carry a CDW modulation.

A second compound that has been studied by STM and exhibits three types of chain per unit cell is TaS_3. Two polytypes of TaS_3 have been identified, one with a monoclinic structure similar to $NbSe_3$ and one with an orthorhombic structure. The monoclinic form has three types of chains, two with short S–S bond lengths of 2.068 and 2.105 Å, and a third chain with a much larger S–S bond length value of 2.83 Å. The orthorhombic structure is a very similar arrangement of three chain types. The exact structure has not been determined, but the unit cell is most likely built up of four slabs of TaS_3 chains, giving rise to 24 chains per unit cell. The relative orientations of the unit cells for the monoclinic and orthorhombic forms of TaS_3 have been discussed by Meerschaut and Rouxel.[2] They have proposed a model that is built up of four slabs with TaS_3 chains related in a zig-zag manner. This model is based on Patterson maps and symmetry considerations and suggests a four-chain grouping somewhat similar to that found in the monoclinic form.

Both forms of TaS_3 show CDW transitions. The monoclinic form exhibits two independent CDW transitions with onset temperature of $T_1 = 240$ K and $T_2 = 160$ K, while orthorhombic TaS_3 shows one CDW transition with an onset of $T = 210$ K. The orthorhombic phase has been studied by STM and results are reviewed later.

8.9.1 STM of $NbSe_3$ at 77 K

The two independent CDWs forming at 144 and 59 K in $NbSe_3$ remain incommensurate[68,69] with the lattice down to the lowest temperature measured (4.2 K). The high-temperature CDW is estimated[70] to obliterate ~20% of the FS, while the low-temperature CDW obliterates ~60% of the remaining FS. The wave vector for the high-temperature CDW has been determined by Tsutsumi *et al.*[71] and Hodeau *et al.*[72] as $\mathbf{q}_1 = (0, 0.243, 0)$ at

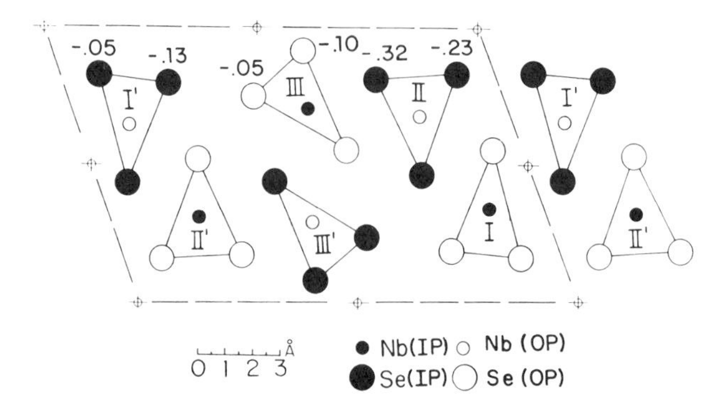

FIG. 37. Cross section of the unit cell of $NbSe_3$ viewed in the **a**–**c** plane. The chain labeling follows that used in Ref. 65 and the numbers refer to the calculated static charge on the surface Se atoms. (From Ref. 65.)

145 K and $\mathbf{q}_1 = (0, 0.241, 0)$ at 59 K. Fleming *et al.*[68] report $\mathbf{q}_1 = (0, 0.243 \pm 0.005, 0)$ with little or no temperature dependence. The low-temperature CDW wave vector was measured as $\mathbf{q}_2 = (0.5, 0.259, 0.5)$ by Tsutsumi *et al.*[71] and Hodeau *et al.*,[72] while Fleming *et al.*[68] measured $\mathbf{q}_2 = (0.5, 0.263 \pm 0.005, 0.5)$. Within experimental error these measurements are in agreement and we take the wavelengths of the two CDWs to be $4.115b_0$ (144 K) and ($2.00\,a_0$, $3.802\,b_0$, $2.00\,c_0$) (59 K), respectively.

At 77 K only one CDW should exist and it should be localized on chains of type III and III′ as identified in the cross section of Fig. 37. Analysis of band structure calculations[65,73] and NMR experiments[74,75] confirm that the high-temperature CDW exists on only one pair of chains. The identification of chain types I, I′; II, II′; and III, III′ follows the designations used in the band structure calculations of Shima and Kamimura.[65] The numbers in Fig. 37 represent the calculated net charge on the Se atoms.

STM scans of $NbSe_3$ at 77 K have been carried out by Slough *et al.*[10,11,62] and the CDW modulation has been clearly resolved. The STM analysis requires identification of the correct chain structure, which must take into account height variations of the surface chains as well as the unequal charges on the Se atoms. As shown in Fig. 38(a) the three surface chains are reasonably well-resolved with one chain exhibiting the maximum z-deflection. However, this is not chain III that contains the row of Se atoms forming the highest surface elevation (see Fig. 37). Rather, it is identified as chain II, which carries the largest negative charge on the Se atoms. Chain III is the chain of intermediate height in the scan of Fig. 38(a). This is confirmed in the profiles plotted in Fig. 39(a), which are taken along the chain axis. The center profile is taken along the chain of intermediate height and shows a modulation of $\sim 4b_0$ as outlined by the dashed curve. The total z-deflection observed

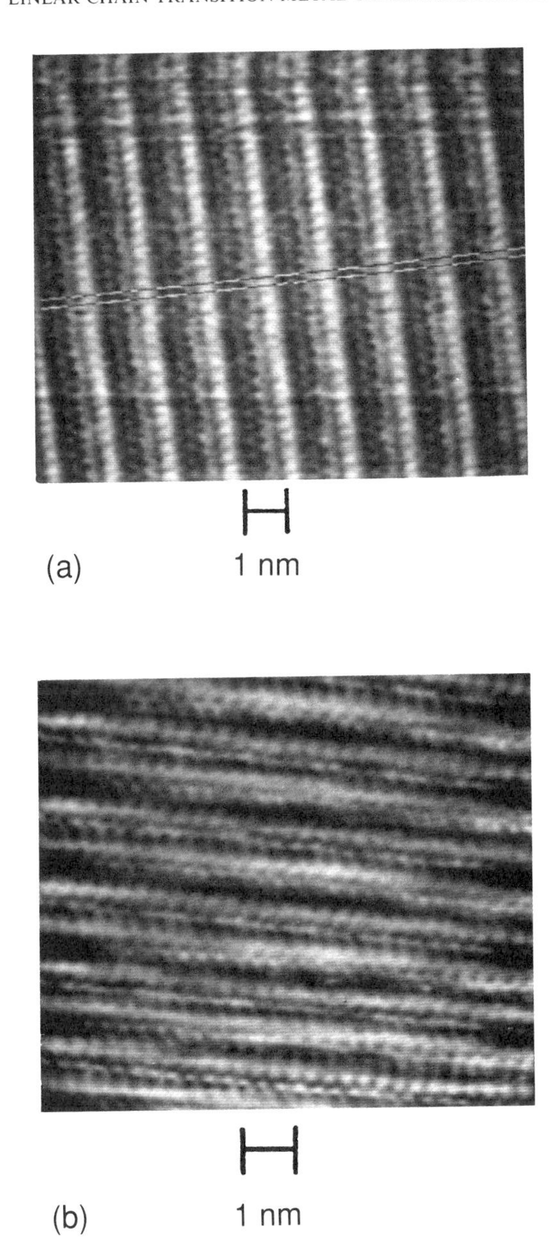

FIG. 38. (a) Grey scale STM image of $NbSe_3$ at 77 K taken with the tip scanning perpendicular to the chains. Three chains per unit cell are resolved in the image ($I = 5\,nA$, $V = 70\,mV$). (From Ref. 10.) (b) Grey scale STM image of $NbSe_3$ at 77 K taken with the tip scanning parallel to the chains ($I = 5\,nA$, $V = 70\,mV$). Three chains per unit cell are resolved. (From Ref. 10.)

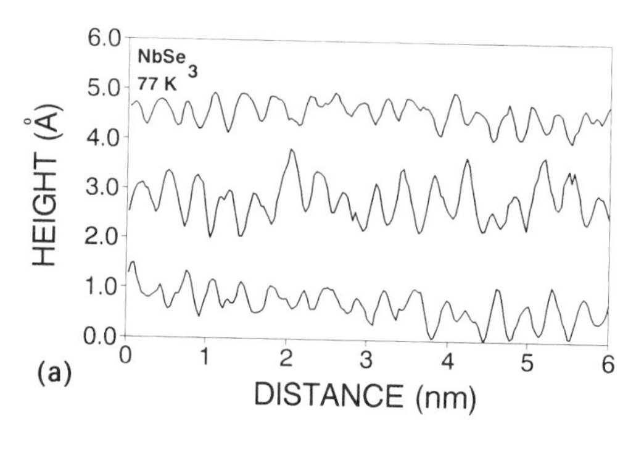

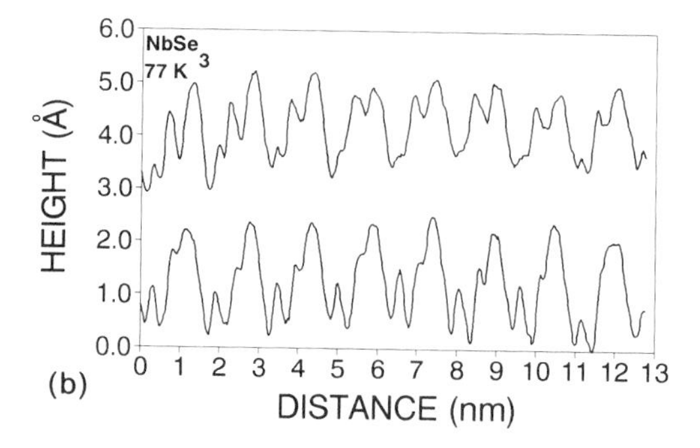

FIG. 39. (a) Profiles of z-deflection taken along the three chains of the unit cell in Fig. 38(a). The atomic modulation of the center profile is out-of-phase with the upper and lower profiles, thereby identifying it as the profile of chain III. It is also the only profile to show a $\sim 4b_0$ CDW modulation, providing further confirmation that it is chain III. (b) Profiles of z-deflecting along the tracks shown in Fig. 38(a), which are separated by $b_0/2$. The chain of intermediate height is out-of-phase as required if it is chain III. (From Ref. 10.)

for the intermediate height chain is ~ 2 Å, while the adjacent chains with little or no CDW modulation show a total z-deflection of ~ 1 Å (upper and lower profiles in Fig. 39(a)). Two profiles taken perpendicular to the chain axis and separated by $b_0/2$ are shown in Fig. 39(b). The chain of intermediate height is out-of-phase with respect to the low and high chains as required if chain III is identified as the chain of intermediate height. (See diagram of Fig. 37.) This phase relation is also confirmed in the contour scan of Fig. 40, where the Se atoms on the chain of intermediate height are clearly out-of-phase with the Se atoms on the other two chains.

The STM scan in Fig. 38(a) was recorded with the scanning tip moving

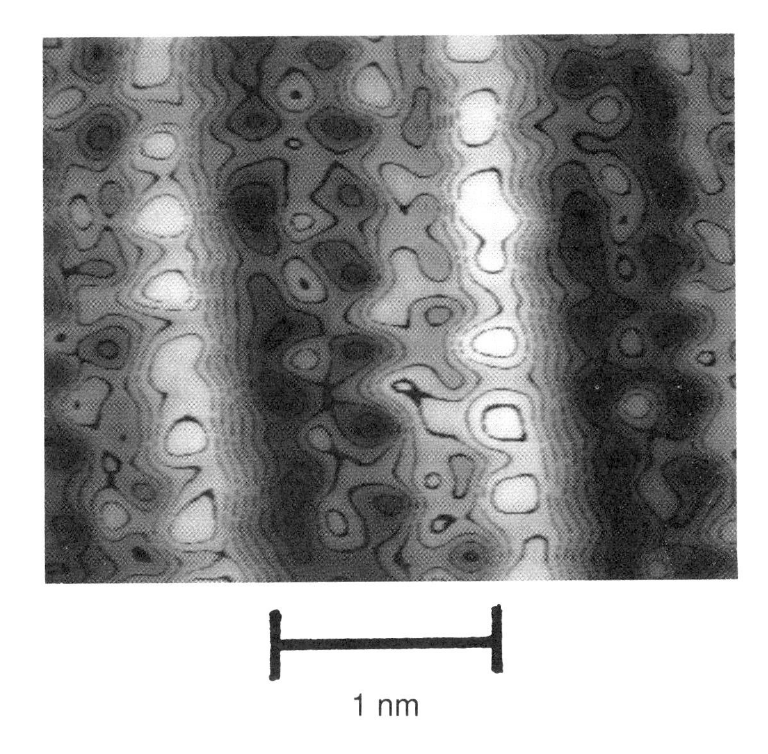

FIG. 40. Magnified plot of contours of constant z-deflection generated from the scan in Fig. 38. Atoms of high chains (white) and low chains (dark) are in phase along the chain axis, while atoms of the intermediate chains are out of phase ($I = 5\,nA$, $V = 70\,mV$). (From Ref. 62.)

perpendicular to the chains. A scan at 77 K with the tip moving parallel to the chains is shown in Fig. 38(b). The total z-deflection is again $\sim 2\,Å$ and there is one high chain and two of lower height. With the tip scanning parallel to the chains, the image shows a "barber pole" effect that may be connected with a slight misalignment between the chain direction and the scan direction. The chain adjacent to the high chain shows a strong CDW modulation of wavelength $\sim 4b_0$ as shown in the profile of Fig. 41(a) taken along this chain, while the other two chains show little or no CDW modulation. Both STM scans in Fig. 38 are therefore in agreement with respect to the existence of one high chain, while the CDW modulation resides on an adjacent chain of intermediate height. As shown in Figs. 39(b) and 41(b), the profiles perpendicular to the chains recorded from STM scans with the tip moving either perpendicular or parallel to the chains show total z-deflections of $\sim 2\,Å$ and similar structures, thereby indicating that the STM response reflects the true surface structure.

The minima occur between chains, and are of variable depth depending on

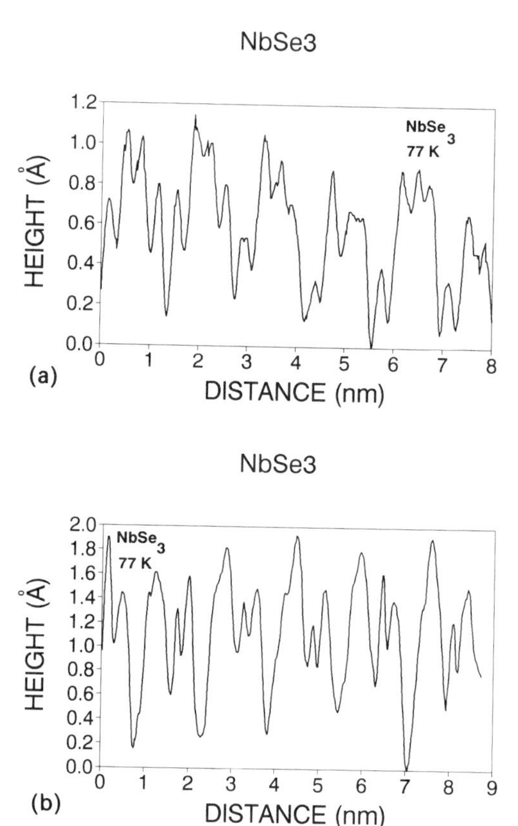

FIG. 41. (a) Profile of z-deflection along the intermediate height chain in Fig. 38(b). CDW modulation of wavelength $\sim 4b_0$ is clearly detected with amplitude of ~ 1 Å. (From Ref. 10.) (b) Profile of z-deflection perpendicular to the chains in Fig. 38(b) (with the tip scanning parallel to the chains). Comparison with the profiles in Fig. 39(b) (with the tip scanning perpendicular to the chains) shows similar structures and z-deflections independent of the scan direction.

the profile track position relative to the Se atoms on a particular chain. In Fig. 39(b) the minima between chains I′ and II, and between chains II and III, appear to have approximately the same depth, and chain I′ lies between these minima with a relatively low z-deflection. The conduction electron density on chain I′ should be large, and comparable to that on chain II and probably larger. Since both chains I′ and II are located at comparable surface heights, we conclude that the difference in STM deflection is due to the difference in static charge on the Se atoms of each chain. The increased static negative charge on the Se atoms of chain II (see Fig. 37) will push the conduction electron wavefunction farther from the surface, and increase the LDOS at the position of the tunneling tip, accounting for the increased

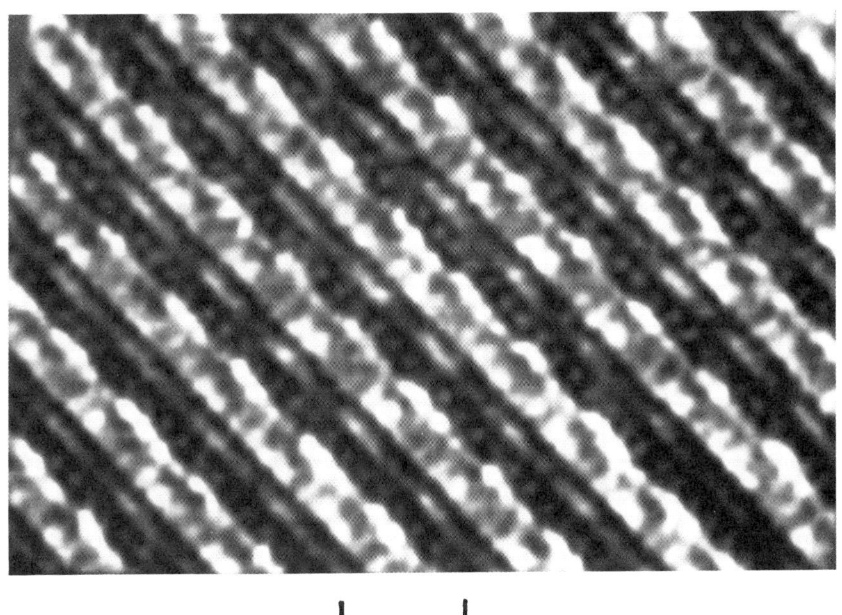

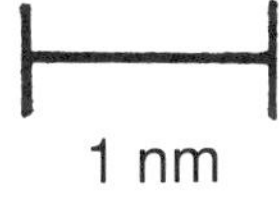

FIG. 42. STM scan of the **b–c** plane of $NbSe_3$ at 4.2 K. The high- and low-temperature CDWs are simultaneously resolved. The low-temperature CDW is identified with the two high white chains. Note that this region is comprised of two modulations that are out-of-phase and represent separate modulations residing on chains I′ and II. Each of these is out-of-phase in adjacent unit cells, consistent with the wavelength component of $2c_0$ for the low-temperature CDW. The high-temperature CDW modulation resides on the intermediate height chain and these CDW maxima are in-phase in adjacent unit cells consistent with the single component of $\sim 4b_0$($I = 2.0$ nA, $V = 50$ mV). (From Ref. 12.)

z-deflection on chain II. At 4.2 K the formation of the low-temperature CDW completely modifies the relative heights observed at 77 K and the detailed results will be discussed later.

8.9.2 STM of $NbSe_3$ at 4.2 K

At 4.2 K the STM scans on $NbSe_3$ show strongly modified profiles due to the formation of the low-temperature CDW. The $\sim 4b_0$ CDW modulation induces extremely large z-deflections on chains of types I′ and II. The high-temperature CDW remains localized on chains of type III, and appears as narrow white maxima surrounded by the dark regions of the scan in Fig. 42. The parallel regions of large white maxima are centered on chains of type I′

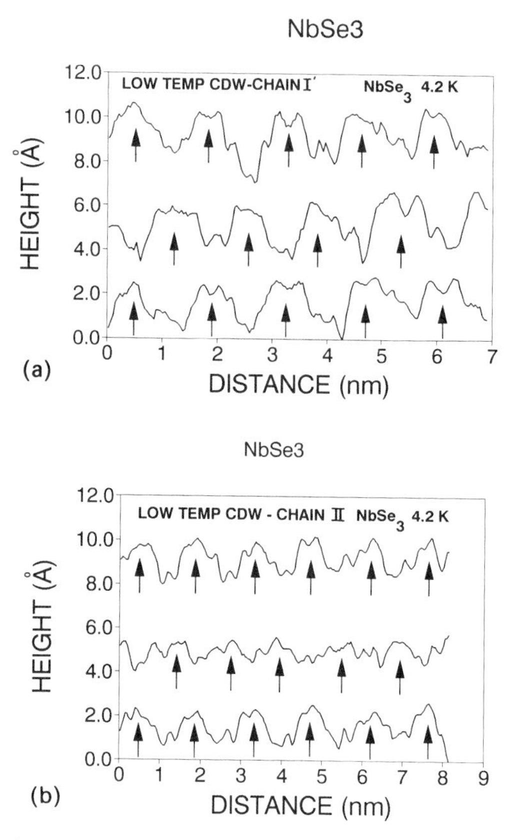

FIG. 43. (a) Profiles of z-deflection taken along the chains of type I′ for three adjacent unit cells in the image of Fig. 42. The amplitude of the low-temperature CDW is ~2 Å and is out-of-phase in adjacent unit cells as required by a $2c_0$ component. (From Ref. 12.) (b) Profiles of z-deflection taken along the chains of type II for three adjacent unit cells in Fig. 42. The low-temperature CDW amplitude is ~2 Å and is again out-of-phase in adjacent unit cells as required by a $2c_0$ component.

and II and are phase shifted by ~ 180° with respect to each other. In adjacent unit cells these modulations alternate in phase by 180° due to the wavelength component of $2c_0$ associated with the low-temperature CDW. Profiles along chains I′ and II in three adjacent unit cells are shown in Figs. 43(a) and 43(b). These remain relatively uniform with a 180° phase shift between unit cells and an amplitude of ~ 2 Å on each chain. The profiles along chains of type III are shown in Fig. 44, and show no phase shift between adjacent unit cells as expected for the single CDW component of $4.115\,b_0$. The amplitude of the CDW modulation is again ~2 Å, and at the positions of these maxima the profile perpendicular to chains, as shown in Fig. 45, indicates that the

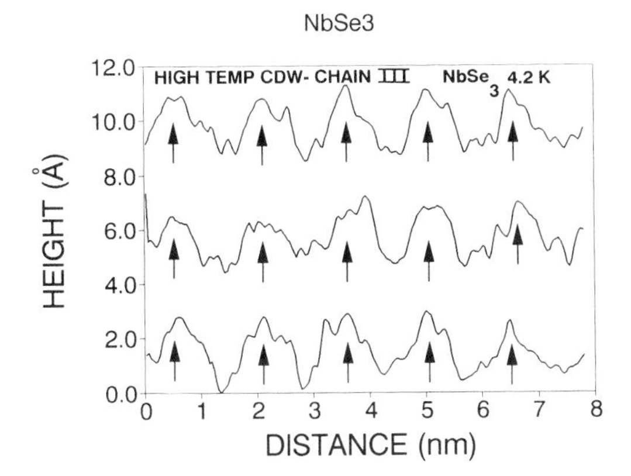

FIG. 44. Profiles of z-deflection taken along chains of type III in three adjacent unit cells of the image in Fig. 42. The amplitude of the high-temperature CDW is ~2 Å and the CDW modulations are in-phase in adjacent unit cells as required for the high-temperature CDW. (From Ref. 12.)

maximum CDW amplitudes are comparable on all three chains. These profiles indicate that the LDOS, at the position of the tip when it is over the CDW maxima, is approximately the same for the maxima appearing on all three chains, although the lateral extent of the CDW maxima on chain III is much less than on chains I′ and II.

The relative phases and amplitudes in these profiles, along with the symmetries, provide convincing evidence that no tip asymmetries or multiple tip effects are playing a role in these scans. Such effects would tend to interrupt the symmetries and phase shifts observed between adjacent unit cells. For

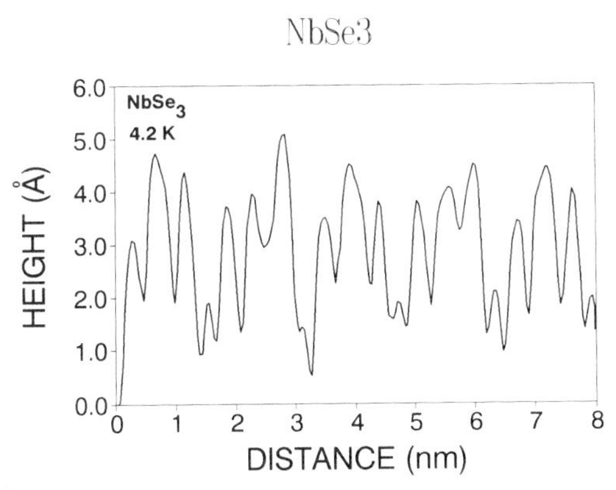

FIG. 45. Profile of z-deflection thaken perpendicular to the chains in Fig. 42. The profile indicates that the CDW amplitudes at the maxima are comparable on all three chains.

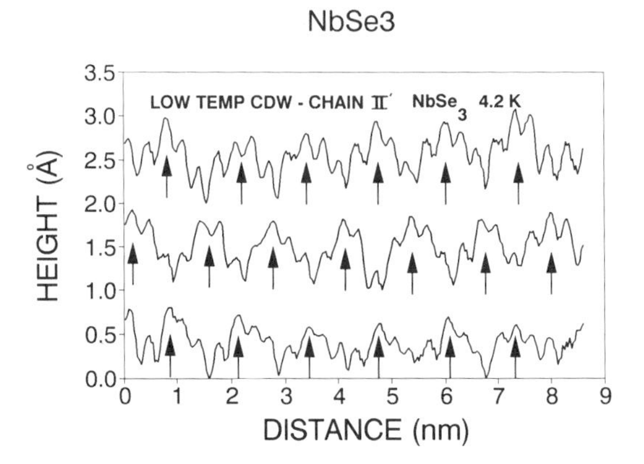

FIG. 46. Profiles of z-deflection taken along chains of type II′ in the deep groove of Fig. 42. The type II′, as well as the surface chains of type II, show a $\sim 4b_0$ CDW modulation with maxima that are out-of-phase across adjacent unit cells. The modulation on chain II′ is out-of-phase with that on the adjacent chain I′ and is therefore in-phase with the modulation on chain II.

example, double tips with spacings coincident with the chain spacing would produce ghosts of chain III on chains I′ and II, which in adjacent unit cells would be alternately in- and out-of-phase with the CDW modulation on chains I′ or II. This would change the amplitude modulation profiles on adjacent chains. Only a phase shift is observed and the CDW amplitude profiles remain approximately the same in adjacent unit cells. The identification of the chain types and the individual CDW modulations in these STM scans are completely consistent with the atomic structure as indicated in the cross section of Fig. 45. The dark strip of minimum z-deflection in the scan of Fig. 42 can be identified with the space between Nb chains I′ and III, which is 6.39 Å versus 5.17 and 4.06 Å for the other two spacings between Nb chains. The existence of this wide spacing also allows the tunneling tip to pick up modulation due to chain II′ in the layer below the surface layer. The presence of this deep groove can be identified in all STM scans showing reasonable resolution of the individual chain structure in $NbSe_3$ at 4.2 K. In some STM scans the profiles recorded parallel to chain II′ in the deep groove can clearly pick up the $\sim 4b_0$ CDW modulation residing on chain II′ giving further confirmation that chains of type II, II′ carry a CDW modulation. Profiles in three adjacent unit cells recorded in the deep groove over chain II′ are shown in Fig. 46. The 180° phase difference between the CDW modulations in adjacent unit cells is clearly resolved with an amplitude of ~ 0.5 Å. This observation provides further evidence that no unusual tip effects or ghosts are contributing to the CDW modulation observed on chains of type

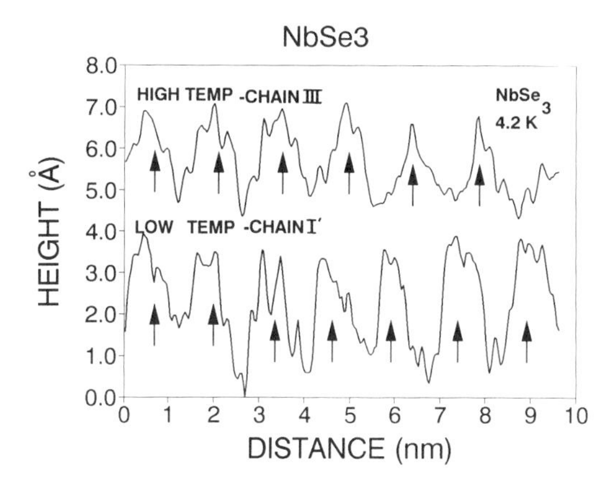

FIG. 47. Profiles of z-deflection along chains of type I′ (lower profile) and type III (upper profile) from the image in Fig. 42. Comparison of the two profiles clearly indicates two different wavelengths as required for identification with the high- and low-temperature CDWs. The two profiles start out in-phase on the left hand side and move continuously out-of-phase as they proceed to the right hand side, with the CDW on chain III having the longer wavelength. (From Ref. 12.)

II, II′. These results have been confirmed in at least five different runs with different crystals and scanning tips.

The differences observed between various high-resolution scans are mainly associated with the intense CDW modulation appearing on chains I′ and II. Although these are observed to be out of phase by 90 to 180° in the same unit cell, the large lateral extent of the CDW modulations on these chains causes them to overlap in many of the scans. However, the deep groove between chains I′ and III can be clearly resolved in these cases, and the CDW modulation on chain II′ in the layer below this groove can be detected. Profiles along the bottom of such a groove are shown for three adjacent unit cells in Fig. 46.

In addition to the profile comparisons given earlier, the identification of the high-temperature and low-temperature CDWs can also be checked by comparing the measured wavelengths. Profiles recorded along chain III and chain I′ in the same unit cell are shown in Fig. 47. They start in-phase at the left of the figure and move continuously out-of-phase, reaching a phase difference of ~108° after ~6 to 7 wavelengths. The two wavelength components along the **b** axis are 4.115 and 3.802 Å, respectively, giving $\Delta\lambda$ of 0.313 Å that would produce a phase difference of ~180° in 6 to 7 wavelengths as observed in Fig. 47. This phase difference also develops continuously in the profiles of Fig. 47 indicating that abrupt discommensurations are not forming, and that both CDWs are continuously incommensurate.

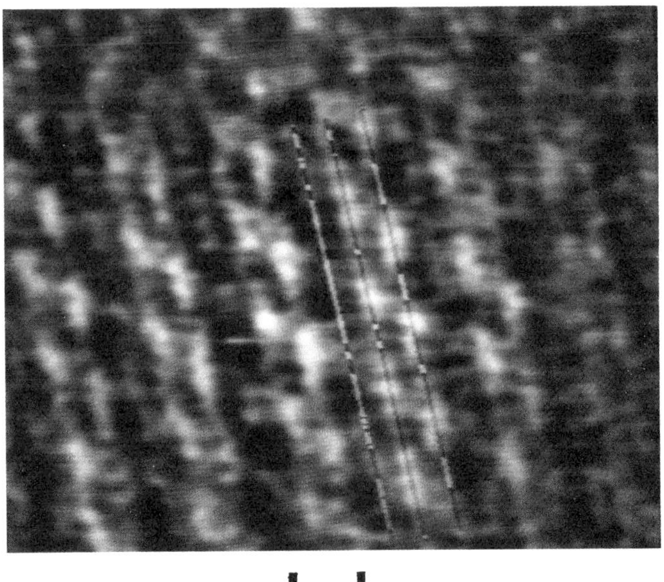

FIG. 48. Grey scale STM scan of orthorhombic TaS_3 at 77 K. CDW modulations are observed on two separate chains with the CDW modulation in adjacent unit cells out-of-phase by 180° ($I = 2$ nA, $V = 164$ mV). (From Ref. 11.)

8.9.3 STM of TaS_3 at 77 K

The orthorhombic form of TaS_3 has been studied using STM at 77 K by Slough *et al.*[11,62] At 77 K the CDW in orthorhombic TaS_3 should be commensurate with the lattice with a wavelength component of $4c_0$ along the chain axis (**c** axis). The **q** vector components along the $\mathbf{a}_0^*$, $\mathbf{b}_0^*$, and $\mathbf{c}_0^*$ directions have been determined by x-ray diffraction experiments to be (0.5, 0.125, 0.250) at temperatures below ~140 K. The wavelength component along the **b** axis perpendicular to the chain direction should therefore be $8b_0$. The STM scan on the **b**–**c** plane should show a predominant modulation with a period of $4b_0$ along the chain direction, with a small phase shift between chains in adjacent unit cells, in order to account for the $8b_0$ component perpendicular to the chain direction.

A grey scale STM scan of orthorhombic TaS_3 at 77 K is shown in Fig. 48. Two strong CDW modulations on separate chains are observed in each unit cell, with a phase shift of 180° between the two modulations within the same

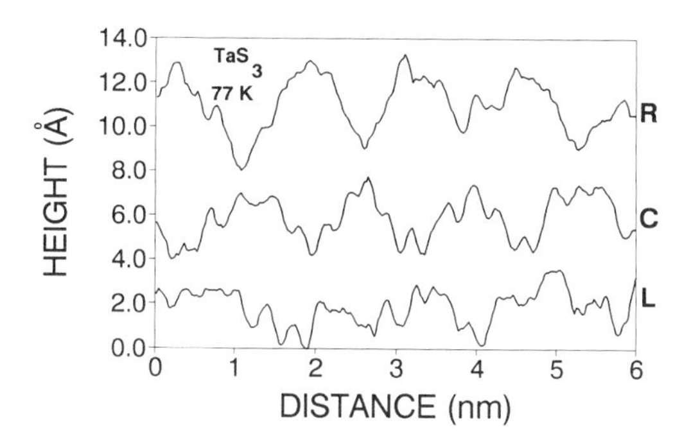

FIG. 49. Profiles of z-deflection parallel to the chains observed for STM scans along the right (R), center (C), and left (L) tracks shown in Fig. 48. The profiles illustrate the phase relation between the CDWs on adjacent chains. (From Ref. 11.)

unit cell. Since only one CDW onset is observed at 215 K in orthorhombic TaS_3, the two modulations are interpreted as coupled to form a single CDW as proposed in a model by Wang *et al.*[76] In this model the single CDW transition is explained in terms of the locking of two CDWs on separate chains and is characterized by a single wave vector. A CDW transition at 215 K on one set of chains induces a CDW on the second set of chains. The single wave vector then adjusts as a function of temperature down to 140 K. Below this temperature the CDW locks into a stable commensurate distortion predominantly along the chain axis. Between 215 and 140 K $\mathbf{b}_0^*$ component changes from 0.1 to $0.125\,b_0^*$.

For the orthorhombic unit cell of TaS_3, the lattice vector along the chain is $c_0 = 3.340$ Å. The CDW modulation wavelength in Fig. 48 is measured as $\lambda_{CDW} = 13.5 \pm 0.2$ Å, in good agreement with the expected value of 13.38 Å, which is equal to 4 times the lattice vector. The unit cell dimension perpendicular to the chains is measured as $b_0 = 14.9 \pm 0.7$ Å in close agreement with the value of $b_0 = 15.173$ Å as determined by x-ray diffraction.

The profiles measured parallel to the chain axis along the tracks shown in Fig. 48 are shown in Fig. 49. The two profiles labeled *R* and *C* in Fig. 49 correspond to the right and center tracks in Fig. 48, and show a strong CDW modulation amplitude of ~ 4 Å. The CDW maxima on these two chains are clearly out of phase by 180°. The track along the third chain labeled *L* in Fig. 49 generates a profile showing very little CDW modulation and exhibits approximately one-half of the total z-deflection that is observed on the other two chains. However, in a number of scans a weak amplitude modulation of wavelength $\sim 4b_0$ can be detected on the third chain. This is a weaker modulation, but may also be induced by the stronger modulations forming on the other two chains. All three modulations clearly shift phase by small equal amounts in adjacent unit cells, as required by the $8b_0$ component perpendicular to the chains.

In the model of Wang *et al.*[76] it is assumed that separate CDWs can form on two different pairs of chains, and these would be characterized by complex order parameters δ_1 and δ_2 for each type of chain pair. In this respect, the CDWs would be very similar to those forming in monoclinic TaS_3 and $NbSe_3$, which form at separate temperatures with two different $\mathbf{q}$ vectors. However, in the orthorhombic phase the separate $\mathbf{q}$ vectors $\mathbf{q}_1$ and $\mathbf{q}_2$, are nearly parallel and the system can choose a unique $\mathbf{q}$ vector between $\mathbf{q}_1$ and $\mathbf{q}_2$ for which a first-order decrease in energy proportional to $\delta_1 \delta_2$ occurs. This can arise from coupling due to electrostatic or elastic interactions connected with the CDW deformation of the lattice. In the monoclinic phase the large $\mathbf{q}$ vector component of $0.5\,\mathbf{c}_0^*$ transverse to the chain direction for the low-temperature CDW would be unfavorable for formation of a single $\mathbf{q}$ vector, and thus the two CDWs form independently at two different temperatures.

The STM scans on orthorhombic TaS_3 indicate that the CDW modulations on the two pairs of chains exhibit substantial lateral extent, with the result that a complex pattern of CDW modulations extends over both pairs of chains as shown in Fig. 50. The pattern shows a 45° slant due to the 180° phase shift between the adjacent chains showing the major CDW modulations. In the grey scale scans of TaS_3 the CDW modulation pattern is oriented at ~84° to the chain direction. This is consistent with a wavelength component of $8b_0$ perpendicular to the chains, although a full supercell with a width of 8 chains is not shown. Wider scans showing 8 or more chains have not provided good resolution of the CDW modulation. Good resolution of the CDW modulation has also only been obtained with the bias voltage set above 160 mV, which is the approximate value of the CDW energy gap structure in the dI/dV versus V curves. STM measurements of the energy gaps are discussed later.

8.9.4 Tunneling Spectroscopy of $NbSe_3$ and TaS_3

The CDW energy gaps in $NbSe_3$ and TaS_3 have been determined with reasonable accuracy using the STM at 4.2 K. In $NbSe_3$ structure in the dI/dV versus V curve can be identified at ~35 mV and at ~101 mV as shown in the example of Fig. 51. These structures can be associated with the low-temperature and high-temperature CDWs respectively. Measurements on four different crystals, and averaging a number of runs from each crystal, gives values for the CDW energy gaps of $\Delta_1 = 35.6 \pm 2.8$ meV and $\Delta_2 = 101.2 \pm 4.1$ meV. The corresponding values of $2\Delta_{CDW}/k_B T_C$ are 14.0 and 16.2 respectively. These are much larger than expected for the weak-coupling, long-coherence

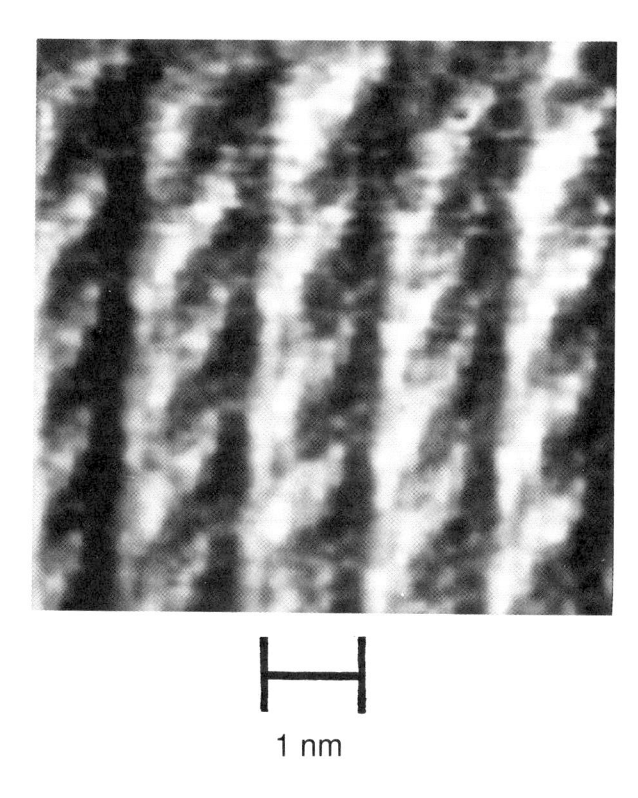

FIG. 50. Grey scale image of orthorhombic TaS_3 at 77 K. The charge maxima are smeared out, creating diffuse maxima elongated at $\sim 45°$ to the chain direction. (From Ref. 12.)

length model and require the short-coherence length model as discussed by McMillan.[48] These values of the CDW energy gaps are in reasonable agreement with a fixed tunnel junction measurement of $\Delta_1 = 35.0 \pm 2.5$ meV carried out by Fournel *et al.*[77] and point contact measurements of $\Delta_1 = 36.5$ meV and $\Delta_2 = 90$ meV carried out by Ekino and Akimitsu.[78]

The energy gaps have been measured for different positions of the tip with respect to the CDW maxima, the chain positions, and the atomic structure. No variation in the position of the energy gap structure in the conductance curves has been detected and none would be expected. The CDW produces partial gapping of the FS, and the DOS change above the gap voltage would be expected to remain uniform as would the Fermi level. Local variations in the CDW gaps would only be expected due to local impurities or defects. The mean free path of the tunneling electron would also be expected to equal a number of chain separation lengths, which would tend to average any local gap variations as well.

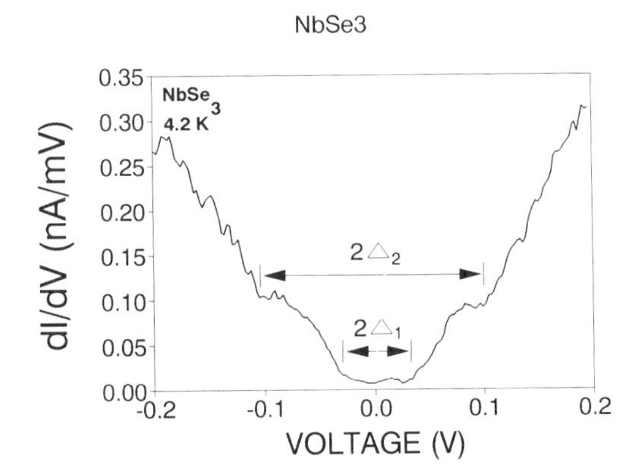

FIG. 51. dI/dV versus V curve for $NbSe_3$ at 4.2 K. Structure identified with both gaps is apparent ~35 mV (low-temperature gap) and ~101 mV (high-temperature gap).

The dI/dV versus V curves obtained for orthorhombic TaS_3 at 4.2 K show a well-defined structure at a voltage of ~80 mV. An example is shown in Fig. 52, where the large change in DOS at the gap edge is consistent with the expected semiconducting behavior of TaS_3 below the CDW transition, which gaps the entire Fermi surface. The value of $2\Delta_{CDW}/k_B T_C$ for orthorhombic TaS_3 is 8.6, a value which indicates a strong coupling, but is a substantially lower value than that observed for $NbSe_3$.

8.9.5 STM of $TaSe_3$

An STM grey scale scan of $TaSe_3$ at 4.2 K is shown in Fig. 53(a) and is

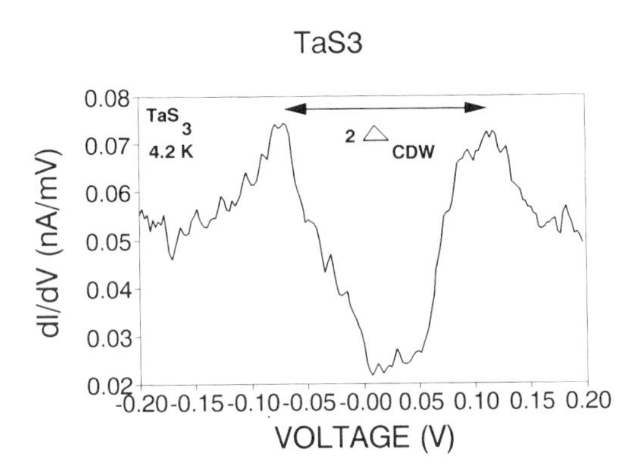

FIG. 52. dI/dV versus V curve for TaS_3 at 4.2 K. A well-defined gap structure is observed with an onset at ~80 mV. (From Ref. 39.) The gap structure is extremely strong, consistent with the semiconducting CDW phase in TaS_3.

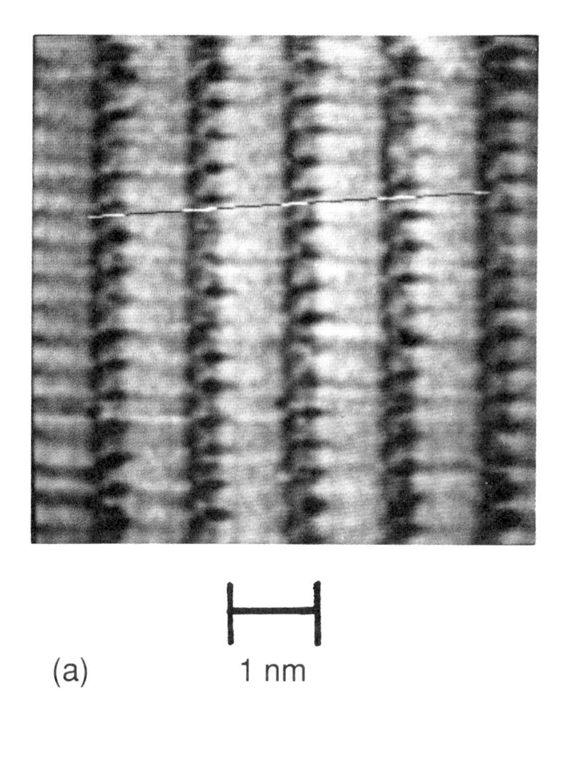

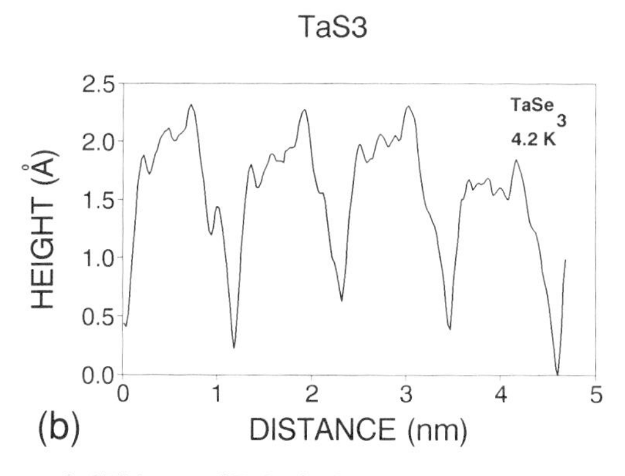

FIG. 53. (a) Grey scale STM scan of $TaSe_3$ in the **b–c** plane at 4.2 K. Four rows of atoms are resolved, consistent with the two-chain surface structure. (b) Profile of z-deflection perpendicular to the chain axis. The amplitude is ~2.0 Å ($I = 2.2$ nA, $V = 200$ mV).

consistent with the atomic structure shown in the unit cell projection of Fig. 36(b). The total z-deflection is ~2.1 Å at 4.2 K. The deep minima correspond to the larger separation between the center chain and the chain on the right in Fig. 36(b). The four Se atom rows per unit cell are also resolved. Three of these form a plateau, while one of these rows lies in a deep crevice. A profile perpendicular to the chains is shown in Fig. 53(b), and follows the expected variation in atomic height. The effective barrier height measured for $TaSe_3$ at 4.2 K is extremely low (< 0.20 eV), and suggests that the STM deflection will be extremely sensitive to the STM operating parameters.

The STM results on $TaSe_3$ show a good resolution of the chain structure. Similar chain structures are resolved in the two compounds exibiting CDWs. The results on $TaSe_3$ add additional evidence that both complex surface atomic structure and CDWs can be resolved with the STM.

8.10 Conclusions

The series of STM experiments reviewed in this chapter have conclusively established the value of the STM in detecting the real space arrangement and the intensity of charge-density wave structure. Both the atomic and CDW structure have been well-resolved, and the STM amplitude response has shown a systematic correlation with the strength of the CDW. CDW structure deduced from other types of experiments has been confirmed on an atomic scale in the real space images, and in a number of cases new and unexpected results on CDW structure have been obtained. For example, $NbSe_3$ has been shown to support a CDW modulation on three pairs of chains rather than two, and orthorhombic TaS_3 shows a correlated induced CDW structure on at least two pairs of chains.

The initial AFM experiments also demonstrate that the AFM can give good resolution of CDW structure, although the amplitude relative to the atomic amplitude is much smaller than observed with the STM. Combined STM and AFM studies should be complimentary.

A large series of spectroscopy measurements have shown that reliable information on CDW energy gaps can also be obtained with the STM. The spectroscopy mode is highly sensitive, but systematic reproducibility has been obtained. The STM can be expected to supply extremely valuable information on many aspects of electronic structure in materials.

Acknowledgments

The research described here has been supported by the U.S. Department of Energy, Grant No. DE-FG05-89ER45072. The authors would like to thank P. K. Hansma, Joe Demuth, B. Giambattista, A. Johnson, and Vittorio Celli for valuable help and discussion.

References

1. J. A. Wilson and A. D. Yoffe, *Adv. Phys.* **18**, 193 (1969).

2. A. Meerschaut and J. Rouxel, p. 205, *Crystal Chemistry and Properites of Materials with Quasi-One-Dimensional Structures*, J. Rouxel, eds, Reidel, Dordrecht, 1986.
3. J. A. Wilson, F. J. Di Salvo, and S. Majahan, *Adv. Phys.* **24**, 117 (1975).
4. *Theoretical Aspects of Band Structures and Electronic Properties of Pseudo-One-Dimensional Solids*, H. Kamimura, ed. Reidel, Dordrecht, 1985.
5. *Electronic Properties of Inorganic Quasi-One-Dimensional Compounds* P. Monceau, ed., Reidel, Dordrecht, 1985.
6. R. V. Coleman, B. Drake, P. K. Hansma, and G. Slough, *Phys. Rev. Lett.* **55**, 394 (1985).
7. R. V. Coleman, B. Giambattista, P. K. Hansma, A. Johnson, W. W. McNairy, and C. G. Slough, *Adv. Phys.* **37**, 559 (1988).
8. *Charge Density Waves in Solids* L. P. Gor'kov and G. Gruner, eds., North-Holland, New York, 1989.
9. G. Gruner, *Rev. Mod. Phys.* **60**, 1129 (1988).
10. C. G. Slough, B. Giambattista, A. Johnson, W. W. McNairy, and R. V. Coleman, *Phys. Rev. B* **39**, 5496 (1989).
11. C. G. Slough and R. V. Coleman, *Phys. Rev. B* **40**, 8042 (1989).
12. Z. Dai, C. G. Slough, and R. V. Coleman, *Phys. Rev. Lett.* **66**, 1318 (1991).
13. J. Tersoff, *Phys. Rev. Lett.* **57**, 440 (1986); private communication (1988).
14. B. Giambattista, C. G. Slough, W. W. McNairy, and R. V. Coleman, *Phys. Rev. B* **41**, 10082 (1990).
15. C. Julian Chen, *Phys. Rev. B* **42**, 8841 (1990).
16. N. D. Lang, *Phys. Rev. Lett.* **56**, 1164 (1986).
17. G. Binnig and D. P. E. Smith, *Rev. Sci. Instrum.* **57**, 1688 (1986).
18. S. Alexander, L. Hellemans, O. Marti, J. Schneir, V. Elings, P. K. Hansma, M. Longmire, and J. Gurley, *J. Appl. Phys.* **65**, 164 (1989).
19. N. V. Smith, S. D. Kevan, and F. J. Di Salvo, *J. Phys. C* **18**, 3175 (1985).
20. Xian Liang Wu and Charles M. Lieber, *Science* **243**, 1703 (1989); *Phys. Rev. Lett.* **64**, 1150 (1990).
21. R. V. Coleman, W. W. McNairy, C. G. Slough (to be published).
22. K. Nakanishi and H. Shiba, *J. Phys. Soc. Jpn.* **43**, 1839 (1977).
23. B. Burk, R. E. Thomson, A. Zettl, and John Clarke (to be published).
24. J. A. Wilson, *J. Phys.: Condens. Matter* **2**, 1683 (1990).
25. C. G. Slough, W. W. McNairy, Chen Wang, and R. V. Coleman, *J. Vac. Sci. Technol. B*, 1036 (1991).
26. R. E. Thomson, U. Walter, E. Ganz, J. Clarke, A. Zettl, P. Rauch, and F. J. Di Salvo, *Phys. Rev. B* **38**, 10734 (1988).
27. J. Garnaes, S. A. C. Gould, P. K. Hansma, and R. V. Coleman (to be published in *J. Vac. Sci. Technol. B*, 1032 (1991).
28. A. M. Woolley and G. Wexler, *J. Phys. C* **10**, 2601 91977).
29. C. G. Slough, W. W. McNairy, R. V. Coleman, J. Garnaes, C. B. Prater, and P. K. Hansma, *Phys. Rev. B* **42**, 9255 (1990).
30. H. W. Myron, *Physica B* **99**, 243 (1990).
31. P. M. Williams, p. 51, *Crystallography and Crystal Chemistry of Materials with Layered Structure* F. Levy, ed., Reidel, Dordrecht, 1976.
32. K. Tsutumi, T. Sambongi, T. Akira, and S. Tanaka, *J. Phys. Soc. Jpn.* **49**, 837 (1980).
33. J. van Landuyt, G. A. Wiegers, and S. Amelinckx, *Phys. Stat. Sol.* **46**, 479 (1978).
34. K. K. Fung, J. W. Steeds, and J. A. Eades, *Physica B* **99**, 47 (1980).
35. K. Tsutsumi, *Phys. Rev. B* **26**, 5756 (1982).
36. Y. Yoshida and K. Motizuki, *J. Phys. Soc. Jpn.* **51**, 2107 (1984).
37. D. J. Eaglesham, R. L. Withers, and D. M. Bird, *J. Phys. C* **19**, 359 (1986).

38. Chen Wang, B. Giambattista, C. G. Slough, R. V. Coleman, and M. A. Subramanian, *Phys. Rev. B* **42**, 8890 (1990).
39. Chen Wang, C. G. Slough and R. V. Coleman, *J. Vac. Sci. Technol. B*, 1048 (1991).
40. K. C. Woo, F. C. Brown, W. L. McMillian, R. J. Miller, M. J. Schaffman, and M. P. Sears, *Phys. Rev. B* **14**, 3242 (1976).
41. F. J. Di Salvo, D. E. Moncton, J. A. Wilson, and J. V. Waszczak, *Bull. Am. Phys. Soc.* **21**, 261 (1976).
42. F. C. Brown, *Physica B* **99**, 264 (1980).
43. N. G. Stoffel, S. D. Kevan, and N. V. Smith, *Phys. Rev. B* **31**, 8049 (1985).
44. K. Motizuki, Y. Yoshida, and Y. Takaoka, *Physica B* **105**, 357 (1981).
45. N. Suzuki, A. Yamamoto, and K. Motizuki, *J. Phys. Soc. Jpn.* **54**, 4668 (1985).
46. D. E. Moncton, J. D. Axe, and F. J. Di Salvo, *Phys. Rev. Lett.* **34**, 734 (1975).
47. B. Giambattista, A. Johnson, R. V. Coleman, B. Drake, and P. K. Hansma, *Phys. Rev. B* **37**, 2741 (1988).
48. W. L. McMillan, *Phys. Rev. B* **16**, 643 (1977).
49. H. F. Hess, R. B. Robinson, R. C. Dynes, J. M. Valles, Jr., and J. V. Waszczak, *Phys. Rev. Lett.* **62**, 214 (1989).
50. R. M. Fleming and R. V. Coleman, *Phys. Rev. B* **16**, 302 (1977).
51. J. E. Graebner, *Solid State Commun.* **21**, 353 (1977).
52. N. J. Doran and A. M. Woolley, *J. Phys. C* **14**, 4257 (1981).
53. G. Wexler and A. M. Woolley, *J. Phys. C* **9**, 1185 (1976).
54. J. P. Tidman, O. Singh, A. F. Curzon, and R. F. Frindt, *Philos. Mag.* **30**, 1191 (1974).
55. G. A. Scholz, O. Singh, R. F. Frindt, and A. E. Curzon, *Solid State Commun.* **44**, 1455 (1982).
56. H. Nishihara, G. A. Scholz, and R. F. Frindt, *Solid State Commun.* **44**, 507 (1982).
57. S. J. Hillenius and R. V. Coleman, *Phys. Rev. B* **18**, 3790 (1978).
58. R. H. Friend, D. Jerome, R. F. Frindt, A. J. Grant, and A. D. Yoffe, *J. Phys. C* **10**, 1013 (1977).
59. W. J. Wattamaniuk, J. P. Tidman, and R. F. Frindt, *Phys. Rev. Lett.* **35**, 62 (1975).
60. G. Wexler, A. Woolley, and N. Doran, *Nuovo Cimento B* **38**, 571 (1977).
61. N. J. Doran, G. Wexler, and A. M. Woolley, *J. Phys. C* **11**, 2967 (1978).
62. C. G. Slough, B. Giambattista, W. W. McNairy, and R. V. Coleman, *J. Vac. Sci. Tech.* **8**, 490 (1990).
63. G. Gammie, J. S. Hubacek, S. L. Skala, R. T. Brockenbrough, J. R. Tucker, and J. W. Lyding, *Phys. Rev. B* **40**, 9529; *Phys. Rev. B*, 11965 (1989).
64. Y. Tajima and K. Yamaya, *J. Phys. Soc. Jpn.* **53**, 495 (1984).
65. N. Shima and H. Kamimura, p. 231, *Theoretical Aspects of Band Structures and Electronic Properties of Pseudo-One-Dimensional Solids*, H. Kamimura, ed., Riedel, Dordrecht, 1985.
66. J. A. Wilson, *Phys. Rev. B* **19**, 6456 (1979).
67. J. A. Wilson, *J. Phys. F* **12**, 2469 (1982).
68. R. M. Fleming, D. E. Moncton, and D. W. McWham, *Phys. Rev. B* **18**, 5560 (1978).
69. P. Monceau, N. P. Ong, A. M. Portis, A. Meerschaut, and J. M. Rouxel, *Phys. Rev. Lett.* **37**, 602 (1972).
70. N. P. Ong and P. Monceau, *Phys. Rev. B* **16**, 3443 (1977).
71. K. Tsutsumi, T. Tagagaki, M. Yamamoto, Y. Shiozaki, M. Ido, T. Sambongi, K. Yamaya, and Y. Abe, *Phys. Rev. Lett.* **39**, 1675 (1977).
72. J. L. Hodeau, M. Marezio, C. Roucau, R. Ayroles, A. Meerschaut, J. Rouxel, and P. Monceau, *J. Phys. C* **11**, 4117 (1978).
73. Myung-Hwan Whangbo, p. 27, *Crystal Chemistry and Properties of Materials with Quasi-One-Dimensional Structures* J. Rouxel, ed., Riedel, Dordrecht, 1986.

74. Shinji Wada, Ryozo Aoki, and Osamu Fujita, *J. Phys. F* **14**, 1515 (1984).
75. F. Devreaux, *J. Phys. (Paris)* **43**, 1489 (1982).
76. Z. Z. Wang, H. Salva, P. Monceau, M. Renard, C. Roucau, R. Ayroles, F. Levy, L. Guemas, and A. Meerschaut, *J. Phys. (Paris)* **44**, L-311 (1983).
77. A. Fournel, J. P. Sorbier, M. Konczykowski, and P. Monceau, *Phys. Rev. Lett.* **57**, 2199 (1986).
78. T. Ekino and J. Akimitsu, *Jpn. J. Appl. Phys. Supplement* **26–3**, 625 (1987).

9. SUPERCONDUCTORS

Harald F. Hess

AT&T Bell Laboratories, Murray Hill, New Jersey

9.1. Introduction

The technique of tunneling has played an important and central role in confirming and extending our knowledge of superconductivity. It was initially used as a tool to understand the bulk properties of these remarkable compounds. By tunneling electrons through thin insulating films into superconductors, one can study how the density-of-states in the superconducting state is modified. This method is routinely used to determine important parameters of superconductors. Traditional thin film tunneling, however, takes place over a broad area and tends to average any spatial variations in the density-of-states.

The STM, on the other hand, allows tunneling to take place only over a very small area with atomic dimensions. This chapter will describe how the STM measures the local density-of-states in a superconductor. The main emphasis will be on those aspects that distinguish the STM from the traditional tunneling methods, and how this new ability to measure local spatial variations in tunneling spectrum (sometimes referred to as scanning tunneling spectroscopy or STS) can reveal new physics of superconductivity.

The recently discovered high temperature superconductors have been the subject of many STM investigations. However, obtaining high quality tunneling data on these compounds remains a challenging goal. This is presently a very active, rapidly developing field and much progress has been made. A review of this topic can be found in Ref. 1, and will not be further discussed here.

9.2. The Superconducting State

9.2.1 Homogeneous Phase

Ever since the existence of superconductivity was established by the pioneering work of H. Kamerlingh Onnes in 1911, many materials were discovered to display the two characteristic properties of the superconducting state: zero resistance and a tendency to expel magnetic fields, the Meissner

METHODS OF EXPERIMENTAL PHYSICS
Vol. 27

ISBN 0-12-475972-6

effect. The conductivity and diamagnetism is perfect below a critical temperature T_C and below a critical magnetic field H_{C1}.

A complete microscopic description of this phase was established when J. Bardeen, L. N. Cooper, and J. R. Schrieffer developed their BCS theory of superconductivity.[2] They predicted a remarkable new ground state where electrons of opposite momentum and spin form pairs. Any excitation of this ground state requires one of these Cooper pairs to be broken into two single particle excitations. A minimum energy of Δ, the energy gap, is required to create one of these quasiparticles. The BCS theory made explicit predictions about how such a gap should evolve with temperature. It also deduced the functional form of the energy distribution for these quasiparticle excitations

$$n(\varepsilon) = 0 \qquad \varepsilon < \Delta(T)$$

$$n(\varepsilon) = n_0 \frac{\varepsilon}{\sqrt{\varepsilon^2 - \Delta(T)^2}} \qquad \varepsilon \geqslant \Delta(T). \tag{1}$$

If an electron is to tunnel from a metallic STM tip into the superconductor, it will enter into one of these quasiparticle states, so the tunneling current provides a direct measure of the density of these states. Summing over all thermally occupied electron states Ψ_t in the metal tip at location R, the quasiparticle states Ψ_s in the sample, and the transition tunneling matrix element $\langle \Psi_s(\varepsilon_i, r) | \Psi_t(\varepsilon_j, r + R) \rangle$, leads to an expression for the differential conductance

$$\frac{dI}{dV}(V) \propto \sum_i \sum_j |\langle \Psi_s(\varepsilon_i, r) | \Psi_t(\varepsilon_j, r + R) \rangle|^2 \frac{df}{d\varepsilon}(\varepsilon_i - \varepsilon_j - eV). \tag{2}$$

Here $df/d\varepsilon(\varepsilon - eV)$ is the derivative of the Fermi distribution which becomes very sharply peaked at $\varepsilon_i - \varepsilon_j = eV$ at low temperatures. An important experimental cornerstone of the BCS theory was the thin film tunneling experiments of I. Giaever,[3] which confirmed this distribution of states.

The STM distinguishes itself from the traditional thin film tunneling technique in several ways. Instead of requiring a high-quality oxide film or other insulating tunneling barrier, a controllable vacuum barrier is used. This opens new possibilities for studying superconductors that otherwise do not support the formation of good, high-quality insulating films. Furthermore, the transition matrix element as a function of tip position becomes proportional to the sample wavefunction amplitudes. The ideal tip wavefunction simply acts like a spatial delta function on the sample quasiparticle wavefunction, and locally maps out its amplitude on the surface. So in the limit of low temperatures and the perfect tip, Eq. (2) simplifies to

$$\frac{dI}{dV}(eV, R) \propto \sum_{\varepsilon_i \simeq eV} |\Psi_s(\varepsilon_i, R)|^2. \tag{3}$$

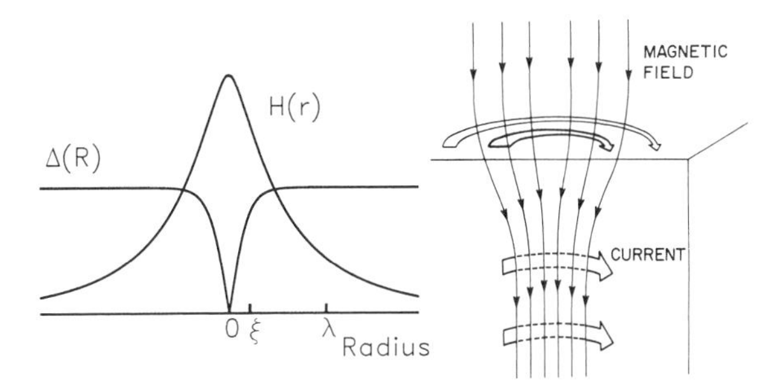

FIG. 1. Basic structure of the vortex showing the profile of the magnetic field, the order parameter, and the relative length scales.

For a superconductor the quasiparticle wavefunctions are given by a spinor with the two components $u(r)$ and $v(r)$. If the wavefunction amplitudes have no spatial structure, the differential conductance is proportional to the density-of-states of Eq. (1).

9.2.2 Inhomogeneous Phase of Superconductivity

The application of a magnetic field to a superconductor can change the homogenous electronic structure into one with marked spatial variations. Below a field of H_{C1} the Meissner effect will initially prevent flux penetration. Above a much larger field H_{C2} the superconductivity is entirely destroyed and the electronic structure reverts to that of a normal metal. For certain superconductors, type II, there exists an intermediate range of magnetic fields between H_{C1} and H_{C2} where a mixed phase exists. This was first predicted by A. A. Abrikosov[4] in 1958. Using the Landau–Ginzberg theory of superconductivity, Abrikosov found that the superconducting order parameter $\Delta(T, \mathbf{r})$ would vanish periodically on a hexagonal lattice of lines. This array of lines is parallel with the field and is known as the Abrikosov flux lattice.

Each one of these singularities is associated with a filament of magnetic field and referred to as a flux line or vortex. One quantum of magnetic flux $hc/2e = 2.07*10^{-7}$ gauss cm^2 defines each one of these filaments. For a magnetic field B, the spacing between flux lines is given by $48900/\sqrt{B}$ Å gauss$^{-1/2}$. The circulating electrical currents that define the field profile of the flux line, decay exponentially with a radial length scale of λ, the magnetic penetration depth. For type II superconductors this length scale is larger than the coherence length ξ, the distance over which superconductivity is quenched. This basic picture of a vortex[5] is illustrated in Fig. 1.

Many elegant experiments have confirmed the existance of these flux lines. Among them are Bitter decoration patterns with fine magnetic particles,[6] neutron scattering,[7] NMR,[8] electron microscopy,[9] and electron holography.[10] Most of these techniques are sensitive to the magnetic field variations.

STM, in contrast, is sensitive to the electronic structure. The motivation to perform local tunneling spectroscopy on the inhomogeneous phases of superconductors has been recognized for some time. Indeed the pioneering development of the STM itself was driven in part by the hope of such an application.

The first direct STM observation of the inhomogeneous phase and the flux lattice was observed in 2H–$NbSe_2$.[11,12,13,14] Before describing these results, we will first review the experimental techniques and procedures that have been useful on superconductors.

9.3. Experimental Techniques for STM on Superconductors

The construction for the STM is standard, following the guidelines of Chapter 2. It has a platinum–iridium tip, and uses a piezoelectric tube scanner that is mounted on a differential spring course approach mechanism. To use this instrument at low temperatures and high magnetic fields requires additional care in the choice of construction materials. Attention must be given to the thermal contraction coefficients and magnetic materials must be avoided.

A high-quality superconducting surface can be produced most easily with 2H–$NbSe_2$. It has a layered structure. The inert surface prevents chemisorption, so clean atomic surfaces are observable with minimal sample preparation. Besides a superconducting phase, this material also supports a charge density wave phase. Refer to the previous chapter for a more complete discussion of this compound and the CDW phase. A number of superconducting parameters have already been determined with $\xi = 77$ Å[15] and $\lambda \simeq 2000$ Å in the plane of the layers,[16] also $\Delta = 1.1$ meV[17] and $T_C = 7.2$ K.

Data on superconductors can be best acquired by measuring the differential conductance, $dI/dV(x, y, z, V)$, at some sample tip bias voltage, V. This is unlike most STM applications where tip deflection $z\ (x, y, V)|_I$ is the experimentally relevant quantity. The differential conductance can be conveniently obtained with a lock-in amplifier technique. A small AC voltage V_{AC} in addition to the bias voltage V is applied across tip and sample. This will result in a small AC current response that is proportional to the differential conductance.

It is important to keep the magnitude of V_{AC} small to obtain high-energy resolution. For a typical normal metal–superconductor barrier the energy resolution is ultimately limited by thermal smearing that is on the order of 3.5 kT. The dither voltage is then best kept comparable or smaller than 3.5 kT/e. For temperatures below 1 K this requires careful shielding of the tip and sample leads, since these voltages are well into the microvolt range.

The distance between tip and sample is set by the constant current feedback loop. For this purpose, a voltage much larger than that of the superconducting gap is used, so that the resistance or spacing is roughly independent of voltage. A resistance of 10^8–10^9 ohms ensures a reasonably large tunnel barrier. dI/ds measurements on 2H–$NbSe_2$ yield 1–2 eV for the barrier height under these conditions. This also forces the DC tunnel current to be small, on the order of tens of picoamperes or less.

A question that often arises is whether such localized currents are sufficiently large to cause local perturbations such as heating, exceeding the local critical current or creating other nonequilibrium effects on superconductivity. Experimental measurements resolve no systematic current-dependent effects in the spectrum. Indeed, one can argue that such effects should not be visible when the individual electron tunneling rates are low compared to the various relevant thermal or quasiparticle relaxation rates. Then the sample can settle back to the equilibrium state between successive electron tunneling events.

The actual signal of the AC tunneling current is correspondingly small, about 100 femtoamperes. A fundamental signal-to-noise ratio is set by shot noise and a reasonable signal-to-noise ratio can be achieved by time averaging. Typical $dI/dV(V)$ curves take a few minutes and complete 2-parameter images can take up to several hours. Fortunately the low-temperature STM does not have thermal drift problems and is a very stable instrument, so long scans are not unreasonable.

Several steps are involved in the data acquisition cycle for collecting the $dI/dV(x, y, z, V)$ data. First the distance z is set with a large tunnel bias voltage V_f while the constant current feedback circuit is active. The feedback loop is next opened while the tip is held steady to better than 0.1 Å for the voltate data acquisition cycle. The first voltage V_N applied across tip and sample should also be much larger than the relevant gap values and $dI/dV(x, y, z, V_N)$ is recorded. This quantity is useful later for normalizing the data. Thereafter, a sequence of voltages V_i are applied and each corresponding $dI/dV(x, y, z, V_i)$ is recorded. The tip–sample voltage is then returned to V_f, the feedback loop switched back on, and the tip moved to a new x, y coordinate. Here the feedback loop is again opened, and the whole voltage data acquisition cycle is repeated for the new location.

A large number of measurements can easily result since three independent parameters x, y and V can be varied. Since signal-to-noise considerations often dictate a long lock-in averaging time for each one of the conductance measurements (sometimes up to one second), it is necessary to limit the number of data points taken for a reasonable length scan. If one wishes to emphasize the two-dimensional spatial structure, the number of voltage values can be limited to just two or three values. The real-space conductance images can show the location and symmetry of the vortices. Alternatively, if spectral information is the goal, the number of voltage samples can be large but the number of spatial coordinates might have to be reduced. The symmetry of the real-space images can guide those choices. For example, for a vortex one may want to sacrifice a complete two-dimensional grid of x, y coordinates, in favor of a one-dimensional sequence of points on a radial line that pierces the vortex at some fixed angle. Such a data set is directly useful to show how the spectrum evolves as a vortex is approached.

There is also a more direct way to image vortex patterns without using the sample and hold cycle. The tip–sample voltage can be reduced to a value comparable to the gap, and the surface scanned with constant current feedback conditions. Vortices are revealed simply by monitoring $z(x, y)|_I$ or $dI/dV(x, y, z(I))|_{V \simeq \Delta}$. The tip, feeding a constant current to the sample, will retract slightly above the higher conductance region areas, such as the vortex cores. Signal amplitudes of several tenths of an angstrom are observed. However, the conductance signal is more direct and also gives a better signal-to-noise ratio.

9.4. Spectrum in Zero Field

An example of how the STM can also measure these quantities is displayed in Fig. 2. The superconducting energy gap of 2H–$NbSe_2$ forms at 7.2°K and opens up to about 1.1 meV at the lowest temperatures. Above 2°K the conductance data agrees roughly with the functional form of Eqs. (1) and (2), provided that finite temperature effects and thermal smearing are included. The evolution of the gap size with temperature is consistent with BCS theory indicated by the solid line in Fig. 2a.

Normally the tunneling current of large-area, thin film junctions is dominated by electrons with the largest normal k_Z vectors, all other directions are exponentially suppressed. When the tunneling takes place in a material such as 2H–$NbSe_2$ with a two-dimensional Fermi surface with little dispersion in the z direction, electrons can tunnel into final states with larger nonzero lateral momentum components. Quasiparticle states on various parts of the Fermi surface are much more evenly weighted in the spectrum.

On closer examination, deviations of the tunneling spectra from the single-valued gap form of Eq. (1) are observed. The top part of Fig. 3 displays the gap structure over a larger voltage range. The superconducting gap is visible as the small structure near zero bias. A second discontinuity in the density-of-states is observed at ± 34 mV. This feature results from the other phase transition, the charge density wave transition. A charge density wave gap opens up on parts of the Fermi surface at $k_Z = 0$ along the six different charge density wavevectors. This anisotropic CDW gap will have important consequences for the superconducting vortex structure. The remaining parts of the

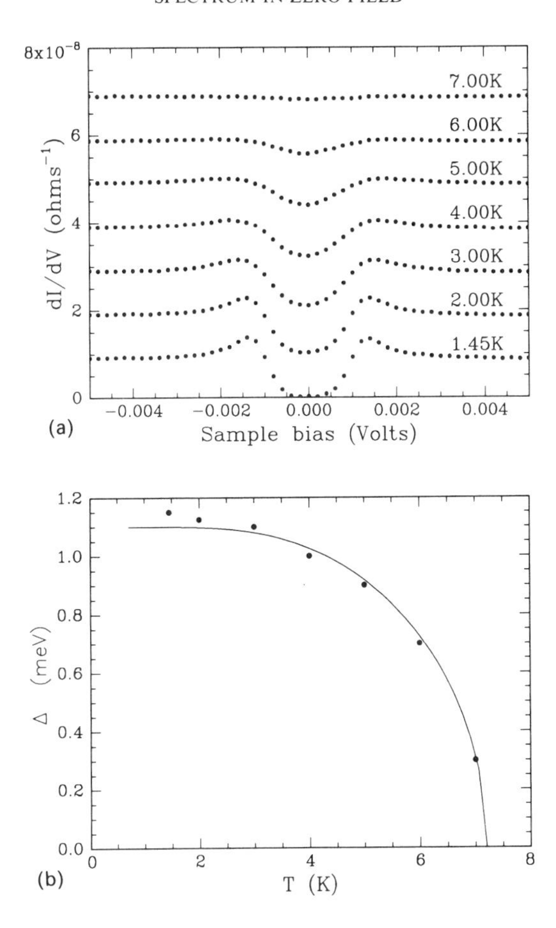

FIG. 2. (a). Behavior of the conductance spectrum of 2H–$NbSe_2$ at different temperatures as it is cooled below the superconducting transition temperature. 2(b). A fit to the spectra yields the temperature dependence of the gap, the solid line is the BCS expectation.

Fermi surface participate in the superconducting transition. A high-resolution spectrum detailing the superconducting gap taken at 50 mK is shown in the bottom part of Fig. 3, where the thermal smearing is now minimal. This spectrum is inconsistent with a density of states with a single valued gap (Eqs. (1) and (2), also shown in Fig. 3 as the dashed line). Such data may well be better described by a distribution of superconducting gaps ranging from 0.7 to 1.4 mV over various parts of the Fermi surface. Residual surface currents may also cause some smearing of the gap.

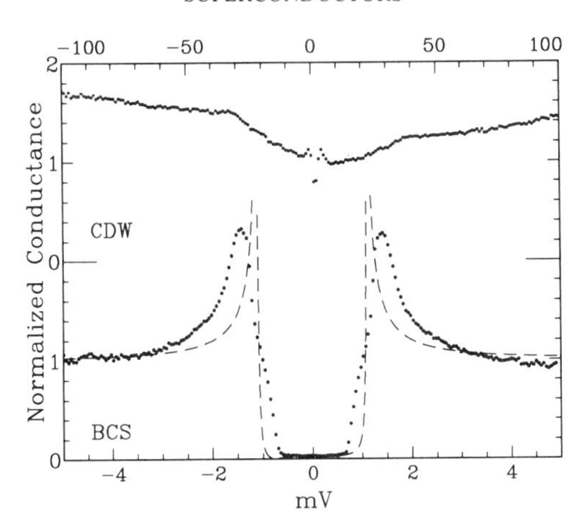

FIG. 3. (top). Spectrum over a larger energy range displaying the charge density wave gap at 33 meV where $2\Delta/kT_c = 24.3$ (bottom). Spectrum taken at 50°mK with reduced thermal smearing displaying nonideal BCS behavior.

9.5. Vortex Data

9.5.1 Spatial Images

The spatial inhomogeneities of the excitation spectrum induced by a vortex can be imaged directly by the STM. This is best seen when the normalized conductance $dI/dV(x, y, V_i)/dI/dV(x, y, V_N)$ is plotted on a grey scale. Such data for a relatively isolated vortex, generated by an applied field of 500 gauss, is shown in Fig. 4. The spatial coordinates range over a 1500 Å square grid of 128 by 128 points. The normalized bias voltage V_N was chosen

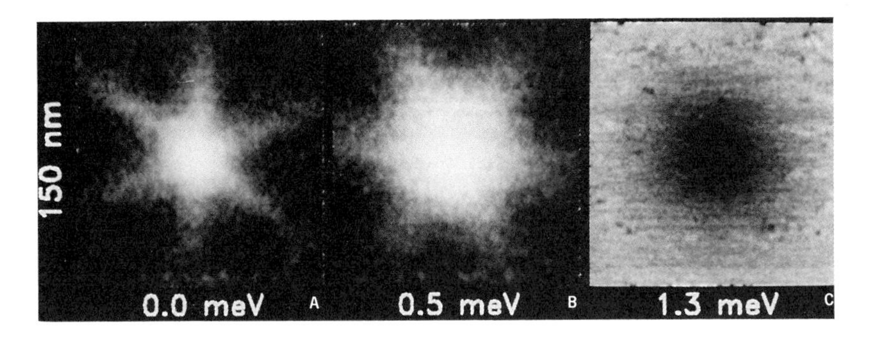

FIG. 4. Spatial images of the normalized differential conductance at three different bias voltages 0.0 mV, 0.5 mV, and 1.3 mV, where the normalization voltage is 5 mV. The grey scale varies from 0 black to 1.59 white, 0.1 to 0.9, and 0.9 to 1.7 for the three respective voltages. The size of the scanned region is 1500 Å.

at $4\Delta/e \simeq 5\,\text{mV}$. Figures 4(a) and 4(b) correspond to the conductance patterns seen at 0.0 mV and 0.5 mV bias, respectively. A bias voltage of 1.3 mV was used for Fig. 4(c). A grey scale representing the conductance ratio ranges from 0 (black) to 1.58 (white), 0.1 to 0.9, and 0.9 to 1.7 for the three applied voltages. The conductance vanishes far from the core for Figs. 4(a) and 4(b) since both voltages are well below the superconducting gap where no states should exist. The high conductance areas in the center correspond to images of the normal metal core. This pattern is reversed in Fig. 4(c) where a bias voltage above the gap is employed. In this case, a somewhat enhanced density-of-states is observed away from the core and decreases at the vortex center.

In the simplest model of a superconductor the size of the normal core region should be of the order of the coherence length. The images present a more complicated, yet more beautiful, picture. The smallest structure is visible for the zero bias images shown in Fig. 4(a) with a radial decay length being of the order of a few coherence lengths. At higher bias, 0.5 meV in Fig. 4(b), the core forms an even larger spatial structure. In both cases the radial decay length depends strongly on the angle and results in a star-shaped pattern. In Fig. 4(a) the decay of conductance is fastest in the crystallographic **a** direction and slowest at 30° away from **a**. This also means that the rays of the zero volt bias star extend along the CDW wavevectors. Surprisingly this situation is reversed for the 0.5 mV bias image in Fig. 4(b). Here the decay is slowest in the crystallographic **a** direction. A comparison of the two images shows an apparent 30° rotation of the star pattern. At bias voltages above the gap this angular structure is no longer as pronounced, and a more circularly symmetric image results. Direct profiles of these vortices at 0.0 meV and 0.5 meV and 2 discrete angles are shown in Fig. 5.

The star patterns observed in the vortex cores might arise from two sources: from two sources: from interactions of nearest neighbor vortices of the hexagonal flux lattice, or possibly it might be intrinsic to a single vortex, produced by the 6-fold anisotropy of the atomic crystalline band structure. Since the star shape persists undiminished even at large vortex separation, the atomic crystalline band structure is most likely responsible for the anisotropy.

When larger fields are applied, the vortex spacing decreases and the effects of vortex–vortex interactions should become more pronounced. Abrikosov predicted that such vortices will organize themselves into a hexagonal lattice. This is present even at the lowest fields of 250 gauss, but can be better observed at higher fields where more vortices are present within the scan range of the microscope. Figure 6 shows this lattice on a 6000 Å square at a field of 10 kG. The grey scale corresponds to $dI/dV(1.3\,\text{mV})$ data taken under constant current feedback conditions with a 1.3 mV bias. The angular orien-

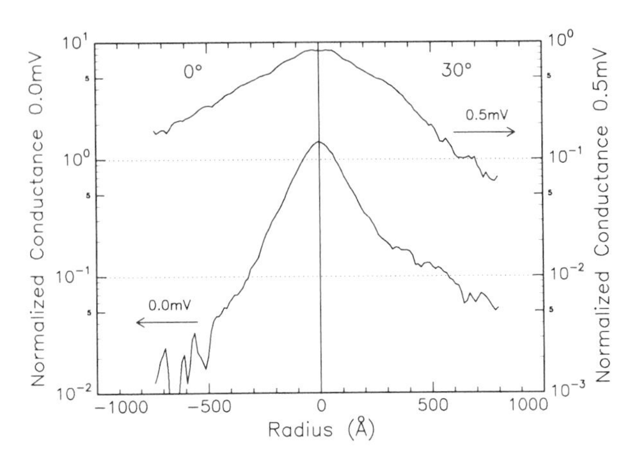

FIG. 5. Conductance profiles of the conductance data of Fig. 4 at two angles with respect to the crystalline lattice.

tation of the flux lattice is always observed to be the same and appears to be locked to the crystalline lattice orientation, where the atomic lattice **a** direction matches the nearest neighbor vortex direction.

9.5.2 Spectral Images

Considerably more insight into the physics of vortices may be obtained from the spectral evolution images. This data corresponds to many spectra taken on a sequence of points along a line that extends radially through a vortex. Figure 7 shows a perspective view of the normalized differential conductance over the range ± 5 mV. The sampling line extends back 800 Å with the vortex located at 600 Å.

A most surprising feature is that the conductance in the core does not show a flat density–of–states that one might expect if it were to be described as a "normal metal." Instead there is a pronounced peak in the conductance at zero bias. This peak has been observed with a maximum normalized conductance of up to 4. Many factors appear to influence its height. It is most pronounced at the lowest temperatures, thermal smearing above a few Kelvin reduce its amplitude. There is also some variation of peak height from one vortex to the next. A potential cause may be impurity or other defects in the vicinity of the core, which can perturb the density–of–states. Some sample-to-sample variation may also be indicative of this. If the sample line misses the core by several Angstroms, the zero-bias amplitude will also be reduced. By taking several parallel, closely spaced, sampling lines one can ensure that one of them is within a reasonable distance of the vortex center.

The spectral evolution is also sensitive to the angle at which the sampling

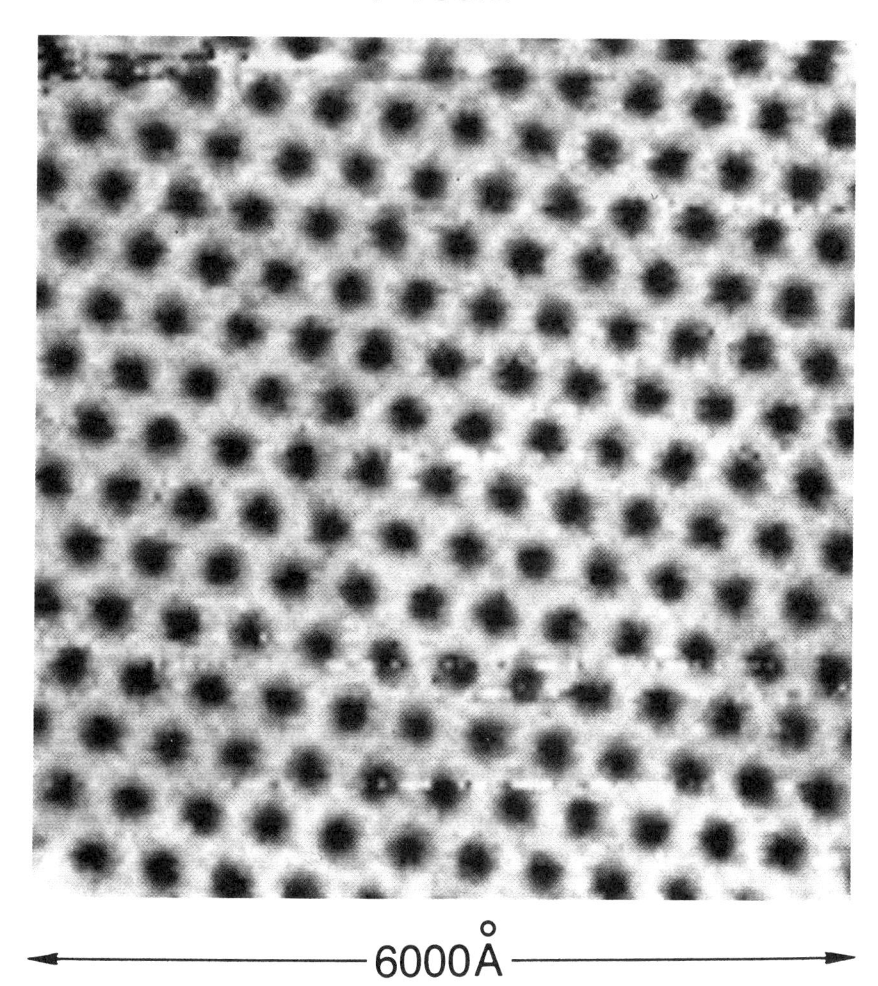

FIG. 6. The Abrikosov vortex lattice as observed in an STM conductance image at 1.3 mV and 1 Tesla field. The grey scale varies from 0.5 to $1.5*10^{-8}$ ohms^{-1}.

line penetrates the core. A sampling of such data at different angles, as sketched in Fig. 8, gives insight into the source of the 6-fold "star" patterns observed in the real space images. Since the anisotropy is most pronounced at energies below the gap, the subgap structure can be best revealed at the lowest temperatures, 50°mK, and focusing on a smaller voltage range, − 1.65

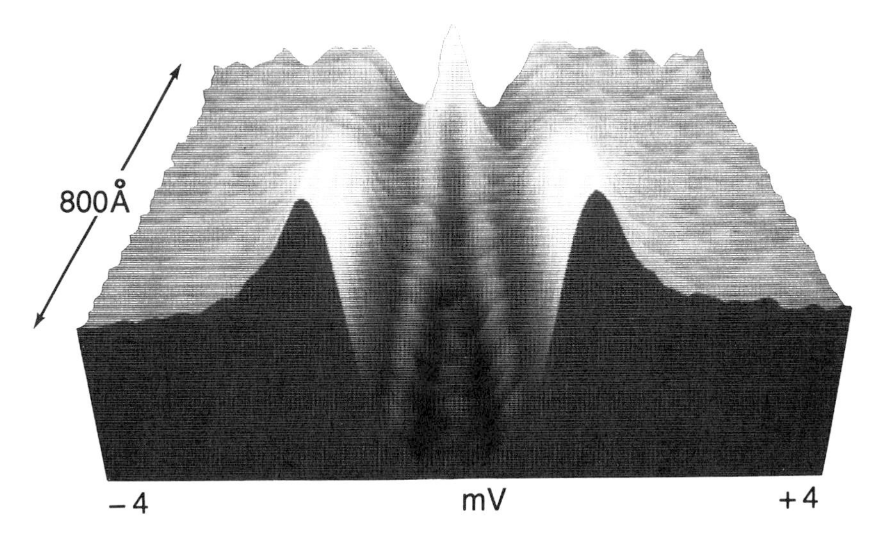

FIG. 7. Perspective view of the superconducting spectra as it evolves on an 800 Å line that penetrates through a vortex. Notice the zero bias peak at the vortex center and how it splits into two subgap peaks at larger radius from the core.

to + 1.35 mV. The data for three different angles measured with respect to the **a** direction is shown in the perspective views of Fig. 9.

Here one can observe that the zero-bias peak splits into not one but two pairs of subgap peaks that merge eventually with the energy gap at large

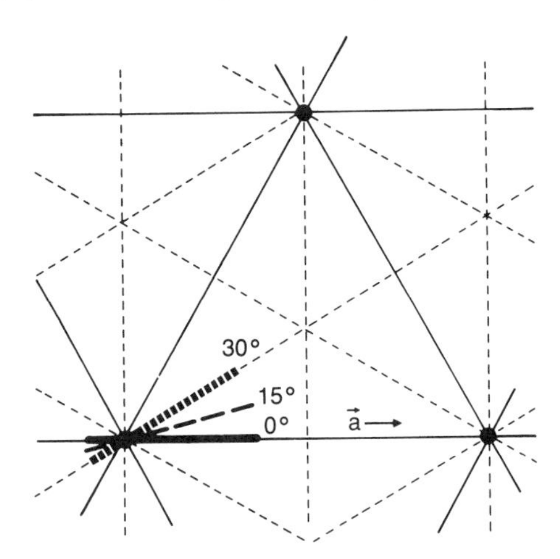

FIG. 8. Schematic of the various sampling lines that pass through the vortex core and are used for the spectral evolution data of Fig. 9. The crystallographic **a** direction is indicated and lines up also with the vortex lattice direction.

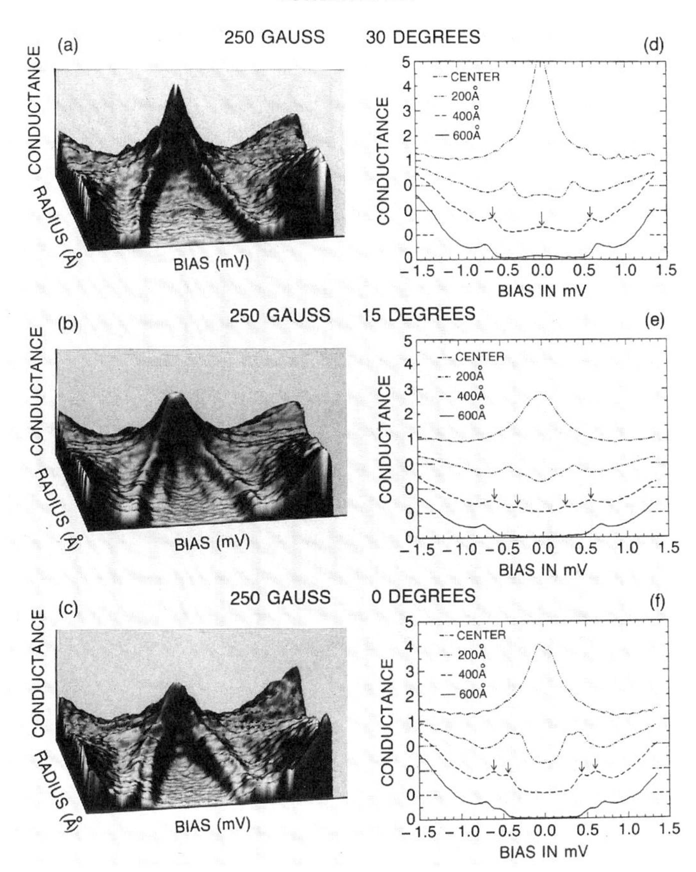

FIG. 9. A more detailed perspective of $dI/dV(V, r)$ showing how it evolves along the three lines sketched in Fig. 8. The energy scale is smaller than Fig. 7 and probes only structure wtihin the gap. The perspective scale corresponds to a 1000 Å sampling line with the vortex positioned at 250 Å from the back. The outer subgap peak is not sensitive to angle, but the inner peak collapses to zero energy at 30°. A few spectra of the perspective data are explicitly plotted in Figs. 9(d), (e), and (f).

radius. The energy evolution of these peaks is displayed more directly in Fig. 10. The outer peaks do not appear to be sensitive to the angle of the sampling line, while the inner pair is. It has its maximum energy at 0° where it is just below that of the outer peaks [see Figs. 9(c) and 9(f)]. At larger angles it decreases to lower energy [see Figs. 9(b) and 9(d)]. Finally at 30° [see Figs. 9(a) and 9(d)] the inner subgap peaks have merged together to form a broad hump at zero bias at all radii. This corresponds also to the rays of the zero mV "star," in Fig. 4(a). The 30° rotation of the 0.5 mV image, in Fig. 4(b),

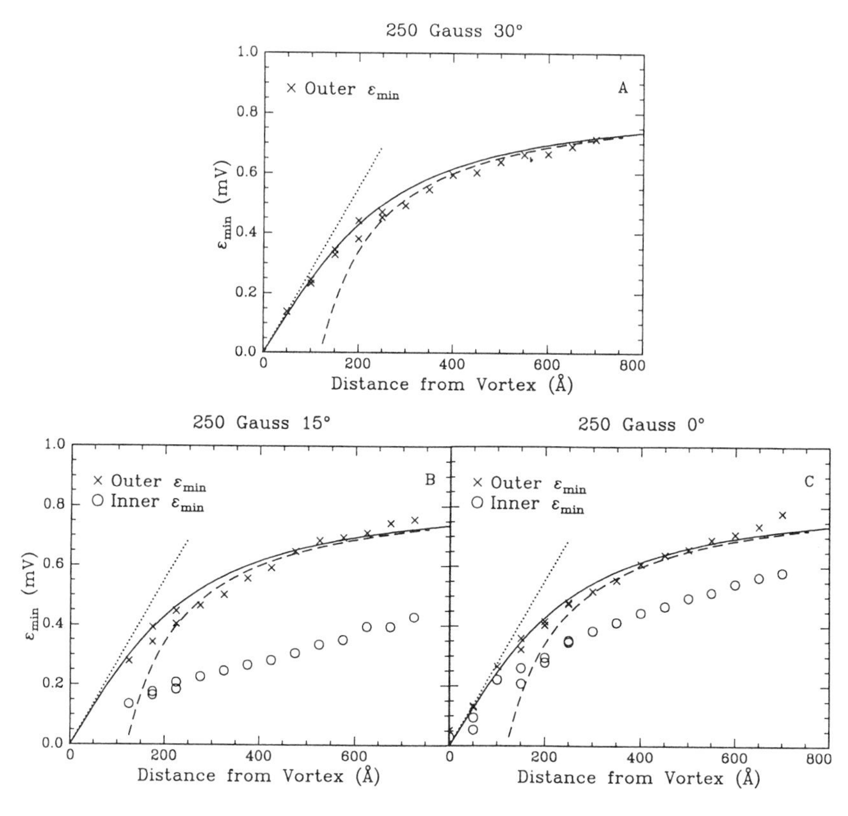

FIG. 10. The subgap peak energies of Fig. 9 as a function of radius. The dotted and dashed lines correspond to asymptotic forms, from Eqs. (8) and (11), respectively. The solid line is a guide corresponding to the outer subgap peak data.

is also consistent with the extra conductance that occurs at this higher bias in the 0° direction associated with the inner subgap peak.

The spectral evolution is also sensitive to the proximity of neighboring vortices. This can be seen clearly in Fig. 11, where the three different applied fields correspond to inter-vortex spacings of 3100 Å, 1550 Å, and 770 Å. The first one of this series approximates that of an isolated vortex, as the separation is larger than a penetration depth. The angle 30° was chosen as the simplest case where the influence on the outer subgap peak can most clearly be observed. The rate at which the zero bias peak splits with radius becomes increasingly abrupt at higher fields. Each peak also tends to broaden and appear less well-defined with increasing fields. On approaching H_{C2} all features disappear and a flat metallic density-of-states is observed everywhere.

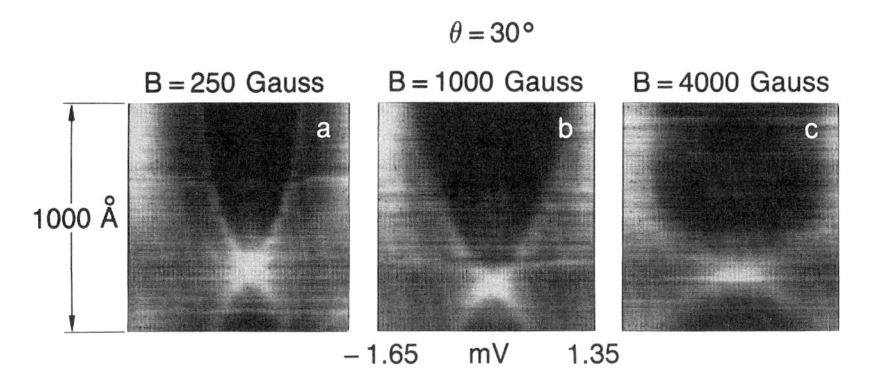

FIG. 11. Grey scale images of the spectral evolution along a 1000 Å line at 30° for a sequence of 3 decreasing vortex–vortex spacings. Notice that the splitting is more rapid at the higher magnetic fields.

9.6. Interpretation

9.6.1 Isolated Vortex Wavefunctions

A simple model of a vortex is that of a circulating supercurrent around a normal metal core. The size of this core where superconductivity is quenched is of order ξ (see Fig. 1). The real-space images of vortices, where a tip–sample bias of 0 mV is employed, as in Fig. 4(a), then represent images of this normal core. The STM data reveals considerably more detail, so to understand more fully the size, shape, and bias voltage dependence of these core images, a more rigorous model is needed.

A more complete insight requires the microscopic solutions of the quasiparticle excitations of this inhomogeneous system. Two approaches have been remarkably successful in predicting and confirming many of the experimental observations. Inhomogeneous superconducting systems can be described by solutions of the Bogoliubov equations. This method was initially applied to vortices by C. Caroli *et al.*[18] and later by J. Bardeen.[19] More recently the solutions of these equations were used by J. Shore *et al.*[20] and F. Gygi *et al.*[21] to evaluate the local density-of-states for a direct comparison with the experimental STM data. For a thorough review of this approach see Ref. 22. Likewise, the Eilenberger equations were solved for vortices in a clean superconductor by L. Kramer and W. Pesch[23] and a recent density-of-states calculation by U. Klein[24] and S. Ullah and A. T. Dorsey[25] also found excellent qualitative agreement with experiment. We will review these results and show how they describe various aspects of the vortex images while emphasizing their physical significance.

The amplitudes of the quasiparticle states are described by the two-

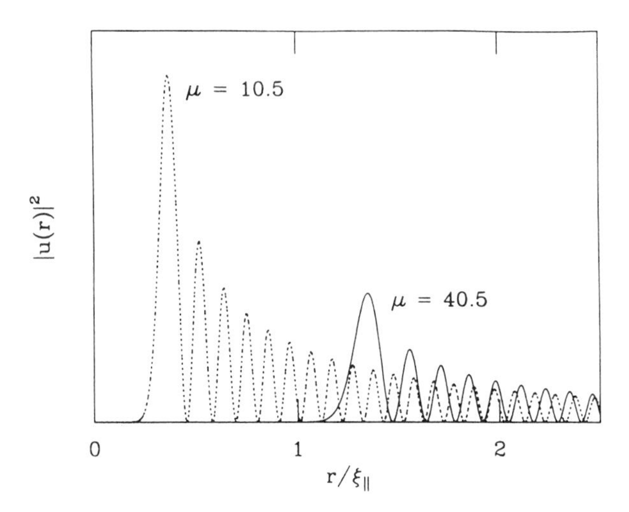

FIG. 12. Bound energy eigenstates of a vortex as calculated from the Bogoliubov equations (from Ref. 20).

component scaler $u(r)$ and $v(r)$. These wavefunctions and their energy eigenvalues, ε_i, can be determined by the Bogoliubov equations,

$$(\varepsilon_i + \varepsilon_f)u_i(r) = \frac{1}{2m}\left(p - \frac{e}{c}A(r)\right)u_i(r) + \Delta v_i(r) \tag{4}$$

$$(\varepsilon_i - \varepsilon_f)v_i(r) = -\frac{1}{2m}\left(p + \frac{e}{c}A(r)\right)v_i(r) + \Delta^* u_i(r).$$

The order parameter, $\Delta(r)$, is determined in turn by the self-consistency condition,

$$\Delta(r) = g \sum u_i(r)v_i^*(r)(1 - 2f(\varepsilon_i)). \tag{5}$$

Both $\Delta(r)$ and the vector potential $A(r)$ act as a radial potential well that define the quasiparticle states. In analogy with the solutions of the Shroedinger equation, the quasiparticle states can be classified as bound states with energy less than Δ_0 and scattering states with greater energy. The bound solutions can be described by the angular momentum quantum numbers μ and the momentum along the axis of the vortex k_z. Examples of such bound states are shown in Fig. 12. These states were calculated using a variational $\Delta(r)$ with a functional form $\Delta_0 \tanh(r/d\xi)$. As might be expected, the lowest angular momentum states have their maximum amplitude closest to the origin. The decay of these wavefunctions far from the core is for the cylindrically symmetric case given by,

$$u(r)^2 + v(r)^2 = \frac{1}{r}\, e^{-r/l} \quad l = \frac{\pi\xi}{\sqrt{1 - (\varepsilon/\Delta)^2}}. \tag{6}$$

Since the tunneling transition matrix depends on the overlap between the tip wavefunction (ideally a spatial delta function on the suface) and the quasiparticle wavefunction, the conductance becomes a direct measure of the wavefunction probability density in Eq. 6. The experimental images in Fig. 4 and the profile in Fig. 5 can be interpreted as the wave function probability density with energies selected by the bias voltage. The data does not have the energy resolution to identify individual energy eigenstates, but rather represents the contribution of several states over a range of order kT of energies about the bias voltage. The fine structure in Fig. 12 at a spatial frequency k_f, is averaged out, yet a number of qualitative trends can be easily identified. The most compact wavefunctions which peak close to the core are observed at zero bias. Higher-bias voltage images, which should highlight the higher angular momentum states, are consistent with a more spatially extended wavefunction. Indeed, the profile in Fig. 5(b), at 0.5 meV, shows a rather flat-topped form with a slight dip at the origin in agreement with high μ wavefunctions, whose maximum should be located at finite radius $R_C = \mu/k_f$. The profiles are consistent with the wavefunction decay of Eq. (6), where l is of the order of 2–4 ξ for low energy states, and significantly larger for the higher energy (0.5 meV), high μ case. With yet higher bias voltages, larger than the gap, the scattering states are observed [see Fig. 4(c)]. In contrast to the bound states, they have a reduced amplitude at the core.

These model states display cylindrical symmetry, yet star-shaped conductance patterns in Fig. 4 are observed in the data. This can also be explained in the framework of the Bogoliubov equations. By adding a 6-fold symmetric, $\cos(6\theta)$, perturbation to the Bogoliubov equations, F. Gygi *et al.*[26] generated modified eigenstates as shown in Fig. 13. The source of this perturbation is from the 6-fold variations in the crystalline band structure. The new states were characterized as bonding states with an amplitude maxima in the angular well minima, and antibonding states where the opposite happens. This approach explains not only the star shape, but also the apparent 30° rotation between high and low energy states, since the bonding and antibonding states must be orthogonal.

9.6.2 Isolated Vortext Spectral Evolution

The same data displayed differently, in terms of the spectral evolution near a vortex, tends to raise different questions and can give additional insight into the physics of vortices. One of the first issues to understand is the size of the

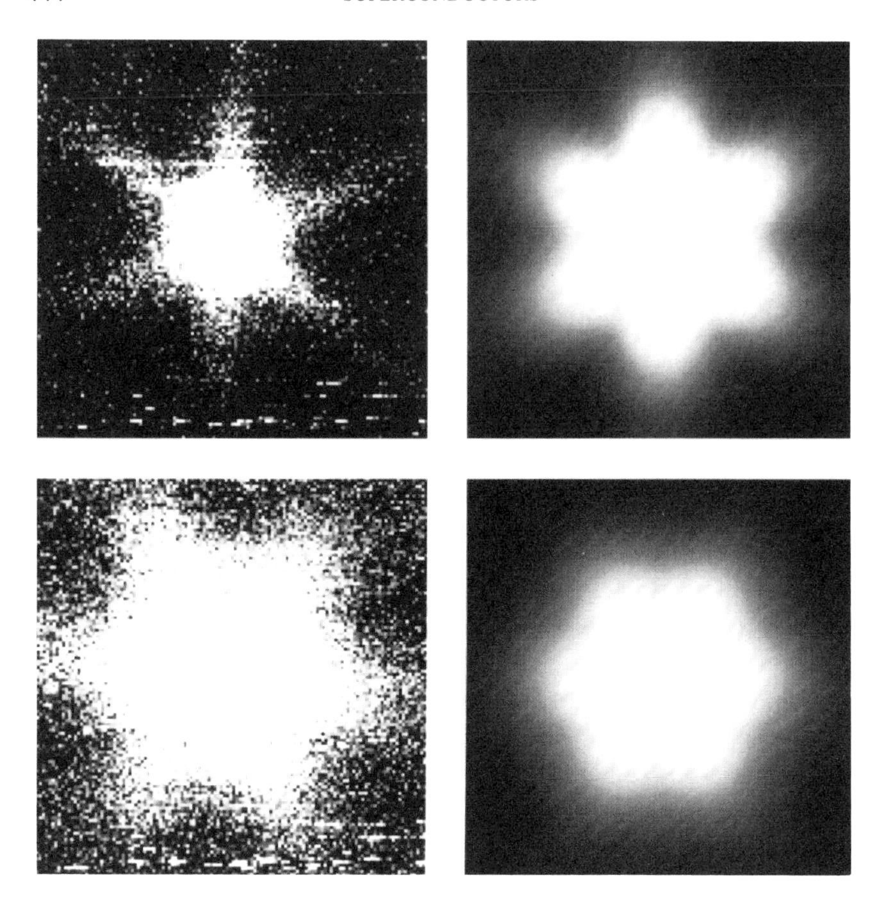

FIG. 13. Star-shaped wavefunction probabilities when an angular perturbation is applied to the Bogoliubov equations (from Ref. 26). Top right image corresponds to the bonding states, the bottom right one, the antibonding states. Left side shows the corresponding experimental data.

zero bias peak that is observed when tunneling into the center of the vortex. The observed normalized conductance is strikingly larger than unity, ranging from 1.5 up to 4.0.

An elegant explanation of the zero bias peak was a macroscopic approach given by A. W. Overhauser and L. L. Daemen.[27] They modeled the core as a cylinder of normal metal surrounded by superconductor, with its characteristic excitation spectrum. This by itself defines a flat density-of-states in the core with a normalized conductance of unity. If the electrons of the two regions are allowed to couple, then the self-energy corrections to the normal electrons of the core will shift their energy toward the Fermi level, and will

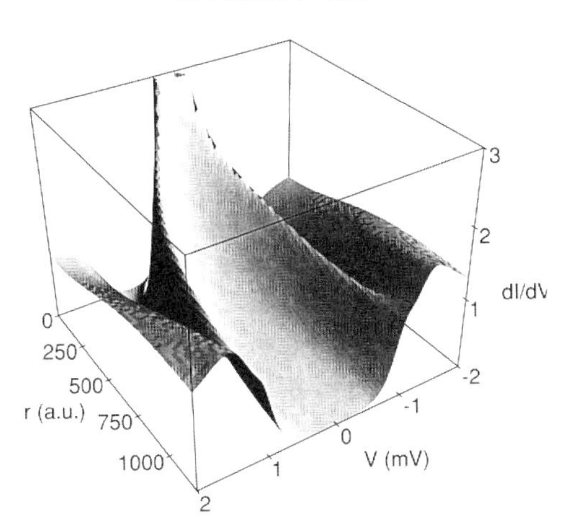

FIG. 14. Perspective view of spectral evolution of a vortex as expected from the Bogoliubov formalism (from Ref. 22).

increase the density-of-states at zero bias. The resulting zero bias peak is similar to that observed by experiment.

The microscopic theories also provide additional insight into the spectra near vortices. The existence of the zero bias peak was initially confirmed in calculations by J. Shore *et al.*[20] for a superconductor in the clean limit on the basis of the Bogoliubov equations. When summing the conductance into the lowest energy-bound states discussed previously, they calculate a peak height somewhat larger than that measured, even when including the appropriate amount of thermal broadening. A similar zero bias peak was obtained by U. Klein[24] on the basis of the Eilenberger equations. However, by including an additional energy-broadening term that describes the effect of impurities, a reduced peak height amplitude consistent with experiment could be explained with reasonable parameters for impurities. In fact, dirty superconductors, where the mean free path is shorter than the coherence length, are not expected to show such an enhanced conductance at zero bias.

The evolution of the spectra with radius from the core also has a number of remarkable properties. Using the Bogoliubov equations, Shore *et al.*[20] predicted that this zero bias peak should split into two separate peaks with increasing radius. The Eilenberger equations used by U. Klein[24] and later by S. Ullah *et al.*[25] gave a result also consistent with the splitting. An example of this spectral evolution as calculated theoretically[22] is shown in Fig. 14 and compared qualitatively to the data of Fig. 7.

This splitting is just another manifestation of the relationship between energy and radius of the bound wavefunctions. In other words, the higher energy wavefunctions have a larger radius, as shown in Fig. 12. The maximum wavefunction amplitude is situated roughly at the radius $R_C = \mu/k_f$. If the order parameter rises linearly from zero in the core, the dispersion is approximately of the form,[28,29]

$$\varepsilon(r) = \frac{\mu}{k_f}\frac{d\Delta}{dr} + \mu\hbar\omega_1(B) \qquad r \ll \xi, \tag{7}$$

where the last Larmor frequency term approximates the influence of the magnetic field. For an extreme type II superconductor this term is small compared to the first one. In the Landau–Ginzberg limit, the gap also arises on the length scale of the order of the coherence length, so that the rate of splitting of the zero bias peak into two peaks is given by,

$$\frac{d\varepsilon(r)}{dr} = \frac{\Delta_0}{c\xi} + k_f\hbar\omega_1(B) \qquad r \ll \xi, \tag{8}$$

where c is a constant of the order unity. (Experiment observes $c \approx 3$–4.) This form should hold within a coherence length radius of the core where the approximations are valid. The low-temperature limit of a clean superconductor $d\Delta/dr$ is predicted to be much larger,[23] of order $k_f\,\Delta_0$. A fully self-consistent calculation[22] agrees with this when $\varepsilon \ll \Delta_0$. Such large values of the splitting rate have not yet been observed in experiment. Other considerations, such as impurity effects and strong anisotropy of the superconductor, make a quantitative comparison to Eq. (8) more involved.

At large radius it is insightful to view the splitting in terms of a quasiclassical picture proposed by M. Cyrot.[30] Consider the quasiparticle states near the gap edge with energy Δ, which are described by Bloch waves with velocity v_f. In the vicinity of vortex they will experience a perturbation determined by the momentum of the circulating superfluid p_s. This will modify the minimum energy of the quasiparticle states to,

$$\varepsilon = \Delta - v_f p_s \qquad r \gg \xi. \tag{9}$$

One can assume a London model vortex, where the expression for the current at the surface[31] is given by,

$$J_s(r) = \frac{c}{4\pi}\frac{\phi_0}{2\pi\lambda^3}\left[\frac{\lambda}{r} - I_1\left(\frac{r}{2\lambda}\right)K_0\left(\frac{r}{2\lambda}\right)\right], \tag{10}$$

where $I_1()$ and $K_0()$ are Bessel functions. Then Eq. (9) can be re-expressed as

$$\varepsilon(r) = \Delta\left(1 - \frac{\pi}{2}\frac{\xi}{\lambda}\left[\frac{\lambda}{r} - I_1\left(\frac{r}{2\lambda}\right)K_0\left(\frac{r}{2\lambda}\right)\right]\right), \qquad r \gg \xi \tag{11}$$

As the vortex is approached, the circulating superfluid momentum gradually increases, resulting in states that are perturbed to lower energies. This describes how the subgap peak splits away and downward from the gap edge.

A comparison of the actual data with the approximate limits in Eqs. (8) and (11) are shown as the dotted and dashed lines, respectively, in Fig. 10(a). The gap and the coherence length were assumed to take the values $\Delta \sim 0.85\,\text{meV}$ and $\xi = 77\,\text{Å}$ and c was evaluated to be about 4 for the small radius limit. For the large radius limit λ was taken to be 2000 Å and a $\Delta \sim 0.85\,\text{meV}$ yielded the best fit.

These simple models are consistent with 2 subgap peaks. The data on 2H–$NbSe_2$ shows four, rather than two, subgap peaks. Other angles of the spectral evolution sampling line show this clearly in Figs. 9(b), 9(e), 9(c), and 9(f). The additional inner pair is angle-sensitive and collapses together at 30° [see Figs. 9(a) and 9(d)], where it forms a broad zero bias hump that exists at all radii. One can specultate that the source of the additional structure originates from the unique band structure[32] of this material. The undulating cylinder shape of the Fermi surface could easily result in two van Hove extrema of the density–of–states and correspond to two separate values of v_f or electron populations with 2 different values of the effective mass. Such extrema are located at $k_z = 0$ and $k_z = \pi/c$. Since the quasiparticle states are formed from the basis states available at the Fermi surface, one might expect two different types of bound states. Each might have a slightly different energy–radius relationship or characteristic splitting rate for its spectral evolution. For example the outer subgap peaks might be associated with the $k_z = \pi/c$ extrema of the Fermi surface. This part does not have any pronounced angular anisotropy and is consistent with a splitting that is independent of angle. The situation is different for the $k_Z = 0$ extrema of the Fermi surface. Quasiparticle states exist only in certain directions in this part of k space. In the six CDW directions (30°) the CDW gap has removed all states from the Fermi surface. In these directions the inner subgap peaks have collapsed to zero and do not split. This may be equivalent to defining a coherence length on this part of the Fermi surface that diverges.

9.6.3 Vortex–Vortex Interactions

On application of higher magnetic fields several effects can provide a qualitative explanation of the behavior of the spectra in Fig. 11. Such interactions set in once the vortices are within a few penetration depths of each other. At this point the magnetic field at each core becomes larger than the isolated vortext value. The Larmour frequency is increased and can become comparable or larger than the $d\Delta/dr$ term of Eq. (7). This will increase the energy of the bound states and result in a more rapid splitting of the subgap peaks. The data is consistent with this trend.

As the vortices are brought in closer proximity to each other, other vortex core state band structure effects[33] become important in describing the spectrum. The wavefunctions of a vortex begins to overlap with its neighbors, and the energy eigenvalues of the bound states are changed, much as atomic orbitals of atoms in a crystal are perturbed into bands in the tight binding model. A new quantum number, the vortex lattice momentum, is now required to fully describe the eigenstates. One consequence of this would be to spoil the unique relationship between radius and energy that is normally evident as a well-defined subgap peak. Indeed, the peaks are broadened in energy at 1000 gauss and even more so at 4000 gauss in Fig. 11. Also the double peak structure observable at 0° and 250 gauss is no longer resolvable at 1000 gauss. The amount of this broadening is a measure of the bound state bandwidths.

9.7. Conclusion

In the STM results described on just one superconductor 2H–$NbSe_2$ one can draw an analogy with many aspects of atomic and condensed matter physics. Bound states and scattering states of a vortex can be directly imaged and suggest qualitative similarities to atoms with their bound and scattering states. Magnetic fields can perturb the energies and radii of these states, an effect that can be uniquely observed via STM in the spatial evolution of the tunneling spectra. Anisotropy in the effective mass from the crystalline band structure has surprising and beautiful consequences in the images and spectra. Finally, bands based on vortex bound states are created at higher fields, forming an analogue to the atomic conduction bands that are normally imaged in atomic scale STM pictures.

The process of extending these techniques to other superconductors is only in its infancy. As further measurements are made, a rich variety of new details about this electronic phase is promised by the STM, which no other instrument can rival.

References

1. J. R. Kirtley, *Intern. J. of Mod. Phys. B* **4**, 201 (1990).
2. M. Tinkham, in *Introduction to Superconductivity*, Robert Krieger Publishing Co., 1980; J. R. Schrieffer, in *Theory of Superconductivity*, Addison-Wesley Publishing Co., 1964; P. G. DeGennes, in *Superconductivity of Metals and Alloys*, Addison-Wesley Publishing Co., 1989.
3. I. Giaever, *Phys. Rev. Lett.* **5**, 147, 464 (1960).
4. A. A. Abrikosov, *Zh. Eksp, Teor. Fiz.* **32**, 1442 (1957); *Sov. Phys. JETP* **5**, 1174 (1957).
5. A. L. Fetter and P. C. Hohenberg, in *Superconductivity*, Vol. 2 R. D. Parks, ed., Marcel Dekker, New York, 1969.
6. U. Essman and H. Traeuble, *Phys. Lett. A* **24**, 526 (1967); N. V. Sarma, *Philos. Mag.* **17**, 1233 (1968).

7. D. Criber, B. Jacrot, L. Madhav Rao, and B. Farnoux, *Phys. Lett.* **9**, 106 (1968); H. W. Weber, J. Scheller, and G. Lippman, *Phys. Status Solidi (b)* **57**, 515 (1973).
8. A. G. Redfield, *Phys. Rev. Lett.* **162**, 367 (1967).
9. J. Bosch, R. Gross, M. Koyanagi, and R. P. Huebeuer, *Phys. Rev. Lett.* **54**, 1448 (1985).
10. T. Matsuda, S. Hasegawa, M. Igarashi, T. Kobayashi, M. Naito, H. Kajiyama, J. Endo, N. Osakabe, A. Tomomura, and R. Aoki, *Phys. Rev. Lett.* **62**, 2519 (1989).
11. H. F. Hess, R. B. Robinson, R. C. Dynes, J. M. Valles Jr., and J. V. Waszczak, *Phys. Rev. Lett.* **62**, 214 (1989).
12. H. F. Hess, R. B. Robinson, R. C. Dynes, J. M. Valles Jr., and J. V. Waszczak, *J. Vac. Sci. Technol. A* **8**, 450 (1990).
13. H. F. Hess, R. B. Robinson, and J. V. Waszczak, *Phys. Rev. Lett.* **64**, 2711 (1990).
14. H. F. Hess, *Physica B,* **169**, 422 (1991).
15. P. de Trey, S. Gygax, and J. P. Jan, *J. Low Temp. Phys.* **11**, 421 (1973).
16. K. Takita and K. Masuda, *J. of Low Temp. Phys.* **58**, 127 (1984).
17. B. P. Clayman, *Can. J. Phys.* **50**, 3193 (1972); R. C. Morris and R. V. Coleman, *Phys. Lett. A* **43**, 11 (1973).
18. C. Caroli, P. G. deGennes, and J. Matricon, *Phys. Lett.* **9**, 307 (1964); C. Caroli and J. Matricon, *Phys. Kondens, Mater.* **3**, 380 (1965).
19. J. Bardeen, R. Kümmel, A. E. Jacobs and L. Tewordt, *Phys. Rev.* **187**, 556 (1969).
20. J. D. Shore, M. Huang, A. T. Doresy, and J. P. Sethna, *Phys. Rev. Lett.* **62**, 3089 (1989).
21. F. Gygi and M. Schluter, *Phys. Rev. B* **41**, 822 (1990).
22. F. Gygi and M. Schluter, *Phys. Rev. B* **43**, 7609 (1991).
23. L. Kramer and W. Pesch, *Z. Phys.* **269**, 59 (1974).
24. U. Klein, *Phys. Rev. B* **41**, 4819 (1990).
25. S. Ullah and A. T. Doresy, *Phys. Rev. B* **42**, 9950 (1990).
26. F. Gygi and M. Schluter, *Phys. Rev. Lett.* **65**, 1820 (1990).
27. A. W. Overhauser and L. L. Daemen, *Phys. Rev. Lett.* **62**, 1691 (1989); *Phys. Rev. B* **40**, 10778 (1989).
28. E. Brun Hansesn, *Phys. Lett. A* **27**, 576 (1968).
29. W. Bergk and L. Tewordt, *Z. Physik* **230**, 178 (1970).
30. M. Cryot, *Phys. Konden. Mater.* **3**, 374 (1965).
31. J. Pearl, *J. App. Phys.* **37**, 4139 (1966).
32. G. Wexler and A. M. Woolley, *J. Phys. C* **9**, 1185 (1976).
33. E. Canel, *Phys. Lett.* **166**, 101 (1965).

Index

A

Adsorbates, metal surfaces,
  scanning tunneling microscopy imaging theory, 14
Ag on Au (111),
  epitaxy, scanning tunneling microscopy of, 296
Ag on Si(111),
  scanning tunneling microscopy of, 199
Apparent barrier height,
  theory of, 20, 26
Apparent size of adatom,
  theory of, 16
As-1 × 1 adsorbed Ge(111),
  scanning tunneling microscopy of, 241, 243
Atom superposition model, 8
Atomic force microscopy, 86
Atomic force microscopy,
  force gradient microscopy, 89
  low-temperature,
    instrument design, 355–7
  thermal noise in, 89
  van der Waal forces in, 87
Atomic force microscopy of charge density waves in,
  1T-TaS_2, 358, 359–72
  1T-$TaSe_2$, 358, 374
  charge density waves (CDW), 350, 385
Au on Au(111),
  surface diffusion, 292
Au on GaAs(110),
  spectroscopy of, 268
Au on Ni(110),
  epitaxy, scanning tunneling microscopy of, 295
Au(100),
  surface reconstruction,
    scanning tunneling microscopy of, 291
  surface reconstruction, missing row,
    scanning tunneling microscopy of, 286
  spectroscopy of, 285
Au(110),
  phase transition, scanning tunneling microscopy of, 287
  roughening transition, scanning tunneling microscopy of, 293
Au(111),
  scanning tunneling microscopy of, 292
  spectroscopy of, 281
Au-adsorbed,
  Si(111),
    scanning tunneling microscopy of, 200
Au/AlAs/GaAs,
  heterostructure interface,
    ballistic electron emission microscopy of, 337, 339
Au/GaAs,
  interface,
    ballistic electron emission microscopy of, 335
    ballistic electron spectroscopy of, 332
    ballistic hole spectroscopy of, 341
    carrier scattering spectroscopy of, 344
Au/Si,
  interface,
    ballistic electron emission microscopy of, 335
    ballistic electron spectroscopy of, 330
    ballistic hole spectroscopy of, 340
    carrier scattering spectroscopy of, 343

B

Ballistic electron emission microscopy, 307
  basic principles, 309
  carrier transport at interfaces,
    critical angle at,
      for electrons, 312
      for holes, 321
    quantum mechanical reflection at, 317, 342
    transverse momentum conservation at, 311, 331
  collector current,
    expression for, 315
    for off-axis conduction band minima, 317
    normalization by tunnel current, 316,

325, 327
electron attenuation length,
in Au, 331
experimental methods, 326
sample preparation, 329
sample electronic properties, 327
imaging, 334
interface resolution of, 313
of Au/AlAs/GaAs, 337, 339
of Au/GaAs, 335
of Au/Si, 335
low temperature, 327
spectroscopy, 316
carrier-carrier scattering, 322
spectroscopy of,
carrier-carrier scattering,
multiple scattering, 324
of Au/GaAs, 344
of Au/Si, 343
theory, 322
threshold behavior, 325
spectroscopy,
electron, 316
of Au/GaAs, 332
of Au/Si, 330
theory, 311
threshold behavior, 316, 325, 331
hole, 319
of Au/GaAs, 341
of Au/Si, 340
theory, 319
threshold behavior, 321
asymmetry compared with electron spectroscopy, 320
tunneling, 314, 320
Band gap states,
scanning tunneling microscopy of, 255, 270, 273, 275
BEEM, ballistic electron emission microscopy, 307

C

C_6H_6 on Rh(111),
chemisorption, scanning tunneling microscopy of, 297
Charge density wave (CDW), 350, 385
Charge density wave gap, 432, 434, 447
Charge density waves in,
2H-$NbSe_2$, 386
1T-TaS_2, 358
2H-TaS_2, 391
4Hb-TaS_2, 393, 400
1T-$TaSe_2$, 358, 374
2H-$TaSe_2$, 388
4Hb-$TaSe_2$, 393
1T-$TiSe_2$, 381
1T-VSe_2, 359
$NbSe_3$, 402, 405
TaS_3, 402, 405, 416
Charge detection microscopy, 90
Chemical potential microscopy, 86, 87
Clean Ge surfaces and alloys,
scanning tunneling microscopy of, 227
Clean semiconductor surfaces,
scanning tunneling microscopy of, 151
Clusters,
scanning tunneling microscopy of, 266, 271, 273
Contact potential microscopy, 92
Corrugation amplitude, 227, 278
anomalous, 289
of metal surface structure, 278
Cross-sectional imaging, 261
of GaAs/AlGaAs superlattices, 263
Cs adsorbed,
GaAs(110),
spectroscopy of, 123
Cu(100),
scanning tunneling microscopy of, 303
Cu(110),
scanning tunneling microscopy of, 300
Cu(111),
scanning tunneling microscopy of, 297
Current imaging tunneling spectroscopy, 129, 159

D

Diffusion,
of O on Cu(110), 302
of Au on Au(111), 292
Dimer structures,
scanning tunneling microscopy of, 176
Dopant profiling microscopy, 90
Dual polarity tunneling images of,
Ge (001), 238
Dynamics,
observation of, 291
of adsorbates on metal surfaces, 298

E

Electronics for scanning tunneling microscopy, 60
Electrostatic force microscopy, 90
Energy gap spectroscopy of,
- 1T-$TaSe_2$, 376–8
- 1T-TaS_2, 363
- 2H-TaS_2, 389, 393
- 4Hb-TaS_2, 401
- 1T-$TaSe_2$, 376
- 2H-$TaSe_2$, 389
- 4Hb-$TaSe_2$, 400
- 1T-$TiSe_2$, 383
- 1T-VSe_2, 378
- $NbSe_3$, 418
- TaS_3, 418
- 1T-TiS_2, 383

Energy gap spectroscopy,
- of charge density waves (CDW), 351, 354, 421, 386

Epitaxy of,
- Ge on Si(001), 247
- Ag on Au(111), 296
- Au on Ni(110), 295
- Ni on Au(111), 296

F

Fe on GaAs(110),
- spectroscopy of, 127, 130, 271

Feedback control theory,
- scanning tunneling microscope design, 33

Feedback electronics, 61
Feedback oscillation,
- experimental problems in scanning tunneling microscope operation, 73

Force gradient microscopy, 89
Forces, role of in STM,
- theory of, 24

Fourier transform of atomic force microscopy (AFM) images, 369, 371, 376

G

GaAs surface reconstruction,
- scanning tunneling microscopy of, 254

GaAs surfaces,
- scanning tunneling microscopy of, 251, 255

GaAs(100),
- scanning tunneling microscopy of, 259

GaAs(110),
- scanning tunneling microscopy of, 110, 255
  - surface buckling, 254, 258
  - O adsorbed, 110
  - Cs adsorbed, 123
  - Fe adsorbed, 127, 130
  - Sb adsorbed, 136
  - General I-V characteristics, 118

GaAs(111),
- scanning tunneling microscopy of, 261

GaAs/AlGaAs superlattices,
- spectroscopy of, 264

GaAs/AlGaAs superlattices,
- cross-sectional imaging of, 263

Ge surfaces,
- scanning tunneling microscopy of, 225

Ge surfaces and alloys,
- scanning tunneling microscopy of, 227

Ge(001),
- spectroscopy of, 237
- dual polarity tunneling images of, 238

Ge(001)-2 × 1,
- scanning tunneling microscopy of, 235

Ge(111),
- high temperature measurements of, 234

Ge(111) c2 × 8,
- spectroscopy of, 230
- scanning tunneling microscopy of, 227

GeSi(111) 5 × 5,
- scanning tunneling microscopy of, 239

Graphite, STM of,
- theory of, 14

H

H-adsorbed,
- Si(111) 7 × 7,
  - scanning tunneling microscopy of, 194

Heterostructure interface,
- Au/AlAs/GaAs
  - ballistic electron emission microscopy of, 337, 339

High temperature scanning tunneling microscopy, 208

I

I on Pt(111),
chemisorption, scanning tunneling microscopy of, 297
Imaging,
dual polarity, 99, 104, 106, 163
interpretation, 164, 167, 177
Imaging band edge states,
theory of, 14
Interfaces,
BEEM, ballistic electron emission microscopy of, 307
critical angle at,
for electrons, 312
for holes, 321
quantum mechanical reflection at, 317, 342
transverse momentum conservation at, 311, 331
Inverse photoemission microscopy, 82

K

Kinks and kink-step interactions,
scanning tunneling microscopy of, 185

L

Local density of states, 7
Low-temperature scanning tunneling microscopy
experimental techniques, 355, 430

M

Magnetic force microscopy, 90
Metal overlayers on Ge(111),
scanning tunneling microscopy of, 241
Metals on GaAs(110),
spectroscopy of, 265
Mo(001),
scanning tunneling microscopy of, 296

N

$NbSe_2$,
2H phase,
scanning tunneling microscopy of, 430, 432, 441
superconducting properties of,
scanning tunneling microscopy of, 430, 432
scanning tunneling microscopy of charge density wave in, 386
spectroscopy of, 386–9
$NbSe_3$,
scanning tunneling microscopy of, 402, 405
spectroscopy of, 418
NH3-adsorbed,
Si(111) 7 × 7,
scanning tunneling microscopy of, 192
Ni Au(111),
epitaxy scanning tunneling microscopy of, 296
Noise,
experimental problems in scanning tunneling microscope operation, 71

O

O adsorbed,
GaAs(110),
spectroscopy of, 110
O adsorbed Ge(111),
scanning tunneling microscopy of, 244
O on Cu(100),
chemisorption, scanning tunneling microscopy of, 303
O on Cu(110),
chemisorption, scanning tunneling microscopy of, 300
surface diffusion, 302
O-adsorbed,
Si(111) 7 × 7,
scanning tunneling microscopy of, 195
Optical absorption microscopy, 86
Optical mixing, 82

P

Pd-adsorbed,
Si(111) 7 × 7,
scanning tunneling microscopy of, 200
Periodic lattice distortion (PLD), 351

Photo emission, 82
Photovoltage microscopy, 91
Point contact, 16
Preamplifier electronics, 60
Pt(111),
  scanning tunneling microscopy of, 297

R

Resolution,
  avve limit for, 77
Re(001),
  scanning tunneling microscopy of, 297
Rh(111),
  scanning tunneling microscopy of, 297
Role of surface electronic structure, 11

S

S on Cu(111),
  chemisorption, scanning tunneling microscopy of, 297
S on Mo(001),
  chemisorption, scanning tunneling microscopy of, 296
S on Re(001),
  chemisorption, scanning tunneling microscopy of, 297
Sample and tip condition,
  experimental problems in scanning tunneling microscope operation, 68
Sample positioner mechanical structure, 57
Sb on GaAs(110),
  spatially resolved spectroscopy of, 266
  surface structure of, 266
Scanner mechanical structure, 53
Scanning electronics, 63
Scanning force microscopy, 86
Scanning noise microscopy, 79, 85
Scanning tunneling microscope design, 31
  common problems, 66
  common problems, feedback oscillation, 73
  common problems, noise, 71
Scanning tunneling microscope design,
  common problems, sample and tip condition, 68
  common problems, thermal drift and piezoelectric hysteresis, 73
  computer interface and data acquisition, 64
  control electronics, 60
  control electronics feedback, 61
  control electronics, preamplifier, 60
  control electronics, scanning, 63
  mechanical structure, 51
  mechanical structure, sample positioner, 57
  mechanical structure, scanner, 53
  mechanical structure, tip, 51
  mechanical structure vibration isolation stage, 58
  theory, 33
  theory, Feedback, 33
  theory, vibration isolation, 43
Scanning tunneling microscopy of charge density waves in,
  2H-$NbSe_2$, 386
  1T-TaS_2, 358
  2H-TaS_2, 391
  4Hb-TaS_2, 393, 400
  1T-$TaSe_2$, 358, 374
  2H-$TaSe_2$, 388
  4Hb-$TaSe_2$, 393
  1T-$TiSe_2$, 381
  1T-VSe_2, 359
Scanning tunneling microscopy of
  adsorbate covered semiconductor surfaces, 191
  adsorbate covered Si, 191
    gases, 191
    NH3, 192
    H, 194
    O, 194
    H, wet chemical preparation, 197
    metals, 199
    Ag, 199
    Au, 200
    Pd, 201
    B, 202
    Al, 202
    Ga, 202
    In, 202
    As, 205
    Ni, 206
    Cu, 206
  adsorbate induced structure, 293
  adsorbates on Ge(111), 244
    O, 244

adsorbates on metal surfaces, 293
  chemisorption, 297
  chemisorption of,
    C6H6 on Rh(111), 297
    I on Pt(111), 297
    O on Cu(100), 303
    O on Cu(110), 300
    S on Cu(111), 297
    S on Re(001), 297
  metallic, 293
band gap states, 255, 270, 273, 275
clean Ge surfaces and alloys, 227
clean Ge surfaces and alloys, 227
clean metal surfaces,
  reconstruction,
    Au(111),
    dynamics, 291
  surface structures, 286
    Au(110), 287
    Au(111), 292
clean semiconductor sufaces, 151
clusters, 266, 271, 273
diffusion, 208
diffusion of Au on Au(111), 292
diffusion of O on Cu(110), 302
dimer structures, 176
dynamics of adsorbates on metal
  surfaces, 298
epitaxial growth of,
  Ag on Au(111), 296
  Au on Ni(110), 295
  Ni on Au(111), 296
epitaxy of,
  Ge on Si(001), 247
  Si(111), 208
  Si(001), 208
GaAs surface reconstruction, 254
GaAs surfaces, 251, 255
GaAs(100), 259
GaAs(110), 110, 255
  surface buckling, 254, 258
GaAs(111), 261
Ge surfaces, 225
Ge(001)-2 × 1, 235
GeSi(111) 5 × 5, 239
kinks and step-kink interactions, 185
metal overlayers on Ge(111), 241
metal overlayers on Ge(111),
  As-1 × 1, 241
  As, 243
  Sn, 243
metal surfaces, 277
  resolution, 288
phase transitions, 212
semiconductors,
  fermi level pinning, 265, 268, 273, 274
Si surfaces, 149
  surface density of states, 156
Si(001), 168
Si(001)-2 × 1,
  low temperature, 173
  room temperature, 170
  smooth (A) and rough (B) steps, 181
Si(110), 187
Si(111), 151
Si(111)-2 × 1, 165
Si(111)-7 × 7, 152
Si(111) 2 × 1, 106
Si(111) laser annealed surfaces, 218
stepped structures, 180
stepped structures at strained surfaces,
  185
superconductors, 427
  experimental techniques, 430
  homogeneous phase, 427
  inhomogeneous phase, 429
surface dimers, 260, 261
surface trimers, 261
vortices in superconductors, 429, 434
Scanning tunneling potentiometry, 79, 81
Semiconductors,
  scanning tunneling microscopy of,
    fermi level pinning, 265, 268, 273, 274
Si surfaces,
  scanning tunneling microscopy of, 149
    surface density of states, 156
Si(001),
  scanning tunneling microscopy of, 168
Si(001)-2 × 1,
  scanning tunneling microscopy of,
    low temperature, 173
    room temperature, 170
Si(111),
  scanning tunneling microscopy of, 151
Si(111)-7 × 1,
  scanning tunneling microscopy of, 152
Si(111)-2 × 1,
  scanning tunneling microscopy of, 106,
    165
  spectroscopy of,

current imaging, 132
I-V measurements, 112
plane wave model of STM contours, 107
surface density of states, 112
voltage dependent imaging of, 106
Si(111)-Al,
scanning tunneling microscopy of, 203
Si(111)-Ga,
scanning tunneling microscopy of, 203
Si(111)-In,
scanning tunneling microscopy of, 204
Si(111): As-1 × 1,
scanning tunneling microscopy of, 205
Si(111): B,
scanning tunneling microscopy of, 204
Si(111): Cu
scanning tunneling microscopy of, 207
Si(111): H,
scanning tunneling microscopy of, 197
Si(111): Ni,
scanning tunneling microscopy of, 206
Sn adsorbed Ge(111),
scanning tunneling microscopy of, 243
Spatially resolved spectroscopy,
experimental techniques, 431
Si(111) 7 × 7, 158, 160
Spatially resolved spectroscopy,
of superconductors, 436
Spectroscopy,
theory of, 22
Spectroscopy of,
1T-TaS_2, 363
1T-$TaSe_2$, 376
1T-TiS_2, 383
1T-$TiSe_2$, 381
1T-VSe_2, 380
2H-$NbSe_2$, 386
2H-TaS_2, 389
2H-$TaSe_2$, 389
4Hb-TaS_2, 400
4Hb-$TaSe_2$, 400
adsorbate covered surfaces, 110
Au(100)-5 × 20, 284
Au(111), 285
band gap states, 270, 273
bulk band gap, 254, 255, 261, 265, 267
clusters, 266, 271, 273
evanescent surface states, 255, 272
GaAs(110),
Cs adsorbed, 123
Fe adsorbed, 127, 130
general I-V characteristics, 118
O adsorbed, 110
Sb adsorbed, 118, 266
GaAs/AlGaAs superlattices, 264
Ge(001), 237
Ge(111)c2 × 8, 230
metal surfaces, 280
metals on GaAs(110), 265
charging effects, 265
NbSe3, 418
Si(111) 2 × 1,
I-V measurements, 112
surface density of states, 112
voltage dependent imaging of, 106
surface state density, 274
TaS_3, 418
transmission resonances, 282
Spectroscopy, 95, 351, 354, 421
barrier resonances in, 103
current imaging spectroscopy, 129
differential conductance, 157
differential conductivity, 282
electronic structure, 282
experimental techniques, 431
fixed separation I-V measurements, 98, 101, 112
instrumentation, 96
normalization methods for,
fixed separation, 112
normalization methods for,
acquisition methods, 135
average conductivity method, 138
band gap effects, 134
general method, 141
of ϕ-S, 281
of I-S, 280
of I-S,
for resonant transmission, 283
spatial semiconductor characteristics, 123
spatially resolved, 123
of superconductor energy gap, 432
of superconductors, 431
in zero field, 432
surface band bending effects, 122, 126
surface density of states, 112
tunneling transmission, 102
of tunneling transmission dependence on

tunnel gap, 284
variable separation measurements, 134
voltage dependent imaging, 99, 104, 106
Stepped structures,
scanning tunneling microscopy of, 180
Stepped structures at strained surfaces,
scanning tunneling microscopy of, 185
Strong coupling regime, 16
Superconductor,
density of states, 428, 432, 436
flux lattice, 429, 435, 437, 447
quasiparticles, 428, 441, 446
scanning tunneling microscopy of, 427
vortex wavefunction of, 441
vortex evolution, 443
vortex-vortex interactions, 447
Superconductor tunneling, 427
Surface dimers,
scanning tunneling microscopy of, 260, 261
Surface state density, 285
Surface trimers,
scanning tunneling microscopy of, 261
Surfaces, metal,
scanning tunneling microscopy theory of, 8
Surfaces, semiconductor,
scanning tunneling microscopy theory of, 11

T

TaS_2,
4Hb phase
scanning tunneling microscopy of charge density wave in, 393, 400
1T phase
scanning tunneling microscopy of charge density wave in, 358
2H phase
scanning tunneling microscopy of charge density wave in, 391
1T phase,
atomic force microscopy of, 358
spectroscopy of, 363
2H phase,
spectroscopy of, 389, 393
4Hb phase,
spectroscopy of, 401
TaS_3,
scanning tunneling microscopy of, 402, 405, 416
spectroscopy of, 418
$TaSe_2$,
1T Phase,
scanning tunneling microscopy of charge density wave in, 358, 374
2H phase,
scanning tunneling microscopy of charge density wave in, 388
4Hb phase,
scanning tunneling microscopy of charge density wave in, 393
1T phase,
atomic force microscopy of, 358, 374
spectroscopy of, 376–8
2H phase,
spectroscopy of, 389
4Hb phase,
spectroscopy of, 400
Theory,
scanning tunneling microscopy imaging of adsorbates, 14
Theory of,
apparent barrier height, 20, 26
apparent size of adatom, 16
atom superposition modeling, 8
forces, role of, i, 24
graphite, imaging, 14
imaging band edge states, 14
local density of states, 7
point contact, 16
role of surface electronic structure, 11
spectroscopy, 22
strong coupling regime, 16
surfaces, metal, 8
surfaces, semiconductor, 11
tip, models for, 4, 6, 14
tip, tip-sample interactions, mechanical, 24
tunneling barrier, 20
tunneling Hamiltonian, 5
vacuum tunneling, 1
vacuum tunneling, transmission probability, 2
voltage dependence in STM imaging, 11, 22
work function, apparent, 20, 26
Thermal drift and piezoelectric hysteresis,
experimental problems in scanning

tunneling microscope operation, 73
thermal microscopy, 83
tip mechanical structure, 51
tip, models for, 4, 6, 14
tip, tip-sample interactions, mechanical,
theory of, 24
TiS_2,
1T phase,
spectroscopy of, 383
$TiSe_2$,
1T phase,
scanning tunneling microscopy of
charge density wave in, 381
spectroscopy of, 383
Transition metal chalogenides,
scanning tunneling microscopy of, 349–55
Transition metal dichalgogenides,
scanning tunneling microscopy of, 352, 357
Transition metal trichalgogenides,
scanning tunneling microscopy of, 352, 358, 402–422
Transmission probability, 284
Transmission resonances, 282
Tunnel current dependence on tunnel gap, 281
Tunneling barrier,
theory of, 20
Tunneling Hamiltonian, 5
Tunneling thermocouple, 85
Tunneling tip,
resolution and corrugation, 278
Tunneling transmission probability, 314

V

Vacuum tunneling, 100, 314, 320
theory of, 1
Vacuum tunneling, transmission probability,
theory of, 2
Vibration isolation stage mechanical structure, 58
Vibration isolation theory,
scanning tunneling microscope design, 43
Voltage dependence in STM imaging,
theory of, 11, 22
Voltage dependent imaging in,
scanning tunneling microscopy, 99, 104, 106
VSe_2,
1T phase,
scanning tunneling microscopy of
charge density wave in, 359
VSe_2,
1T phase,
spectroscopy of, 378

W

Work function, apparent,
theory of, 20, 26

Z

$ZrSe_3$,
scanning tunneling microscopy of, 403–4

ISBN 0-12-475972-6

90040>

9 780124 759725